Diffuse Waves in Complex Media

NATO Science Series

A Series presenting the results of activities sponsored by the NATO Science Committee. The Series is published by IOS Press and Kluwer Academic Publishers, in conjunction with the NATO Scientific Affairs Division.

A. Life Sciences	IOS Press
B. Physics	Kluwer Academic Publishers
C. Mathematical and Physical Sciences	Kluwer Academic Publishers
D. Behavioural and Social Sciences	Kluwer Academic Publishers
E. Applied Sciences	Kluwer Academic Publishers
F. Computer and Systems Sciences	IOS Press

1. Disarmament Technologies	Kluwer Academic Publishers
2. Environmental Security	Kluwer Academic Publishers
3. High Technology	Kluwer Academic Publishers
4. Science and Technology Policy	IOS Press
5. Computer Networking	IOS Press

NATO-PCO-DATA BASE

The NATO Science Series continues the series of books published formerly in the NATO ASI Series. An electronic index to the NATO ASI Series provides full bibliographical references (with keywords and/or abstracts) to more than 50000 contributions from international scientists published in all sections of the NATO ASI Series.
Access to the NATO-PCO-DATA BASE is possible via CD-ROM "NATO-PCO-DATA BASE" with user-friendly retrieval software in English, French and German (© WTV GmbH and DATAWARE Technologies Inc. 1989).

The CD-ROM of the NATO ASI Series can be ordered from: PCO, Overijse, Belgium.

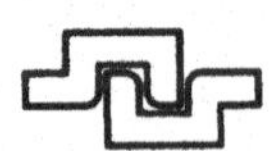

Series C: Mathematical and Physical Sciences – Vol. 531

Diffuse Waves in Complex Media

edited by

Jean-Pierre Fouque

CNRS, Ecole Polytechnique, France
and
Department of Mathematics,
North Carolina State University,
Raleigh, NC, U.S.A.

Kluwer Academic Publishers

Dordrecht / Boston / London

Published in cooperation with NATO Scientific Affairs Division

Proceedings of the NATO Advanced Study Institute on
Diffuse Waves in Complex Media
Les Houches, France
March 17–27, 1998

A C.I.P. Catalogue record for this book is available from the Library of Congress

ISBN 0-7923-5679-9 (HB)
ISBN 0-7923-5680-2 (PB)

Published by Kluwer Academic Publishers,
P.O. Box 17, 3300 AA Dordrecht, The Netherlands.

Sold and distributed in North, Central and South America
by Kluwer Academic Publishers,
101 Philip Drive, Norwell, MA 02061, U.S.A.

In all other countries, sold and distributed
by Kluwer Academic Publishers,
P.O. Box 322, 3300 AH Dordrecht, The Netherlands.

Printed on acid-free paper

Printed in the Netherlands

Contents

Preface

The NATO Advanced Study Institute on *Diffuse Waves in Complex Media* was held at the "Centre de Physique des Houches" in France from March 17 to 27, 1998.

The Schools' scientific content, wave propagation in heterogeneous media, has covered many areas of fundamental and applied research. On the one hand, the understanding of wave propagation has considerably improved during the last thirty years. New developments and concepts such as, speckle correlations, weak and strong localization, time reversal, near-field propagation are under active research. On the other hand, wave propagation in random media is now being investigated in many different fields such as applied mathematics, acoustics, optics, atomic physics, geophysics or medical sciences. Each community often uses its own langage to describe the same phenomena. The aim of the School was to gather worldwide specialists to illuminate various aspects of wave propagation in random media.

This volume presents fourteen expository articles corresponding to courses and seminars given during the School. They are arranged as follows.

The first three articles deal with the phenomena of localization of waves: B. van Tiggelen (p. 1) gives a critical review of the physics of localization, J. Lacroix (p. 61) presents the mathematical theory and A. Klein (p. 73) describes recent results for randomized periodic media.

The next two contributions are concerned with spectral gap materials: C. Soukoulis present an overview of photonic band gap materials and ways to obtain them. Asymptotic models of high-contrast photonic crystals are studied by A. Figotin (p. 109).

In his contribution, C. Beenakker (p. 137) presents the relation between the photon statistics of a random medium and its scattering matrix. A formalism for the study of spectra of large random matrices is discussed by V. Freilikher and E. Kanzieper (p. 165).

The next two articles deal with atomic physics. R. Loudon (p. 213) describes the optical properties of ordered and disordered dielectric media and R. Kaiser (p. 249) introduces us to the scattering of light by laser cooled atoms.

Wave scattering from randomly rough surfaces is treated by M. Nieto-Vesperinas and A Madrazo (p.289). In particular they discuss the backscattering enhancement due to multiple scattering phenomena.

The next two papers are concerned with acoustic waves propagating in randomly layered media. J.P. Fouque (p. 319) presents recent results obtained in the regime of separation of scales in the case of plane layered media while W. Kholer, G. Papanicolaou and B. White (p. 347) discuss the robustness of this theory by studying the reflection and transmission properties of locally-layered media.

The last two contributions deal with waves propagating in the earth. M. Campillo, L. Margerin and N. Shapiro (p. 383) discuss the radiative transfer equation and the diffusive regime. They compare numerical simulations and real observations. In his contribution, R. Snieder (p. 405) discuss the problem of imaging in complex media in the context of seismic signals. Time reversal techniques and a comparison between waves and particles are presented.

The French research group POAN (Propagation d'Ondes en milieux Aléatoires et/ou Nonlinéaires) has organized several meetings on Waves in Random Media. It is during one of these meetings, the Cargèse conference in May 1996, that it was decided to organize an international interdisciplinary School covering this subject.

The Scientific Direction of the School, J.P. Fouque, A. Lagendijk, R. Maynard and G. Papanicolaou, takes this opportunity to thank the NATO Science Committee whose financial support made this School possible. We also thank the French CNRS and research group POAN for their supports.

The excellent organization of the School is due to C. Prada and C. Vanneste with the help of J. Garnier, J.P. Montagner, B. van Tiggelen. The Direction and the Personel of the "Centre de Physique des Houches" have made our stay very enjoyable. We thank all of them.

Special thanks are due to S. Whitaker at NCSU for her valuable assistance in the preparation of this manuscript for publication.

I personally take this opportunity to thank the authors of this volume and all the participants to the School who have greatly contributed to its success.

Raleigh NC, December 1998 *Jean-Pierre Fouque*

Participants

Bal Guillaume, Stanford, USA.
Beenakker Carlo, Leiden, The Netherlands.
Boulgakov Sergei, Madrid, Spain.
Bressoux Richard, Grenoble, France.
Busch Kurt, Toronto, Canada.
Campillo Michel, Grenoble, France.
Carminati Rémi, Chateney Malabry, France.
Chabanov Andrey, New York, USA.
Chapurin Igor, Kishinev, Moldova.
Chevrot Sébastien, Paris, France.
Cohen-Addad Sylvie, Marne-la-Vallée, France.
Cowan Michael, Manitoba, Canada.
Cwilich Gabriel, New York, USA.
Delande Dominique, Paris, France.
Dorn Oliver, Stanford, USA.
Faure Frédéric, Grenoble, France.
Figotin Alexander, Irvine, USA.
Fink Mathias, Paris, France.
Fouque Jean-Pierre, Raleigh, USA.
Freilikher Valentin, Ramat-Gan, Israël.
Garcia-Martin Antonio, Madrid, Spain.
Garnier Josselin, Palaiseau, France.
Gomez-Rivas Jaime, Amsterdam, The Netherlands.
Gopal Venkatesh, Bangalore, India.
Greffet Jean-Jacques, Chateney Malabry, France.
Herrmann Felix, Stanford, USA.
Hohler Reinhard, Marne-la-Vallée, France.
Hovenier Joop, Amsterdam, The Netherlands.
Kafesaki Maria, Crete, Greece.
Kaiser Robin, Nice, France.
Khater Antoine, Le Mans, France.
Klein Abel, Irvine, USA.
Kuske Rachel, Minneapolis, USA.
Lacombe Cécile, Grenoble, France.
Lacoste David, Grenoble, France.
Lacroix Jean, Paris, France.
Lagendijk Ad, Amsterdam, The Netherlands.
Lenke Ralf, Konstanz, Germany.
Loudon Rodney, Colchester, UK.
Maret Georg, Konstanz, Germany.
Margerin Ludovic, Grenoble, France.
Maurel Agnes, Paris, France.
Maynard Roger, Grenoble, France.
Mertelj Alenka, Ljubljana, Slovenia.
Mishchenko Michael, New York, USA.
Montagner Jean-Paul, Paris, France.
Mortessagne Fabrice, Nice, France.
Muzzi Alessandro, Florence, Italy.
Nieto-Vesperinas Manuel, Madrid, Spain.
Orlowsky Arkadiusz, Warszawa, Poland.
Page John, Manitoba, Canada.
Pagneux Vincent, Le Mans, France.
Papanicolaou George, Stanford, USA.
Patra Michael, Leiden, The Netherlands.
Peters François, Nice, France.
Pitter Mark, Nottingham, UK.
Prada Claire, Paris, France.
Quartel John, London, UK.
Röhm Axel, Utrecht, The Netherlands.
Ripoll Jorge, Madrid, Spain.
Rose James, Iowa, USA.
Ryzhik Leonid, Chicago, USA.
Saenz Juan Jose, Madrid, Spain.
Scheffold Frank, Konstanz, Germany.
Sebbah Patrick, Nice, France.
Shchegrov Andrei, Irvine, USA.
Skipetrov Serguei, Moscow, Russia.
Snieder Roel, Utrecht, The Netherlands.
Soukoulis Costas, Iowa, USA.
Sparenberg Anja, Stuttgart, Germany.
Spetzler Jerper, Utrecht, The Netherlands.
Spott Thorsten, Trondheim, Norway.
Stam Daphne, Amsterdam, The Netherlands.
Stark Holger, Stuttgart, Germany.
Stoytchev Marin, New York, USA.
Tourin Arnaud, Paris, France.
Tualle Jean-Michel, Villetaneuse, France.
Turner Joseph, Nebraska, USA.
Tweer Ralf, Konstanz, Germany.
Vanneste Christian, Nice, France.
Van Tiggelen Bart, Grenoble, France.
Van Soest Gijs, Amsterdam, The Netherlands.
Wiersma Diederik, Florence, Italy.
Yodh Arjun, Philadelphia, USA.

LOCALIZATION OF WAVES

B.A. VAN TIGGELEN
Laboratoire de Physique et Modélisation des Systèmes Complexes
CNRS/Maison des Magistères, Université Joseph Fourier
B.P. 166, 38042 Grenoble Cedex 9, France
E-mail: tiggelen@belledonne.polycnrs-gre.fr

Protect me from knowing what I don't need to know
Protect me from even knowing that there are things to know that I don't know
Protect me from knowing that I decided not to know about
the things that I decided not to know about
Douglas Adams, 1992 in: *Mostly Harmless*

1. Introduction

Writing a short and critical review about localization of waves is an ambiguous project, doomed to be incomplete, and doomed to be unfair to the many good papers that have been written on this fascinating subject. At least every subject raised in this modest contribution merits, I think, an entire review and, fortunately, some exist already. By no means is this review going to be a replacement for the excellent works by Thouless [1], Ramakrishnan [2], Mott [3], Souillard [4], John [5, 6], and Vollhardt and Wölfle [7]. Answering the question whether or not reported experiments reveal wave localization is not the aim of this review either. The emphasis in this review will be on definitions, features and consequences.

The phenomenon of localization appears in many disguises, sometimes an "unrecognizable monster" according to its creator Anderson in looking back at 25 years of localization [8]. Originally, the concept of localization came forward out of the wish to understand metal-insulator transitions [3], where "disorder" prevents the electrons in a semiconductor to move freely. Like all great discoveries it initiated a new way of thinking, with impact in domains far from the one it originated, like radiative transfer [5, 9], seismology [10], and even atomic physics [11] and high-energy physics [12]. Localization has been at the origin of a new "mesoscopic" physics, i.e. a physics between microscopic and macroscopic, with best of both worlds,

J.-P. Fouque (ed.), Diffuse Waves in Complex Media, 1–60.

namely disorder and phase. Finding convincing evidence for localization has been a constant challenge for experimental physicists working with waves in disordered media. Only a handful of experiments claim observation of localization in a very direct way. Nevertheless, many new phenomena - coherent backscattering, weak localization, long-range correlations, resonant multiple scattering to mention a few - have been discovered on the way. In all reports of localization, "alternative" explanations are possible. This has largely contributed to the actual confused state that localization seems to be in, at least for the younger generation. Several lectures given in Les Houches in march 1998, two generations after the discovery, made this very clear.

Most theoretical approaches are extremely intuitive, especially the good ones. A good example is the scaling theory of localization [13], inspired by the successful renormalization theory of statistical physics. More technical methods, using Green's functions, diagrams, spontaneous symmetry breaking and all those things that many always wanted to avoid, have been guided by a good deal of physical intuition, ranging from the first "locator expansion" by Anderson [14], to the microscopic "diagrammatic" theories [15, 16] developed in the end of the seventies, the "non linear sigma" models [17, 18, 19] borrowed from quantum field theory. They all aim to find and verify "great principles" and "simple ideas" in a mess of complicated physics. In the eighties, rigorous mathematical proofs [20] started to appear that showed that localization really exists in the regime where physical intuition believed it to occur, briefly that Anderson was right after all. As a result, they did not shock the physics community that much. Yet to me, mathematical theories showed how subtle the concept is when trying to make really strong statements, a subtlety that is not evident from theoretical treatments.

Two models have learned us a lot about Anderson localization. Both are idealizations of the real world, but contain the physics that is believed to be essential. The most recent one is random matrix theory. This theory is now in rapid development and applies to open "quasi one dimensional" random systems [21], an elegant theory where symmetry plays a dominant role, the rest "just" being calculation. I found it funny to learn that this theory actually assumes a disordered medium to be a stack of Sinai billiards, to each of which one applies chaos theory developed by Dyson. I suppose an experimentalist would never have thought of his sample this way. The second model is the fundamental Anderson Hamiltonian, which models the hopping of a tightly bound electron from one site to the other, either in an infinite lattice or in a finite closed lattice. Many thorough numerical studies have been carried out with this Hamiltonian.

Giving up any of these simplifications rapidly makes the problem unsolvable, leaving room for all kinds of wild speculations. Altogether, the subject is hard and perhaps too vast to comprehend for an interested newcomer in the field, not knowing what treatment is best suited to study her or his particular problem, but eager to understand the impact of localization in her or his field.

2. What is Localization?

The phenomenon of localization can occur when waves propagate in some kind of random medium. However, the exact definition of localization does not seem to be unique. Perhaps it is easier to tell what localization is certainly not. Localization is not the same thing as "weak localization" or "coherent backscattering". These mesoscopic phenomena refer to interference effects in multiple scattering that do even persist for low disorder. Localization is not trapping either, which would occur by simply surrounding the wave with a bunch of ideal mirrors, an effect that even persists classically. Equivalently, a state in the gap of a semi-conductor or a photonic band gap material, created by some local defect, is not called "localization" in the sense of the present paper. It is just a bound state. As I will point out below, localization requires a *finite number of bound states per unit volume.* This definition is strong in infinite media, but ambiguous in a finite medium since even one bound state has finite density in a finite volume. In an open system, the spectrum is continuous and microstates cannot be counted. Localization in open systems is often associated with an exponentially small ensemble-averaged transmission, which is - I think - only one aspect of localization. Fluctuations are at least as important.

2.1. LOCALIZATION IN INFINITE SYSTEMS

The easiest definition is the one suggested by the title of the original paper by Anderson, namely the "absence of diffusion". Many authors have accepted the criterion

$$D(E) = 0\,, \tag{1}$$

as a working criterion for localization for waves at energy E. Indeed, the diffusion constant $D(E)$ at energy E seems to be a relevant quantity because it is strongly related to ensemble-averaged conductance, and that is a basic quantity measured in the laboratory, at least for conductors at that time. I want to stress here immediately that the diffusion constant is typically a transport quantity in an *infinite* system, whereas "conductance" is a property of an *open finite* system. We now know that this distinction is crucial in any discussion on localization.

Some people go even one step further by defining localization as a "small" diffusion constant, e.g. by comparing to the diffusion coefficient anticipated from classical transport. This puts the phenomenon on the same footing as "weak localization". The latter is sometimes seen as a precursor of strong localization, but the exact relation is far from evident. Small optical diffusion constants have been reported [22]. An interpretation in terms of localization is dangerous as reasons other than localization exist why diffusion can be suppressed [23].

The restrictions of the criterion (1) can be revealed by recalling the Einstein random walk formula for the diffusion constant,

$$D = \frac{1}{2d} \lim_{t\to\infty} \left\langle \frac{\mathbf{r}^2}{t} \right\rangle , \tag{2}$$

where $\langle\cdots\rangle$ stands for ensemble averaging over disorder, d is the dimension, $\mathbf{r}$ is the center-of-mass of the wave packet in space and t denotes time. Apparently, criterion (1) imposes that $\mathbf{r}^2 < \sqrt{t}$ for an ensemble-averaged medium, but that does definitely not "localize" the wave packet for a typical realization of the disorder. To have that we must impose the much stronger constraint that

$$\mathbf{r}^2 < \text{constant}, \qquad \text{almost surely.} \tag{3}$$

Mathematicians call this dynamical localization. I would like to warn the reader that this terminology is not equal to what a physicist usually calls dynamical localization (section 3.1.1)

A definition for localization exists that circumvents time evolution and ensemble averaging from the start. Localization in a certain energy interval ΔE is said to occur when, *for "almost every realization"* $\{V\}$ of the random potential V, the random Hamiltonian $H = H_0 + V$ has, in the energy range ΔE only point spectrum with (at least) exponentially localized eigenfunctions $\psi_n(\mathbf{r})$, i.e.

$$|\psi_n(\mathbf{r})| \leq C_n(\{V\}) \exp\left(-A|\mathbf{r} - \mathbf{r}_n(\{V\})|\right) \quad \text{almost surely.} \tag{4}$$

Here A is independent of the realization and identified with the maximal localization length in the interval ΔE. The "almost surely" requirement is very strong, but not too strong to be impossible. Because the eigenfunctions decay exponentially, the spectrum of the Hamiltonian in the interval ΔE will be "pure point". The difference with normal bound states is that the interval ΔE contains an *infinite* number of eigenvalues. The spectrum is said to be "dense pure point". In an infinite system, bound states with infinitely close energies are possible because they correspond to localized eigenstates that are infinitely far apart and hardly overlap. It is well known from e.g. quantum tunneling problems that any overlap of wavefunctions would cause eigenvalue repulsion. We will make this important notion more precise in the following section.

One consequence of the dense point spectrum is that the localized regime has, just like the conventional extended regime, a finite number of eigenstates $N(E, \Delta E)$ *per unit volume* in some finite energy interval ΔE, a number that can actually shown to be equal for "almost any" realization of the disorder [25]. This is rather unfortunate: The density of states $\rho(E) \equiv \lim_{\Delta E \to 0} N(E, \Delta E)/\Delta E$ is an important quantity in transport theory and is also experimentally accessible, but is - in an infinite system - not sensitive to localization. The absence of localization effects in the density of states implies that no signatures can be found in thermodynamic quantities such as the specific heat.

It is constructive to investigate the relation of definition (4) to the more dynamical one involving the time dependence of $\mathbf{r}(t)$. It is known that criterion (4) implies that (almost surely) $|\mathbf{r}(t)|/t = 0$ for large times t. Consequently, criterion (4) is stronger than the one in Eq. (1), but still not strong enough to obey criterion (3). It is also known that if $\mathbf{r}^2(t)$ is bounded than criterion (4) must be obeyed. The reverse is not true: Exotic models exist, that obey criterion (4) but have un unbounded $\mathbf{r}^2(t)$, and are thus not localized in the strict sense. To guarantee this property too, one must impose that $C_n(\{V\}) \leq C_\epsilon(\{V\}) \exp(\epsilon|\mathbf{r}_n(\{V\}|)$ for any small positive ϵ [26], i.e. the localization length of states far apart must be "more or less" equal for almost any realization.

A common way to study localization is by means of the so-called *return probability* $P(\mathbf{r})$, defined as,

$$P(\mathbf{r}) = \lim_{T \to \infty} \frac{1}{T} \int_0^T dt \, |\langle \mathbf{r}| \exp(-iHt) \, |\mathbf{r}\rangle|^2 \, . \tag{5}$$

It gives the probability density - or when integrated over some volume a probability - for a particle leaving at position $\mathbf{r}$ to finally come back at the same point. We expect this quantity to vanish for an extended state, and to be nonzero for a localized state. By using the spectral decomposition,

$$\exp(-iHt) = \sum_n \exp(-iE_n t) \, |\psi_n\rangle \, \langle\psi_n| \tag{6}$$

it is not difficult to see that

$$P(\mathbf{r}) = \sum_n |\psi_n(\mathbf{r})|^4 \, . \tag{7}$$

This brings the return probability on the same footing as the *inverse participation number* $\mathcal{P}^{-1}(E)$ for an eigenstate at energy E, defined as

$$\mathcal{P}^{-1}(E) = \sum_i |\psi_E(\mathbf{r}_i)|^4 \, , \tag{8}$$

where the summation runs over all sites, or in the case of a continuum, involves an integral over space. The participation number measures the number of sites covered by the wave function at energy E, or equivalently the size of a state at energy E. The relation between participation ratio and the return probability evidently is $\sum_i P(\mathbf{r}_i) = \sum_n \mathcal{P}^{-1}(E_n)$. For an exponentially localized state with localization length ξ we estimate $\mathcal{P} \sim (\xi/a)^d$, with a the lattice spacing.

Another convenient way to study localization is by means of the spatial correlation functions. The localization length $\xi(E)$ can be defined ¿from the intensity-intensity correlation function

$$\left\langle |\psi_E(\mathbf{r})|^2 |\psi_E(\mathbf{r}')|^2 \right\rangle \sim \exp\left(-\frac{|\mathbf{r}-\mathbf{r}'|}{\xi(E)}\right) . \tag{9}$$

I want to stress here that the field-field correlation function also decays exponentially, and defines the scattering mean free path or more exactly the extinction length $\ell(E)$

$$\langle \psi_E(\mathbf{r}) \psi_E^*(\mathbf{r}') \rangle \sim \exp\left(-\frac{|\mathbf{r}-\mathbf{r}'|}{\ell(E)}\right) . \tag{10}$$

The ensemble-averaging is crucial here, because that's the mechanism responsible for the exponential decay. According to criterion (4) for the localized regime, exponential decay occurs for "almost any realization", and not only for the ensemble-average of the intensity correlation. Contrary to the field correlation function, the intensity correlation in Eq. (9) would decay algebraically in the extended regime.

2.2. CLOSED SYSTEMS

By definition, a closed system has boundary conditions $\psi = 0$ or $\partial_n \psi = 0$ on its boundaries. In both cases the current density, whose component normal to the surface is given by $J_n(\mathbf{r}) \sim \mathrm{Im}\, \psi^*(\mathbf{r}) \partial_n(\mathbf{r}) \psi$, vanishes. Periodic boundary conditions are also encountered, transforming a finite system effectively into a periodic infinite one, ¿from which no current can escape.

In a closed system all states are genuine bound states with *finite* level spacing , and the question is how this evolves as the system becomes bigger. I emphasize that numerically exact "*ab initio*" studies can only deal with finite-size systems, either closed or open.

We note that the inverse participation ration (8) is well defined in a closed system. In Figure 1 we show the participation ratio as calculated by Schreiber [27] using a careful numerical technique to solve the Anderson model on a 2D square lattice with N sites. What is important to note here is that the participation ratio declines as the disorder W increases. This is

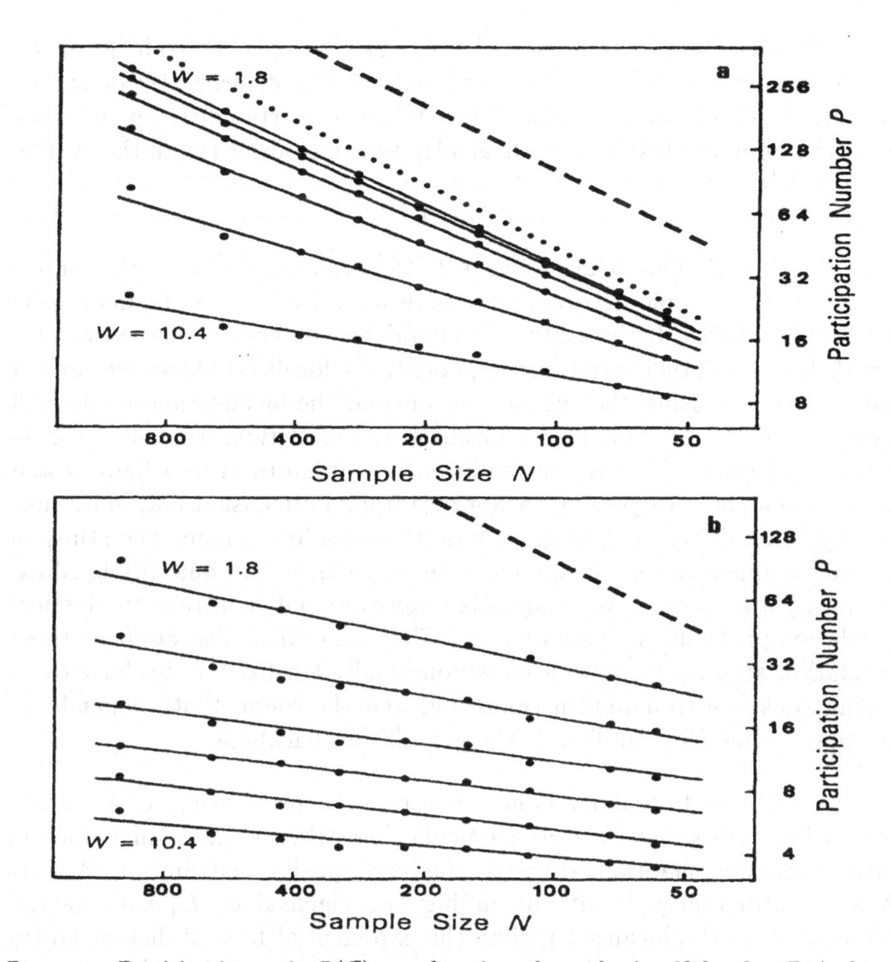

Figure 1. Participation ratio $\mathcal{P}(E)$ as a function of sample size N for the 2D Anderson model on a cubic lattice, for random energies $W/V = 1.8$ to 10.4. The top figure corresponds to an energy interval in the band center, the bottom figure to an energy interval just beyond the band edge. The broken line has $\mathcal{P}(E) = N$, the maximally possible value obtained for an extended Bloch plane wave. Taken from Schreiber [27], with kind permission from the author.

consistent with the fact that eigenstates in 2D are expected to be localized with a localization length ξ that shortens as the disorder W increases. Secondly, the participation ratio seems to be proportional to some power law of the number of sites, or equivalently to some power law of the system size $L \sim N^{1/d}$,

$$\mathcal{P}(E) \sim L^{\beta} . \tag{11}$$

where $0 \leq \beta \leq d$. This relation suggests that the eigenstates sort of interconnect in the sample according to some fractal structure, familiar from percolation studies. If the states are completely extended, one anticipates $\beta = d$. One the other hand, for exponentially localized states we do not expect that increasing the system size beyond the localization length will change a lot: $\beta = 0$. Finally, for localized eigenfunctions that decay algebraically $|\phi_E()| \sim 1/r^{\alpha}$ the tails sufficiently overlap to yield a finite β and one finds $\beta = 2d - 4\alpha$ [28, 29]. A nonzero fractal dimension may thus indicate algebraic decay of the eigenfunctions over a long range. The study of the fractal dimension β can therefore be very efficient to quantify localization in a *finite* system. We emphasize however that, contrary to classical percolation problems, it is ambiguous to define a critical disorder from these percolation arguments, since even exponentially localized states have overlapping tails due to quantum tunneling, and the connectivity depends on the "grey" scale used to display the percolation backbone.

A quantity that regularly shows up in recent considerations of Anderson localization in closed systems, in particular in relation with Dyson's random matrix theory of chaotic systems, is the level spacing distribution $P(s)$. It gives the probability density of finding two eigenvalues E_n with energy difference s. In the localized regime the exponential tails of distant states are hardly aware of each other. As a result, elementary statistical arguments suggest that the level spacing distribution is "Poisson",

$$P(s) = \frac{1}{\Delta} \exp\left(-\frac{s}{\Delta}\right) \quad \text{(localized regime)} \tag{12}$$

where Δ is the average level spacing and finite as long as a finite system is considered. This formula confirms that infinitely close levels are possible. We remark that Eq. (12) has assumed that the localization length of the levels ξ is much smaller than the sample dimension L. If not, we would not really speak of the localized regime, and a finite level spacing $\Delta E \approx 1/N(E) \sim L^{-d}$ would persist.

What can be expected in the extended regime? In principle all states overlap in space and one expects Eq. (12) to break down. It is tempting to apply ideas developed in chaos theory to this situation. The basic hypothesis in random matrix theory is that the energy E is the only constant of motion during the multiple scatterings and all other variables are completely scrambled. The only severe constrain is energy and the presence of time-reversal symmetry (for particles with spin there is one more). Time-reversal symmetry forces the matrix -element $\mathcal{H}_{nm} = \langle\psi_n|H|\psi_m\rangle$ to be real-valued. The rest is just a matter of evaluating the proper volume element (Jacobian) for the quantity to be considered. Time-reversal symmetry is characterized by a level spacing distribution,

$$P(s) = \frac{\pi s}{2\Delta^2} \exp\left(-\frac{\pi s^2}{4\Delta^2}\right) \quad \text{(extended regime; time reversal)} \tag{13}$$

whereas the absence of time-reversal symmetry is characterized by [30],

$$P(s) = \frac{32 s^2}{\pi\Delta^3} \exp\left(-\frac{4s^2}{\pi\Delta^2}\right) \quad \text{(extended regime; no time reversal)} \tag{14}$$

Expressions (13) and (14) differ considerably from the Poisson distribution expected in the localized regime. In particular, the extended regime is characterized by $P(0) = 0$, a phenomenon called *level repulsion*, and by Eq. (12) absent in the localized regime.

The level spacing distribution functions have been thoroughly investigated for simplified models such as the Anderson model [83, 84]. Some results will be discussed in the next section. Equations (13) and (14), and in particular their behavior as $s \to 0$ are often taken as a definition of "quantum chaos", since the classical definition of "sensibility to initial conditions" turns out to be too strong for quantum systems. On the basis of their level spacing distribution, closed disordered systems can be called "chaotic" only in the extended regime.

2.3. OPEN SYSTEMS: TRANSPORT OF WAVES

The characteristics of localization discussed above assume that the system is either infinite or finite but closed. In 3D, both are not very realistic experimentally. After all, to see 3D localization one has to open somehow the system and relate the observed leaks to localization. Open systems are characterized by a nonvanishing current density on their boundary. The study of localization in open systems is a study of wave transport and transmission matrix of the system, and not one of eigenfunctions, which become continuum eigenfunctions. In an open system, the eigenstates of the closed system achieve a finite width, due to leakages through the boundaries. As first realized by Thouless this rather innocent notion enables to formulate a criterion for localization in open systems, which is by many - including myself - regarded as the most important result in localization theory after its discovery. This criterion will be discussed in Chapter 4.

In the following I will first discuss localization of electrons, considering originally only ensemble-averaged transport. Next, I will outline what random matrix theory has to say about localization. This theory focuses on fluctuations and probability distributions, setting a new trend for modern mesoscopic physics. These studies have seen experimental applications in microwave tubes and quantum dots. Finally, I discuss a promising numerical method that actually studies the resonant remnants of the localized states in open media. Especially in 2D, experiments directly probing the wave function are possible. Experimental studies of 2D localized wave functions of microwaves [31], acoustic waves [32] and bending waves [33] have been reported.

2.3.1. *Metal-Insulator Transition*

In electronic systems (metals) the whole concept of localization was devised to explain metal-insulator transitions. "Classical" transport theory predicts a disordered electronic system to be conducting. The word classical should - of course - be between quotes, because deep quantum-mechanical wave concepts such as "Bloch states" and "incompletely filled bands" have been necessary to understand why only deviations from perfect periodicity induce a nonzero electronic conductivity. Such impurities (phonons, vacancies or thermal fluctuations,) had been given a classical treatment.

A conceptual difference seems to exist between localization in *uncompensated* and *compensated* doped semi-conductors. Compensated means that the number of acceptor (donor) levels exceeds the number of valence electrons (holes). In uncompensated semi-conductors electronic transport is governed by "free" electrons colliding with impurities, having much in common with classical wave localization. In compensated semi-conductors, conduction is largely determined by impurity band propagation, i.e. quantum tunneling due to overlapping "tightly bound" quantum states of impurity levels. This mimics the physics of the Anderson model discussed later, although one should not underestimate the role of electron-electron interactions and spin scattering that have not been taken into account by this model. All aspects play an important role in the solution of the critical exponents puzzle: Uncompensated materials, of which Si:P is the most famous, seem to have different critical exponents than compensated materials [34]. Recently, Shmilak etal. [35] suggested that the critical exponent of 1/2 is actually due to an incorrect extrapolation of the electronic conductivities towards zero temperature, and that all materials have a critical exponent close to one.

Let me summarize in a nutshell what Anderson localization implies for electron conduction. I stress once more that answering the important question whether or not experiments reveal localization is not the aim of this review.

The important transport coefficients, not only the electronic conductivity σ, but also thermal conductivity K and thermoelectric power S depend on temperature T and electron concentration n_e, and one would like to understand how. They can all be expressed in terms of the kinetic coefficients [36],

$$L(T, n_e) = \int_{-\infty}^{\infty} dE \, \frac{\partial f}{\partial E} D(E) \mathcal{L}[E - \mu(T)] \,. \tag{15}$$

Here $f(E,T)$ is the Fermi function. At finite temperature, transport coefficients always involve a finite energy width $\delta E \approx k_B T$, a crucial difference with classical wave transport. The generally accepted picture says that in the conducting regime, the diffusion constant vanishes according to a power law $D(E) \sim (E - E_c)^s$ with s some critical exponent, and E_c the mobility edge. Unfortunately, there is no consensus on the values for s. Different variants of the Anderson model (with and without broken time reversal, with and without spin, diagonal and off-diagonal spin interactions) exist that have different critical exponents. Nevertheless, the existence of critical exponents implies the transition to be continuous, as also predicted by the scaling theory of localization published in 1980, and not first-order, despite arguments put forward by Mott (see Chapter 4) in favor of the latter. The paper by Mott in 1972 [37] gives his point of view on the existence of a mimimum conductivity given the localization arguments put forward by Anderson and Thouless. In 3D he finds the rather small value $\sigma_{\min} \approx 0.025 e^2/\hbar a$. In reading the paper by Kaveh and Mott [38], published 10 years later and two years after the scaling theory had appeared, I got the impression that Mott never really gave up his minimum-conductivity version of localization.

As $T \to 0$, the Fermi level E_F is inversely proportional to the electron density and one sees that

$$\sigma(T = 0, n_e) \sim (E_F - E_c)^s \sim (n_e - n_c)^s \,. \tag{16}$$

This equation is, of course, the basis of electron localization. At the critical electron density one finds that $\sigma(T, n_c) \sim T^s$ and $K(T, n_c) \sim T^{s+1}$ [36]. Not all transport coefficients show critical behavior: at the critical density n_c the thermoelectric power reaches a large but finite value that is independent of temperature [36]. The complication in electronic systems is the presence of a second mechanism for an electronic phase transition, called Mott-Hubbard localization, and driven by electron-electron interactions. In addition, other mechanisms (phonons) may influence the thermal conductivity and thermopower.

In the insulating phase the dielectric constant $\varepsilon(E) \sim (E_c - E)^{-\nu}$ has the same critical exponent ν as the localization length $\xi(E)$ near the mobility edge E_c. At finite temperatures, conductance is still possible by thermally excited hopping from one localized state to the other. Mott showed that this "variable range hopping" leads to a temperature dependence [39]

$$\sigma(T) \sim \exp(-T_0/T)^{1/(d+1)} \tag{17}$$

in d dimensions. This law is widely observed [39].

Perhaps the most beautiful and direct manifestation of localized states in electronic 2D disordered media is the integer quantum Hall effect, observed first in 1980 [40]. Localization theory is not only capable to predict the plateau values of the Hall conductivity very precisely, it also explains the scaling in between, where the eigenstates go from localized to extended, as described by the scaling theory of localization [41, 42, 43].

2.3.2. *Mesoscopic Physics*

The study of classical wave localization is interesting because of the relatively large wavelengths involved. For classical waves it is much easier to measure the transmission coefficient from one angle (channel) to another (see Figure 2), as well as frequency and channel correlations. For a finite tube of length L with N transverse modes, classical diffusion theory tells that the ensemble-averaged transmission coefficient $T_{ab}(L)$ equals,

$$\langle T_{ab}(L)\rangle \sim \frac{\ell^*}{NL}, \tag{18}$$

where the transport mean free path ℓ^* is closely related to the diffusion constant D according to $D = v_E \ell^*/d$. Keeping definition (1) in mind, a "small" transmission may be the first signature of localization, exactly like in one dimension. In the localized regime, one may expect that

$$\langle T_{ab}(L)\rangle \sim \exp(-L/\xi), \tag{19}$$

with ξ the localization length. Such exponential decay has been reported both for microwaves [44] and light [45]. Important is, of course, to exclude absorption, which trivially leads to an exponential decay of the kind (19). In an open finite system, one can define localization as,

$$\xi(E) < L \quad \text{(localization in open system).} \tag{20}$$

We have seen earlier that - at least for an infinite medium - strict localization, in the sense of Eq. (4), is a property "for almost all" realizations

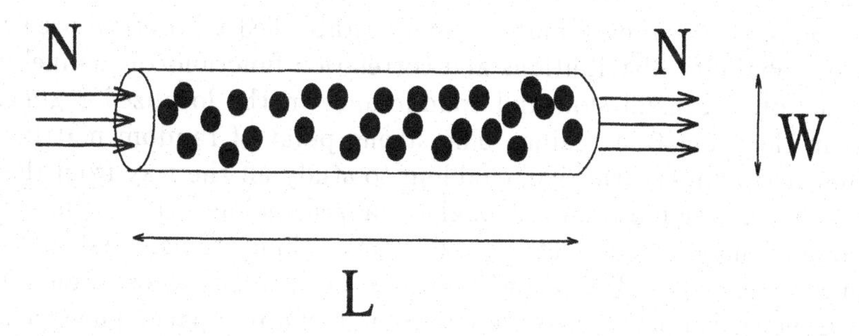

Figure 2. Geometry of a quasi one-dimensional system. The transverse size W is comparable to one mean free path; The length L is arbitrary. The tube has N input channels and N output channels.

of the disorder, and not only of a "typical" one. Of course, we would like to understand how this notion translates to the transmission coefficients. To this end it is constructive to consider not only ensemble-averages, as in Eq. (19), but the whole probability distribution. Fortunately, microwaves and acoustic waves facilitate such measurements. Except for the channel transmission coefficient T_{ab} one can consider the total transmission coefficient for a mode a,

$$T_a = \sum_b T_{ab} , \tag{21}$$

and the (dimensionless) conductance,

$$g \equiv \sum_{ab} T_{ab} . \tag{22}$$

This quantity makes the link with the conduction problem studied in electronic systems, since by the Landauer formula [21], g equals the conductance of a mesoscopic disordered system, except for a trivial factor $2e^2/h$ with the dimension of conductance. The presence of nonlocal mode correlations makes the total transmission and the conductance already nontrivial in the extended regime.

In closed systems, the energy eigenvalue distribution was seen to be a very characteristic feature. In open systems, the N transmission eigenvalues T_n of the transmission matrix T_{ab} take over this role. The transmission eigenvalue density $\rho(T)$ gives the probability density to find the eigenvalue T among the N transmission channels. Random matrix theory predicts $\rho(T)$ to be significantly different in extended and localized regime, as discussed below.

So far only random matrix theory established theoretical predictions for the probability distributions and correlation functions of channel transmission, total transmission and conductance in the localized regime, as well as in the transition regime. The strong point of random matrix theory is that it is nonperturbative, enabling to study all the way from the metallic (extended) regime to the strongly localized regime. At the time of writing, random matrix theory still has two severe limitations. First, it "only" applies to quasi-one dimensional systems, i.e. systems whose transverse width is at most comparable to the transport mean free path. The generalization to genuine 2D or 3D systems does not seem easy. Secondly, random matrix theory still is a stationary theory and is not yet capable to study spectral correlations or time tails. This last aspect is certainly going to be solved in the very near future.

2.3.3. *Statistics of Transmission in Extended Regime*
Elementary arguments show that, in the extended regime [46]

$$P(T_{ab}) = \frac{1}{\langle T_{ab}\rangle} \exp\left(-\frac{T_{ab}}{\langle T_{ab}\rangle}\right) \quad \text{(extended regime)} \tag{23}$$

i.e. the celebrated Rayleigh statistics for speckle intensities. On the basis of the law of large numbers one would expect the total transmission T_a to be Gaussian distributed with mean $\langle T_a\rangle \sim \ell^*/L$ and variance $(\Delta T_a)^2 = N\langle T_{ab}\rangle^2 \sim (\ell^*)^2/NL^2$. However, studies in the eighties have established that substantial nonlocal correlations exist between different channels [47]. This leads to the relatively large variance

$$(\Delta T_a)^2 \approx \frac{\ell^*}{NL}\,, \tag{24}$$

for the total transmission coefficient in the extended regime, i.e. a factor L/ℓ^* larger than what would have been expected if the N channels were independent. The long-range correlations also give rise to a nongaussian distribution for T_a [48].

The ensemble average $\langle g\rangle \sim N\ell^*/L$ can be related to the Ohmic conductance of a copper wire of length L and width A. This can be seen by realizing that $N \approx Ak^{d-1}$ in terms of the cross-section A of the wire and the (Fermi) wave number. The quantity $\ell^* k^{d-1}$ is nothing more than the Drude expression for the electronic conductivity, except again for a factor $2e^2/h$, containing only fundamental constants, and with the dimension of conductance. Again, due to long range correlations, fluctuations in g are much bigger than the naive result $\Delta g \sim \ell^*/L$ expected for N independent channels. The result [49],

$$\Delta g \approx 1 \tag{25}$$

is known as a Universal Conductance Fluctuation, because it hardly depends on details of the system such as dimension, size or mean free path. Random matrix theory gives the exact result $\Delta g = \sqrt{2/15\beta}$, where $\beta = 1$ for time reversal symmetry, and $\beta = 2$ for broken time reversal symmetry [21]. Thus, broken time reversal symmetry kills the universal conductance fluctuations by a factor of *exactly* two. The factor $\Delta g = \sqrt{2/15}$ is confirmed by diagrammatic calculations for quasi 1D systems. For cubic 3D systems the somewhat larger factor $\sqrt{0.296}$ is found [50].

Random matrix theory predicts a fundamental relation between the probability distributions of transmission T_a and one-channel transmission T_{ab} [52],

$$P(T_{ab}) = \int_0^\infty \frac{dT_a}{T_a/N} \exp\left(-\frac{T_{ab}}{T_a/N}\right) P(T_a) \,. \tag{26}$$

This equation relates the moments of total transmission and one channel transmission according to $\langle T_{ab}^n \rangle = n! \langle T_a^n \rangle$. One concludes immediately that any finite ΔT_a necessarily implies a deviation from the law (23). These deviations - of order $1/\langle g \rangle$ - have been observed [51]. When $\langle g \rangle = 1$, fluctuations in T_a become as large the average. In Chapter 4 we shall discuss that $\langle g \rangle = 1$ marks the onset of localization in a finite, open system. Far in the extended regime $\langle g \rangle \gg 1$, we expect both T_a and g to be normally distributed. Yet, a rigorous general theory that is also able to include the above long range correlations, is still absent. Only a perturbation theory for T_a exists [53], and a nonperturbational theory for broken time-reversal [54].

The distribution of transmission eigenvalues $\rho(T)$ in the extended regime was given by Mello and Pichard [55], using random matrix theory. In the limit $N \to \infty$,

$$\rho(T) = \frac{\ell^*}{L} \frac{1}{2T\sqrt{1-T}} \qquad \text{for } 4\mathrm{e}^{-2\mathrm{L}/\ell^*} < \mathrm{T} < 1 \,. \tag{27}$$

This distribution implies that $\langle T \rangle = \ell^*/L$, i.e. the channel average equals the ensemble-average given by Ohm's law (18). It also implies that many channels have *exponentially* small transmission, i.e. are "closed", whereas the rest has $T \approx 1$, i.e. are open.

2.3.4. *Open Chaotic Cavities*

Today, considerable attention is devoted to the analogies between disordered systems and chaotic cavities, stimulated by the many successes achieved for both. Though physically entirely different, conceptually they look very similar. In both cases, the transmission eigenvalue distribution $\rho(T)$ as well as transmission fluctuations $P(T_{ab})$ play a crucial role. For instance, for a chaotic system with N reflection channels and N transmission channels, the equivalent of Eq. (27) is [56]

$$\rho(T) = \frac{1}{\pi\sqrt{T}\sqrt{1-T}}, \tag{28}$$

which has less weight at small transmission values as compared to disordered wires in the extended regime. As a result, the mean $\langle T \rangle = 1/2$ is much bigger.

A second difference concerns the eigenvalue repulsion. For chaotic systems the Dyson ensembles predict that

$$P(T_1, T_2) \sim |T_1 - T_2|^\beta = \exp\left(-\beta \log |T_1 - T_2|\right). \tag{29}$$

i.e. the probability that two different channels have the same transmission is small. This implies the existence of strong channel correlations, just as we have seen in the extended regime of disordered systems. The eigenvalue repulsion is often written as an exponential in order to make contact with Gibb ensembles in statistical mechanics. For chaotic systems the pair potential $-\log |T_1 - T_2|$ responsible for repulsion is seen to be logarithmic.

For disordered systems, the pair potential has been calculated by Beenakker and Rejaei [57] for the case $\beta = 2$ (broken time reversal). They concluded that it coincides with the one for chaotic cavities only when $T_1, T_2 \approx 1$. For small transmission however, the "pair potential" approaches $-\frac{1}{2}\log |T_1 - T_2|$, i.e. a factor 2 smaller but still logarithmic. This result looks innocent and academical. One important consequence is that the universal conductance value in a chaotic cavity is not equal to the one ($\Delta g = \sqrt{2/15\beta}$) calculated for disordered wires in the extended regime. For an open chaotic billiard one finds the somewhat smaller value [21]

$$\Delta g = \sqrt{1/8\beta}. \tag{30}$$

2.3.5. *Statistics of Transmission in Localized Regime*

Pichard and Sanquer [83], followed by Van Langen, Brouwer and Beenakker [54] predict both channel transmission and total transmission to be distributed log-normally in the localized regime, i.e.

$$P(T) = \frac{1}{\sqrt{4\pi}T}\frac{\xi}{L}\exp\left[-\frac{\xi}{4L}\left(\frac{L}{\xi} + \log T\right)^2\right]. \tag{31}$$

Here ξ is the localization length and in random matrix theory given by $\xi = \beta N\ell^*/2$, i.e. essentially the number of modes times the transport mean free path. I recall that the localized regime is defined by Eq. (20), which leads to $N\ell^* > L$. From Eq. (31) it follows that

$$\langle \log T \rangle^2 = -2 \langle \log T \rangle \,. \tag{32}$$

As a result $-(\log T)/L$ converges "almost surely" to $1/\xi$ as the length of the sample exceeds the localization length. This notion provides a crucial difference between absorption and localization, who both yield an exponentially small average transmission. In the localized regime the lognormal distribution predicts a peak at a very small value but, in sharp contrast to absorption, a very significant tail towards large values. It is interesting to note that by breaking time-reversal symmetry ($\beta = 1 \to 2$), the localization length actually doubles. The distribution of the conductance g has also been argued to become lognormal [58, 21], implying that the universal conductance fluctuations (25) disappear in the localized regime.

The transmission eigenvalue distribution $\rho(T)$ in the localized regime is predicted to be quite different ¿from the one in the extended regime. It is convenient to introduce the variable x according to $T = 1/\cosh^2 x$ so that, according to Eq (27), x is uniformly distributed between 0 and $L/\ell^* \gg 1$ in the extended regime. In the localized regime the distribution $\rho(x)$ is predicted to "crystallize": It exhibits maxima at periodic positions of x [21]. In this respect the transition from the extended to the localized regime looks like a liquid-solid phase transition.

2.3.6. *Ab initio studies*

Ab-initio studies try to solve a "real" disordered system numerically exactly. Consider the scattering of an incident plane wave $\psi_0(\mathbf{r})$ ¿from M identical scatterers, with scattering matrix t, distributed randomly in d dimensions. In the simplest case that these can be represented by point particles (a severe simplification), the solution of this problem requires the diagonalization of a complex-valued $M \times M$ matrix. In that case, the equation to be solved is [59]

$$\psi(\mathbf{r}_i) = \psi_0(\mathbf{r}_i) + t \sum_{j \neq i}^{M} G(\mathbf{r}_{ij})\, \psi(\mathbf{r}_j) \,. \tag{33}$$

The matrix $G(\mathbf{r}_{ij})$ describes a scattered spherical wave emitted by scatterer i and propagating to scatterer j. One wishes to solve this equation in a medium with typical size L and typical density M/L^d. To guarantee multiple scattering, the system size must exceed the mean free path ℓ. Assuming that $\ell \approx 1/n\sigma$ and assuming optimal scattering $\sigma \sim \lambda^{d-1}$ one learns that $L < \lambda\, M^{1/(d-1)}$. In 3D and for $M = 1000$ one can consider a medium which is at most 30 times bigger than the wavelength. Of resonance, in particular for pure Rayleigh scattering, the maximal size is even smaller.

Equation (33) can be written as [60, 61]

$$\Psi(\{\mathbf{r}_i\}) = \mathbf{M}(\{\mathbf{r}_i\})^{-1} \cdot \Psi_0(\{\mathbf{r}_i\}), \tag{34}$$

where $\{\mathbf{r}_i\}$ stands for one realization of the scatterers, $\mathbf{M} = \mathbf{I} - \mathbf{t}\mathbf{G}$ and Ψ is the M-dimensional vector containing the wave function at the scatterer positions. From this equation it is clear that an eigenvalue of $\mathbf{M}$ close to zero corresponds to a large scattered field, typically the case for a resonant frequency. An eigenvalue exactly equal to zero would actually correspond to a localized state, that would even persist in the absence of incident field. However, scattering outside the system prevents states to be really localized. The clustering of eigenvalues near zero in a certain frequency interval may thus be a signature of localized states.

The t-matrix of the particles depends on frequency. By energy conservation (optical theorem) it has the form $t = c[\exp(2i\phi)-1]$ with c a real-valued constant and $\phi(\omega)$ the scattering phase shift. The eigenvalues of $\mathbf{M}$ and $\mathbf{G}$ are related by $\lambda_{\mathbf{M}} = 1 - t\lambda_{\mathbf{G}}$. An eigenvalue $\lambda_{\mathbf{G}}$ with $\mathrm{Re}\,\lambda_{\mathbf{G}} = -1$ allows exactly one choice for ϕ such that $\lambda_{\mathbf{M}} = 0$, the choice depending on $\mathrm{Im}\,\lambda_{\mathbf{G}}$. The frequency associated with this phase shift would then correspond to a localized state.

In Figure 3 we show the outcome of numerical work done by Rusek, Orlowski and Mostowski [60, 61]. It shows contour plots of the density of eigenvalues $\lambda_{\mathbf{G}}$ for randomly positioned point cylinders i.e the numerical solution of Eq. (33) in 2D. It can be seen that as the size of the system grows, eigenvalues start to cluster more and more around the point $\mathrm{Re}\,\lambda_{\mathbf{G}} = -1$, the convergence being fastest for $\mathrm{Im}\,\lambda_{\mathbf{G}} = 0$. This value corresponds to a phase shift $\phi = \pi/2$ i.e. resonating scatterers. In principle one would expect all states to become localized in 2D, even the ones with very small scattering phase shift ϕ (which in Fig. 3 would correspond to $\mathrm{Im}\,\lambda_{\mathbf{G}} \gg 1$). But to see that one has to increase the system - at constant number density - to astronomical lengths.

3. Anderson Model

The simplest model that exhibits basic features of localization is the Anderson model. This model describes the hopping of a tightly bound electron from one site to the other. Disorder is introduced by a random potential ε_n at each site, so that the Hamiltonian becomes,

$$H = \sum_{n \in S} \varepsilon_n \,|n\rangle\langle n| + V \sum_{n,m \in S}{}' |n\rangle\langle m| \,. \tag{35}$$

$E = 0$ corresponds to the tightly bound state in the absence of quantum tunneling ($V = 0$). The set S indicates the location of the sites $|n\rangle$ in real

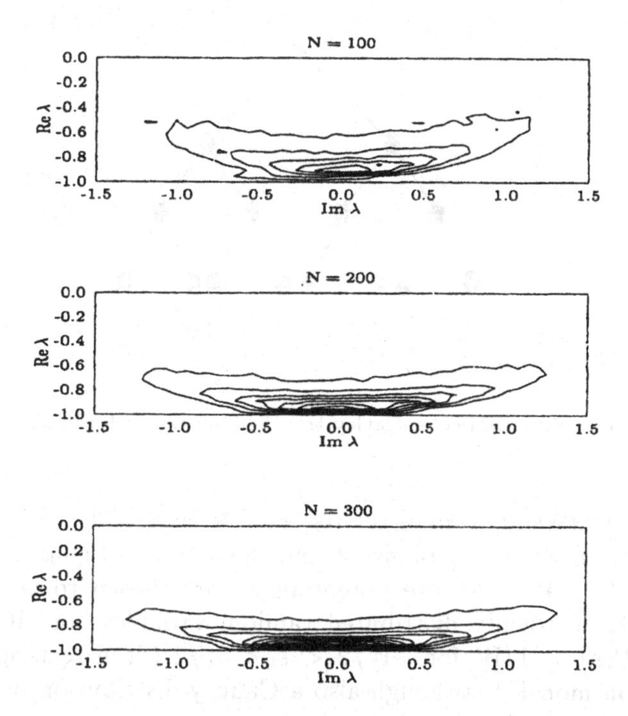

Figure 3. Contour plot of the density of complex-valued eigenvalues λ_G calculated for 10^3 different distributions of randomly positioned point cylinders. The number of particles N increases from top to bottom, at constant number density n, chosen to be one scatterer per wavelength squared. The clustering of eigenvalues near $\text{Re}\,\lambda = -1$ marks the onset of localization for the infinite system. Taken from Rusek, Orlowski and Mostowski [61], with kind permission from the authors.

space, which can e.g. be Z^d for the simple cubic lattice in d dimensions. Usually, the summation $\sum'_{n,m}$ runs over nearest neighbours only. This kinetic "off-diagonal" part describes the hopping ¿from one site to the other, and is in the standard Anderson model equal to a deterministic hopping matrix element V. Also models with off-diagonal disorder have been discussed, where V is random rather than ε, modeling situations with either broken time-reversal symmetry or spin-orbit scattering. They are believed to be part of a different universality class, characterized by different critical exponents at the transition [62, 63], and sometimes even different transitions.

A crucial property in the model is the number of nearest neighbours Z. The simple cubic lattice in dimension d ($Z = 2d$) is the simplest choice, but - despite its exotic structure - also the Cayley tree (or Bethe Lattice, see Figure 4) with connectivity K ($Z = K + 1$) is of interest because for this geometry, like in one dimension, the Anderson model can be solved exactly.

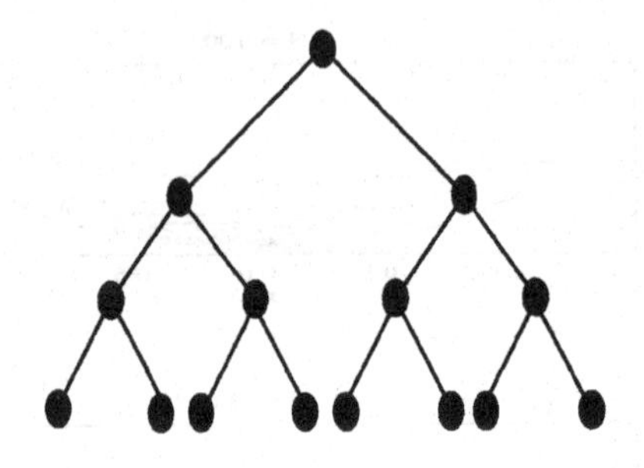

Figure 4. The Bethe lattice or Cayley tree, here drawn with connectivity $K = 2$.

In both cases, the spectrum of the regular system ($\varepsilon = 0$) is determined by the number of nearest neighbours Z and the amount of quantum tunneling V: $E \in [-ZV, ZV]$. The site potentials ε_n are chosen to be statistically independent, identically distributed random variables, and it is common to choose $P(\varepsilon) = 1/W$ for $-W/2 < \varepsilon < W/2$ ("the rectangular diagonal Anderson model"), although also a Cauchy distribution or a Gaussian distribution for ε is frequenctly encountered.

The Anderson model with diagonal disorder and its numerical solutions have been reviewed recently in the book by Ping Sheng [64]. Frohlich and Spencer [20] proved strong localization for the Anderson model with diagonal disorder on a cubic lattice in any dimension, in the sense of criterion (4), provided that either the disorder W is sufficiently large, or, for any disorder, at energies sufficiently outside the spectrum of the regular model: $|E - 2dV| \gg 0$. This proof was later refined by Aizenman and Molchanov [65].

3.1. ANDERSON MODEL IN 1D

The extensive literature justifies to spend a few extra words on the one dimensional Anderson model. Some people consider 1D localization not interesting because there is no mobility edge. The critical behavior at the mobility edge is often considered as the most important aspect of Anderson localization.

The study of 1D localization goes back to the early sixties where random matrix techniques by Furstenberg and Osseledec were first applied. The basic starting point is the product of n statistically independent random transfer matrices, $\mathbf{M}_1(E) \cdot \mathbf{M}_2(E) \cdots \mathbf{M}_n(E)$. Transfer matrices typically show up in one dimensional problems and are characterized by $\det \mathbf{M}_n = 1$.

The eigenvalue equation for the 1D Anderson model can be written as,

$$V\left(\psi_{n-1}+\psi_{n+1}\right)+\varepsilon_n \psi_n = E\psi_n\,, \tag{36}$$

where $\psi_n = \langle n|\psi_E\rangle$ is the eigenfunction at site n. Note that $\psi_{n+1}+\psi_{n-1}-2\psi_n$ is a discrete version for the Laplacian in 1D. Equation (36) is equivalent to,

$$\begin{pmatrix} \psi_n \\ \psi_{n+1} \end{pmatrix} = \begin{pmatrix} 0 & 1 \\ -1 & (E-\varepsilon_n)/V \end{pmatrix} \cdot \begin{pmatrix} \psi_{n-1} \\ \psi_n \end{pmatrix}, \tag{37}$$

where the matrix is identified as $\mathbf{M}_n$. Evidently, ψ_n for large n is determined by a product of the kind $\mathbf{M}_1(E)\cdot\mathbf{M}_2(E)\cdots\mathbf{M}_n(E)$.

Furstenberg's theorem says that the norm of this matrix is, for "almost every" realization, proportional to $\exp[n\gamma(E)]$, where for any energy E, the Lyapunov exponent $\gamma(E) > 0$. A simple proof, though restricted to 2×2 matrices has been given by Delyon, Kunz and Souillard [66]. Ossedelecs theorem says that if $\gamma(E) > 0$, for" almost every" realization of the system, one *unique* vector $\psi_0(E)$ exists such that the product $\mathbf{M}_1(E)\cdot\mathbf{M}_2(E)\cdots\mathbf{M}_n(E)\cdot_0$ decays as $\exp[-n\gamma(E)]$. Note that, by Furstenberg's theorem applied to $\mathbf{M}_n(E)^{-1}\cdots\mathbf{M}_2(E)^{-1}\cdot\mathbf{M}_1(E)^{-1}$, there is at least one such a vector.

Osseledec's Theorem excludes - for almost all realizations - a solution of the 1D Anderson model to be part of the absolute continuous spectrum, since solutions either decay or grow exponentially. The solution at energy E will be localized when the vectors $\psi_0^{\pm}$ at some site for left- $(n\to-\infty)$ and righthand $(n\to\infty)$ decay coincide, assuring exponential decay at both sides. If not, E is not part of the spectrum. The subtle point here is that the set of realizations with probability zero for which Osseledec's Theorem does not hold, depends on the energy. Statements for almost all realizations and all energies cannot be made. As a result one can only state that "for almost all realizations", and then for "almost any" energy, the 1D Anderson model has a pure point spectrum and is thus localized. It is still possible that the whole spectrum coincides with the set of measure zero that we don't know anything about! One example is by simply putting $\varepsilon_n = \lambda\cos(2\pi\nu)$, with ν some irrational number and $\lambda > 2$. Except for exotic potentials, the 1D Anderson model is - under very broad conditions - localized in the sense of definition (4).

We shall briefly state some major findings of the 1D Anderson model. One of the most important fundamental results - formulated first by Thouless [67] - is the Kramers-Kronig relation between Lyapunov exponent $\gamma(E)$ and density of states (per site) $\rho(E)$, showing that $\gamma(E)+i\rho(E)$ is actually an analytic function when continued in the complex sheet $\mathrm{Im\,E} > 0$. Kappus and Wegner [68] and Lambert [71] showed that the Lyapunov exponent near the band center $E = 0$ takes the form,

$$\gamma(E) = \frac{\langle \varepsilon^2 \rangle}{V^2} F\left(\frac{EV}{\langle \varepsilon^2 \rangle}\right) \quad (W, |E| \ll V). \tag{38}$$

with $F(x)$ a smooth function, obeying $F(0) \approx 8.7537$ and $F(x \to \pm\infty) = 8$. The result $\gamma = 8\langle \varepsilon^2 \rangle / V^2$ coincides with the perturbational result first obtained by Thouless [69]. The presence of the function $F(x)$ denotes that perturbation theory breaks down for $|E| < \langle \varepsilon^2 \rangle / V$, which is called an *anomaly*. The anomaly is weak and the localization length basically scales as $\xi(E) \approx V^2/8\langle \varepsilon^2 \rangle$.

Near the band edges $E = \pm 2V$ of the regular system one obtains a rather nontrivial scaling relation as $\langle \varepsilon^2 \rangle \ll V^2$,

$$\gamma(E) = \frac{\langle \varepsilon^2 \rangle^{1/3}}{V^{2/3}} H\left(\frac{V^{1/3}(|E| - 2V)}{\langle \varepsilon^2 \rangle^{2/3}}\right). \tag{39}$$

The function $H(x)$ has been explicitly calculated by Derrida and Gardner [70], and satisfies $H(0) = \sqrt{\pi}6^{1/3}/2\Gamma(\frac{1}{6}) \approx 0.2893$, $H(x \to \infty) = \sqrt{x}$ and $H(x \to -\infty) = -1/8x$. At the band edge we thus find the nonanalytic scaling $\gamma(E) \sim (\delta\varepsilon/V)^{2/3}$. The scaling relations (38) and (39) can actually be derived using the elegant renormalization procedure developed by Bouchaud and Daoud [72].

It is worth mentioning that for the Cauchy distribution, the Anderson model with diagonal disorder allows for an exact solution of the Lyapunov exponent for all energies and for all disorders. This model is often called the Lloyd model. Assuming the distribution to be $P(\varepsilon) = W^2/\pi(W^2 + \varepsilon^2)$, one finds [73]

$$\gamma(E, W) = \log|z(E, W)|, \text{ with } \mathrm{z} + \frac{1}{\mathrm{z}} = \mathrm{E} + \mathrm{iW}. \tag{40}$$

which exhibits a different scaling $\gamma = \sqrt{W/2}$ on the band edges, as compared to distributions with finite variance.

The 1D Anderson model has an analogy in the scattering of classical waves from random layers, where the relations (38) and (39) have been seen to apply. The anomalies in the band center of the optical model are much more pronounced, leading sometimes to giant localization lengths [74]. In Fig. 5 I show the anomaly function $F(x)$ defined by Eq. (38), calculated light propagation in a random dielectric stack. The band center is here chosen to be right in the middle of a pass band. The scaling (38) is confirmed for small fluctuations in the dielectric constant, with an almost vanishing Lyapunov exponent in the band center. The solid line denotes the prediction of a

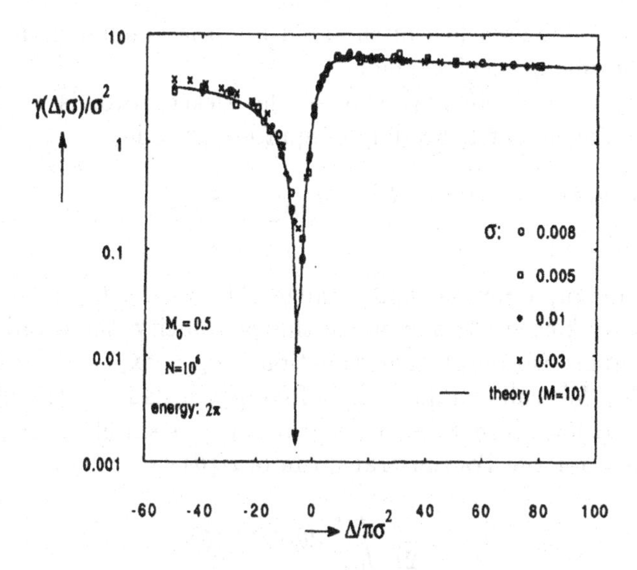

Figure 5. Numerical simulation of the Lyapunov exponent in a stack composed of 10^6 random dielectric layers with a contrast $m_0 = 0.5$ in their index of refraction. Δ denotes the distance to the band center of a transmission band; σ is the standard deviation of the fluctuations in the dielectric constant. The solid line denotes the prediction of degenerate perturbation theory. Taken from Van Tiggelen and Tip [74].

modified version of the perturbation theory by Lambert [71]. The agreement is very good.

It has been shown that 1D localization of waves is unstable against small isotropic scattering [75]. This is bad news for seismology, where 1D localization has been investigated in view of the layered structure of the Earth crust [10]. Localization of light in 1D layered media has also been discussed in relation with gain [76]. It turned out that gain, just like absorption, *enhances* the Lyapunov exponent, somewhat surprising since gain may have been expected to enhance transmission.

3.1.1. *Kicked Rotor*

The Anderson model in 1D has seen an interesting application in the quantum mechanical treatment of a kicked rotor. The classical picture of a rotor with moment of inertia I that is kicked abruptly at times t_n with a torque $\partial V/\partial\theta$ is easy. The equations of motion for the classical rotor after kick $\sum_n \Delta t\, \delta(t - t_n)$ easily show that the angular momentum $J_n = J(t_n + 0)$ will finally increase in time as,

$$\lim_{n\to\infty} \frac{J_n^2}{n} = (\Delta t)^2 \left\langle \left(\frac{\partial V}{\partial\theta}\right)^2 \right\rangle . \tag{41}$$

This looks like a diffusion law of the kind (2), the angular momentum J_n taking over the role of displacement $\mathbf{r}$.

The quantum mechanical version of the kicked rotor is given by the following time-dependent Schrödinger equation ($\hbar = 1$),

$$i\frac{\partial\psi(\theta,t)}{\partial t} = \frac{1}{2I}\frac{\partial^2\psi(\theta,t)}{\partial\theta^2} + V(\theta)\Delta t\sum_n \delta(t-t_n)\psi(\theta,t)\,. \tag{42}$$

Following Fishman, Grempel and Prange [11] we set $t_n = nt_0$, i.e. the rotor is regularly kicked. As a result of this periodicity, the solution can be written as a superposition of "eigenfunctions" $\psi_\omega = \exp(-i\omega t)u(\theta,t)$, with $u(t) = u(t+t_0)$. We call ω a quasi eigenfrequncy because $u(t)$ still depends on time. Let $U(\theta) = \frac{1}{2}[u(\theta,t_n-0)+u(\theta,t_n+0)]$. A simple closed equation can be obtained for the Fourier transform of $U(\theta)$,

$$U_k = \frac{1}{2\pi}\int_0^{2\pi} d\theta\, \mathrm{e}^{\mathrm{i}k\theta}\, \mathrm{U}(\theta) \tag{43}$$

which is nothing more than Fourier component of the eigenfunction corresponding to a quantized angular momentum $J_k = k$, which reads

$$\sum_{l\neq 0} V_l U_{k+l} + \varepsilon_k U_k = EU_k\,, \tag{44}$$

This equation looks like the 1D Anderson model (36). The "random site potential" is given by $\varepsilon_k = \tan(\frac{1}{2}\omega t_0 - \frac{1}{4}t_0k^2/I)$. The hopping element V_l is given by the Fourier transform (43) applied to $\tan[\frac{1}{2}\Delta t V(\theta)]$. Finally, the "energy" $E = -V_0$. Compared to the standard Anderson model, Eq. (44) allows for "hopping" (discrete transitions in the angular momentum) beyond nearest neighbours. Only the somewhat special potential $V(\theta) = (2/\Delta t)\arctan(\mathrm{v}\cos\theta - \mathrm{w})$ has nearest neighbour hopping only, with strength v.

Without kicking (V = constant), quasi eigenfrequencies $\omega_k = -\arctan \mathrm{w} + \frac{1}{2}\mathrm{k}^2\mathrm{t}_0/2\mathrm{I} \bmod(2\pi)$ exist with eigenfunction $u_n = \delta_{nk}$ "localized" at site k, i.e. with constant angular momentum k. In the presence of kicking we notice that, during the kicking from one site to the other, the variable $\frac{1}{2}\omega t_0 - \frac{1}{4}t_0k^2/I \pmod{\pi}$ becomes a pseudo-random variable uniformly distributed between $-\pi/2$ and $\pi/2$ provided that $t_0/4\pi I$ is an *irrational* number. As a result, ε_k is pseudo-random with a Cauchy distribution and the model resembles the Lloyd model discussed above. We may expect all wave functions to localized in n space, with a localization length given by Eq. (40). Localization occurs in angular momentum space, i.e. the probability of finding high angular momenta is exponentially small *at all times*. Hence the name "dynamical localization". This statement clearly violates the classical result (41). The validity of the Lloyd model for the exponential decay of the wave function has been verified [11].

The localization principle tells us that the quasi eigenfrequency spectrum is discrete. All localized states that overlap a given site k are separated in frequency. Assuming the rotor to be at rest at $t = 0$, one expects the time dependent wave function,

$$\psi_k(t) = \sum_n \exp(-i\omega_n t) U_k^{(n)} \bar{U}_0^{(n)} , \tag{45}$$

to be a superposition of some countable set of incommensurate frequencies, making it quasi periodic in time. As a result, the probability of finding the oscillator at the original site, i.e. with the same kinetic energy is finite at any time. In section 2.1 a finite "return probability" was argued to be a hallmark for localization.

A number of questions come up by this analogy. The first is mathematically interesting but difficult to answer in general: What kind of "pseudo-randomness" in the Anderson model is sufficient to make it equivalent to one with real randomness? The second question touches the heart of modern quantum mechanics. By the correspondence principle, the classical solution must be some limit of quantum mechanics. Yet, the last one gives non-diffuse behavior, whereas the first yields diffuse behavior for the angular momentum of the kicked rotor. The solution of this "coherence" problem is believed to reside in the coupling of the quantum object with the environment. As this coupling increases, the classical solution is believed to apply more and more.

Ammann etal. recently reported the observation of decoherence in a gas of ultracold cesium atoms, subject to a pulses wave of light [77]. The model of the quantum kicked rotor applies here. The decoherence is in this case caused by the spontaneous emission, which destroys the phase of the Cesium atoms. A significant modification to Eq. (41) was observed as the amount of spontaneous emission decreased. Dynamical localization in more than one dimension has also been reported in literature [78].

3.2. ANDERSON MODEL IN 2D AND 3D

The original paper of Anderson dealt with the Hamiltonian (35) assuming many nearest neigbours, and a rectangular distribution $P(\varepsilon) = 1/W$ for $-W/2 < \varepsilon < W/2$. Anderson showed that eigenstates become localized for sufficiently large value for W/V. The critical value for W/V increases with the number of nearest neighbours, implying that localization becomes more and more difficult as the dimension of the system increases. The dependence of localization upon dimensionality is thus very basic.

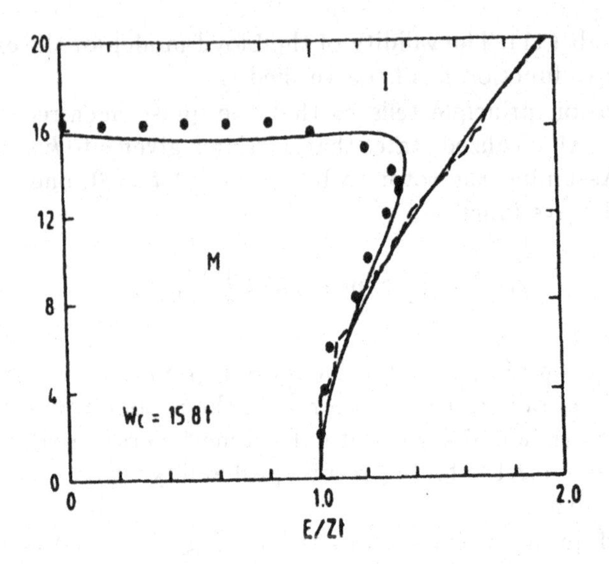

Figure 6. Phase diagram of the rectangular Anderson model with diagonal disorder on a 3D cubic lattice. The solid triangles are exact numerical results for the mobility edge. The solid line denotes the outcome of the selfconsistent theory of localization, discussed in Chapter 5. The thin solid line denotes the prediction of effective medium theory for the shift of the band edge beyond $E = 6V$. Taken from Vollhardt and Wölfle [7], with kind permission from the authors.

We now know that the Anderson model is fully localized in one and two dimensions ($Z = 2$ and $Z = 4$) for *any* value of W/V. Only the off-diagonal Anderson model with spin-orbit scattering seems to have a transition in 2D [62]. For $d > 3$ the spectrum exhibits a mobility edge E_c separating two regimes with exponentially localized and extended eigenfunctions, separated by a "mobility edge". This mobility edge was first suggested by Mott. The localized states start to appear at the edges $E = \pm ZV$ of the spectrum, and move gradually towards the band center as the disorder increases. In Fig. 6 we show the phase diagram on the basis of a sophisticated numerical solution of the Anderson model on a simple cubic lattice in three dimensions. It can be inferred that the extended regime completely disappears for $W/V > 16.5$ (or equivalently $\Delta\varepsilon/V > 4.7$), which is not far from the value of 26 estimated by the back-of-the-envelope calculation of Anderson. For the Cauchy distribution $P(\varepsilon) = (1/\pi)\, W/(\varepsilon^2 + W^2)$ the critical value in 3D is $W/V = 3.8$ whereas the Gaussian distribution has $W/V = 20.9$ [79].

An interesting consequence of the random potential is the formation of localized states outside the spectrum of the regular system. This phenomenon has been studied extensively, and it has turned out that the den-

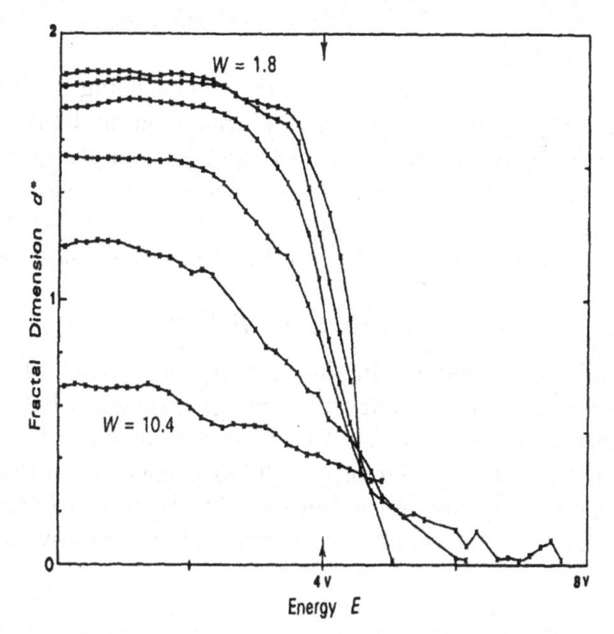

Figure 7. Fractal dimension, defined in Eq. (11) of the 2D Anderson model with diagonal disorder, as a function of energy. $E = 4V$ is the band edge of the model without disorder. Taken from Schreiber [27], with kind permission from the author.

sity of states $\rho(E)$ follows a universal curve $\rho(E) \sim \exp[-|E - E_b|^{-d/2}]$, called a Lifshitz tail [24, 25], provided the spectrum is bounded (with bound E_b i.e. the random potentials ε are bounded. This is not true if their distribution is Gaussian. In that case one finds [24, 79] $\rho(E) \sim \exp[-C|E|^{2-d/2}]$. In 1D, one finds more precisely $\rho(E) \sim |E|^{3/4}W^{-5/3})\exp[-C|E|^{3/2}/W^2]$ [70].

In Figure 7 we show the fractal dimension β, defined in Eq. (11), for the 2D Anderson model on a simple 2D square lattice. In the center of the band ($E = 0$) β is somewhat smaller than 2, indicating that these states are nearly extended. Beyond the band edge $E = 4V$ of the regular system, the exponentially localized Lifshitz tail makes that $\beta = 0$, even at modest disorders. As the disorder W increases, the localization length decreases, and the appearance of exponentially localized states ($\beta = 0$) becomes more and more evident. In 2 dimensions, in principle all states are exponentially localized, but the localization length ξ may be macroscopically large. Economou etal. [82] give the approximate formula,

$$\xi(E = 0) = 2.72\ell(0)\exp\left(\frac{\sqrt{2}\pi\ell(0)}{a}\right) \tag{46}$$

for the localization length in the band center in terms of the mean free path ℓ

and the lattice spacing a. In 3D one finds a fractal dimension $\beta = 1.54 \pm 0.08$ at the mobility edge [80].

As mentioned already in section 2.1, the level spacing distribution function is sensitive to localization. For the 3D Anderson model it has explicitly been verified that the distributions (13) and (14) apply in the extended regime, respectively with and without broken time reversal, whereas a totally uncorrelated (Poisson) distribution emerges in the localized regime. At the critical disorder $W = 16.5\,V$ a critical distribution shows up [84],

$$P(s) \approx As \exp(-Bs^{1.25}), \tag{47}$$

that seems to be independent on the presence or absence of time reversal symmetry. Near the mobility edge the critical exponent of the localization length is inferred to be $\nu \approx 1.4$ [81], largely independent of the chosen distribution for ε_n and significantly different from the mean field value $\nu = 1$ obtained by approximate (mean field) theories. Surprisingly, this critical exponent is seen to be left unchanged if time-reversal is broken [84, 85].

3.3. CAYLEY TREE

The Anderson model on the Cayley tree has first been discussed by Abou-Chacra, Anderson and Thouless [86], with important contributions later by Kunz and Souillard [87] and Kawarabayashi and Suzuki [88]. It was realized by Abou-Chacra etal. that a significant approximation made in the original Anderson paper necessary to solve the tight-binding problem, is actually not an approximation for the Cayley tree. This means that the original argument by Anderson for localization is rigorous for this system. The Cayley tree is special because it has no closed paths and the sites on the boundary are as numerous as inside the system. In comparing the number of nearest neighbours $K+1$ to the number $2d$ for a simple cubic lattice, the connectivity K can be identified as an effective dimension d by means of the relation $K+1 = 2d$.

It can be shown that the exponential decay rate $\gamma(E)$ of the Green's function (in units of the lattice spacing), defined as,

$$\gamma(E) = -\lim_{j\to\infty} \frac{\log|G(0,j,E)|}{j}, \tag{48}$$

is self-averaging and given by,

$$\gamma(E) = -\log|V| + \left\langle \log|G_{>}(i,i,E)|^{-1} \right\rangle . \tag{49}$$

The symbol > indicates that the return Green's function $G(i,i)$ is calculated without counting paths on earlier levels of the tree. The absence of closed paths on the Cayley tree guarantees that the $G_>(i,i)$ with i on the same level of the tree are all mutually independent. This notion enables to solve for the whole distribution of $\tilde{G}(i,i)$, and to evaluate the decay (49) of the Green's function straightforwardly.

However, on the Cayley, Eq. (48) does not yet guarantee strict localization. The number of sites at distance L grows exponentially with L according to $(K+1)K^{L-1}$, and not algebraically L^{d-1} as in Euclidean space. In order to have strict localization, the exponential decay of the Green's function must overcome this exponential growth. Hence, the criterion for localization becomes [88],

$$\frac{1}{\xi(E)} \equiv \gamma(E) - \log K > 0\,. \tag{50}$$

This immediately confirms the conjecture of Anderson that localization becomes more difficult as the dimensionality $K \sim d$ increases. It can be shown that the critical exponent of the localization length $\xi(E)$ near the mobility edge, is unity, i.e. $\xi(E) \sim |E - E_c|^{-1}$.

The calculation of $\gamma(E)$ looks very much like the one of the Lyapunov exponent of the one dimensional Anderson model discussed earlier in Eq. (40). In fact, for $K = 1$ it *is* exactly the 1D Anderson model. Just like in 1D, an analytical solution for $\xi(E)$ exists, valid for all energies and any disorder , for a Cauchy distribution of the potential ε, $P(\varepsilon) = W/\pi(\varepsilon^2 + W^2)$. By criterion (50) the mobility edge separating localized and delocalized states is given by [88]

$$\frac{E^2}{(K+1)^2} + \frac{W^2}{(K-1)^2} = V^2\,. \tag{51}$$

In particular, all states become localized when $W/V > K - 1$, a conclusion that can be compared qualitatively to Anderson's "best estimate" (Figure 3 of Ref. [14]). For $K = 1$ all states are localized for any disorder, as we know already from other work. For $K = 3$, the equivalent of two dimensions on the Cayley tree, we notice the existence of a mobility edge, contrary to the 2D Anderson model on a cubic lattice.

In Fig. 8 we show a comparison of the phase diagram for the Anderson Model on the Cayley tree ($K = 5$) and the one on a 3D simple cubic lattice, obtained by finite size scaling [79], both using a Cauchy distribution of the site potentials. The agreement is remarkable. In higher dimensions the agreement is expected to be even better.

4. Great Principles

Simplified models give a good impression of what is going on and when localization can be expected to occur. More realistic models, hopefully closer

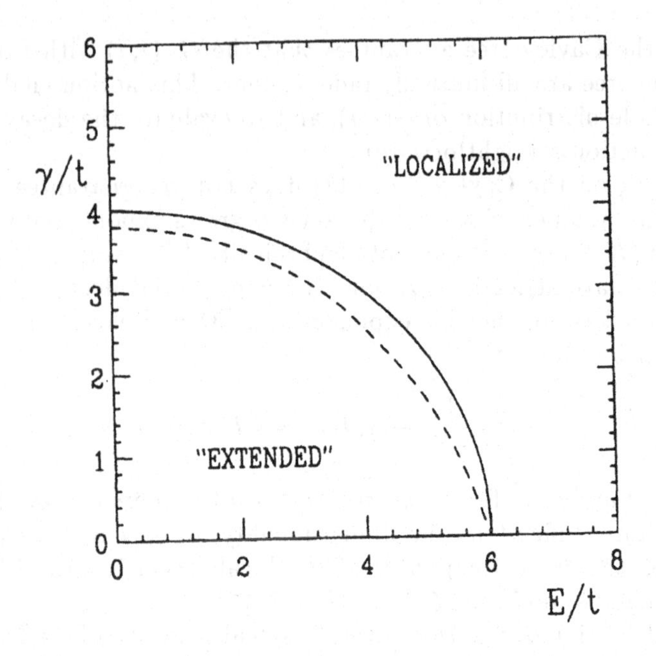

Figure 8. Mobility edge trajectory obtained for a Anderson model with diagonal Cauchy disorder. Solid line line is Eq. (51) for the Cayley tree with $K = 5$, dashed is the exact numerical solution for the 2D cubic lattice. Taken ¿from Kawarabayashi and Suzuki [88], with kind permission from the authors.

to experiments, are hardly ever exactly solvable. The question is whether we can formulate "great principles" that provide basic features of localization, and that apply to more complicated systems as well, such as ones with absorption, with short or long-range correlations, anisotropic and continuous models (described by spatial and/or time-dependent correlation function rather than with discrete random potentials). This section addresses three great principles.

4.1. IOFFE-REGEL CRITERION

The mean free path (or extinction length) ℓ denotes a typical length scale of a particle or wave packet between two subsequent collisions. For a low number density n of particles with total scattering cross-section σ one derives the familiar expression $\ell = 1/n\sigma$. A long mean free path denotes that "disorder is weak".

The important length scale for interference effect is - of course - the wavelength $\lambda = 2\pi/k$. What happens if the mean free path becomes comparable to the wavelength? It means that interference continues to dominate during scattering. It signals a regime in which scattering can no longer be described classically, and some kind of catastrophe has to occur. This kind of reasoning led Ioffe and Regel to a criterion for localization in disordered semi-conductors,

$$k\ell < \text{constant} \approx 1\,, \tag{52}$$

now called the Ioffe-Regel criterion. Actually, this reasoning is only a very small part of the extensive paper by Ioffe and Regel [89]. In my opinion, their intention was more to identify the "strongly scattering" regime, where one cannot longer identify multiple scattering with a subsequent propagation of plane waves, but rather one with overdamped waves. The criterion for this regime look like a Ioffe-Regel criterion: $k\ell < 0.5$ [90]. Equation (52) seems to state that localization and "strong scattering" are both complicated, and must therefore be related. It was who Mott who really made the link between the Ioffe-Regel criterion (52) and localization of waves. He also estimated the constant in Eq. (52) to be of order unity.

Many physicists have criticized the Ioffe-Regel criterion. Their main objection is the fact that ℓ is a length scale of the field, according to Eq. (10), whereas localization is (at least) a property of the intensity, as indicated in Eq. (9). In spite of these arguments, the Ioffe-Regel criterion has been verified to hold quite accurately in the 3D Anderson model, the constant ranging from 0.8 to 1.0 [64]. In addition, it also emerges from approximate theories such as the self consistent theory of localization [15, 16] and quantum field theories [19]. It is now generally accepted as an approximate criterion for localization of waves in infinite 3D random media. I stress that it is not a *universal* criterion, in the sense that the constant is known to vary from one model to the other, and perhaps even from experiment to experiment, as opposed to critical exponents around the transition [91].

It is important to note is that the mean free path figuring in the Ioffe-Regel criterion is the *scattering* mean free path (extinction length), and not the transport mean free path ℓ^*. I recall that the latter is defined by means of the relation $D = v_E \ell^*/d$. The DC electronic conductivity can easily be shown to be

$$\sigma \approx \frac{1}{d}(e^2/h)\ell^* k^{d-1}\,. \tag{53}$$

If we accept definition (1) as then $\ell^* = 0$ at the mobility edge. The velocity v_E is not believed to vanish.

The confusion of transport and scattering mean free path can lead to erroneous conclusions, of which I mention one of historical importance. In his discussion of the Ioffe-Regel criterion, Mott concluded that if $\sigma = 0$ in the localized regime, and finite in the extended regime given by criterion

(52), a *minimum conductivity* $\sigma_{\text{min}} \approx (e^2/h)k^{d-2}$ must exist, suggesting the Anderson transition to be first-order. In 2D this minimum conductivity would even be a fundamental constant. Although this argument still does not rule out a minimum conductivity, it is now widely accepted that the transition is continuous, and that the Mott minimum conductivity does not exist.

4.1.1. *Generalizations*

Several authors have tried to find a microscopic base of the Ioffe-Regel criterion. Both the selfconsistent theory of localization and non-linear field methods suggest that if one accepts Eq. (1) as a working definition for localization, the Ioffe-Regel criterion follows, the constant being somewhat dependent on the model but always of the order of unity. It would be tempting to apply the criterion in dimensions less then three. Yet, we know that in principle all states are localized. Non-linear sigma models are theories that expand around the critical dimension $d = 2$, and extrapolate towards higher dimensions [18, 19]. The outcome is a generalized criterion for d dimensions,

$$k\ell = \left(\frac{1}{2\pi(d-2)}\right)^{1/(d-1)} . \tag{54}$$

This formula confirms that the criterion for localization becomes more severe in higher dimensions, a conclusion that followed already from the work of Anderson, and clearly identifies $d = 2$ as the critical dimension for the presence of a mobility edge.

For the 3D Anderson model, the Ioffe-Regel criterion was derived by Zdetsis, Soukoulis and Economou [82, 92], starting from the self consistent theory of localization, to be discussed later. They noticed that the formula for the mobility edge has a lot in common with the well-known condition for a 3D potential barrier to have a bound state. On the basis of this "potential well analogy" they concluded to mobility edge to be given by,

$$S(E)\ell^2 \approx 8.96 \,, \tag{55}$$

where $S(E)$ is the constant-energy (or -frequency) surface in wave number space. One can assign an average effective wave number $k(E)$ using the relation in free space, $S(E) = 4\pi k(E)^2$, leading to a Ioffe-Regel criterion with a constant equal to 0.844. A completely different model, using randomly distributed point scatterers in 3D yields the constant 0.972 [93].

4.1.2. *Applications: electrons versus classical waves*

What does the Ioffe-Regel criterion learn us? One should first realize that the mean free path ℓ is a complicated function of both energy (or frequency) and disorder. But the mean free path may be expected to decrease with increasing disorder (or increasing impurity density). As a result, the mobility edge occurs at a minimum amount of disorder that depends on k, and thus on energy. This agrees qualitatively with what we know for the Anderson model.

Let us consider an electron at energy $E = k^2/2m$ that propagates in a medium with randomly positioned particles with cross-section $\sigma(E)$ and number density n. For sufficiently low density expects the mean free path to be given by $\ell(E) = 1/n\sigma(E)$. It is well known that the cross-section for ordinary potential scattering is finite at low energies. Hence $\ell(E = 0) < \infty$ for any density n. Since $k \to 0$ at low energies one arrives at the conclusion that all states are localized at sufficiently small energy, i.e near the continuum edge of the regular medium.

Unfortunately, this argument breaks down for classical waves. For light or sound the cross-section at low frequencies has the familiar Rayleigh form, $\sigma(\omega) \sim \omega^4$. As a result $k\ell \sim \omega^{-3} \gg 1$ at low frequencies, and no localization is expected there.

The relation $\ell = 1/n\sigma$ suggests that localization might be possible if the cross-section is large, i.e. at resonances. However, the cross-section can never exceed the "unitary limit" $\sigma \approx \lambda^2/\pi$. Localization may thus be possible for sufficiently high density $n\lambda^3 \approx 1$. The possibility of localization near resonances was first argued in a paper by Sornette and Souillard [94], which was later followed by extensive calculations [95, 96, 97]. One obvious deficiency of the argument is that it is based on a formula for the mean free path that holds only if the number density is small. Both spatial correlations and "dependent scattering" [93] - i.e. recurrent scattering from the same particle - can make the mean free path bigger than anticipated from a low-density extrapolation.

In the band center of the Anderson model $E = 0$ the wavenumber is inversely proportional to the lattice spacing, meaning that localization sets in when the lattice constant exceeds the mean free path, a reasonable conclusion. Near the band edges one finds $k^2 \sim 6V - |E| \to 0$, whereas the mean free path stays finite. This means that localization can set in even when $\ell \gg a$. I note that by Eq. (50) the Cayley tree has its mobility edge, in any dimension larger than one, always determined by the criterion $\ell \equiv 1/\gamma \approx a$.

According to the above, localization seems easier near band edges. It was first suggested by John [5] that periodic structures, subject to small disorder, are very good candidates to observe localization, in particular for classical waves. In a slightly disordered photonic band gap material, localization of light can be expected near the band gap where the density of states of the regular system vanishes, just like electron localization was predicted to occur at $k \approx 0$. The Ioffe-Regel criterion near the band edge ω_b takes the form $(\omega - \omega_b)\ell(\omega)/c_0 \approx 1$. The possibility of light localization still is one of the drivers to manufacture of photonic band gap materials. Recently, exact proofs have been published for classical wave localization - in the sense of criterion (4) - near a frequency gap of the host system [98, 99].

4.2. THOULESS CRITERION

The opening of a disordered system has one important consequence, as first discussed by the classical paper of Thouless [1]. Consider a closed sample of size L, much bigger than the mean free path and in the extended regime, and open it up at two sides in an ideal way, i.e. the boundaries do not prevent in any way the waves to leak through. In the extended regime, the typical time for a wave to traverse the sample is L^2/D. As a result, the eigenstates of the closed system become resonances with typical width $\delta E \approx D/L^2$, called the *Thouless energy*. For De Broglie waves we have assumed that $\hbar = 1$. For classical waves one should look at δE as a typical frequency. Crucial is how the Thouless energy compares to the average level spacing ΔE, which is approximately given by the inverse density of states $1/\rho(E)L^d$. We can define,

$$g \equiv \frac{\delta E}{\Delta E} \approx \rho(E)D(E)L^{d-2} . \tag{56}$$

The dimensionless variable g is called the dimensionless conductance, because the second equality actually makes it coincide with the Drude conductance of the medium.

According to Thouless $g > 1$ implies that the diffusion process can be supported by the microstates of the disordered system implying it to be in the metallic (extended) regime. For $g < 1$, the diffusion process cannot be supported microscopically, because the typical level width necessary for diffusion would contain only one eigenstate, which is insufficient to support a dynamical process. Hence the system is in the localized regime. The criterion for localization in open systems becomes,

$$g = \text{constant} \approx 1 . \tag{57}$$

It is instructive to compare the Thouless criterion to the one formulated earlier in Eq. (20) for open systems. For a tube with N transverse modes, length L and mean free path ℓ^*, we argued that localization sets in when $L > \xi$, with the localization length ξ defined by $\xi \approx N\ell^*$. In the extended regime the ensemble -average of the conductance, defined in Eq. (22) is $\langle g \rangle \approx N\ell^*/L$. Hence, near cross-over to the localization regime one recovers the Thouless criterion $\langle g \rangle \approx 1$.

Some find it convenient to rephrase the Thouless criterion in terms of time scales. The time $t_D\hbar/\delta E \sim L^2/D$ is a typical time for an excitation to move through the entire system, whereas $\hbar/\Delta E$, often called the Heisenberg time t_H, represents the maximal time scale possible in the medium. Localization means that the diffuse traversal time exceeds the Heisenberg time. One may apply this to classical wave propagation in a medium with resonant scatterers. Due to resonant scattering the transport velocity v_E can be small [23], i.e. much smaller than c_0. As a result, the diffusion constant $D = \frac{1}{d} v_E \ell^*$ will be small, so that the propagation time L^2/D is relatively long. One is tempted to conclude that g becomes small and localization may occur. Yet, resonant scattering has no direct relation with localization. Indeed, one can show that the Heisenberg time is *equally* enlarged, because the density of states $\rho(E)$ per unit volume becomes large near resonant scattering [100, 101]. As a result, delay in resonant scattering does not affect the criterion for localization.

An important question is what happens to localization when absorption enters the problem. I will discuss some aspects of this question in Chapter 6. Here I want to define the Thouless dimensionless conductance g for an open sample with absorption. This problem has been investigated experimentally by Genack etal. [51, 102] for microwaves. According to random matrix theory, discussed in Chapter 2, and without dissipation, the probability distribution of the total transmission T_a depends functionally on the Thouless parameter $\langle g \rangle = N\ell^*/L$, which is essentially the second moment of the total transmission: $(\Delta T_a)^2/\langle T_a \rangle^2 \sim 1/g$, c.f. Eq. (24). In the presence of absorption, this relation can still serve to define a new Thouless parameter g_a. It turns out empirically that the fluctuations in both T_a and T_{ab} are still correctly modeled by the random matrix results provided that the conductance is taken as [51].

$$\frac{(\Delta T_a)^2}{\langle T_a \rangle^2} \sim \frac{1}{g_a}. \tag{58}$$

In the presence of absorption this conductance no longer relates to the real conductance. If $g_a < 1$, one enters a regime with anomalous fluctuations, due to localization. In the last Chapter we shall discuss the role of absorption on localization. At the time of writing there is no elegant Thouless argument of the kind outlined above, why the conductance g_a emerges as a universal parameter in the presence of absorption.

4.2.1. *Conductance and Boundary Conditions*

The Thouless relation (56) implies a deep connection between conductance and typical energy scales. The ambiguity of the Thouless parameter g is that it addresses energy-levels in an open system. On the other hand, an open system cannot have exact energy eigenvalues, because the spectrum is absolute continuous.

Indeed, the original argument by Thouless was slightly different. Thouless actually considered a *closed* system with some boundary conditions. He defined δE as the sensitivity of the eigenstate to a change of boundary conditions. In the localized regime, the wave functions decay exponentially with a length much smaller than the system size so that a modification on the boundary will in general not change a lot. This leads again to criterion (57).

A simple but instructive model exists in which this Thouless argument can be made more precise. Following Akkermans and Montambaux [103] we consider a ring of length L enclosing a magnetic flux Φ. The advantage of this model is that exact eigenvalues exist because the systems is closed. In spite of this, the "Ohmic" conductance G can be defined as the factor of proportionality between induced Lenz current I and the time-derivative of the flux $d\Phi/dt$: $I = G\, d\Phi/dt$, as if the system would be open. The Hamiltonian for the ring is

$$H = \frac{\mathbf{p}^2}{2m} + V(\mathbf{r}), \tag{59}$$

where $V(\mathbf{r})$ is some random potential in the ring. Periodic boundary conditions are chosen for the wave function: $\psi(x + L) = \psi(x)\exp(i\phi)$, where x is the coordinate along the ring.

The first question is how the eigenvalues change for small phase. The equation for $u(\mathbf{r}) = \psi(\mathbf{r})\exp(-i\phi x/L)$ reads,

$$H = \frac{(\mathbf{p} - \phi\,\hat{\mathbf{x}}/L)^2}{2m} + V(\mathbf{r}), \tag{60}$$

¿From this equation it follows that the phase shift ϕ can be associated with an Aharonov-Bohm flux Φ, and it not just a mathematical trick on the boundary. For small phase, the "perturbing" potential is $\delta V = -\phi p_x/mL + \phi^2/2mL^2$. Second-order perturbation theory immediately gives,

$$\frac{\partial E_n(\phi)}{\partial \phi^2} = \frac{1}{2mL^2} + \frac{1}{4m^2L^2}\sum_{k\neq n}\frac{|\langle\psi_n|\,p_x\,|\psi_k\rangle|^2}{E_n - E_k}. \tag{61}$$

The ensemble average of this quantity, divided by the mean level spacing can be defined as the dimensionless conductance,

$$g \equiv \langle N(E)\rangle \left\langle \frac{\partial E_n(\phi = 0)}{\partial \phi^2} \right\rangle . \tag{62}$$

This definition is a more precise variant of definition (56). It has been proven by Akkermans and Montamboux [103] that in the metallic regime $g \gg 1$, definition (62) coincides *rigorously* with the Ohmic conductance G defined above. In the localized regime this relation breaks down.

4.3. SCALING THEORY

After the work of Thouless, an important development was the scaling theory of localization, a phenomenological theory put forward in 1979 by the so-called "gang of four" [13], and inspired by the theory of phase transitions. For this reason, the onset of localization is often called an "electronic" phase transition [7]. This theory is believed to be somewhat dated as it considers only the ensemble-average of the conductance, and disregards fluctuations. Nevertheless, it has been of historical importance, as it set a trend for all work that followed. In addition to the clear, pioneering paper by the gang of four, good descriptions of the ideas behind scaling theory can be found in the works by Thouless [1], Ramakrishnan [2] Vollhardt and Wölfle [7], and recently Ping Sheng [64]. For that reason, I will restrict myself to a brief outline.

The basic question addressed by the scaling theory is the dependence of the conductance (transmission, summed over all channels in and out) on the volume of the system. Classically, that is without considering interference, the total conductance of the system depends on the transverse transverse A and its length L according to

$$g \sim \sigma(E)\frac{A}{L} = \rho(E)D(E)L^{d-2}\,. \tag{63}$$

The second equality assumes a disordered system with equal size in all directions (a cube or sphere). In three dimensions this means that the conductance increases when the system size finally increases, whenever the diffusion constant is nonzero. In one dimension, g would decrease. In the previous section we mentioned that if the conductance becomes small enough, one enters the localized regime.

¿From the previous argument it follows that increasing the systems size will in principle not help to approach localization. Only in one dimension this will be the case, and indeed all states are localized in one dimension. Quasi one-dimensional systems, i.e three dimensional systems where $0 < A < \ell^2$ is kept fixed, are subject to the same conclusion that all states are localized if the system is sufficiently long, consistent with criterion (20). In three dimensions increasing the systems size will only help if one is already in the localized regime. In that case

$$g \sim \exp\left(-\frac{L}{\xi}\right). \tag{64}$$

It looks like if the size dependence of the conductance, its derivative in particular, depends on the conductance itself. The scaling theory of localization asserts the existence of a length ξ, depending on the amount of disorder in the system but *independent* of the sample size L, such that the dimensionless conductance depends only on the ratio ξ/L,

$$g = F\left(\frac{\xi}{L}\right). \tag{65}$$

This equation implies the existence of a one-parameter scaling function

$$\beta(g) = \frac{\mathrm{d}\log g}{\mathrm{d}\log L}, \tag{66}$$

which depends *only* on the dimensionless conductance g. This identifies g as the universal scaling parameter. The scaling ansatz asserts $\beta(g)$ to be a smooth and continuous function of g. From Eqs. (63) and (64) one anticipates the limits,

$$\beta(g) = \begin{cases} d - 2 - a/g & g \gg g_c \\ \log g + \text{constant} & g \ll g_c \end{cases} \tag{67}$$

The term a/g is the weak localization correction that can be calculated from perturbation theory. In three dimensions the scaling ansatz together with both asymptotic limits leads us to the conclusion that some critical point g_c exists that obeys $\beta(g_c) = 0$. This point is identified as the critical value for the conductance in the Thouless criterion for localization (57). For $g > g_c$ one has $\beta > 0$ meaning that the conductance increases with sample size. In that case ξ is interpreted as a correlation length. If $g < g_c$ the conductance decreases with sample size, and ξ is identified as the localization length. These notions are consistent with the existence of extended and localized states respectively. Hence, the scaling theory provides an additional - though rather intuitive - support for Thouless' criterion. I want to stress that the assertion of a *continuous* scaling function immediately excludes a discontinuous conductivity near g_c as claimed by Mott. It is however, not a proof since continuity has been asserted, and not established. The derivative

$$\left.\frac{\mathrm{d}\beta}{\mathrm{d}\log g}\right|_{g_c} = \frac{1}{\nu}, \tag{68}$$

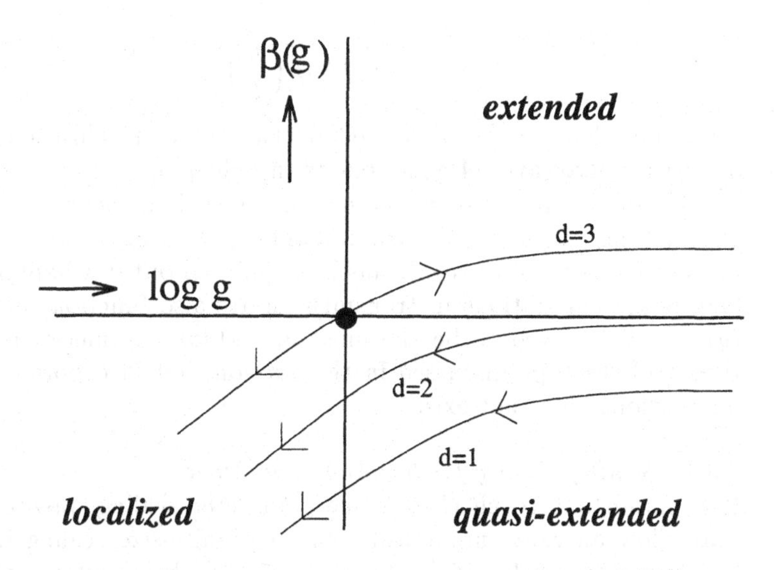

Figure 9. The scaling function $\beta(g)$, in dimensions $d = 1$, 2 and 3. In dimensions $d > 2$ a critical point exists defined as a zero of β. The three regimes have been indicated. The left upper sheet is forbidden by Thouless' criterion. The arrows denote the sense of scaling when the system size increases. A quasi 1D system (recall Figure 2) would scale here like an ordinary 1D system.

can be seen to determine the critical exponents on both sides of the transition. Near the critical point, Eq. (68) implies that $\log g = \log g_c + (L/\xi)^{1/\nu}$. Only the choice $\xi \sim |\omega - \omega_c|^{-\nu}$ makes the conductance g analytic near the mobility edge ω_c. In the extended regime we expect that $D \sim |\omega - \omega_c|^s$ and relation (63) indicates that $\xi \sim |\omega - \omega_c|^{-s/(d-2)}$. One-parameter scaling imposes,

$$s = \nu(d-2). \tag{69}$$

This hyperscaling equation was first derived using an elegant scaling theory [104]. In particular, in 3D the exponents ν and s are equal. Fig. 9 shows the expected qualitative behavior of the scaling function $\beta(g)$ in different dimensions.

The fixed point g_c has one important consequence for the total transmission T_a. In three dimensions, the relation between (ensemble-averaged) conductance g and transmission T_a is $T \approx g/(kL)^2$, where $(kL)^2$ is proportional to the number of conducting modes. At the transition, $g = g_c$ is independent of the size of the sample. At the same time we expect $k \approx 1/\ell$. Hence,

$$T(L) \approx \left(\frac{\ell}{L}\right)^2 , \tag{70}$$

a relation that has been claimed by two recent localization experiments, one with microwaves [44] and one with light [45].

One of the major shortcomings of the scaling theory is the absence of fluctuations. Actually, the conductance g figuring in it is the ensemble-averaged conductance $\langle g \rangle$. It has been pointed out in Chapter 2 that many long-range fluctuations in transmission are also functions of $\langle g \rangle$. As such, $\langle g \rangle$ is still believed to be the most important parameter. Nevertheless, a theory of the type discussed in this section, but incorporating mesoscopic fluctuations, does not exist.

4.3.1. *Scaling Theory for the Anderson Model*

It is important to check the one-parameter scaling ansatz in situations that allow an exact numerical solution. Finite-size scaling in the 2D and 3D Anderson model, given by Eq. (35) has been extensively studied by MacKinnon and Kramer [105].

Consider the Anderson model on a 2D cubic lattice with $M \times N$ sites. We shall impose periodic boundary conditions in the M direction, so that we are studying a cylinder of length N and radius M. For $i = 1, \cdots N$ the M-dimensional vector wave function $\mathbf{\Psi}_i$ can be defined as $\mathbf{\Psi}_i \equiv (\psi_{i1}, \psi_{i2}, \cdots \psi_{iM})$, and obeys the equation,

$$V\left(\mathbf{\Psi}_{i+1} + \mathbf{\Psi}_{i+1}\right) + \mathbf{E}_i \cdot \mathbf{\Psi}_i = E\mathbf{\Psi}_i \,. \tag{71}$$

This equation looks like the 1D Anderson model (36). The diagonal $M \times M$ matrix $\mathbf{E}_i$ contains the random site energies ε_{ij}. As the longitudinal size $N \to \infty$ we expect all states to be localized along the cylinder, with localization length ξ_M. As M increases, this system approaches the 2D Anderson model. If ξ_M converges to a final value ξ_∞ we can conclude that the 2D model is localized with localization length ξ_∞. If ξ_M stays larger than M, states must be extended in the 2D Anderson model.

Like has been done in ansatz (65) we can postulate the existence of a length ξ, independent of M such that

$$\Lambda \equiv \frac{\xi_M}{M} = F\left(\frac{\xi}{M}\right) . \tag{72}$$

Similarly a scaling function $\beta(\Lambda)$ can be defined, following Eq.(67). Numerically one now faces the considerable job to find, for each value of the disorder parameter W/V, an appropriate ξ such that all calculations fall on the same curve, which then determines b the still unknown function

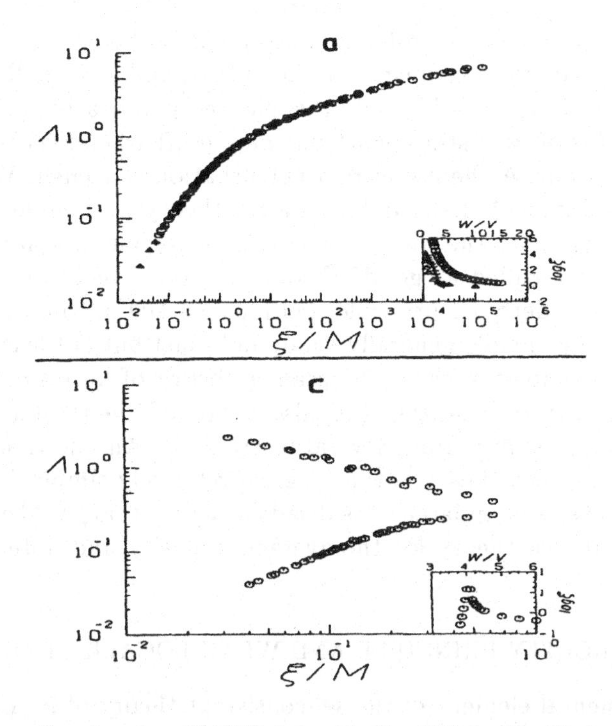

Figure 10. The scaling function $F(\xi/M)$ defined in Eq. (65) for the 2D Anderson model on a cylinder. Top: Diagonal disorder, which gives completely localized behavior. Bottom: Off-diagonal (sympletic) disorder, which shows a mobility edge at $W/V \approx 4$. Taken from MacKinnon [62].

$F(x)$ in Eq. (72). In Fig. 10 we show the results of MacKinnon and Kramer [105]. The top figure shows the 2D Anderson model with diagonal disorder, the bottom figure shows an "off-diagonal" variant used to model spin-orbit scattering (in random matrix theory this would correspond to "$\beta = 4$"). Both graphs confirm the scaling hypothesis (72) beautifully. We can infer that diagonal disorder results in $\Lambda(M)$ that strictly decreases as M increases. Thus, $\beta(\Lambda) < 0$ so that according to the previous section all states are localized. The off-diagonal disorder seems to have a mobility edge even in 2D, as apparent from the upper branch where $\beta(\Lambda) > 0$.

5. Self-Consistent Theory of Localization

Localization is the phenomenon that wave functions become localized due to disorder. As may have become clear from the previous Chapters, one of the most important issues has been how localization influences the transport of waves through disordered media. One would like to identify a microscopic mechanism that induces localization by destroying explicitly the familiar picture of diffusion. This problem is not going to be easy. Classical transport theory predicts the transport to be diffuse, and any small perturbation caused by interference will modify this picture quantitatively, but not qualitatively: diffusion will still persist but may be characterized by a different diffusion constant. A theory beyond perturbation is needed. We mentioned already in Chapter 2 that random matrix theory is capable of describing the diffuse regime all the way to the localized regime in a nonperturbative way and may be called a "good" theory. A "good" theory is characterized by a rigorous underlying principle, perhaps worked out in an approximate manner but in principle generally valid, and - last but not least - consistent with great principles such as the scaling theory of localization. The selfconsistent theory of localization is also a "good" theory, but like random matrix theory is suffers from several restrictions. An additional advantage of the selfconsistent theory is that it gives relatively simple formulas, from which the physics is apparent. Its disadvantage is that, unlike random matrix theory, it is a theory for the average intensity and diffusion constant only.

5.1. RECIPROCITY PRINCIPLE AND WEAK LOCALIZATION

The fundamental element of the selfconsistent theory of localization is the reciprocity principle. As outlined below, classical transport theory can never be a "good" theory for waves because it doesn't obey the reciprocity principle. I will first outline how transport theory is set up and how reciprocity comes is.

The basic observable in a transport experiment is the specific intensity (in reflection or transmission) which is proportional the complex field amplitude squared: $I(\mathbf{r},t) = |\psi(\mathbf{r},t)|^2$. Transport theory tries to formulate a relation between the ensemble-average of incident intensity and an outgoing intensity. More generally one can try to connect space-time correlation functions of incident and outgoing field. Since we consider media with linear response, the relation must be,

$$\langle\psi(\mathbf{r}_1,t_1)\psi^*(\mathbf{r}_3,t_3))\rangle = \int_2\int_4 \Gamma(\mathbf{r}_1,\mathbf{r}_2;\mathbf{r}_3,\mathbf{r}_4,t_1,t_2,t_3,t_4)\,\langle\psi(\mathbf{r}_2,t_2)\psi^*(\mathbf{r}_4,t_4))\rangle\,, \tag{73}$$

where the integrals are done over time and space. This relation identifies the vertex $\Gamma(\mathbf{r}_1,\mathbf{r}_2;\mathbf{r}_3,\mathbf{r}_4,t_i)$ as the fundamental object independent of incident field. Let us for simplicity assume monochromatic waves, so that the time-dependence $\exp(-i\omega t_i)$ becomes trivial. The reciprocity relation for Γ interchanges detector and source, i.e. incoming and outgoing position vectors,

$$\Gamma(\mathbf{r}_1,\mathbf{r}_2;\mathbf{r}_3,\mathbf{r}_4) = \Gamma(\mathbf{r}_2,\mathbf{r}_1;\mathbf{r}_4,\mathbf{r}_3) = \Gamma(\mathbf{r}_1,\mathbf{r}_2;\mathbf{r}_4,\mathbf{r}_3)\,. \tag{74}$$

The first equality is a reciprocity relation for intensity, the second one concerns the complex field field only.

Transport theory aims to find a simple expression for Γ in terms of microscopic properties of the scatterers and their statistics. A multiple scattering sequence can be identified in terms of the so-called irreducible vertex $U(\mathbf{r}_1, \mathbf{r}_2; \mathbf{r}_3, \mathbf{r}_4)$, defined by the matrix relation,

$$\Gamma = U + U \cdot G \times G^* \cdot \Gamma \equiv U + R \ , \tag{75}$$

in which G is the averaged amplitude Green's function. R is a new object that contains only geometrical multiplications of the vertex U. Sometimes it is convenient to think as U in terms of a super single scattering, and R in terms of super multiple scattering. This is the closest one can get to the familiar multiple scattering picture without making any approximations. The adjective "super" is used to remind that U represents "one" scattering from a still cpmplicated object that cannot be disentangled further without giving up the ensemble averaging. As such, U can be called a collision operator. Boltzmann transport theory replaces U by the scattering ¿from one scatterer. This low-density approximation turns R into a genuine incoherent multiple scattering series, which obeys the familiar radiative transport equation. This equation disregards interference in multiple scattering.

It is important to realize that a price has to be paid by disentangling the exact solution into two operators R and U. The object R alone does not obey the reciprocity relation (74). The ones that have to be added to R to restore reciprocity are contained in a set C defined by interchanging bottom indices of R,

$$C(\mathbf{r}_1, \mathbf{r}_2; \mathbf{r}_3, \mathbf{r}_4) = R(\mathbf{r}_1, \mathbf{r}_2; \mathbf{r}_4, \mathbf{r}_3) \ . \tag{76}$$

This set is not part of the set of "reducible" events contained in R, but is part of U. The object C is the rigorous mathematical definition for the set of "most-crossed diagrams", a jargon that is used to refer to diagrams that generate all kinds of interference effects in multiple scattering, such as coherent backscattering and weak localization. The collision operator U can now be decomposed into

$$U(\mathbf{r}_1, \mathbf{r}_2; \mathbf{r}_3, \mathbf{r}_4) = C(\mathbf{r}_1, \mathbf{r}_2; \mathbf{r}_3, \mathbf{r}_4) + S(\mathbf{r}_1, \mathbf{r}_2; \mathbf{r}_3, \mathbf{r}_4). \tag{77}$$

By construction, the subset S is closed under the operation carried out Eq. (76), just like the total set Γ. Contrary to C it also exists classically. Physically it contains "loop" events that finally return to where they came from. In the theory of coherent backscattering, R describes "background" and C the "coherent peak". It is relation (76) that guarantees a coherent backscattering peak with an enhancement factor of approximately 2. It is only approximately 2, because the vertex S does not give a peak, and contributes to background [106]. In Figure 11 we show typical diagrammatic representations of the vertices R, S and C, evaluated in wave number space.

We now come to the basic point of the selfconsistent theory for localization. The vertex $\Gamma = R + S + C$ finally determines the ensemble-averaged intensity. The reciprocity relation (74) states that the solution of Eq. (75) is necessarily a selfconsistent equation for the vertex R, since C is basically the same object as R and serves as input of Eq. (75). In particular, the Boltzmann approximation disregards C and thus violates the reciprocity principle! The reciprocity principle forbids the existence a transport equation with separated microscopic input and mesoscopic output.

5.2. MICROSCOPIC THEORY FOR DIFFUSION CONSTANT

The next issue is how the conclusion of the previous section influences wave transport, the diffusion constant in particular. The first thing to notice is that energy conservation "proves" that wave propagation can be described by a diffusion constant at long time and length scales (mathematicians would not agree with this statement). More precisely, the object $\Gamma((\mathbf{r}_1 \approx \mathbf{r}_3; \mathbf{r}_2 \approx \mathbf{r}_4) = \Gamma(\mathbf{r}_{12})$ obeys a diffusion equation. The diffusion constant can be calculated from the following exact Kubo formula [107],

$$\rho(\omega)D(\omega) = \frac{1}{3\pi}\int \frac{d\mathbf{p}}{(2\pi)^3}\left[p^2|G(\omega,p)|^2\gamma(\omega,p) - \frac{\partial \mathrm{Re}\, G(\omega,p)}{\partial p^2}\right], \qquad (78)$$

where the object $\gamma(\omega, p)$ must satisfy the equation

$$\gamma(\omega,p) = 1 + \int \frac{d\mathbf{p}'}{(2\pi)^3}\,\frac{\mathbf{p}\cdot\mathbf{p}'}{p^2}|G(\omega,p')|^2 U_{\mathbf{pp}'}(\omega)\,\gamma(\omega,p'). \qquad (79)$$

In these equations, ω is the frequency (we consider classical waves here, with vacuum velocity 1) and $\mathbf{p}$ the momentum. The vertex $U_{\mathbf{pp}'}(\omega)$ and the Green's function $G(\omega, p)$ have been introduced formally in the previous section and are here presented as matrix elements in wavenumber space, using the convention of Figure 11; $\rho(\omega)$ is the density of states per unit volume. We have mentioned that U and Γ are related by reciprocity. As a result, U obeys a diffusion equation too, *with the same diffusion constant.* In this way one can see that Eqs. (78) and (79) are in fact coupled equations for the diffusion constant $D(\omega)$.

This selfconsistency was first pointed out by Götze [108] and worked out in detail by Vollhardt and Wölfle [7, 15, 16]. In the diffuse regime the Fourier transform of $R(\mathbf{r})$ is (see Fig. 11),

$$R_{\mathbf{pp}'}(\Omega, \mathbf{q}) \approx \frac{4\pi}{\ell^2}\,\frac{1}{-i\Omega + Dq^2}. \qquad (80)$$

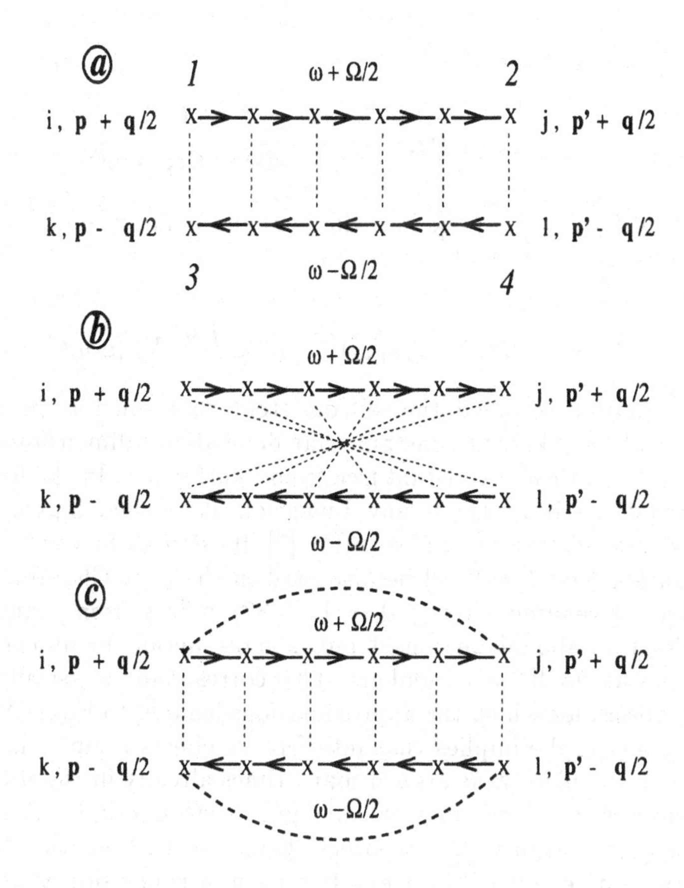

Figure 11. Typical diagrammatic representations of two correlated events $\langle\psi(a)\psi(b)^*\rangle$. By convention, the top line denotes $\psi(a)$ and bottom line the complex conjugate $\psi^*(b)$ (more precisely the propagation at energy $\omega - \Omega/2 - i0$, which explains why it propagates in the opposite direction); a, b denote paths with different frequencies, different wave numbers and different polarizations at in- and output, as indicated in the graphs; Crosses denote scattering events. Dashed lines connect identical scatterers. The numbers 1,2,3 and 4 denote the convention in Eqs. (73) and (74). A). Typical contribution to the reducible vertex R. The present graph represents a totally incoherent event (a "ladder" diagram), where the phase cancels. B). Typical contribution to the vertex C ("most-crossed" diagram), obtained by time-reversing the lower path of the vertex R. C.) Typical contribution to the vertex S which is here an incoherent path with equal start and end ("loop").

The relation (76) can be reformulated in wave number space, and the object

$C_{\mathbf{pp}'}$ takes the form

$$C_{\mathbf{pp}'}(\Omega, \mathbf{q}) \approx \frac{4\pi}{\ell^2} \frac{1}{-i\Omega + D(\mathbf{p} + \mathbf{p}')^2} \tag{81}$$

Inserting $U_{\mathbf{pp}'} = S + C_{\mathbf{pp}'}$ into Eq. (79) and making some intuitive approximations, one arrives at, for $\Omega = 0$,

$$\frac{1}{D(\omega)} = \frac{1}{D_0(\omega)} + \frac{C_d}{\rho(\omega)\ell} \int \mathrm{d}^d\mathbf{q} \frac{1}{D(\omega)q^2} \,. \tag{82}$$

This equation is called the selfconsistent equation for the diffusion coefficient; C_d is a known constant that depends on dimensionality. The first term is the diffusion constant free from interference. In the form it has been written down, it applies to any dimension. It can be adapted for anisotropy or inelastic scattering or absorption [7]. Its straightforward dynamical generalization, i.e. $\Omega \neq 0$ will be discussed in the next Chapter.

The wavenumber integral in Eq. (82) suffers from a number of divergencies that should be considered in more detail. In dimensions $d \leq 2$ it diverges at small wave numbers. This corresponds physically to long wave trajectories, for which the approximations leading to Eq. (82) should apply. This catastrophe implies that interference effects inhibit classical diffusion in $d \leq 2$ completely, as argued many times already in the present paper. In dimensions $d \geq 2$ the integral diverges at large q, indicating that some of our approximations start to break down at small lengths. We expect this length to be equal to the mean free path, a conclusion that emerges from a detailed analysis [93]. In dimensions $d > 2$ we put $q_{\max} \approx 1/\ell$ so that

$$D(\omega) = D_0(\omega) \left(1 - \frac{C_d}{\rho(\omega)\ell^{d-1}}\right) \,. \tag{83}$$

This equation predicts localization - in the sense of criterion (1) - to occur when $\rho(\omega)\ell^{d-1} = C_d$, which agrees with the generalized Ioffe-Regel criterion (54) formulated earlier, realizing that the density of states scales as $\rho(\omega) \sim k^{d-1}$. A precise consideration of C_d demonstrates that $(k\ell)^{d-1} \sim (d-2)/d$ at the mobility edge.

The second interesting aspect of Eq. (82) is the relative easy to investigate the effect of a finite open medium in an approximate way. In that case we anticipate trajectories longer than the system size L to be absent. Hence a minimum value $q_{\min} \approx 1/L$ may be considered. In 3D, the selfconsistent equation can be rewritten as ,

$$\frac{1}{D(\omega, L)} \left(1 - \frac{C_3}{\rho(\omega)\ell^2}\right) = \frac{1}{D_0(\omega)} \left(1 - \frac{C_3'}{g(\omega, L)}\right) \,. \tag{84}$$

Here $g(\omega, L) \sim \rho(\omega)D(\omega, L)L$ denotes the dimensionless conductance, defined earlier in Eq. (63). The advantage of this form is that both the Ioffe-Regel criterion (55) and the Thouless criterion (56) appear. Because the system is finite, the diffusion constant never really vanishes but at the mobility edge of the infinite medium, described by the Ioffe-Regel criterion, the dimensionless conductance defined in Eq. (56) equals some constant of order unity. As a result, the diffusion constant D scales as $1/L$, a conclusion that also emerged from the scaling theory of localization. The selfconsistent equation for $g(\omega, L)$ satifies the scaling ansatz (65) with a correlation length that diverges if $k\ell \approx 1$. For more details I refer to the excellent review paper by Vollhardt and Wölfle [7].

In the localized regime and for an infinite system the diffusion constant D vanishes, its role being taken over by the localization length ξ. It is now necessary to consider the dependence of the diffusion constant on the hydrodynamic frequency Ω. We set $D(\Omega) = -i\Omega\xi^2(\omega)$, so that Eq. (80) for the reducible vertex R takes the form,

$$R(\Omega, \mathbf{q}) \approx \frac{4\pi}{-i\Omega\ell^2} \frac{1}{1 + \xi^2(\omega)q^2} . \tag{85}$$

In real space and in the time domain this functional dependence describes a stationary, and exponentially small (ensemble averaged) intensity correlation, as specified earlier in Eq. (9). The selfconsistent equation for the localization length is easily seen to be,

$$\frac{1}{\xi^2(\omega)} = \frac{C_d}{\rho(\omega)} \int_{q<\ell} \mathrm{d}^d\mathbf{q} \frac{1}{1 + q^2\xi^2(\omega)} . \tag{86}$$

The solution of this equation heavily depend on dimensionality. In one dimension, $\xi \approx \ell$, a not too surprising conclusion in view of Chapter 3. In 2D, all states are still localized with localization length $\xi \approx \ell \exp(\rho\ell)$, consistent with the relation (46) obtained for the 2D Anderson model. In three dimensions the localization length diverges at the mobility edge according to,

$$\xi(\omega) \sim \frac{\ell}{(k\ell)_c - k\ell} , \tag{87}$$

i.e. with a critical exponent $\nu = 1$. In the selfconsistent theory, the case of four dimensions is a critical upper dimension. For $d > 4$ it can be inferred that

$$\xi(\omega) \sim \frac{\ell}{[(k\ell)_c - k\ell]^{1/2}} \tag{88}$$

i.e. a critical exponent $\nu = \frac{1}{2}$ *independent* of the dimensionality. Note that such an upper critical dimensionality was not obtained for the Cayley tree.

Figure 6 of Chapter 3 shows a comparison of the phase diagram as predicted by the self consistent equation and the exact numerical solution for the 3D Anderson model. It can be seen that the agreement is excellent. This agreement suggests that localization is for a great deal a catastrophe in weak localization (because that is the physics that goes in), a statement on which not everybody agrees. One of the problems with the selfconsistent theory is that it predicts typical mean-field critical exponents around the transition, whereas numerical simulations (of the Anderson model) indicate the critical exponents to be different. As a result, the selfconsistent equation is believed to break down very close to the transition. Also when considering magnetic fields, in which case the basic reciprocity principle (76) underlying the selfconsistent theory breaks down, the selfconsistent theory disagrees with the now generally accepted picture that wave localization is not destroyed - at most modified - by broken time-reversal symmetry.

6. Localization in the Time Domain

Some time-dependent features of localization have been briefly addressed in Chapter 2. The time-evolution of the center-of mass $\mathbf{r}(t)$ of a wave packet plays an important role and was seen to be involved already in the very definition of the phenomenon. Time dependent features of wave propagation are - by Fourier transformation - related to frequency correlations.

Many more dynamical features of localization have been studied. For electronic systems the basic dynamic observable is the dynamic AC (electronic) conductivity, i.e. the conductivity of electrons driven by an oscillating electric field $E(\Omega) \sim \exp(-i\Omega t)$ implying that electronics mainly deals frequency formulations. In ultrasonics or seismology, observations are carried out directly in the time domain. In this Chapter we will have a closer look how localization is assumed to affect dynamical wave propagation. Time dependent wave localization in one-dimensional systems has been studied out loud by Sheng, Papanicolaou etal. , in particular in relation to its potential applications in seismology. Finally, time-dependent acoustic wave propagation has been investigated numerically by Weaver etal. A dynamic version of random matrix theory is now in rapid development.

6.1. FREQUENCY CORRELATIONS IN 1D

Localization of acoustic waves in one dimension has been extensively studied by Sheng, Papanicolaou, White and Zhang [10, 109, 110, 111]. It may have applications to seismology in view of the layered structure of the Earth. The source initiating wave propagation, such as Earth quakes and dynamite explosions, are typically limited in time. Observations are therefore intrinsically carried out in the time-domain. The 1D solution is important because it provides a rigorous treament of time-dependent fluctuations in the localized regime, that may also be relevant to other domains, and perhaps even to higher dimensions.

Consider a semi-infinite layered medium. Different layers may have different size or different acoustic properties, such as different wave velocity and different density. The background velocity and density are given by v_0 and ρ_0. The wave equation reads

$$i\partial_t \mathbf{\Psi}(z,t) = \mathbf{H}(z)\cdot\mathbf{\Psi}(z,t)\,. \tag{89}$$

The wave function $\mathbf{\Psi}(z,t) = (p, v)$ contains pressure and velocity variations over the background. The "Hamiltonian" is given by

$$\mathbf{H(z)} = \begin{pmatrix} 0 & -K(z)p_z \\ \rho^{-1}(z)p_z & 0 \end{pmatrix} \tag{90}$$

which may be symmetrized by a change of variables. Both the elastic bulk modulus $K(z)$ and the density $\rho(z)$ are 1D random functions. The statement is that "almost" all states are exponentially localized. If we consider monochromatic solutions of this wave equation for the semi-infinite stack, the localization implies that the the reflection coefficient has modulus one for "almost all" frequencies ω, i.e. $R(\omega) = \exp(i\psi)$. In this 1D model the absolute value is not subject to fluctuations, unless of course we allow some leaks at the bottom. The phase, however, fluctuates and one may expect that $\langle R \rangle = 0$. Information may be retrieved from the spectral correlation function,

$$U(\omega_1,\omega_2) = \langle R(\omega_1)R^*(\omega_2)\rangle\,. \tag{91}$$

This correlation function can be deduced from the fluctuations of the pressure field $p(t, z = 0)$ at the top. Consider an initial pulse at $z = 0$ whose frequency distribution is described by $F(\omega)$. The pressure field $p(t, z = 0)$ is given by,

$$p(t) = \int_{-\infty}^{\infty} \frac{\mathrm{d}\omega}{2\pi} \exp(\mathrm{i}\omega \mathrm{t})\mathrm{F}(\omega)\mathrm{R}(\omega)\,. \tag{92}$$

Carring out some Fourier transformations we find the simple relation,

$$U\left(\omega - \frac{1}{2}\Omega, \omega + \frac{1}{2}\Omega\right) = \frac{1}{|F(\omega)|^2}\int_{-\infty}^{\infty} \mathrm{d}\tau\, \exp(-\mathrm{i}\Omega\tau) \int_{-\infty}^{\infty} \mathrm{dt}\, \exp(-\mathrm{i}\omega \mathrm{t})\, \langle \mathrm{p}(\tau+\mathrm{t})\mathrm{p}(\mathrm{t})\rangle \tag{93}$$

We assumed the typical value for Ω to be much smaller than ω, so that $F(\omega)$ can be considered constant over the band Ω. Correlations over a band width Ω typically probe wave paths less than v_0/Ω.

Sheng etal. [10] have calculated the spectral correlation function U in a special but relevant limit. This limit applies when $a \ll \lambda \ll \xi(\omega)$, i.e. in case of separation of the length scales a (the thickness of the layers), the wave length $\lambda \sim v_0/\omega$ and the 1D localization length $\xi(\omega)$. This limit has been studied by various authors [10, 112]. It is found that,

$$U\left(\omega - \frac{1}{2}\Omega, \omega + \frac{1}{2}\Omega\right) \approx \int_0^\infty ds \frac{s}{s + iY} \exp(-s). \tag{94}$$

The correlation function depends on one parameter $Y = \Omega\xi(\omega)/v_0$ only . An important conclusion is that the time-dependent correlation function give direct access to the frequency dependence of the localization length, since the dephasing frequency Ω is typically equal to $v/\xi(\omega)$, i.e. the inverse time to travel one localization length. This also confirms that $|\Omega| \ll |\omega|$ as has been assumed earlier. At low frequencies $\omega \to 0$ the localization length $\xi \sim \omega^{-2}$ is large and dephasing occurs rapidly.

Equation (94) can be transformed into one for the power spectrum $S(\omega, \tau)$ of fluctuations with frequency ω in a time window τ. The result is

$$S(\omega, \tau) = \int_{-\infty}^{\infty} \mathrm{dt} \exp(-\mathrm{i}\omega\mathrm{t}) \langle \mathrm{p}(\tau + \mathrm{t})\mathrm{p}(\mathrm{t})\rangle = |\mathrm{F}(\omega)|^2 \times \frac{1}{|\tau|} \frac{\chi}{(1+\chi)^2}. \tag{95}$$

where $\chi = |\tau| v_0/\xi(\omega)$. The fluctuations at frequency ω are suppressed once the time window of observation τ exceeds the time to traverse one localization length. A very useful conclusion, which I think, is worth investigating in higher dimensional or quasi-one dimensional systems.

6.2. TIME-DEPENDENT TIGHT BINDING MODELS

Ultrasonic localization experiments have initiated the study of a time-dependent variant of the Anderson model (35),

$$\partial_t^2 \psi_n + \varepsilon_n \psi_n - V \sum_m{}' \psi_m = F_n(t). \tag{96}$$

Here $\{\varepsilon_n\}$ are again identical, independently distributed stochastic variables, and V is a hopping element ¿from a site to a nearest neighbour. They have a uniform distribution between $-\frac{1}{2}W$ and $\frac{1}{2}W$. The second-order time derivative suggests that we are dealing with classical waves although one should not forget that light and ultrasound would have a time dependent interaction of the form $(1 + \varepsilon_n)\partial_t^2 \psi_n$. The function $F_n(t)$ denotes a source term and rapidly decays to zero for large times and at sites far away from the source.

Equation (96) has been studied numerically in 2D by Weaver [113]. For the 2D Anderson model all states are known to be localized. It is found that beyond a certain time the *ensemble-averaged* wave energy takes a stationary profile that is decays exponentially away from the source. This is consistent with localization and in particular with criterion (3). For all values of the microscopic parameter W/V, and for all distances from the source, the time-evolution towards the stationary state seems to be a universal function, empirically given by,

$$|\psi_n(t)|^2 \sim \exp\left[\frac{-r_n}{\xi} - \left(\frac{r_n^{2.46}}{4D_p t \xi^{0.46}}\right)^{0.76}\right], \tag{97}$$

where ξ is the localization length and D_p is a quantity with the dimension of diffusivity. So far, no microscopic theory is able to explain this numerical result. The exponent $2.46 > 2$ indicates that the spread of energy is already sub-diffuse $\langle \mathbf{r}^2(t)\rangle \sim t^{0.81}$ well before the wave packet becomes localized.

An important recent development for *ab initio* time-dependent studies is the wave automaton. This is an efficient numerical tool that considers random S-matrices on a large lattice. These S-matrices control the scattering at each site. If they are taken random, but unitary, they model the random hopping of a wave packet ¿from one site to the other, in much the same way as the Anderson model. In fact, it can be shown that the wave automaton mimics a time-dependent version of the Anderson model with diagonal and second-nearest neighbour coupling [114].

The appearance of a subdiffusive regime, eventually leading to a bounded $\langle \mathbf{x}^2(\mathbf{t})\rangle$ at large times has been confirmed by Vanneste, Sebbah and Sornette in numerical studies using the 2D wave automaton [114]. They report a behavior $\langle \mathbf{r}^2(t)\rangle \sim t^{0.90}$, not far from the subdiffuse law reported by Weaver for the Anderson model. The wave automaton exhibits amazingly long times, up to 10^6 times the intersite travel time, to become saturated due to the onset of localization. To my knowledge, the wave automaton has never been studied in the case of 3D. Another interesting aspect may be to investigate broken time-reversal symmetry, which according to random matrix theory should modify localization.

6.3. AC CONDUCTIVITY & TIME-DEPENDENT DIFFUSION CONSTANT

The selfconsistent theory of localization can be straightforwardly generalized for the AC diffusion constant $D(\omega, \Omega)$. Here ω is the microscopic cycle frequency, for electrons typically equal to k_F/v_F, i.e. the "Fermi" frequency and for classical waves simply the central frequency of the pulse; Ω is the carrier frequency of the envelope, determining the dynamics of the wave packet. In electronic language, a conductance or diffusivity dependent on Ω can be thought of as a parallel capacitance.

In general one has $|\Omega| \ll |\omega|$. One might anticipate that for small Ω the diffusion pole still dominates, i.e. $R(\Omega, \mathbf{q}) \sim 1/[-i\Omega + D(\Omega)q^2]$, so that

Eq. (82) generalizes to

$$\frac{1}{D(\omega,\Omega)} = \frac{1}{D_0(\omega)} + \frac{C_d}{\rho(\omega)\ell} \int d^d\mathbf{q} \frac{1}{-i\Omega + D(\omega,\Omega)q^2}. \tag{98}$$

Like always with frequency response functions, $R(\Omega, \mathbf{q})$ must be an analytic function in the (physical) sheet $\mathrm{Im}\,\Omega > 0$, since at times $t < 0$ there was no pulse at all. The diffusion pole is determined by the analytic continuation of this function in the sheet $\mathrm{Im}\,\Omega < 0$.

I will focus on the time dependent behavior of an infinite 3D random medium. In the extended regime of can identify three regimes,

$$D(\Omega) = \begin{cases} D_0 & \Omega\tau > 1 \\ (i\Omega\tau)^{1/3} & (D(0)/D_0)^3 < \Omega\tau < 1 \\ D(0) & \Omega\tau < (D(0)/D_0)^3 \end{cases} \tag{99}$$

with $\tau = 3D_0/\ell$ the mean free time. This qualitative behavior of $D(\Omega)$ agrees with the time-dependent scaling by Berkovits and Kaveh [115]. Exactly at the mobility edge one finds $D(\Omega) \approx D_0(i\Omega\tau)^{1/3}$, in agreement with the elegant scaling theory first set up by Wegner [104]. In real space and in the time domain such AC diffusion would yield a time tail of the kind,

$$R(t \to \infty, r) \sim \frac{1}{rt^{2/3}}, \tag{100}$$

rather than the familiar diffuse result $R(t \to \infty, r) \sim 1/(D_0 t)^{3/2}$ for an infinite diffuse medium. The time-dependent reflection coefficient $R(t)$ of a semi-infinite medium at the mobility edge was estimated by Berkovits and Kaveh [115] to vary as $R(t) \sim 1/t^{4/3}$. The time dependent transmission through a slab of length L at the mobility edge is estimated to be [116]

$$T(t \to \infty) \sim \frac{1}{t^{2/3}} \exp\left(-\gamma t^{2/3}/L^2\right), \tag{101}$$

to be compared to the outcome $T(t) \sim \exp(-Dt/L^2)$ of the conventional diffusion equation. It would be fascinating to guide time-dependent localization experiments in this spirit. Also 3D seismic localization may possibly be deduced from anomalous time tails in this way.

6.3.1. *Role of Dissipation*

So far we have discussed localization without considering dissipation. As a result, Hamiltonians are symmetric operators and S-matrices are unitary. Experiments with classical waves always deal with some sort of dissipation. It is important to think the concept of localization over again in the presence of absorption. Absorption is easiest to understand in the time domain.

The definitions of localization given in section 2.1 all break down for one reason or another when absorption comes into play. In the case of light, absorption is described by a conductivity $\sigma(\mathbf{r})$ (at optical frequencies). The wave equation for the electrical field reads,

$$\varepsilon(\mathbf{r})\partial_t^2\mathbf{E}(\mathbf{r},t) + \nabla \times \nabla \times \mathbf{E}(\mathbf{r},\mathbf{t}) = -\sigma(\mathbf{r})\partial_t\mathbf{E}(\mathbf{r},\mathbf{t}) . \tag{102}$$

I shall assume both dielectric constant $\varepsilon(\mathbf{r})$ and conductivity $\sigma(\mathbf{r})$ to be independent of frequency. The simplest situation occurs when the absorption time $\tau_a = \varepsilon(\mathbf{r})/\sigma(\mathbf{r})$ is independent of $\mathbf{r}$, i.e. the absorption is homogeneous. In that case, one has, on top of the time evolution due to multiple scattering, an extra exponential decay in time,

$$\mathbf{E}(\mathbf{r},t) = \mathbf{E}_0(\mathbf{r},t)\,\mathrm{e}^{-t/2\tau_a} . \tag{103}$$

The field $E_0(\mathbf{r},t)$ satisfies the wave equation,

$$\varepsilon(\mathbf{r})\partial_t^2\mathbf{E}_0(\mathbf{r},t) - \frac{\varepsilon(\mathbf{r})}{4\tau_a^2}\mathbf{E}_0(\mathbf{r},t) + \nabla \times \nabla \times \mathbf{E}_0(\mathbf{r},\mathbf{t}) = \mathbf{0} . \tag{104}$$

This is essentially the wave equation for an electric field without absorption. The second, new term can be seen to be very small since for small absorption $\omega\tau_a \gg 1$. Therefore, Eq. (104) can serve to define localization (of light) in the presence of dissipation, using the result of Chapter 1. In particular, the center-of-mass $\mathbf{r}_0(t)$ is introduced as,

$$\mathbf{r}_0^2(t) \equiv \langle \mathbf{E}_0(t)|\,\varepsilon(\mathbf{r})\,\mathbf{r}^2\,|\mathbf{E}_0(t)\rangle = \mathrm{e}^{t/\tau_a}\,\langle \mathbf{E}(\mathrm{t})|\,\varepsilon(\mathbf{r})\,\mathbf{r}^2\,|\mathbf{E}(\mathrm{t})\rangle . \tag{105}$$

A "diffusion constant" can then be introduced as,

$$D_a = \frac{1}{d}\lim_{t\to\infty}\left\langle \frac{\mathbf{r}_0^2}{t}\right\rangle = \frac{1}{d}\lim_{t\to\infty}\mathrm{e}^{t/\tau_a}\left\langle \frac{\mathbf{r}^2}{\mathrm{t}}\right\rangle . \tag{106}$$

Even without making the assumption of homogeneous absorption, one might *speculate* about the existence of some minimal time τ_a such that Eqs. (105) and (106) are finite. I don't know of any exact results for this problem. Localization in infinite absorbing media can now be defined as a bounded $\mathbf{r}_0^2(t)$ in time (for "almost all realizations"). More weakly one can impose $D_a = 0$.

It is straightforward to apply the selfconsistent theory of localization, discussed in the previous Chapter, to Eq. (104), as done by Josefin [117]. The conclusion is almost trivial: this equation does not suffer from absorption, and therfore, the final equation for the diffusion constant D_a will also be free from absorption, so that all localization phenomena will remain

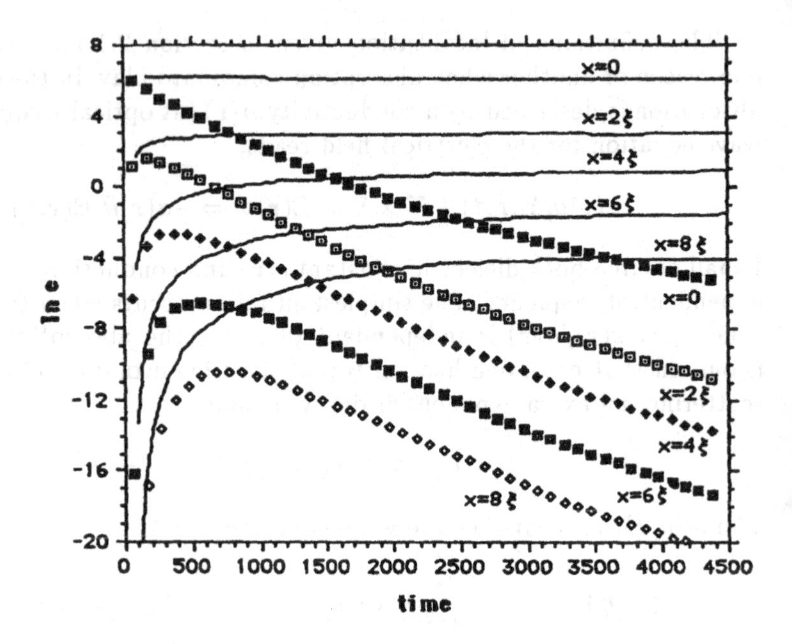

Figure 12. The evolution of the energy density in a viscous 2D Anderson model with $W/V = 11$, for different distances from the source. Solid: undamped; symbols denote damped system. Taken from Weaver [113], with kind permission from the author.

unaltered, except for the "trivial" exponent $\exp(-t/\tau_a)$. Numerical experiments carried out by Weaver [113, 118] for the 2D Anderson model with absorption, confirm this. In Figure 12 I show numerical results by Weaver [118] for the energy density $\psi(n,t)^2$ on the 2D Anderson model with diagonal disorder and a viscous term $\sigma_n \partial_t \psi(n,t)$ in much the same way as in Eq. (102), with the difference that also σ_n is chosen random ("diagonal random absorption"). The graph shows that except for an overall time decay of the kind $\exp(-t/\tau_a)$, the (ensemble-averaged) wave packet remains exponentially localized in space, as discussed in Chapter 2. The physical idea behind is that absorption only kills the amplitude but not the phase. In this respect absorption of classical waves is fundamentally different from inelastic scattering processes in the solid state. Inelastic scattering does not affect the amplitude but destroys the phase. The electron is not destroyed but is reset. Genack etal [102] multiplied their microwave data in transmission with the same factor $\exp(t/\tau_a)$ to arrive again at the conclusion that absorption is an almost trivial complication in the experiment that can be restored in the computer.

Despite these many strong arguments I believe that statements like "localization is not modified by absorption" and "absorption is a trivial aspect of localization" are too strong. When τ_a is chosen too small in Eqs. (105) and (106) one is actually putting gain into the medium, and the limits tend to infinity. At the time of writing it is unclear to me how this statement relates to the known and perhaps unexpected result that "gain" actually lowers the localization length in just the same way as absorption does. This was first shown for one-dimensional media [76] and recently for quasi one-dimensional media [119].

Secondly, the problem is that definition (106) does not always describe the measurement. Average intensities and field correlation functions are always described by the vertex R defined in Chapter 5. In the presence of absorption, the expression for R changes into,

$$R(\Omega, \mathbf{q}) \sim \frac{1}{-i\Omega + D(\Omega)\mathbf{q}^2 + 1/\tau_a}. \tag{107}$$

It can easily be checked that $D(\Omega = -i\tau_a) = D_a$, defined in Eq. (106). But stationary measurements, among which the DC conducivity featuring in the scaling theory of localization, are properties for $\Omega = 0$. In that case, the selfconsistent equation becomes,

$$\frac{1}{D(\omega)} = \frac{1}{D_0(\omega)} + \frac{C_d}{\rho(\omega)\ell}\int \mathrm{d}^d\mathbf{q}\frac{1}{D(\omega)q^2 + 1/\tau_a}. \tag{108}$$

We remark that the argument of reciprocity leading to this selfconsistent equation for diffusion is not violated in the presence of absorption. It can easily be shown from Eq. (108) that the absorption term destroys the mobility edge and $D > 0$ for all values of the mean free path, and in any dimension. Near the mobility edge of a nonabsorbing 3D system Eq. (108) predicts a highly nontrivial dependence of the diffusion constant on the absorption time τ_a, $D \sim 1/\tau_a^{1/3}$, which is the counterpart of Eqs. (99). It is well known that the length $L_a \equiv \sqrt{D\tau_a}$ describes the exponential fall-off of the ensemble-averaged transmission [47]. Thus, near $k\ell = 1$,

$$T(L) \sim \exp(-L/L_a) \sim \exp(-\gamma L/\tau_a^{1/3}). \tag{109}$$

This relation agrees with the "theory of white paint" by Anderson [9].

Acknowledgements. This work has been made possible by collaborations with Frédéric Faure, Azriel Genack, Ad Lagendijk, Roger Maynard, Patrick Sebbah, Marin Stoytchev, Adriaan Tip and Diederik Wiersma. Discussions with Erik Akkermans, Carlo Beenakker, Kurt Busch, Michel Campillo, Valentin Freilikher, Sajeev John, Abel Klein, Jean Lacroix, Tom Lubensky,

Theo Nieuwenhuizen, George Papanicolaou, Ping Sheng, Costas Soukoulis, Christian Vanneste, Pedro de Vries, Richard Weaver and Peter Wölfle are highly appreciated.

References

1. D.J. Thouless, Electrons in Disordered Systems and the Theory of Localization, Phys. Rep. **13**, 93 (1974).
2. T.V. Ramakrishnan, Electron Localization, in: *Chance and Matter*, Les Houches, session XLVI, edited by J. Souletie, J. Vannimenus and R. Stora (Elsevier, 1987).
3. N.F. Mott, Metal Insulator Transitions, Physics Today November 1978.
4. B. Souillard, Waves and Electrons in Inhomogeneous Media, in: *Chance and Matter*, Les Houches, session XLVI, edited by J. Souletie, J. Vannimenus and R. Stora (Elsevier, 1987).
5. S. John, The Localization of Light and Other Classical Waves in Disordered Media, Comments Cond. Phys. **14**, 193 (1988).
6. S. John, Localization of Light, Physics Today, May 1991.
7. D. Vollhardt and P. Wölfle, Selfconsistent Theory of Anderson Localization, in: *Electronic Phase Transitions* (Elsevier Science, Amsterdam, 1992).
8. P.W. Anderson, Some Unresolved Questions in the Theory of Localization, in: *Localization, Interaction and Transport Phenomena*, ed. B. Kramer and Y. Bruynserade (Springer-Verlag, Berlin, 1985).
9. P.W. Anderson, The Question of Classical Localization: A Theory of White Paint, Phil. Mag. B**52**, 505 (1985).
10. P. Sheng, B. White, Z.Q. Zhang, and G. Papanicolaou, Wave Localization and Multiple Scattering in Randomly-Layerered Media, in: *Scattering and Localization of Classical Waves in Random Media* edited by Ping Sheng (World Scientific, Singapore, 1990).
11. S. Fishman, D.G. Grempel and R.E. Prange, Chaos, Quantum Recurrences and Anderson Localization, Phys. Rev. Lett. **49**, 509 (1982).
12. R. Tawel and K.F. Canter, Observation of a Positron Mobility Threshold in Gaseous Helium, Phys. Rev. Lett. **56**, 2322 (1986).
13. E. Abrahams, P.W. Anderson, D.C. Licciardello and T.V. Ramakrishnan, Scaling Theory of Localization, Absence of Diffusion in Two Dimensions, Phys. Rev. Lett. **42**, 673 (1979).
14. P.W. Anderson, Absence of Diffusion in Certain Random Lattices, Phys. Rev. **109**, 1492 (1958).
15. D. Vollhardt and P. Wölfle, Scaling Relations from a Selfconsistent Theory for Anderson Localization, Phys. Rev. Lett. **48**, 699 (1982).
16. D. Vollhardt and P. Wölfle, Diagrammatic, Selfconsistent Treatment of the Anderson Localization Problem in $d \leq 2$ Dimensions, Phys. Rev. B. **22**, 4666 (1980).
17. S. Hikami, Anderson Localization in a Nonlinear σ-model Representation, Phys. Rev. B **24**, 2671 (1981).
18. A.J. McKane and M. Stone, Localization as an Alternative to Goldstone's Theorem, Annals of Physics **131**, 36 (1981).
19. S. John, H. Sompolinsky and M.J. Stephen, Localization in a Disordered Elastic medium near two Dimensions, Phys. Rev. B **27**, 5592 (1983).
20. J. Fröhlich and Th. Spencer, Absence of Diffusion in the Anderson Tight Binding Model for Large Disorder and Low Energy, Commun. Math. Phys. **88**, 151 (1983).
21. C.W.J. Beenakker, Random matrix Theory of Quantum Transport, Rev. Mod. Phys. **69**, 731 (1997).
22. J.M. Drake and A.Z. Genack, Observation of Nonclassical Optical Diffusion, Phys. Rev. Lett. **63**, 259 (1989).

23. M.P. van Albada, B.A. van Tiggelen, A. Lagendijk, A. Tip, Speed of Propagation of Classical Waves in Strongly Scattering Media, Phys. Rev. Lett. **66**, 3132 (1991).
24. A.B. Harris and T.C. Lubensky, Mean-field Theory and ϵ-expansions, Phys. Rev. B **23**, 2640 (1981).
25. W. Kirsch, Random Schrödinger Operators, in: *Schrödinger Operators*, edited by H. Holden and A. Jensen (Springer-Verlag, 1988).
26. R. del Rio, S. Jitomirskaya, Y. Last and B. Simon, What is Localization?, Phys. Rev. Lett. **75**, 117 (1995).
27. M. Schreiber, Numerical Characterisation of Electronic States in Disordered Systems, in: *Localisation 1990* (Inst. Phys. Conf. Ser. No 108, Imperial College, London, 1990).
28. M. Schreiber, Fractal Character of Eigenstates in Weakly Disordered 3D Systems, Phys. Rev. B **31**, 6146 (1985).
29. P. de Vries, *Trottering Through Quantum Physics*, Ph.D. Thesis (1991, University of Amsterdam) unpublished.
30. O. Bohigias, J.M. Giannoni and C. Schmit, Characterisation of Chaotic Quantum Spectra and Universality of Level Fluctuation Laws, Phys. Rev. Lett. **52**, 1 (1984).
31. R. Dalichaouch, J.P. Armstrong, S. Schultz, P.M. Platzman and S.L. McCall, Microwave Localization by 2D Random Scattering, Nature **354**, 53 (1991).
32. R.L. Weaver, Anderson Localization of Ultrasound, Wave Motion **12**, 129 (1990).
33. L. Ye, G. Cody, M. Zhou and P. Sheng, Observation of Bending Wave Localization and Quasi Mobility Edge in two Dimensions, Phys. Rev. Lett. **69**, 3080 (1992).
34. G.A. Thomas and M.A. Paalanen, Recent Developments in the Metal-Insulator Transition, in: *Localization, Interaction and Transport Phenomena*, ed. B. Kramer and Y. Bruynserade (Springer-Verlag, Berlin, 1985).
35. I. Shlimak, M. Kaveh, R. Ussyshkin, V. Ginodman and L. Resnick, Determination of the Critical Conductivity Exponent for the Metal-Insulator Transition at Nonzero Temperatures: Universality of the Transition, Phys. Rev. Lett. **77**, 1103 (1996).
36. J.E. Enderby and A.C. Barnes, Electron Transport at the Anderson Transition, Phys. Rev.B **49**, 5062 (1994).
37. N.F. Mott, Conduction in Non-Crystalline Materials: IX. The Minimum Conductivity, Phil. Mag. **26**, 1015 (1972).
38. M. Kaveh and N.F. Mott, The Metal-Insulator Transition in Disordered 3D Systems: A New View, J. Phys. C: Solid sate Phys. **15**, L697 (1982).
39. N.F. Mott, Conduction in Non-Crystalline Materials: III. Localized States in a Pseudo-Gap and Near Extremities of Conduction and Valence Bands, Philos. Mag. **19**, 835 (1969).
40. K. von Klitzing, G. Dorda and M. Pepper, New Method for High Accuracy Determination of Fine-structure Constant based on Quantized Hall Resistance, Phys. Rev. Lett. **45**, 494 (1980).
41. A.A.M. Pruisken, Localization and the Integer Quantum Hall Effect, in: *Localization, Interaction and Transport Phenomena*, ed. B. Kramer and Y. Bruynserade (Springer-Verlag, Berlin, 1985).
42. B.I. Halperin, Helv. Phys. Acta **56**, 75 (1983).
43. H.P. Wei, D.C. Tsui and A.A.M. Pruisken, Metal-Insulator Transition in the Integer Quantum Hall Effect, in: *Localization and Confinement of Electrons in Semiconductors*, ed. F. Kuchar, H. Heinrich and G. Bauer (Springer-Verlag, Berlin, 1990)
44. N. Garcia and A.Z. Genack, Anomalous Photon Diffusion at the Threshold of the Anderson Localization Transition, Phys. Rev. Lett. **66**, 1850 (1991).
45. D.S. Wiersma, P. Bartolini, A. Lagendijk and R. Righini, Localization of Light in a Disordered Medium, Nature **390**, 671 (1997).
46. J.W. Goodman, *Statistical Optics* (Wiley, 1985).
47. A.Z. Genack, Fluctuations, Correlations and Average Transport of Electromagnetic Radiation in Random Media, in: *Scattering and Localization of Classical Waves in random media*, edited by Ping Sheng (World Scientific, Singapore, 1990).

48. J.F. de Boer, M.C.W. van Rossum, M.P. van Albada, Th.M. Nieuwenhuizen and A. Lagendijk, Probability Distribution of Multiple Scattered Light measured in Total Transmission, Phys. Rev. Lett. **73**, 2567 (1994).
49. P.A. Lee, Universal Conductance Fluctuations in Disordered Metals, Physica **140A**, 169 (1986).
50. P.A. Lee, A.D. Stone, and H. Fukuyama, Phys. Rev. B **35**, 1039 (1987).
51. M. Stoytchev and A.Z. Genack, Measurement of the Probability Distribution of Total Transmission in Random Waveguides, Phys. Rev. Lett. **79**, 309 (1997).
52. E. Kogan and M. Kaveh, Random matrix Theory Approach to the Intensity Distributions of Waves Propagating in Random Media, Phys. Rev. B **52**, R3813 (1995).,
53. Th.M. Nieuwenhuizen and M.C.W. van Rossum, Intensity Distributions of Waves Transmitted through a Multiple Scattering Medium, Phys. Rev. Lett. **74**, 2674 (1995).
54. S.A. van Langen, P.W. Brouwer and C.W.J. Beenakker, Nonperturbative Calculation of the Probability Distribution of Plane-Wave Transmission through a Disordered Waveguide, Phys. Rev. E **53**, R1344 (1996).
55. P.A. Mello and J.L. Pichard, Maximum Entropy Approaches to Quantum Electronic Transport, Phys. Rev. B **40**, 5276 (1989).
56. H.U. Baranger and P.A. Mello, Mesoscopic Transport Through Chaotic Cavities: A random *S*- matrix Theory Approach, Phys. Rev. Lett. **73**, 142 (1994).
57. C.W.J. Beenakker and B. Rejaei, Exact Solution for the Distribution of Transmission Eigenvalues in a Disordered Wire and Comparison to RMT, Phys. Rev. B **49**, 7499 (1994).
58. E.R. Muciolo, R.A. Jalabert and J.L. Pichard, Parametric Studies of the Scattering Matrix; From Metallic to Insulating Quasi 1D Disordered Systems, J. Phys. France **10**, 1267 (1997).
59. B.A. van Tiggelen, A. Lagendijk and A. Tip, Multiple Scattering Effects for the Propagation of Light in 3D Slabs, J.Phys.C: Condens. Matter **2**, 7653 (1990).
60. M. Rusek, A. Orlowski and J. Mostowski, Localization of Light in 3D Random Dielectric Media, Phys. Rev. E **53**, 4122 (1996)
61. M. Rusek, A. Orlowski and J. Mostowski, Band of Localized Electromagnetic Waves in random Arrays of Dielectric Cylinders, Phys. Rev. E **56**, 4892 (1997).
62. A. MacKinnon, The Scaling Theory of Localisation, in: *Localization, Interaction and Transport Phenomena*, ed. B. Kramer and Y. Bruynserade (Springer-Verlag, Berlin, 1985).
63. A. MacKinnon, Critical Behavior of the Metal-Insulator Transition, in: *Localization and Confinement of Electrons in Semiconductors*, ed. F. Kuchar, H. Heinrich and G. Bauer (Springer-Verlag, Berlin, 1990).
64. Ping Sheng, *Introduction to Wave Scattering, Localization, and Mesoscopic Phenomena* (Academic, San Diego, 1995).
65. M. Aizenmann and S. Molchanov, Localization at Large Disorder and at Extreme Energies: An Elementary Derivation, Commun. Math. Phys. **157**, 245 (1993).
66. F. Delyon, H. Kunz and B. Souillard, 1D Wave Equations in Disordered Media, J. Phys. A **16**, 25 (1983).
67. D.J. Thouless, A Relation between Density of States and Range of Localization for 1D Random Systems, J. Phys. C: Solid State Phys. **5**, 77 (1972).
68. M. Kappus and F. Wegner, Anomaly in the Band Center of the 1D Anderson Model, Z. Phys. B - Condensed Matter **45**, 15 (1981).
69. D.J. Thouless, Percolation and Localization, in: Les Houches *Ill-Condensed Matter*, edited by R. Balian, R. Maynard, G. Toulouse (Amsterdam, North-Holland, 1979).
70. B. Derrida and E. Gardner, Lyapounov Exponent of the 1D Anderson Model: Weak Disorder Expansions, J. Physique **45**, 1283 (1984).
71. C.J. Lambert, Anomalies in the Transport Properties of a Disordered Solid, Phys. Rev. B **29**, 1091 (1984).
72. E. Bouchaud and M. Daoud, Reflection of Light by a Random Layered System, J. Physqiue **47**, 1467 (1986).

73. R. Carmona and J. Lacroix, *Spectral Theory of Random Schrödinger Operators* (Birkhäuser, Boston, 1990).
74. B.A. van Tiggelen and A. Tip, Photon Localization in Disorder- induced Periodic Multilayers, J.Phys. I France **1**, 1145 (1991).
75. P. Sheng and Z.Q. Zhang, Is a Layerered Medium One Dimensional, Phys. Rev. Lett. **74**, 1343 (1995).
76. Z.Q. Zhang, Light Amplification and Localization in Randomly Layered Media with Gain, Phys. Rev. B **52**, 7960 (1995).
77. H. Amman, R. Gray, I. Shvarchuck and N. Christensen, Quantum Delta-Kicked Rotor: Experimental Observation of Decoherence, Phys. Rev. Lett. **80**, 4111 (1998).
78. A. Buchleitner and D. Delande, Dynamical Localization in More than One Dimension, Phys. Rev. Lett. **70**, 33 (1992).
79. B. Bulka, M. Schreiber and B. Kramer, Localization and Quantum Interference and the Metal-Insulator Transition, Z. Phys. B **66**, 21 (1987).
80. M. Schreiber, Fractal Eigenstates in Disordered Systems, Physica A **167**, 188 (1990) (Special Issue on the Anderson Transition and Mesoscopic Fluctuations).
81. A. MacKinnon, Critical Exponents for the Metal-Insulator Transition, J. Phys. Cond. Matter **6**, 2511 (1994).
82. A.D. Zdetsis, C.M. Soukoulis, E.N. Economou and G.S. Grest, Localization in Two- and Three Dimensional Systems away from the Band Center, Phys. Rev. B **32**, 7811 (1985).
83. J.L. Pichard and M. Sanquer, Quantum Conductance Fluctuations and Maximum Entropy Ensembles for the Transfer matrix, Physica A **167**, 66 (1990). (Special issue on Anderson Transition and Mesoscopic Fluctuations).
84. E. Hofstetter and M. Schreiber, Does Broken T-Symmetry modify the Critical Behavior at the Metal-Insulator Transition in 3D Disordered Systems?, Phys. Rev. Lett. **73**, 3137 (1994).
85. M. Henneke, B. Kramer and T. Ohtsuki, Anderson Transition in a Strong Magnetic Field, Europhys. Lett. **27**, 389 (1994).
86. R. Abou-Chacra, P.W. Anderson and D.J. Thouless, A Selfconsistent Theory for Localization, J. Phys. C: Solid State Phys. **6**, 1734 (1973).
87. H. Kunz and B. Souillard, The Localization Transition on the Bethe Lattice, J. Physique Lett. **44**, L411 (1983).
88. T. Kawarabayashi and M. Suzuki, Decay Rate of the Green Function in a Random Potential on the Bethe Lattice and a Criterion for Localization, J.Phys. A: Math. Gen. **26**, 5729 (1993).
89. A.F. Ioffe and A.R. Regel, Non-Crystalline, Amorphous and Liquid Electronic Semiconductors, Progress in Semiconductors **4** 237 (1960).
90. A. Lagendijk and B.A. van Tiggelen, Resonant Multiple Scattering of Light, Phys. Rep. **270**, 143 (1996).
91. A. Lagendijk, private communication.
92. E.N. Economou, C.M. Soukoulis and A.D. Zdetsis, Localized States in Disordered Systems as bound States in Potential Wells, Phys. Rev. B **30**, 1686 (1984).
93. B.A. van Tiggelen, A. Lagendijk, A. Tip and G.F. Reiter, Effect of Resonant Scattering on Localization of Waves, Europhys. Lett. **15**, 535 (1991).
94. D. Sornette and B. Souillard, Strong Localization of Waves by Internal Resonances, Europhys. Lett. **7**, 269 (1988).
95. C.A. Condat and T.R. Kirkpatrick, Resonant Scattering and Anderson Localization of Acoustic Waves, Phys. Rev. B **36**, 6782 (1987).
96. T.R. Kirkpatrick, Localization of Acoustic Waves, Phys. Rev. B **31**, 5746 (1985).
97. P. Sheng and Z.Q. Zhang, Scalar-Wave Localization in a Two-Component Composite, Phys. Rev. Lett. **57**, 1879 (1986).
98. A. Figotin and A. Klein, Localization of Classical Waves. I: Acoustic Waves, Comm. Math. Phys. **180**, 439 (1996).
99. A. Figotin and A. Klein, Localization of Classical Waves. II: Electromagnetic Waves,

Comm. Math. Phys. **184**, 411 (1997)
100. B.A. van Tiggelen and E. Kogan, Analogies between Light and Electrons: Friedels' Identity and Density of States, Phys. Rev. A **49**, 708 (1994).
101. B. Elattari, V. Kagalovsky and H.A. Weidenmüller, Non-linear Supersymmetric σ-Model for Diffuse Scattering of Classical Waves with Resonant Enhancement, Europhys. Lett. **42**, 13 (1998).
102. M. Stoytchev and A.Z. Genack, Measurements of Intensity Distributions in the Approach to Localization, preprint.
103. E. Akkermans and G. Montambaux, Conductance and Statistical Properties of Metallic Spectra, Phys. Rev. Lett. **68**, 642 (1992).
104. F.J. Wegner, Electrons in Disordered Systems: Scaling near the Mobility Edge, Z. Phys. B **25**, 327 (1976).
105. A. MacKinnon and B. Kramer, The Scaling Theory of Electrons in Disordered Solids,, Z. Phys. B **53**, 1 (1983).
106. B.A. van Tiggelen, D.S. Wiersma and A. Lagendijk, Selfconsistent Theory for the Enhancement Factor in Coherent Backscattering, Europhys. Lett. **30**, 1 (1995).
107. G.D. Mahan, *Many particle Physics* (Plenum, New York, 1981).
108. W. Götze, A Theory for the Conductivity of a Fermion Gas Moving in a Strong 3D Random Potential, J. Phys. C **12**, 1279 (1979).
109. R. Burridge, G. Papanicolaou and B. White, SIAM J. Appl. Math. **47**, 146 (1987).
110. P. Sheng, Z.Q. Zhang and G. Papanicolaou, Multiple Scattering Noise in 1D, Universality through Localization-Length Scaling Phys. Rev. Lett. **57**, 1000 (1986).
111. B. White, P. Sheng and Z.Q. Zhang, Wave Localization Characteristics in the Time Domain, Phys. Rev. Lett. **59**, 1918 (1987).
112. J.P. Fouque, Transmission and Reflection of Acoustic Pulses by Randomly Layered Media, in: *New Aspects of Electromagnetic and Acoustic Wave Diffusion* (Springer-Verlag, Heidelberg, 1998)
113. R.L. Weaver, Anderson Localization in the Time-Domain: Numerical Studies of Waves in 2D Media Phys. Rev. B **49**, 5881 (1994).
114. C. Vanneste, P. Sebbah and D. Sornette, A Wave Automaton for Time-dependent Wave Propagation in Random Media, Europhys. lett. **17**, 715 (1992).
115. R. Berkovits and M. Kaveh, Backscattering of Light near the Optical Anderson Transition, Phys. Rev. B **36**, 9322 (1987).
116. R. Berkovits and M. Kaveh, Propagation of Waves Trough a Slab near the Anderson Transition: A Local Scaling Approach, J. Phys. C: Cond. Matter **2**, 307 (1990).
117. M. Josefin, Localization in Absorbing Media, Europhys. Lett. **25**, 675 (1994).
118. R.L. Weaver, Anomalous Diffusivity and Localization of Classical Waves in Disordered Media: Effect of Dissipation, Phys. Rev. B **47**, 1077 (1993).
119. J.C.J. Paasschens, T.Sh. Misirpashaev and C.W.J. Beenakker, Localization of Light, Dual Symmetry between Absorption and Amplification, Phys. Rev. B **54**, 11887 (1996).

MATHEMATICAL THEORY OF LOCALIZATION

J. LACROIX
Laboratoire de Probabilités, Université de Paris VI
4 Place Jussieu, F-75252 Paris Cedex 5
E. mail: lacroix@proba.jussieu.fr

1. THE MODEL

Localized states appear in a lot of models related to wave propagation in disordered media. This property cannot be explained by a perturbation of the free case and is somewhat unexpected and intricate. The major drawback in the mathematical approach to this phenomenon is that the physical background of the problem is "almost surely lost" (faithful readers will understand soon...) and this certainly indicates that a "good proof" of localization is still missing. For historical reasons (the pioneer work of Anderson in 1958 [1]) and in order to avoid too much technicalities we will essentially consider the Schrödinger equation on $\mathbb{Z}^\nu$:

$$i\frac{\partial \psi}{\partial t} = H\psi, \qquad H = K + U + \lambda V$$

where:

- V is a random potential associated with an ergodic dynamical system. Actually we will restrict ourselves to the case of independent potentials V_x at each site $x \in \mathbb{Z}^\nu$ and assume that their common probability distribution has a bounded density. The tunable parameter λ expresses the strength of the disorder.
- U is a periodic potential.
- The operator K is the Laplace operator $K\psi(x) = \sum_{|x-y|=1} \psi(y)$ (Or a translation invariant and exponentially decaying symmetric kernel)

The operator H is then self adjoint on the Hilbert space $\mathcal{H} = \ell^2(\mathbb{Z}^\nu)$.

J.-P. Fouque (ed.), Diffuse Waves in Complex Media, 61–71.

It is also possible to consider models with off diagonal disorder or having an additional constant magnetic field of the form

$$K\psi(x) = \sum_{|x-y|=1} \exp(-iA(x,y))\psi(y)$$

where A is an antisymmetric function of the oriented bonds (x,y). The simplest model corresponding to the Laplace operator, without periodic potential, is called the "Anderson model". In this review we are essentially concerned with the multidimensional case, hence a lot of earlier and important papers are not cited. Much more information on the spectral theory of Schrödinger and almost periodic operators can be found in example in two textbooks [12] [13] or in a presentation of the one dimensional case [20].

2. DEFINITION OF LOCALIZATION

One says that the operator H is strongly localized if any spectral measure of H is pure point. In other words, let P_B be the spectral projector on a Borel subset $B \subset \mathbb{R}$, then for any $f \in \mathcal{H}$ the spectral measure $\sigma_f(B) = ||P_B(f)||^2$ is pure point. This is equivalent to the fact that there exists an orthonormal basis of $\mathcal{H}$ made of eigenfunctions of H. As a consequence, for any function $\psi_0(x) \in \mathcal{H}$ then $\psi(t,x) = (\exp(-itH)\psi_0)(x)$ exhibits the localized behaviour:

$$\lim_{r\to\infty} \sup_t \sum_{|x|>r} |\psi_t(x)|^2 = 0$$

This behaviour is in contrast with the free operator situation ($\lambda = 0$) in which one gets an absolutely continuous spectrum formed of conductivity closed bands separated by gaps.

In general one also considers two "stronger" forms of localization:

1. Exponential localization: the operator H is strongly localized and any eigenfunction of H is actually exponentially decaying.
2. Dynamical localization: Let $r(t)$ be the mean distance traveled in time t by a wave packet associated with function $\psi_0(x)$ of finite support, namely:

$$r^2(t) = \sum_x |x|^2 |\psi_t(x)|^2$$

   Then the function $r(t)$ is bounded.

Dynamical localization implies strong localization (RAGE theorem) but there are exponentially localized models for which $r(t)$ grows faster than t^δ for any $0 < \delta < 1$. Localization only implies abscence of balistic motion that is $\lim_{t\to\infty} r(t)/t = 0$.

For random operators, strong localization, exponential localization or dynamic localization are understood to hold for almost all realization ω of the potentials, thas is for almost all operators $H(\omega)$. Actually, in the case of independent potentials, we will see below that in regimes where strong localization has been proved then one also gets exponential and dynamical localization but this is no longer true for almost periodic models.

3. A KEY PROPERTY TOWARD LOCALIZATION

In most of the recent proofs of localization, the main step is achieved by proving exponential decay of the Green function under various forms. We will below refer to the strong (S.E.D.G) and weak (W.E.D.G) exponential decay of the Green function.

One says that (S.E.D.G) holds on a bounded interval $[a, b]$ if:

$$\mathbb{E}\{\sup_{\epsilon>0} |G(E + i\epsilon, 0, x)|^s\} \leq C \exp(-\gamma|x|)$$

for some $0 < s < 1$, $\gamma > 0$, $C < \infty$ and for any $E \in [a, b]$.

For Anderson models, such a bound has been obtained in the one dimensional case for any interval $[a, b]$ and any $\lambda > 0$ as a consequence of the positivity of the Lyapunov exponent associated with a product of random transfer matrices and the theory of Laplace's transform on $SL(2, \mathbb{R})$. In dimension greater than one, such an estimate is only obtained at large disorder (uniformly with respect to $E \in \mathbb{R}$) or at any non zero disorder for intervals $[a, b]$ contained in $|E| > E_0$ for some large value E_0.

Using the fact that the probability distributions of $G(E + i\epsilon, x, y)$ and $G(E + i\epsilon, 0, y - x)$ are indentical, (S.E.D.G.) implies that for any fixed energy E in $[a, b]$ then for almost all realizations of the potentials ω one has:

$$\sup_{\epsilon>0} |G(E + i\epsilon, x, y)| \leq C'(E, \omega) \exp(-\gamma'|x - y|)$$

for somme $\gamma' > 0$, $C'(E, \omega) < \infty$ and for any $x, y \in \mathbb{Z}^\nu$. This weaker form of exponential decay is called (W.E.D.G). Up to now, such an estimate has been proved to be valid in the same regimes as for the stronger form (S.E.D.G).

We will below give some hints about the proofs of these exponential bounds but we first review some consequences:

- Exponential localization.
- Dynamical localization.
- Absence of level repulsion in the Anderson model.

- Vanishing of the electrical conductivity given by the Kubo formula in the Anderson Model.

We also want to point out that exponential localization has been proved to hold in different regimes (near band edges of the unperturbed spectrum) or different models (acoustic and electro-magnetic waves, continuous models) by means of extensions of the multi-scale analysis that we present in the next section.

4. PROOFS OF (W.E.D.G.) and (S.E.D.G)

4.1. PROOF OF (W.E.D.G): THE MULTISCALE ANALYSIS

J. Fröhlich T. Spencer [4], *J. Fröhlich F. Martinelli E. Scoppola T. Spencer* [6], *H. von Dreifus, A. Klein* [11]

The multiscale analysis has been developed first for Anderson models but can be extended to magnetic models. Let ℓ be a positive real number. The box $\Lambda(x)$ centered at the point x and of size ℓ is defined as:

$$\Lambda_\ell(x) = \{y \in \mathbb{Z}^\nu \,;\, |x-y| \leq \frac{\ell}{2}\}$$

The symbol Λ denotes a box without reference to its center. The boundary $\partial\Lambda$ of the box Λ is the subset of $\mathbb{Z}^\nu \times \mathbb{Z}^\nu$ defined by:

$$\partial\Lambda = \{(u,v)\,;\, |u-v| = 1\ ,\ u \in \Lambda \text{ and } v \notin \Lambda\}$$

One defines the interior boundary $\partial\Lambda^{in}$ by:

$$\partial\Lambda^{in} = \{u \in \Lambda\,;\ \text{ there exists }\ v \notin \Lambda \text{ such that } (u,v) \in \partial\Lambda\}$$

For a complex number z and a box Λ we denote by $G^\Lambda(z,x,y)$ the Green's kernel of the restriction of the operator H to the box Λ, extended to the whole of $\mathbb{Z}^\nu$ by 0 whenever x or y are not in the box Λ.
Let E be a fixed energy and γ be a positive number. The box $\Lambda_\ell(x)$ is said to be a γ-"good box" if for any $\epsilon \neq 0$ one has:

$$\sum_{y \in \partial\Lambda_\ell^{int}(x)} |G^{\Lambda_\ell(x)}(E+i\epsilon,x,y)| \leq e^{-\gamma\ell}$$

The box Λ_ℓ is said to be "γ adapted" if:

$$\mathbb{P}\{\omega\,;\ \Lambda_\ell \text{ is a } \gamma\text{-"good box" }\} \geq 1 - \frac{1}{\ell^{2\nu+4}}$$

The essential argument in multiscale analysis is the following:
If there exists $\gamma_0 > 0$ and ℓ_0 "sufficiently large" such that the box Λ_{ℓ_0} be "γ_0 adapted" then it is possible to find a scale $1 < \alpha < 2$ such that the sequence of boxes of width $\ell_{n+1} = \ell_n^\alpha$ be γ_n adapted with $inf_n(\gamma_n) > 0$. In order to implement this machinery one needs an extra ingredient called "Wegner estimate" which implies that the probability of resonances between large boxes far apart is very small. This assumption is always satisfied for potentials having a probability distribution with a bounded density.
Using this increasing sequence of adapted boxes it is possible to recover the Green function on the whole space hence the desired result. In order to make the multi-scale analysis to work, one first needs to prove that there exists an adapted box to start with. In dimension one this is always possible but in dimension greater than one one needs large disorder (and this works for any energy E) or large energies, more precisely there exist constants E_0 and λ_0 such that if we are in one of the following situation:

- The energy is large i.e. $|E| \geq |E_0|$
- The disorder is large i.e. the potential is scaled by a factor λ such that $|\lambda| \geq |\lambda_0|$

then there exists an initial γ adapted box of sufficiently large size to initiate the multiscale procedure.

4.2. PROOF OF (S.E.D.G): THE MOMENT ANALYSIS

M. Aizenman, S. Molchanov [14], *M. Aizenman G.M. Graf* [18]

This method can be applied to different lattices (in particular the Bethe lattice) and is much less technical than the multi-scale analysis. One first proves that for $0 < s < 1$ then:

$$\mathbb{E}\{|G(z,0,x)|^s\} \leq C(s) < \infty, \text{ for } \Im m(z) \neq 0$$

This is a consequence of the assumptions on the distribution of potentials and of the resolvent identity:

$$G(z,x,y) = \frac{\widetilde{G}(z,x,y)}{1 + \lambda V_x \widetilde{G}(z,x,x)}$$

where $\widetilde{G}(z,.,.)$ is the Green function of the operator H with the potential V_x replaced by 0. Actually the presence of the exponent s is due to the fact that the probability distribution of $|G(E+i0,x,x)|$ has a Cauchy tail:

$$G(z,x,x) = \frac{1}{\widetilde{G}(z,x,x)^{-1} + \lambda V_x} = \frac{1}{u + \lambda V_x}$$

Hence for $t > 0$:

$$\mathbb{P}\{|G(z,x,x)| > t\} \leq \mathbb{P}\{|\Re e(u) + \lambda V_x| < 1/t\} \leq \frac{2\lambda||\varphi||_\infty}{t}$$

where φ is the density of the distribution of potentials. This relation entails:

$$\mathbb{E}\{|G(z,x,x)|^s\} \leq C(s) \quad \text{for } 0 < s < 1$$

The proof of the same bound for $\mathbb{E}\{|G(z,x,y)|^s\}$ is more involved. Using the definition of the Green function:

$$|z - \lambda V_x - U_x|^s |G(z,0,x)|^s \leq \sum_y |K(x,y)|^s |G(z,0,y)|^s, \quad x \neq 0$$

(In our situation $|K(x,y)| = 1$ but it is possible to include exponentially decreasing kernels). It only remains to prove the following "decoupling lemma":

$$\mathbb{E}\{|z - \lambda V_x - U_x|^s |G(z,0,x)|^s\} \geq a(\lambda,s)\mathbb{E}\{|G(z,0,x)|^s\}$$

It is clear that this can only be true when the factor $|z - \lambda V_x - U_x|$ is bounded below with a large probability. Assuming λ large enough this can be done uniformly with respect to z, $\Im m z \neq 0$. Hence one gets:

$$a(s,\lambda)\mathbb{E}\{|G(z,0,x)|^s\} \leq \sum_y \mathbb{E}\{|G(z,0,y)|^s\}, \quad x \neq 0$$

For $a > 2\nu$ this is an uniform subhamonicity statement for the function $x \hookrightarrow \mathbb{E}\{|G(z,0,x)|^s$ which is bounded. Exponential decay of $\mathbb{E}\{|G(z,0,x)|^s\}$ is an easy consequence of this fact.

For small values of λ the decoupling lemma remains valid for large energies and the bound is uniform on a bounded interval of energy. One can remark that the domain of validity of this approach is the same as in the multiscale analysis.

5. CONSEQUENCES OF (W.E.D.G.) AND (S.E.D.G)

5.1. ABSENCE OF ABSOLUTELY CONTINUOUS SPECTRUM

F. Martinelli E. Scoppola [8]

Let $\sigma(\omega,x)$ be the spectral measure of $H(\omega)$ associated with the delta function at the point x. From the relation:

$$\Im m G(E+i\epsilon,x,x) = 2\epsilon \sum_y |G(E+i\epsilon,x,y)|^2$$

one can conclude from (W.E.D.G) and de la Vallée Poussin's theorem that for almost all realizations of the potentials, the absolutely continuous part of $\sigma(\omega, x)$ is null. This property is identical to the Pastur's theorem in the one dimensional case. But it remains to exclude the singular continuous part of the spectrum and this is the hard part of the work!

5.2. STRONG AND EXPONENTIAL LOCALIZATION

S. Kotani [5], *B. Simon T. Wolff* [9],*F. Delyon, Y. Levy, B. Souillard* [7], *F. Martinelli E. Scoppola* [10]

We first give the argument of Simon & Wolff which is easy to state but somewhat far from the physical nature of the problem... Let define the set $W \subset (\Omega, \mathbb{R})$ by:

$$W = \{(\omega, E) \; ; \; \lim_{\epsilon \to 0} \sum_{x} |G(E + i\epsilon, 0, x)|^2 < \infty \; , E \in [a, b]\}$$

Under the condition (W.E.D.G) the theory of Simon & Wolff immediately yields the strong localization property since it only requires that the set W be of full $\mathbb{P} \otimes \ell$ measure (where ℓ is the Lebesgue measure on $[a, b]$), but one does not get exponential localization...In order to obtain this extra property one needs an other approach, nearer from the original ideas of Borland [2] and related to the multiscale analysis.

Some facts about the non-random operator

Let ℓ_n be an increasing sequence to be chosen later and satisfying $\ell_0 > 1$, $\ell_{n+1} = \ell_n^\alpha$ for some $1 < \alpha < 2$. One denotes by $\Lambda_n(x)$ the box of center x and size ℓ_n and by A_n the annulus $\{x \; ; \; \ell_n \leq |x| \leq \ell_{n+1}\}$. The energy E is said to be hyperbolic if there exists a positive number $\gamma(E)$ such that for n sufficiently large the box $\Lambda_n(x)$ is a $\gamma(E)$- good box for any $x \in A_n$. Let us denote by σ_x the spectral measure of the self-adjoint operator H at the point $x \in \mathbb{Z}^\nu$. Then one has the following properties:

1. Let m be a nonnegative continuous measure on $\mathbb{R}$ which is carried by hyp(H). Then m is orthogonal to any spectral measure σ_x.
2. If the spectral measure σ_x is carried by hyp(H), then σ_x is pure point. Moreover any eigenvector $\psi_E(x)$ corresponding to an eigenvalue E is exponentially decaying.

In order to prove these statements let E be an hyperbolic energy and ψ be a non trivial "slowly growing" generalized eigenfunction i.e. such that $|\psi(x)| \leq C|x|^2$. Let choose $x \in A_n$ and n sufficiently large. Writing the Poisson formula (where we set $\Lambda = \Lambda_n(x)$ in order to simplify the notations)

$$< \psi \, , (H^\Lambda - E) G^\Lambda(E, x, .) > \;\; = \;\; < \psi \, , (H^\Lambda - H) G^\Lambda(E, x, .) >$$

$$\psi(x) = \sum_{(u,v)\in\partial\Lambda} G^{\Lambda}(E,x,u)\psi(v)$$

(all the dot products are well defined thank to the the slow growth of ψ and the exponential decay of thee Green's kernel) we get that for some finite constant constant C' one has:

$$|\psi(x)| \leq C'\ell_{n+1}^2 \exp(-\ell_n\gamma(E)) \leq C'|x|^4 \exp(-\gamma(E)|x|^{1/\alpha})$$

It follows that ψ is exponentially decaying hence that E is an eigenvalue of H. The measure m being continuous the countable set of eigenvalues of H is m negligible. On the other hand we know that for σ_x almost energy there exists a non trivial slowly growing generalized eigenfunction. Then one can conclude that m is orthogonal to σ_x. If σ_x is supported by the hyperbolic energies, the same argument shows that σ_x is actually supported by the eigenvalues of H with exponentially decaying eigenfunctions.

The random operator

Let define the set $W \subset (\Omega, \mathbb{R})$ by:

$$W = \{(\omega, E)\,;\, E \in \text{hyp}(H(\omega)) \cap [a,b]\}$$

Let E be a fixed energy in $(a,b]$ and ℓ_n be the increasing sequence of size of boxes obtained in the Frölich and Spencer analysis. Let set

$$\Omega_n = \{\omega\,;\, \Lambda_n(x) \text{ is a } \gamma \text{ good box for any } x \in A_n\}$$

Then one has:

$$\mathbb{P}(\Omega_n) \geq 1 - \frac{|A_n|}{\ell_n^{2\nu+4}} \geq 1 - \frac{\ell_{n+1}^{\nu}}{\ell_n^{2\nu+4}} \geq 1 - \frac{1}{\ell_n^4}$$

A direct application of the Borel-Cantelli lemma yields that for any energy E in $[a,b]$ the section W_E of W is of full $\mathbb{P}$ probability measure. Let $\sigma(\omega)$ denotes a maximal spectral measure for $H(\omega)$ and m be a given positive measure on $\mathbb{R}$. Using the Fubini's theorem, one gets that for almost all ω then m almost all energy E in $[a,b]$ is hyperbolic for $H(\omega)$. Let suppose for a while that it is possible to replace " *m almost all energy E in $[a,b]$* "in the previous sentence, by " *$\sigma(\omega)$ almost all energy E in $[a,b]$* " then the criterion of localization of the non random case yields almost sure exponential localization. Unfortunately the above proposed replacement is far from to be obvious since for almost all ω the spectral measures $\sigma(\omega)$ is orthogonal to any fixed m, and we are stuck! Fortunately Kotani gave some sort of extension of the Fubini's Theorem ("Kotani's trick") which allows us to justify this replacement assuming that the probability distribution of potentials is absolutely continuous with a bounded density. It essentially uses the fact that if we denote by $\mathcal{F}_n$ the sigma-algebra generated by the potentials $V(k)$ for $|k| \geq n$ then W is measurable with respect to the sigma field $\mathcal{F}_n \otimes \mathcal{B}$ where $\mathcal{B}$ is the Borel σ field of the real line.

5.3. DYNAMICAL LOCALIZATION

M. Aizenman [15], *R. del Rio S. Jitomirskaya Y. Last B. Simon* [16]
S. De Bièvre, F. Germinet [19]

It is possible to deduce from (S.E.D.G.) the following estimate:

$$\mathbb{E}\{\sup_{t>0} |\exp(-itH)P_{[a,b]}\delta_0(x)|\} \leq C\exp(-\gamma|x|)$$

where $P_{[a,b]}$ is the spectral projection on the interval $[a, b]$ and for somme real constants $\gamma > 0$, $C < \infty$.

Dynamical localization follows immediately for large disorder, and otherwise for the restriction of the operator to the energies in $[a, b]$.

It is also possible to get dynamical localization by a direct analysis of the behaviour of eigenfunctions. One says that H has semi-uniformly localized eigenfunctions (SULE) if H has a complete set φ_n of orthonormal eigenfunctions, there is $\gamma > 0$ and a sequence m_n, such that for any $\epsilon > 0$ there exists $C_\epsilon < \infty$ with:

$$|\varphi_n(x)| \leq C_\epsilon \exp(\epsilon|m_n| - \gamma|x - m_n|)$$

Then SULE implies:

$$\sup_{t>0} |\exp(-itH)\delta_0(x)| \leq C\exp(-\gamma|x|)$$

hence dynamical localization. It has been recently remarked that a weaker (and easier to prove) property than SULE also implies dynamical localization:

$$|\varphi_n(x)\varphi_n(y)| \leq C_(n, y)\exp(-\gamma_n|x - y|)$$

Such a bound has been proved to hold for some almost periodic models or for random dimer models.

5.4. ABSENCE OF LEVEL REPULSION

S. Molchanov [3], *N. Minami* [17]

Let H^L be the operator H restricted to a box Λ_L of size L and centered at the origin, with Dirichlet boundary conditions. The empirical distribution of eigenvalues of H^L is defined as:

$$K^L = \frac{1}{|\Lambda_L|} \sum_j \partial_{E_j} , \qquad E_j \in \text{eigenvalues of } H^L$$

It follows from the ergodic theorem that for almost all realization of the potential these distributions converge weakly when $L \to \infty$ to a non random probability measure called the "Integrated density of states". Since that the common law of the potentials is assumed to have a bounded density, this probability measure has a continuous density $n(E)$ called the "density of states". In order to investigate the local structure of the states, one defines for a given energy E the point process:

$$N_E^L = \sum_j \partial_{|\Lambda|(E-E_j)} , \qquad E_j \in \text{eigenvalues of } H^L$$

In other words for any $a > 0$ one has:

$$N_E^L([-a,a]) = \{\#j \ ; \ E_j \in [E - \frac{a}{|\Lambda_L|}, E + \frac{a}{|\Lambda_L|}]\}$$

Then it can be proved that (S.E.D.G.) implies that this point process converges in distribution when $L \to \infty$ to a Poisson Process on $\mathbb{R}$ with intensity $n(E)\ell$ where ℓ is the Lebesgue measure on $\mathbb{R}$.

5.5. ELECTRICAL CONDUCTIVITY

J. Fröhlich T. Spencer [4], *M. Aizenman G.M. Graf* [18]

In the Anderson model, the electrical conductivity at an energy E is given by the Kubo's formula:

$$\chi(E) = \lim_{\epsilon \to 0} \frac{\epsilon^2}{\pi} \sum_x |x|^2 \mathbb{E}\{|G(E+i\epsilon,0,x)|^2\}$$

The obvious bound $\epsilon|G(E+i\epsilon,0,x)| \leq 1$ yields:

$$\chi(E) \leq \liminf_{\epsilon \to 0} \frac{\epsilon^s}{\pi} \sum_x |x|^2 \mathbb{E}\{|G(E+i\epsilon,0,x)|^s\}$$

for any $0 \leq s \leq 2$.

Then (S.E.D.G.) implies $\chi(E) = 0$ for any $E \in [a,b]$

The same result have been obtained in [4] only using (W.E.D.G).

In presence of a magnetic field, the Hall conductance is given by a trace formula and using (E.D.G.) it can be proved that it is a constant integral multiple of e^2/h on different intervals of energies.

References

1. Ph. Anderson: *"Absence of diffusion in certain random lattices"*, Physical Review **109**,1492-1505 (1958)
2. R.E. Borland: *" THe nature of the electronic states in disordered one-dimensional systems"*, Proc. Royal Soc. London, **A274**, 529-545, (1963)
3. S.A. Molcanov: *The structure of eigenfunctions of one dimensional unordered structures"*, Math. USSR Izvestjia **12**, 69-101, (1978)
4. J. Frölich, T. Spencer: *"Abscence of diffusion in the Anderson tight binding model for large disorder or low energy "*, Comm. Math. Phys. **88**,151-184 (1983)
5. S. Kotani: *"Lyapunov exponents and spectra for one-dimensional random Schrödinger operators"*, Proc. Conf. on Random Matrices and their Applications. Contemporary Math. A.M.S. Providence R.I.
6. J. Frölich , F. Martinelli, E. Scoppola, T. Spencer: *"A constructive proof of localization in the Anderson tight binding model"*, Comm. Math. Phys. **101**,21-46 (1985)
7. F. Delyon, Y. Levy, B. Souillard: *" An approach à la Borland to multidimensional localization"*, Phys. Rev. Lett. **55**,618-621 (1985)
8. F. Martinelli, E. Scoppola: *Remark on the absence of absolutely continuous spectrum...*, Comm. Math. Phys., **97**, 465-471 (1985)
9. B. Simon, T. Wolff: *"Singular continuous spectrum under rank one perturbations and localization for random Hamiltonians"*,
   Comm. Pure. Appl. Math. **39**,75-90 (1986)
10. F. Martinelli, E. Scoppola: *Introduction to the mathematical theory of Anderson Localization*, La Rivista del Nuovo Cimento, **10**, (1987)
11. H. von Dreifus, A. Klein: *" A new proof of localization in the Anderson tight binding model"*, Comm. Math. Phys. **124**,285-299 (1989)
12. R. Carmona, J. Lacroix: *"Spectral Theory of random Schrödinger Operators"*, Birkhaüser, Boston (1990).
13. L. Pastur, A. Figotin: *"Spectra of Random and Almost-Periodic Operators"*, Springer-Verlag, Berlin (1991).
14. M. Aizenman, S. Molchanov: *"Localization at large disorder and at extreme energies: an elementary derivation"*, Comm. Math. Phys. **157**,245- (1993)
15. M. Aizenman: *Localization at weak disorder: Some elementary bounds"*, Rev. Math. Phys. **6**, 1163-1182, (1994)
16. R. del Rio, S. Jitomirskaya, Y. Last, B. Simon: *"Operators with singular continuous spectrum, IV. Hausdorff dimensions, rank one perturbations, and localization"*, J. Analyse Mathématique **69** 153- (1996)
17. N. Minami: *"Local fluctuation of the spectrum of a multidimensional Anderson tight binding model"*, Comm. Math. Phys. **177**,709- (1996)
18. M. Aizenman, G.M. Graf: *"Localizations Bounds for an Electron Gas"* Preprint. To appear in J. Phys. A (1998)
19. F. Germinet, S. De Bievre: *" Dynamical Localization for Random Schrödinger Operators"* Comm. Math. Phys, (1998)
20. J. Lacroix: *" Localization Theory"*, New Aspects of Electromagnetic and Acoustic Wave Diffusion, Springer Tracts in Physics, **144**, (1998)

LOCALIZATION OF LIGHT IN RANDOMIZED PERIODIC MEDIA

ABEL KLEIN
University of California, Irvine
Department of Mathematics
Irvine, CA 92697-3875, USA
aklein@math.uci.edu

Abstract. In this article we describe a mathematically rigorous proof of Anderson localization of light in randomized periodic media, based on the macroscopic Maxwell equations.

1. Introduction

There is now experimental evidence for Anderson localization of light in a disordered medium [44, 45]. *In this article we describe a mathematically rigorous proof of Anderson localization of light in randomized periodic media, based on the macroscopic Maxwell equations* [14].

We start with a perfectly periodic lossless dielectric medium, called a *photonic crystal.* The propagation of electromagnetic (EM) waves in photonic crystals has been the subject of intensive study in recent years [46, 7, 39, 43, 26, 24, 37, 40, 18, 19, 42, 10, 41]. The most significant manifestation of coherent multiple scattering in the periodic medium is the rise of a gap in the spectrum, called a *photonic band gap.* Frequencies in the gap are forbidden: EM waves with such frequencies cannot propagate in the photonic crystal. If a periodic dielectric medium with a band gap is perturbed by either a single defect (impurity) or a random array of defects, localized EM waves can arise under some conditions. The frequencies of these localized waves lie in the gap. In the case of a single defect the localized eigenmodes are often called defect (or impurity) midgap eigenmodes. In the case of a random medium the phenomenon of localization has the same nature as the Anderson localization of electrons [1, 38, 33, 2, 27, 28], which is now well understood in the mathematical literature [20, 23, 21, 9, 5, 36, 6, 11].

J.-P. Fouque (ed.), Diffuse Waves in Complex Media, 73–92.

Wave localization, due to either a single defect or a random array of defects, is a general wave phenomenon. In addition to electron and EM waves, this phenomenon is also relevant to acoustic waves [35, 12, 13, 14, 15, 16], elastic waves [32, 30], acoustic phonons [34], and more complicated excitations involving coupled waves such as polaritons [8].

Our approach to the mathematical study of localization of classical waves, developed in collaboration with A. Figotin [12, 13, 14, 15, 16, 17], is operator theoretic and reminiscent of quantum mechanics. It is based on the fact that Maxwell equations (and other classical waves equations) can be recast as a first order conservative linear equation, i.e., as a Schrödinger-like equation. The corresponding self-adjoint operator, which governs the dynamics, is a first order partial differential operator, but its spectral theory can be studied through an auxiliary second order operator. This second order partial differential operator can be treated in an analogous way to Schrödinger operators in quantum mechanics.

This approach, originally developed for acoustic and Maxwell equations, has been extended to a general framework which includes elastic waves and anisotropic media [30]. It has also been used to prove *dynamical localization* for light and other classical waves [25], using methods of [22].

This article is organized as follows: Section 2 contains the operator formulation of Maxwell equations in inhomogeneous media. The Maxwell operator is introduced, as well as its restriction to finite cubes with periodic boundary condition. The concept of a localized wave is discussed. In Section 3 we study spectral gaps, defects and midgap eigenvalues. The existence of *bound states of light* is discussed. Section 4 concerns periodic dielectric media. In Section 5 we describe a mathematically rigorous proof of Anderson localization in randomized periodic media. Finally, in Section 6 we discuss a general framework for the study of localization of classical waves, which includes acoustic, elastic and electromagnetic waves, and allows anisotropic media.

2. The Operator Formulation of Maxwell Equations

2.1. MAXWELL EQUATIONS IN INHOMOGENEOUS MEDIA

We start from Maxwell equations in a linear, lossless dielectric medium:

$$\begin{aligned} \mu\tfrac{\partial}{\partial t}\mathbf{H} &= -\nabla\times\mathbf{E} \qquad & \nabla\cdot\mu\mathbf{H} &= 0 \\ \varepsilon\tfrac{\partial}{\partial t}\mathbf{E} &= \nabla\times\mathbf{H} \qquad & \nabla\cdot\varepsilon\mathbf{E} &= 0, \end{aligned} \tag{1}$$

where $\mathbf{E}=\mathbf{E}(x,t)$ is the electric field, $\mathbf{H}=\mathbf{H}(x,t)$ is the magnetic field, ε is the dielectric constant, and μ is the magnetic permeability. We use the Giorgi system of units.

For inhomogeneous media the dielectric constant $\varepsilon = \varepsilon(x)$ and the magnetic permeability $\mu = \mu(x)$ are in general position dependent. Since the medium is lossless, $\varepsilon(x)$ and $\mu(x)$ are real valued. We neglect their frequency dependence. We always have

$$\begin{aligned} 0 < \varepsilon_- \leq \varepsilon(x) \ &\leq \ \varepsilon_+ < \infty, \\ 0 < \mu_- \leq \mu(x) \ &\leq \ \mu_+ < \infty, \end{aligned} \tag{2}$$

for some constants $\varepsilon_\pm$ and $\mu_\pm$.

Typically, $\varepsilon(x)$ and $\mu(x)$ take on a finite number of values; we only assume they are real valued measurable functions satisfying (2).

The energy density $\mathcal{E}(x,t) = \mathcal{E}_{\mathbf{H},\mathbf{E}}(x,t)$ and the (conserved) energy $\mathcal{E} = \mathcal{E}_{\mathbf{H},\mathbf{E}}$ of a solution $(\mathbf{H}, \mathbf{E})$ of the Maxwell equations (1) are given by

$$\begin{aligned} \mathcal{E}(x,t) &= \frac{1}{2}\left[\varepsilon(x)|\mathbf{E}(x,t)|^2 + \mu(x)|\mathbf{H}(x,t)|^2\right], \\ \mathcal{E} &= \int_{\mathbb{R}^{\#}} \mathcal{E}(x,t)\, dx \ \text{ at any time } t. \end{aligned} \tag{3}$$

2.2. MAXWELL EQUATIONS AS A SCHRÖDINGER-LIKE EQUATION

Our approach is based on the fact that Maxwell equations may be recast as a Schrödinger-like equation (i.e., a first order conservative linear equation):

$$-i\frac{\partial}{\partial t}\mathbf{\Psi}_t = \mathbb{M}\mathbf{\Psi}_{\approx}, \tag{4}$$

with

$$\mathbf{\Psi}_t = \begin{pmatrix} \mathbf{H}_t \\ \mathbf{E}_t \end{pmatrix} \in \mathbb{H}, \qquad \mathbb{M} = \begin{bmatrix} 0 & \frac{i}{\mu}\nabla^\times \\ \frac{-i}{\varepsilon}\nabla^\times & 0 \end{bmatrix}. \tag{5}$$

Here $\mathbb{H} = \mathbb{H}_{\mu.\varepsilon} = \mathbb{S}_\mu \oplus \mathbb{S}_\varepsilon$ is the Hilbert space of finite energy solutions. For a given $\varrho = \varrho(x) > 0$, bounded from above and away from 0, we set $\mathbb{S}_\varrho$ to be the closure in $L^2(\mathbb{R}^{\#}, \varrho(\frown)\frown; \mathbb{C}^{\#})$ of the linear subset of functions Ψ with $\varrho\Psi \in C_0^1(\mathbb{R}^{\#}; \mathbb{C}^{\#})$, $\nabla \cdot \varrho \gneqq = \nvdash$.

The matrix operator $\mathbb{M} = \mathbb{M}_{\mu.\varepsilon}$, where $\nabla^\times$ denotes the operator given by $\nabla^\times\Psi = \nabla\times\Psi = \operatorname{curl}\Psi$, has a natural definition as a self-adjoint operator on $\mathbb{H}$. This operator governs the dynamics of EM fields; finite energy solutions to (4) are of the form $\mathbf{\Psi}_t = e^{it\mathbb{M}}\mathbf{\Psi}_0$, with $\mathbf{\Psi}_0 \in \mathbb{H}$. Their energy is given by

$$\mathcal{E} = \frac{1}{2}\|\mathbf{\Psi}_t\|_{\mathbb{H}}^2 = \frac{1}{2}\|\mathbf{\Psi}_0\|_{\mathbb{H}}^2. \tag{6}$$

Notice that if $\mathbf{J}$ denotes the antiunitary involution corresponding to complex conjugation on $\mathbb{H}$, i.e., $\mathbf{J}\mathbf{\Psi} = \overline{\mathbf{\Psi}}$, we have $\mathbf{J}\mathbb{M}\mathbf{J} = -\mathbb{M}$. It follows that $\mathbb{J}\mathbb{M}_+\mathbb{J} = \mathbb{M}_-$, with $\mathbb{M} = \mathbb{M}_+ - \mathbb{M}_-$ being the decomposition of $\mathbb{M}$ into its positive and negative parts. In particular, the spectrum of $\mathbb{M}$, $\sigma(\mathbb{M})$, is symmetric, i.e., $\sigma(\mathbb{M}) = -\sigma(\mathbb{M})$.

2.3. THE SECOND ORDER PARTIAL DIFFERENTIAL OPERATORS

If Ψ_t is a solution of equation (4), it must satisfy the second order equation $\frac{\partial^2}{\partial t^2}\Psi_t = -\mathbb{M}^2\Psi_t$, so the magnetic and electric fields solve the second order equations

$$\frac{\partial^2}{\partial t^2}\mathbf{H}_t = -\frac{1}{\mu}\nabla\times\frac{1}{\varepsilon}\nabla\times\mathbf{H}_t, \quad \mathbf{H}_t \in \mathbb{S}_\mu, \tag{7}$$

$$\frac{\partial^2}{\partial t^2}\mathbf{E}_t = -\frac{1}{\varepsilon}\nabla\times\frac{1}{\mu}\nabla\times\mathbf{E}_t, \quad \mathbf{E}_t \in \mathbb{S}_\varepsilon. \tag{8}$$

The second order partial differential operators $\mathbf{M_H} = \frac{1}{\mu}\nabla\times\frac{1}{\varepsilon}\nabla\times$ and $\mathbf{M_E} = \frac{1}{\varepsilon}\nabla\times\frac{1}{\mu}\nabla\times$ have natural definitions as nonnegative self-adjoint operators on $\mathbb{S}_\mu$ and $\mathbb{S}_\varepsilon$, respectively. The two operators are unitarily equivalent:

$$\mathbf{M_E} = U\mathbf{M_H}U^*, \tag{9}$$

where $U : \mathbb{S}_\mu \to \mathbb{S}_\varepsilon$ is the unitary operator given by

$$U\mathbf{H} = \frac{-i}{\varepsilon}\nabla\times\mathbf{M_H}^{-\frac{1}{2}}\mathbf{H} \text{ if } \mathbf{H} \in \operatorname{Ran}\mathbf{M_H}^{\frac{1}{2}}. \tag{10}$$

Note $\sigma(\mathbf{M_H}^{\frac{1}{2}}) = \sigma(\mathbb{M}_+)$. We obtain solutions of (4) by setting

$$\Psi_{\pm,t} = \left(e^{\pm it\mathbf{M_H}^{\frac{1}{2}}}\mathbf{H}_{0,\pm}, \pm Ue^{\pm it\mathbf{M_H}^{\frac{1}{2}}}\mathbf{H}_{0,\pm}\right), \quad \mathbf{H}_{0,\pm} \in \mathbb{S}_\mu. \tag{11}$$

Conversely, any solution Ψ_t of (4) can be written as $\Psi_t = \Psi_{+,t} + \Psi_{-,t}$ for appropriate $\mathbf{H}_{0,\pm} \in \mathbb{S}_\mu$.

It follows that to find all eigenvalues and eigenmodes for $\mathbb{M}$, it is necessary and sufficient to find all eigenvalues and eigenmodes for $\mathbf{M_H}$. For if $\mathbf{M_H}\mathbf{H}_{\omega^2} = \omega^2\mathbf{H}_{\omega^2}$, with $\omega > 0$, $\mathbf{H}_{\omega^2} \in \mathbb{S}_\mu$, $\mathbf{H}_{\omega^2} \neq 0$, we have

$$U\mathbf{H}_{\omega^2} = \frac{-i}{\omega\varepsilon}\nabla\times\mathbf{H}_{\omega^2} \tag{12}$$

and

$$\mathbb{M}\left(\mathbf{H}_{\omega^2}, \pm\frac{-i}{\omega\varepsilon}\nabla\times\mathbf{H}_{\omega^2}\right) = \pm\omega\left(\mathbf{H}_{\omega^2}, \pm\frac{-i}{\omega\varepsilon}\nabla\times\mathbf{H}_{\omega^2}\right). \tag{13}$$

Conversely, if $\mathbb{M}(\mathbf{H}_{\pm\omega}, \mathbf{E}_{\pm\omega}) = \pm\omega(\mathbf{H}_{\pm\omega}, \mathbf{E}_{\pm\omega})$, with $\omega > 0$, $(\mathbf{H}_{\pm\omega}, \mathbf{E}_{\pm\omega}) \in \mathbb{H}$, we have $\mathbf{M_H H}_{\pm\omega} = \omega^2 \mathbf{H}_{\pm\omega}$ and $\mathbf{E}_{\pm\omega} = \pm U\mathbf{H}_{\pm\omega} = \pm\frac{-i}{\omega\varepsilon}\nabla^{\times}\mathbf{H}_{\pm\omega}$.

2.4. LOCALIZED WAVES

Intuitively, a localized electromagnetic wave is a finite energy solution of Maxwell equations with the property that as much as desired of the wave's energy remains in a fixed bounded region of space at all times. A precise definition of a localized wave is to require

$$\lim_{R\to\infty} \inf_t \frac{1}{\mathcal{E}} \int_{|x|\le R} \mathcal{E}(x,t)\, dx = 1. \tag{14}$$

If the operator $\mathbb{M}$ has an eigenvalue ω with eigenmode Ψ_ω, i.e., $\mathbb{M}\Psi_\omega = \omega\Psi_\omega$, with $\Psi_\omega \in \mathbb{H}$, $\Psi_\omega \neq 0$, then $\Psi_{\omega,t} = \mathrm{e}^{it\omega}\Psi_\omega$ is a localized electromagnetic wave, i.e., it satisfies (4) and (14). Since in this case $-\omega$ is also an eigenvalue of $\mathbb{M}$ with eigenmode $\overline{\Psi}_\omega$, we have that $\overline{\Psi}_{\omega,t} = \mathrm{e}^{-it\omega}\overline{\Psi}_\omega$ is also a localized wave. Moreover, linear combinations of eigenmodes are also localized waves.

Thus we can search for localized waves by looking for eigenvalues and eigenmodes of the second order partial differential operator $\mathbf{M_H}$.

2.5. THE MAXWELL OPERATOR

It is convenient to study the spectral theory of the operator $\mathbf{M_H}$ on the Hilbert space $L^2(\mathbb{R}^3, dx; \mathbb{C}^3)$ instead of $L^2(\mathbb{R}^3, \mu(dx)dx; \mathbb{C}^3)$. To do so, note that multiplication by the function $\sqrt{\mu(x)}$ is a unitary map from the Hilbert space $L^2(\mathbb{R}^3, \mu(dx)dx; \mathbb{C}^3)$ to $L^2(\mathbb{R}^3, dx; \mathbb{C}^3)$, which takes $\mathbf{M_H}$ to the self-adjoint operator $\mathbf{M}$ on $\mathbb{S}$, where $\mathbf{M}$ is formally given by

$$\mathbf{M} = \mathbf{M}_{\mu,\varepsilon} = \frac{1}{\sqrt{\mu}}\nabla^{\times}\frac{1}{\varepsilon}\nabla^{\times}\frac{1}{\sqrt{\mu}}, \tag{15}$$

and $\mathbb{S}$ is the closure in $L^2(\mathbb{R}^3, dx; \mathbb{C}^3)$ of the linear subset of functions Ψ with $\sqrt{\mu}\Psi \in C_0^1(\mathbb{R}^3; \mathbb{C}^3)$ and $\nabla \cdot \sqrt{\mu}\Psi = 0$.

We call $\mathbf{M}_{\mu,\varepsilon}$ *the Maxwell operator for the dielectric medium described by* μ *and* ε. For the rigorous definition, we start by defining the unrestricted Maxwell operator

$$M = M_{\mu,\varepsilon} = \frac{1}{\sqrt{\mu}}\nabla^{\times}\frac{1}{\varepsilon}\nabla^{\times}\frac{1}{\sqrt{\mu}} \quad \text{on} \quad L^2(\mathbb{R}^3, dx; \mathbb{C}^3), \tag{16}$$

as the nonnegative self-adjoint operator uniquely defined by the nonnegative quadratic form given as the closure of

$$\mathcal{M}(\Psi, \Phi) = \langle \nabla \times \frac{1}{\sqrt{\mu}}\Psi, \frac{1}{\varepsilon}\nabla \times \frac{1}{\sqrt{\mu}}\Phi \rangle, \quad \Psi, \Phi \in \sqrt{\mu}\, C_0^1(\mathbb{R}^3; \mathbb{C}^3). \tag{17}$$

By Weyl's decomposition (see [4]), we have

$$L^2(\mathbb{R}^{\nVdash}, \frown; \mathbb{C}^{\nVdash}) = \mathbb{S} \oplus \mathbb{G}, \tag{18}$$

where $\mathbb{G}$ is the closure in $L^2(\mathbb{R}^{\nVdash}, \frown; \mathbb{C}^{\nVdash})$ of the linear subset of functions $\Phi = \sqrt{\mu}\nabla\varphi$, $\varphi \in C_0^1(\mathbb{R}^{\nVdash})$. The spaces $\mathbb{S}$ and $\mathbb{G}$ are left invariant by M, with $\mathbb{G} \subset \mathcal{D}(\mathbf{M})$, the domain of the operator M, and $M|_{\mathbb{G}} = 0$. We define $\mathbf{M}$ as the restriction of M to $\mathbb{S}$, i.e., $\mathcal{D}(\mathbf{M}) = \mathcal{D}(M) \cap \mathbb{S}$ and $\mathbf{M} = M|_{\mathcal{D}(M)\cap\mathbb{S}}$. Thus

$$\mathbf{M} = P_{\mathbb{S}} M I_{\mathbb{S}} = M I_{\mathbb{S}}, \tag{19}$$

with $P_{\mathbb{S}}$ the orthogonal projection onto $\mathbb{S}$ and $I_{\mathbb{S}} : \mathbb{S} \to \mathbb{L}^{\nvdash}(\mathbb{R}^{\nVdash}, \frown; \mathbb{C}^{\nVdash})$ the restriction of the identity map. Notice that $M = \mathbf{M} \oplus 0_{\mathbb{G}}$ and $0 \in \sigma(\mathbf{M})$, so

$$\sigma(\mathbf{M}) = \sigma(M). \tag{20}$$

We can thus work with M to answer questions about the spectrum of $\mathbf{M}$. Note that $\sigma(\mathbf{M}) \subset [0, \infty)$, with 0 always in the essential spectrum of $\mathbf{M}$, and that eingenmodes of $\mathbf{M}$ correspond to localized waves.

2.6. THE HILBERT-SCHMIDT CONDITION FOR MAXWELL OPERATORS

The following result [14] plays an important role in the study of Maxwell operators. It provides the crucial estimate for proving important properties of Maxwell operators, including the existence of *polynomially bounded generalized eigenfunctions* [31].

Theorem 1 (Hilbert-Schmidt condition) *Let M be an operator as in (16), and let W denote the bounded operator given by multiplication by $(1 + |x|^2)^{-\frac{n}{2}}$, with $\delta > \frac{3}{2}$. Then*

$$P_{\mathbb{S}}(M + I)^{-1} W = \left[(\mathbf{M} + I_{\mathbb{S}})^{-1} \oplus 0_{\mathbb{G}}\right] W \tag{21}$$

is a Hilbert-Schmidt operator, ie.,

$$\operatorname{Tr}\left\{W \left[(\mathbf{M} + I_{\mathbb{S}})^{-2} \oplus 0_{\mathbb{G}}\right] W\right\} < \infty. \tag{22}$$

2.7. FINITE VOLUME MAXWELL OPERATORS

We will need to work with finite volume Maxwell operators. We write

$$\Lambda_L(x) = \{y \in \mathbb{R}^{\nVdash};\ \|\frown - \frown\| \le \frac{\mathbb{L}}{\nvdash}\}, \tag{23}$$

for the (closed) cube of side L centered at $x \in \mathbb{R}^3$, and $\chi_{x,L}$ for its characteristic function. Here

$$\|x\| = \max\{x_1, x_2, x_3\} \quad \text{for} \quad x = (x_1, x_2, x_3) \in \mathbb{R}^3. \tag{24}$$

Given a (closed) cube Λ in $\mathbb{R}^3$ and M as in (16), we will denote by M_Λ the restriction of M to Λ with periodic boundary condition, i.e., M_Λ is the nonnegative self-adjoint operator on $L^2(\Lambda, dx; \mathbb{C}^3)$, uniquely defined by the nonnegative quadratic form given as the closure of

$$\mathcal{M}_\Lambda(\Psi, \Phi) = \langle \nabla \times \frac{1}{\sqrt{\mu}}\Psi, \frac{1}{\varepsilon}\nabla \times \frac{1}{\sqrt{\mu}}\Phi \rangle, \quad \Psi, \Phi \in \sqrt{\mu}\, C^1_{\text{per}}(\Lambda; \mathbb{C}^3), \tag{25}$$

the inner product being in $L^2(\Lambda, dx; \mathbb{C}^3)$. Here C^1_{per} denotes periodic C^1 functions, i.e., C^1 functions on the torus we obtain by identifying opposite faces of the cube.

3. Spectral Gaps, Defects and Midgap Eigenmodes

It is a well known fact in solid state physics that, in three dimensions, a deep potential well creates a bound state. Following the analogy between the operator formulation of Maxwell equations and quantum mechanics, we may look for bound states of light, i.e., eigenmodes of the Maxwell operator. They will correspond to localized EM waves.

In spite of the similarity between the creation of eigenmodes for classical and electron waves, there are some important differences. First of all, for the electron it suffices to perturb locally a homogeneous medium to generate a bound state (eigenmode). For classical waves a local perturbation of a homogenous medium cannot generate an eigenmode.

The reason for this difference between classical waves and electrons can be explained as follows. The motion of the electron in a homogenous medium is described by the Schrödinger operator $H_0 = -\Delta + V_0$ with a constant potential $V_0(x) \equiv v_0$. Clearly the spectrum $\sigma(H_0)$ of the operator H_0 is the interval $[v_0, \infty)$, so we may consider the infinite interval $(-\infty, v_0)$ as a gap in the spectrum of the operator H_0. Note that the edge v_0 of the gap depends on the homogeneous medium. Hence, if we perturb this homogeneous medium by a defect, say a potential well, the spectrum can expand into the interior of the gap $(-\infty, v_0)$, and if this happens the corresponding eigenmodes will be exponentially localized.

For EM waves in a homogeneous medium, described by the Maxwell operator $\mathbf{M}$ with constant μ and ε, we always have $\sigma(\mathbf{M}) = [0, \infty)$, so, as for Schrödinger operators, we may consider the infinite interval $(-\infty, 0)$ as a gap in the spectrum. But for EM waves the bottom of the spectrum does not depend on the medium at all, it is always 0 and belongs to the essential spectrum. A local perturbation of the medium cannot expand the spectrum into the gap $(-\infty, 0)$.

Thus, in order to employ a mechanism for the creation of eigenmodes of Maxwell operators, similar to the one for Schrödinger operators, we have to start with a dielectric medium such that the corresponding Maxwell operator has a gap inside its spectrum. The edges of the spectral gap must depend on the medium, i.e., on the functions $\mu(x)$ and $\varepsilon(x)$ that describe the medium. Such media can be perturbed locally by a defect, giving rise to exponentially localized eigenmodes with corresponding eigenvalues in the interior of the gaps. Such midgap eigenmodes of the Maxwell operator correspond to localized EM waves.

3.1. SPECTRAL GAPS

A spectral gap for the operator A is an interval (a,b), disjoint from the spectrum $\sigma(A)$ of A (i.e., $(a,b) \cap \sigma(A) = \emptyset$), with $a, b \in \sigma(A)$.

Note that an interval (a,b) is a spectral gap for a Maxwell operator $\mathbf{M}$ if and only if it is a spectral gap for the corresponding unrestricted Maxwell operator M.

The physical significance of the existence of a spectral gap lies in the fact that a wave with frequency in the gap cannot propagate in the medium. This is shown by the exponential decay of the Green's functions corresponding to frequencies in the gap.

To see that, for z not in the spectrum of M, we write $R(z) = (M - zI)^{-1}$ for the resolvent. We also introduce operator-valued Green's functions:

$$G_\ell(x, y; z) = \chi_{x,\ell} R(z) \chi_{y,\ell} \tag{26}$$

for $x, y \in \mathbb{R}^{\not k}$, where $\chi_{x,\ell}$ is the characteristic function of the cube of side ℓ centered at x. The following theorem exhibits the exponential decay of these Green's functions when z is in a spectral gap [14, 30, 3].

Theorem 2 (Exponential decay in spectral gaps) *Let the interval (a,b) be a spectral gap for a Maxwell operator M as in (16). Then for $E \in (a,b)$ and $\ell > 0$ we have*

$$\|G_\ell(x, y; E)\| \leq \frac{C}{\min\{E - a, b - E\}} e^{\sqrt{3}\ell} e^{-c\sqrt{(E-a)(b-E)}\|x-y\|} \tag{27}$$

for all $x, y \in \mathbb{R}^{\not k}$, where C and c are finite constants depending only on ε_-, μ_-, a and b.

The finite volume operator-valued Green's function is

$$G_{\ell,\Lambda}(x, y; z) = \chi_{x,\ell} R_\Lambda(z) \chi_{y,\ell}, \quad x, y \in \Lambda, \tag{28}$$

where $R_\Lambda(z) = (M_\Lambda - zI)^{-1}$ is the finite volume resolvent. Here $\chi_{x,\ell}$ is the characteristic function of the cube $\Lambda_\ell(x)$ in the torus obtained from identifying opposite faces in Λ. *There is a version of Theorem 2 for finite volume Maxwell operators with periodic boundary condition, the decay of the finite volume Green's function being with respect to the distance in the torus* [14].

3.2. DEFECTS AND MIDGAP EIGENVALUES

A *defect* is a modification of a given medium in a bounded domain. Two dielectric media, described by $\mu_0(x)$, $\varepsilon_0(x)$ and $\mu(x)$, $\varepsilon(x)$, are said to *differ by a defect*, if they are the same outside some bounded region, i.e., $\mu(x) = \mu_0(x)$ and $\varepsilon(x) = \varepsilon_0(x)$ outside the bounded region.

We recall that the essential spectrum $\sigma_{ess}(A)$ of an operator A consists of all the points of its spectrum, $\sigma(A)$, which are not isolated eigenvalues with finite multiplicity. Essential spectra of Maxwell operators are not changed by defects [15].

Theorem 3 (Stability of essential spectrum) *Let* $\mathbf{M}_0$ *and* $\mathbf{M}$ *be the Maxwell operators for two dielectric media which differ by a defect. Then*

$$\sigma_{\text{ess}}(\mathbf{M}) = \sigma_{\text{ess}}(\mathbf{M}_0). \tag{29}$$

If (a,b) *is a gap in the spectrum of* $\mathbf{M}_0$, *the spectrum of* $\mathbf{M}$ *in* (a,b) *consists at most of isolated eigenvalues with finite multiplicity, with the corresponding eigenmodes decaying exponentially fast, with a rate depending on the distance from the eigenvalue to the edges of the gap.*

Theorem 1 provides the key ingredient for the proof of Theorem 3 .

Is it possible to create at least one eigenvalue in a gap of $\mathbf{M}_0$ *by introducing a defect?* By Theorem 3, such an eigenvalue will be isolated with finite multiplicity, with the corresponding eigenmodes decaying exponentially fast. The next theorem shows that one can introduce simply defined defects which create eigenvalues in any given subinterval of a spectral gap of $\mathbf{M}_0$ [15]. In particular, *it shows that localized waves are allowed by Maxwell equations.*

Theorem 4 (Creation of midgap eigenmodes) *Let* $\mathbf{M}_{\mu_0,\varepsilon_0}$ *and* $\mathbf{M}_{\mu,\varepsilon}$ *be the Maxwell operators for two dielectric media which differ by a defect. Let* (a,b) *be a gap in the spectrum of* $\mathbf{M}_{\mu_0,\varepsilon_0}$, *select* β *in* (a,b), *and pick* $\theta > 0$ *such that the interval* $[\beta-\theta, \beta+\theta]$ *is contained in the gap, i.e.,* $[\beta-\theta,\beta+\theta] \subset (a,b)$. *If* $\mu(x) \equiv \hat{\mu}$ *and* $\varepsilon(x) \equiv \hat{\varepsilon}$ *in a cube of side* ℓ, *with*

$$\ell^2 \hat{\mu}\hat{\varepsilon} > \frac{24\beta}{\theta^2}\left(1 + \sqrt{1 + \frac{33\theta^2}{8\beta^2}}\right), \tag{30}$$

then the Maxwell operator $\mathbf{M}_{\mu,\varepsilon}$ *has at least one eigenmode with corresponding eigenvalue inside the interval* $[\beta - \theta, \beta + \theta]$.

Note that the simpler criterion

$$\ell^2 \hat{\mu}\hat{\varepsilon} > \frac{79\beta}{\theta^2} \tag{31}$$

suffices, since it implies (30).

4. Periodic Dielectric Media

The most natural way to obtain Maxwell operators with spectral gaps is to consider periodic dielectric media (photonic crystals), i.e., media described by periodic dielectric constant $\varepsilon_0(x)$ and magnetic permeability $\mu_0(x)$. If $\mathbb{L}$ is the lattice of periods, we have

$$\varepsilon_0(x+\tau) = \varepsilon_0(x) \tag{32}$$

$$\mu_0(x+\tau) = \mu_0(x) \tag{33}$$

for any τ in $\mathbb{L}$. We assume for simplicity that $\mathbb{L} = q\mathbb{Z}^3$. In this case we say that the dielectric medium and the functions $\varepsilon_0(x)$ and $\mu_0(x)$ are q-periodic. We take q to be a positive integer without loss of generality.

We will use $\mathbf{M}_0 = \mathbf{M}_{\mu_0,\varepsilon_0}$ and $M_0 = M_{\mu_0,\varepsilon_0}$ for the corresponding Maxwell operators, which we will also call q-periodic. In view of the periodicity, the Floquet-Bloch theory can be applied, so their spectrum has band structure (e.g., [14]). It may happen that these spectral bands do not cover the whole semiaxis $[0,\infty)$, in which case there will be a spectral gap.

The band structure of the spectrum of a periodic Maxwell operator is a generic property due to the periodicity. But the existence of spectral gaps is not a generic property. It depends in a subtle way on the geometry and properties of the dielectric materials in the periodic medium (see [10]).

The spectrum of a periodic Maxwell operator can be approximated by the spectra of its restriction to finite volume with periodic boundary condition. In particular, *spectral gaps of periodic Maxwell operators are preserved by restricting the operator to finite volume with periodic boundary condition.*

To state the precise result, we need some definitions. If $k, n \in \mathrm{N}$, we say that $k \preceq n$ if $n \in k\mathrm{N}$ and that $k \prec n$ if $k \preceq n$ and $k \neq n$.

Theorem 5 *Let* M_0 *be a* q*-periodic Maxwell operator. Let* $\{\ell_n;\ n = 0,1,2,\ldots\}$ *be a sequence in* N *such that* $\ell_0 = q$ *and* $\ell_n \prec \ell_{n+1}$ *for each* $n = 0,1,2,\ldots$. *Then*

$$\sigma\left(M_{0,\Lambda_{\ell_n}(0)}\right) \subset \sigma\left(M_{0,\Lambda_{\ell_{n+1}}(0)}\right) \subset \sigma(M_0) \text{ for all } n = 0,1,2,\ldots, \tag{34}$$

and

$$\sigma(M_0) = \overline{\bigcup_{n\geq 1} \sigma\left(M_{0\Lambda_{\ell_n}(0)}\right)}. \tag{35}$$

It follows from (34) that taking the restriction of a q-periodic Maxwell operator M_0 to a cube $\Lambda_\ell(x)$, $q \preceq \ell$, with periodic boundary condition, can only shrink the spectrum. An important consequence is that a gap in the spectrum of M_0 sits inside a gap in the spectrum of $M_{0,\Lambda_\ell(x)}$ for any $x \in \mathbb{R}^{\not k}$ and $\ell \in q\mathrm{N}$.

5. Localization in Randomized Periodic Media

According to Theorem 4, a strong enough single defect in a periodic dielectric medium with a spectral gap creates exponentially localized electromagnetic waves. *If we have a random array of such defects, then, under some natural conditions, the localized waves created by individual defects do not couple (i.e., the electromagnetic wave tunneling becomes inefficient), so we get an infinite number of localized waves whose frequencies are dense in an interval contained in the spectral gap of the Maxwell operator of the underlying periodic medium.* This phenomenon is analogous to the Anderson localization of electron waves in random media.

We start with a q-periodic dielectric medium (the background medium), described by $\varepsilon_0(x)$ *and* $\mu_0(x)$*, such that the corresponding Maxwell operator* M_0 *has a spectral gap* (a, b). We recall that the period q is taken to be a positive integer without loss of generality.

5.1. RANDOMIZED PERIODIC MEDIA

The randomized periodic medium is modeled by random dielectric constant and magnetic susceptibility of the form

$$\varepsilon_{g,\omega}(x) = \varepsilon_0(x)\gamma_{g,\omega}(x)\ , \textit{with}\ \gamma_{g,\omega}(x) = 1 + g\sum_{i\in\mathbb{Z}^{\not k}} \omega_i u_i(x), \tag{36}$$

$$\mu_{g,\omega}(x) = \mu_0(x)\beta_{g,\omega}(x)\ , \textit{with}\ \beta_{g,\omega}(x) = 1 + g\sum_{i\in\mathbb{Z}^{\not k}} \omega_i v_i(x), \tag{37}$$

where

(i) $u_i(x) = u(x-i)$, $v_i(x) = v(x-i)$ *for each* $i \in \mathbb{Z}^{\not k}$, $u(x)$ *and* $v(x)$ *being nonnegative measurable real valued functions with compact support, say* $u(x) = v(x) = 0$ *if* $\|x\| \geq r$ *for some* $r < \infty$*, such that*

$$0 \leq U_- \leq U(x) \equiv \sum_{i\in\mathbb{Z}^{\not k}} u_i(x) \ \leq\ U_+ < \infty, \tag{38}$$

$$0 \leq V_- \leq V(x) \equiv \sum_{i \in \mathbb{Z}^{\not{k}}} v_i(x) \leq V_+ < \infty, \tag{39}$$

for some constants $U_\pm$ and $V_\pm$, with $U_- + V_- > 0$.

(ii) $\omega = \{\omega_i;\, i \in \mathbb{Z}^{\not{k}}\}$ is a family of independent, identically distributed random variables taking values in the interval $[-1, 1]$, whose common probability distribution has a bounded density $\rho > 0$ a.e. in $[-1, 1]$.

(iii) g, satisfying $0 \leq g < (\max\{U_+, V_+\})^{-1}$, is the disorder parameter.

The same random variables $\{\omega_i\}$ are used in the random dielectric constant and in the random magnetic permeability, modeling the randomness in the dielectric medium. Note that a change in the material that constitutes the medium causes a simultaneous change in the dielectric constant and in the magnetic permeability.

We always have $\varepsilon_{g,\omega}(x)$ and $\mu_{g,\omega}(x)$ satisfying (2). Note also that, since we only require that either $U_- > 0$ or $V_- > 0$, we may have either $\mu_{g,\omega}(x) \equiv \mu_0(x)$ or $\varepsilon_{g,\omega}(x) \equiv \varepsilon_0(x)$.

5.2. THE RANDOM MAXWELL OPERATOR AND ITS SPECTRAL GAP

We study the random Maxwell operators (see [13, Appendix A] for the definition of a random operator)

$$M_g = M_{g,\omega} = M_{\mu_{g,\omega}, \varepsilon_{g,\omega}}; \quad \mathbf{M}_g = \mathbf{M}_{g,\omega} = \mathbf{M}_{\mu_{g,\omega}, \varepsilon_{g,\omega}}. \tag{40}$$

It is a consequence of ergodicity (measurability follows from [13, Theorem 38]) that there exists a nonrandom set Σ_g, such that $\sigma(\mathbf{M}_{g,\omega}) = \sigma(M_{g,\omega}) = \Sigma_g$ with probability one. Thus Σ_g can be called the spectrum of the random operator. In addition, the decompositions of $\sigma(\mathbf{M}_{g,\omega})$ and $\sigma(M_{g,\omega})$ into pure point spectrum, absolutely continuous spectrum and singular continuous spectrum are also independent of the choice of ω with probability one [29, 36], so the pure point, absolutely continuous, and singular continuous spectrum of the random operator are well defined, nonrandom sets.

Note that the pure point spectrum is defined as the *closure* of the set of eigenvalues. This closed set is the same for different realizations of the random operator with probability one, but the set of eigenvalues is random. In fact, the probability of a fixed point being an eigenvalue is zero.

The random Maxwell operator $\mathbf{M}_g$ was obtained by perturbing the periodic Maxwell operator $\mathbf{M}_0$. Since $\mathbf{M}_0$ has the spectral gap (a, b), one may ask what happens to this gap under the random perturbations. The answer, given in the following theorem [14, 30], is that if the disorder parameter g is not too large, then the gap shrinks, but does not close. If g is allowed to become too large, the gap closes.

Theorem 6 (Spectral gap of the random Maxwell operator) *There exists* $g_0 \leq (\max\{U_+, V_+\})^{-1}$, *with*

$$\frac{1}{U_+ + V_+}\left(1 - \left(\frac{a}{b}\right)^{\frac{1}{2}}\right) \leq g_0 \tag{41}$$
$$\leq \min\left\{\frac{1}{U_+}\left(\left(\frac{b}{a}\right)^{\frac{U_+}{2U_-}} - 1\right), \frac{1}{V_+}\left(\left(\frac{b}{a}\right)^{\frac{V_+}{2V_-}} - 1\right)\right\},$$

and strictly increasing, Lipschitz continuous real valued functions $a(g)$ *and* $-b(g)$ *on the interval* $[0, (\max\{U_+, V_+\})^{-1})$, *with* $a(0) = a$, $b(0) = b$ *and* $a(g) \leq b(g)$, *such that:*

(i)

$$\Sigma_g \bigcap [a, b] = [a, a(g)] \bigcup [b(g), b]. \tag{42}$$

(ii) *For* $g < g_0$, *we have* $a(g) < b(g)$ *and* $(a(g), b(g))$ *is a gap in the spectrum of the random operator* $\mathbf{M}_g$, *located inside the gap* (a, b) *of the unperturbed periodic operator* $\mathbf{M}_0$. *Moreover, we have*

$$a \leq a(1 + gU_+)^{\frac{U_-}{U_+}}(1 + gV_+)^{\frac{V_-}{V_+}} \leq a(g) \leq \frac{a}{(1 - gU_+)(1 - gV_+)}, \tag{43}$$

and

$$b(1 - gU_+)(1 - gV_+) \leq b(g) \leq \frac{b}{(1 + gU_+)^{\frac{U_-}{U_+}}(1 + gV_+)^{\frac{V_-}{V_+}}} \leq b. \tag{44}$$

(iii) *If* $g_0 < (\max\{U_+, V_+\})^{-1}$, *we have* $a(g) = b(g)$ *when* $g_0 \leq g < (\max\{U_+, V_+\})^{-1}$, *and the random operator* $\mathbf{M}_g$ *has no gap inside the gap* (a, b) *of the unperturbed periodic operator* $\mathbf{M}_0$, *i.e.,* $[a, b] \subset \Sigma_g$.

This theorem is proven by first approximating the (nonrandom) spectrum of the random operator by spectra of (nonrandom) periodic operators, which are then approximated by spectra of operators on finite cubes with periodic boundary condition using Theorem 5. The latter operators have compact resolvents (i.e., Green's functions) on the orthogonal complement of their kernel, and bounds on their eigenvalues are obtained by the min-max principle.

5.3. ANDERSON LOCALIZATION

We say that the random operator $\mathbf{M}_g$ *exhibits (Anderson) localization in an interval* $I \subset \Sigma_g$, *if* $\mathbf{M}_g$ *has only pure point spectrum in* I *with probability one. We have exponential localization in* I *if we have localization and, with* *probability one, all the eigenmodes corresponding to eigenvalues in I are exponentially decaying (in the sense of having exponentially decaying local L^2-norms).*

Note that the energy densities (3) corresponding to exponentially decaying eigenmodes of $\mathbf{M}_g$ also have exponentially decaying local L^2-norms, uniformly in the time t [14].

The randomization of the periodic medium creates exponentially localized eigenmodes near the edges of the gap. Our method of proof requires low probability of extremal values for the random variables. The results below are formulated for the left edge of the gap, with similar results holding at the right edge [14, 30]. The quantities g_0, $a(g)$, $b(g)$ are defined in Theorem 6.

Theorem 7 (Localization at the edge) *Suppose the probability density ρ of the random variables ω_i satisfies*

$$\int_{1-\gamma}^{1} \rho(t)dt \leq K\gamma^{\eta} \text{ for } 0 \leq \gamma \leq 1, \tag{45}$$

where $K < \infty$ and $\eta > \frac{3}{2}$. For any $g < g_0$ there exists $\delta(g) > 0$, depending only on the constants g, q, $\varepsilon_{0,\pm}$, $\mu_{0,\pm}$, $U_\pm, V_\pm$, r, K, η, an upper bound on $\|\rho\|_\infty$, and on the edges a, b, such that the random operator $\mathbf{M}_g$ exhibits exponential localization in the interval $[a(g) - \delta(g), a(g))$.

Theorem 7 can be extended to the situation when the gap is totally filled by the spectrum of the random operator [14, 30]. In this case we establish the existence of an interval (inside the original gap) where the random Maxwell operator exhibits exponential localization. Under somewhat different assumptions on the density ρ we can arrange for localization in any desired fraction of the original spectral gap [14].

Theorem 8 (Localization at the meeting of the edges) *Suppose the probability density ρ of the random variables ω_i satisfies*

$$\int_{1-\gamma}^{1} \rho(t)dt, \int_{-1}^{-1+\gamma} \rho(t)dt \leq K\gamma^{\eta} \tag{46}$$

for $0 \leq \gamma \leq 1$, where $K < \infty$ and $\eta > 3$. Suppose $g_0 < (\max\{U_+, V_+\})^{-1}$ (this can be arranged using estimate (41)), so the random operator $\mathbf{M}_g$ has no gap inside (a, b) for g in $\left[g_0, (\max\{U_+, V_+\})^{-1}\right)$. Then there exist $0 < \epsilon < (\max\{U_+, V_+\})^{-1} - g_0$ and $\delta > 0$, such that the random operator $\mathbf{M}_g$ exhibits exponential localization in the interval $[a(g_0) - \delta, a(g_0) + \delta]$ for all $g_0 \leq g < g_0 + \epsilon$.

Theorems 7 and 8 are proven in [13, 14] by a multiscale analysis [20, 21, 9, 13], which reduces the proof to the verification of appropriate decay of the (random) Green's function in a given finite scale, with high probability. This decay with high (appropriate for the scale) probability is then shown to imply exponential decay for larger and larger scales with appropriate probabilities. Finally, the exponential decay of the Green's function in all scales is used to show exponential localization.

We will now discuss some of the key steps in the proof. We work with unrestricted Maxwell operators as in (16), since we are only interested in spectrum away from 0.

5.3.1. *A Wegner-type Estimate*

The multiscale analysis requires control of the norm of $R_{g,\omega,\Lambda}(E)$, the resolvent of the random operator $M_{g,\omega,\Lambda}$, with high probability. This is given by a Wegner-type estimate [14, 30], which says that the probability that the norm of such resolvents is bigger than a given number $\frac{1}{\eta}$ is at most of the order of $\eta\,|\Lambda|^2$.

Theorem 9 (Wegner-type estimate) *There exists a constant $Q < \infty$, depending only on the constants $\varepsilon_{0,+}$, $\mu_{0,+}$, r, such that*

$$\mathbb{P}\left\{\|\mathbb{R}_{\eth,\not\geq}(\mathbb{E})\| \geq \frac{\not\Bbbk}{\eta}\right\} \leq \frac{\mathbb{Q}\|\rho\|_\infty\sqrt{|\mathbb{E}|}}{\eth\,(\not\Bbbk - \eth\max\{\mathbb{U}_+,\mathbb{V}_+\})\,(\mathbb{U}_- + \mathbb{V}_-)}\eta|\not\geq|^{\not\Bbbk} \tag{47}$$

for all $E > 0$, cubes Λ in $\mathbb{R}^{\not\Bbbk}$, and all $0 < \eta < E$.

This estimate is typically used with Λ a cube of side L and $\eta = L^{-s}$ for suitable $s > 6$; for large L we have $\eta\,|\Lambda|^2 = L^{-s+6}$ small. It shows that, with high probability, the eigenvalues of $M_{g,\omega,\Lambda}$ do not want to be too close to any given E, a precursor of Anderson localization.

5.3.2. *The Multiscale Analysis*

The multiscale analysis [20, 21, 9, 13, 14] starts from a statement of decay of the random Green's functions in a finite cube, with appropriate high probability. If the initial scale, the length of a side of the cube, is sufficiently large, the multiscale analysis shows decay of the random Green's functions in cubes of increasing scale, with suitably increasing probabilities. This collection of statements about decay of the random Green's functions in cubes of increasing scales, with suitably increasing probabilities, is then used to show Anderson localization. This will be now made precise.

Our finite volumes will be cubes $\Lambda_L(x)$, with $x \in q\mathbb{Z}^{\not\Bbbk}$; below L will always be a multiple of $2q$. We write $G_{x,L}(y, y'; E)$ for the operator-valued Green's function $G_{q,\Lambda_L(x)}(y, y'; E)$ as in (28); we will always take $y, y' \in$

$q\mathbb{Z}^d \cap \Lambda_L(x)$. For the random operator M_g as in (40), we write $G_{g,x,L}(y,y';E) = G_{g,\omega,x,L}(y,y';E)$ for the corresponding finite volume Green's function.

Given M as in (16) and $E > 0$, we make the following definitions:

- *Let* $m > 0$, *we say the cube* $\Lambda_L(x)$ *is* (m,E)-regular, *if* $E \notin \sigma(M_{x,L})$ *and*

$$\|G_{x,L}(y,x;E)\| \leq \mathrm{e}^{-m\frac{L}{2}} \text{ if } \|y-x\| = \frac{L}{2} - q. \tag{48}$$

- *Let* $\nu > 0$, *we say the cube* $\Lambda_L(x)$ *is* (ν,E)-suitable, *if* $E \notin \sigma(M_{x,L})$ *and*

$$\|G_{x,L}(y,x';E)\| \leq \frac{1}{L^\nu} \text{ if } \|y-x\| = \frac{L}{2} - q \text{ and } \|x'-x\| \leq \frac{L}{4}. \tag{49}$$

Note that if $\Lambda_L(x)$ is (ν,E)-suitable, it is also $(2\nu\frac{\log L}{L},E)$-regular.

Theorem 10 (Multiscale analysis) *Let* $E_0 > 0$ *and* $\nu > 40$. *Suppose*

$$\limsup_{L\to\infty} \mathbb{P}\{\Lambda_L(0) \text{ is } (\nu,E_0)\text{-suitable}\} = 1. \tag{50}$$

Then there exists $\delta > 0$, *such that, with probability one,* M_g *has only pure point spectrum in* $(E_0-\delta, E_0+\delta)$, *and the corresponding eigenfunctions decay exponentially fast.*

Note that, by the definition of lim sup, the periodicity of the background medium, and the stationarity of the probability distributions, condition (50) is the same as requiring that we find $x \in \mathbb{Z}^d$ and a sequence $L_n \to \infty$, such that

$$\lim_{n\to\infty} \mathbb{P}\{\Lambda_{L_n}(x) \text{ is } (\nu,E_0)\text{-suitable}\} = 1. \tag{51}$$

Theorem 10 is proven by a double multiscale analysis [13]. The first multiscale analysis shows that, given $p > 3$, we can find L_0 and $\beta > 0$, such that

$$\mathbb{P}\{\Lambda_{L_k}(0) \text{ is } (\nu,E_0)\text{-suitable}\} \geq 1 - \frac{1}{L_k^p}, \tag{52}$$

at all scales L_k, where $L_{k+1} = \beta L_k$ for all $k = 0,1,\ldots$. In particular, it proves that

$$\mathbb{P}\{\Lambda_{L_k}(0) \text{ is } (2\nu\tfrac{\log L_k}{L_k},E_0)\text{-regular}\} \geq 1 - \frac{1}{L_k^p} \tag{53}$$

for all $k = 0,1,\ldots$. A second multiscale analysis then shows that if we pick k_0 large enough, we can choose $\alpha > 1$ such that we have

$$\mathbb{P}\{\Lambda_{\ell_j}(0) \text{ is } (m_0,E_0)\text{-regular}\} \geq 1 - \frac{1}{\ell_j^p}, \tag{54}$$

at all scales ℓ_j, $j = 0, 1, \ldots$, with $\ell_0 = L_{k_0}$, $\ell_{j+1} = \ell_j^{\alpha}$, and $m_0 = \nu \frac{\log L_{k_0}}{L_{k_0}}$. (See [13] for the details.)

Exponential localization follows from a stronger version of (54), which is also proven from (53), plus the fact that Maxwell operators have polynomially bounded generalized eigenfunctions.

Theorem 10 reduces the proof of Theorem 7 to the verification of condition (50) with $E_0 = a(g)$. To do so we use Theorems 6 and 5, a finite cube version of Theorem 2, and assumption (45). Theorem 8 is proven in a similar way.

6. The General Framework for Classical Waves

The operator approach, described in this article for electromagnetic waves, is also applicable to other classical waves. The relevant second order partial differential operators for acoustic, elastic and electromagnetic waves are as follows. Recall ∇ denotes the gradient operator, $\nabla^{\times}$ denotes the curl operator, and $-\nabla^*$ is the divergence operator, i.e., $\nabla^*\Psi = -\nabla \cdot \Psi$. We use I_ν for the $\nu \times \nu$ identity matrix.

Acoustic waves:

$$A = \frac{1}{\sqrt{\kappa(x)}} \nabla^* \frac{1}{\rho(x)} \nabla \frac{1}{\sqrt{\kappa(x)}} \tag{55}$$

on $L^2(\mathbb{R}, \curvearrowleft)$, where $\kappa(x)$ is the compressibility and $\varrho(x)$ is the mass density.

Elastic waves:

$$E = \frac{1}{\sqrt{\rho(x)}} \left[\nabla\left(\lambda(x) + 2\mu(x)\right)\nabla^* + \nabla^{\times}\mu(x)\nabla^{\times}\right] \frac{1}{\sqrt{\rho(x)}} \tag{56}$$

on $L^2\left(\mathbb{R}^{\nVdash}, \curvearrowleft; \mathbb{C}^{\nVdash}\right)$, where $\rho(x)$ is the mass density and $\mu(x)$ and $\lambda(x)$ are the Lamé moduli.

Electromagnetic waves:

$$\mathbf{M} = \frac{1}{\sqrt{\mu(x)}} \nabla^{\times} \frac{1}{\varepsilon(x)} \nabla^{\times} \frac{1}{\sqrt{\mu(x)}} \tag{57}$$

on $\mathbb{S} = \left[\operatorname{Ker}\left(\nabla^{\times} \frac{\nVdash}{\sqrt{\mu(\curvearrowleft)}}\right)\right]^{\perp}$ in $L^2\left(\mathbb{R}^{\nVdash}, \curvearrowleft; \mathbb{C}^{\nVdash}\right)$, where $\varepsilon(x)$ is the dielectric constant and $\mu(x)$ is the magnetic permeability.

These operators are special cases of the following operator:

$$\mathbf{W} = \sqrt{\mathcal{K}(x)}\mathbb{D}^* \mathcal{R}(x) \mathbb{D}\sqrt{\mathcal{K}(x)} \tag{58}$$

on $\mathbb{H} = \left[\operatorname{Ker}\left(\mathbb{D}\sqrt{\mathcal{K}(\curvearrowleft)}\right)\right]^{\perp}$ in $L^2(\mathbb{R}, \curvearrowleft; \mathbb{C}^{\prime})$. Here

(i) $\mathbb{D}$ is a $\nu' \times \nu$ matrix whose entries are of the form $\mathbb{D}_{i,j} = a^{(i,j)} \cdot \nabla$, $i = 1, 2, \ldots, \nu'$, $j = 1, 2, \ldots, \nu$, where $a^{(i,j)} \in \mathbb{C}$ and ∇ is the d-dimensional gradient. $\mathbb{D}^*$ is its formal adjoint. There is a constant $C > 0$ such that

$$\mathbb{D}^*\mathbb{D}|_{[\mathrm{Ker}\mathbb{D}]^\perp} \geq C\left[(-\Delta) \otimes I_\nu\right]|_{[\mathrm{Ker}\mathbb{D}]^\perp}. \tag{59}$$

(ii) $\mathcal{K}(x)$ is a position dependent positive $\nu \times \nu$ matrix such that

$$0 < \mathcal{K}_- I_\nu \leq \mathcal{K}(x) \leq \mathcal{K}_+ I_\nu \tag{60}$$

for some constants $\mathcal{K}_\pm$.

(iii) $\mathcal{R}(x)$ is a position dependent positive $\nu' \times \nu'$ matrix such that

$$0 < \mathcal{R}_- I_{\nu'} \leq \mathcal{R}(x) \leq \mathcal{R}_+ I_{\nu'} \tag{61}$$

for some constants $\mathcal{R}_\pm$.

The general operator $\mathbf{W}$ has a natural definition as a self-adjoint operator on the subspace $\mathbb{H}$ of $L^2(\mathbb{R}, \frown; \mathbb{C}^\nu)$, as in (16)-(19). The results described in this article have been extended to $\mathbf{W}$, including a rigorous proof of Anderson localization for general acoustic, elastic and electromagnetic waves in randomized periodic anisotropic media [30].

Acknowledgement. This research was partially supported by the National Science Foundation under grants DMS-9500720 and DMS-9800883.

References

1. P. W. Anderson, "Absense of Diffusion in Certain Random Lattice," Phys. Rev. **109**, 1492-1505 (1958).
2. P. W. Anderson, "A Question of Classical Localization. A Theory of White Paint," Philosophical Magazine B **53**, 505-509 (1985).
3. J. M. Barbaroux, J. M. Combes and P. D. Hislop, " Localization near Band Edges for Random Schrodinger Operators," Helv. Phys. Acta, to appear.
4. M. Sh. Birman and M. Z. Solomyak, "L^2Theory of the Maxwell Operator in Arbitrary Domains," Russian Math. Surveys, **42:6**, 75-96, 1987.
5. R. Carmona and J. Lacroix, *Spectral Theory of Random Schrödinger Operators* (Birkhäuser, Boston, USA, 1990).
6. J. Combes and P. Hislop, "Localization for some Continuous, Random Hamiltonians in d-dimensions," J. Funct. Anal. **124**, 149-180 (1994).
7. *Development and Applications of Materials Exhibiting Photonic Band Gaps*, Journal of the Optical Society of America B **10**, 280-413 (1993).
8. L. Deych and A. Lisyansky, "Impurity Localization of Electromagnetic Waves in Polariton Region," Phys. Rev. Lett., to appear.
9. H. von Dreifus and A. Klein, "A New Proof of Localization in the Anderson Tight Binding Model," Commun. Math. Phys. **124**, 285-299 (1989).
10. A. Figotin, "High-contrast Photonic crystals," in this volume.
11. A. Figotin and A. Klein, "Localization Phenomenon in Gaps of the Spectrum of Random Lattice Operators," J. Stat. Phys. **75**, 997-1021 (1994).

12. A. Figotin and A. Klein, "Localization of Electromagnetic and Acoustic Waves in Random Media. Lattice Model," J. Stat. Phys. **76**, 985-1003 (1994) .
13. A. Figotin and A. Klein, "Localization of Classical Waves I: Acoustic Waves," Commun. Math. Phys. **180**, 439-482 (1996).
14. A. Figotin and A. Klein, "Localization of Classical Waves II: Electromagnetic Waves," Commun. Math. Phys. **184**, 411-441 (1997).
15. A. Figotin and A. Klein, "Localized Classical Waves Created by Defects," J. Stat. Phys. **86**, 165-177 (1997).
16. A. Figotin and A. Klein, "Midgap Defect Modes in Dielectric and Acoustic Media," SIAM J. Appl. Math. (1998).
17. A. Figotin and A. Klein, "Localization of Light in Lossless Inhomogeneous Dielectrics," J. Opt. Soc. Am. A **15**, 1423-1435 (1998).
18. A. Figotin and P. Kuchment, "Band-Gap Structure of Spectra of Periodic Dielectric and Acoustic Media. I. Scalar Model," SIAM J. Appl. Math. **56**, 68-88 (1996).
19. A. Figotin and P. Kuchment, "Band-Gap Structure of Spectra of Periodic Dielectric and Acoustic Media. II. 2D Photonic Crystals," SIAM J. Appl. Math. **56**, 1561-1620 (1996).
20. J. Fröhlich and T. Spencer, "Absence of Diffusion in the Anderson Tight Binding Model for Large Disorder or Low Energy," Commun. Math. Phys. **88**, 151-184 (1983).
21. J. Fröhlich, F. Martinelli, E. Scoppola and T. Spencer, "Constructive Proof of Localization in the Anderson Tight Binding Model," Commun. Math. Phys. **101**, 21-46 (1985).
22. F. Germinet and S. De Bievre, "Dynamical Localization for Discrete and Continuous Random Schrödinger Operators," Commun. Math. Phys., to appear.
23. H. Holden and F. Martinelli, "On Absence of Diffusion near the Bottom of the Spectrum for a Random Schrödinger Operator on $L^2(\mathbb{R}^\nu)$," Commun. Math. Phys. **93**, 197-217 (1984).
24. P. M. Hui and N. F. Johnson, "Photonic Band-Gap Materials," in *Solid State Physics*, vol. **49**, H. Ehrenreich and F. Spaepen, eds., 151-203 (Academic Press, New York, USA, 1995).
25. S. Jitomirskaya and A. Klein, in preparation.
26. J. Joannopoulos, R. Meade and J. Winn, *Photonic Crystals* (Princeton University Press, Princeton, USA, 1995).
27. S. John, "Localization of Light," Phys. Today, May 1991, 32-40.
28. S. John, "The Localization of Light," in *Photonic Band Gaps and Localization*, C.M. Soukoulis, ed., NATO ASI Series B., Vol. **308**, 1-22 (Plenum, New York, 1993).
29. W. Kirsch and F. Martinelli, "On the Ergodic Properties of the Spectrum of General Random Operators," J. Reine Angew. Math. **334**, 141-156 (1982).
30. A. Klein and A. Koines, "A General Framework for Localization of Classical Waves," in preparation.
31. A. Klein and M. Seifert, in preparation.
32. W. Kohler, G. Papanicolaou and B. White, "Localization and Mode Convertion for Elastic Waves in Randomly Layered Media," Wave Motion **23**, 1-22 and 181-201 (1996).
33. I. M. Lifshits, S. A. Greduskul and L. A. Pastur, *Introduction to the Theory of Disordered Systems* (Wiley, New York, USA, 1988).
34. A. Maradudin, E. Montroll and G. Weiss, *Theory of Lattice Dynamics in the Harmonic Approximation* (Academic Press, New York, 1963).
35. J. Maynard, "Acoustic Anderson Localization," in *Random Media and Composites*, B. V. Kohn and G. W. Milton, eds., 206-207 (SIAM, Philadelphia, USA, 1988).
36. L. Pastur and A. Figotin, *Spectra of Random and Almost-periodic Operators* (Springer-Verlag, Heidelberg, USA, 1991).
37. J. Rarity and C. Weisbuch, eds., *Microcavities and Photonic Bandgaps: Physics and Applications* (Kluwer Academic Publishers, Dordrecht, Holland, 1995).
38. B. Shklovskii and A. Efros, *Electronic Properties of Doped Semiconductors*

(Springer-Verlag, Heidelberg, Germany, 1984).
39. C. Soukoulis, ed., *Photonic Band Gaps and Localization* (Plenum, New York, 1993).
40. C. Soukoulis, ed., *Photonic Band Gap Materials* (Kluwer Academic Publishers, Dordrecht, Holland, 1996).
41. C. Soukoulis, "Photonic Band Gap Materials," in this volume.
42. P. R. Villeneuve and J. Joannopoulos, "Working at the Speed of Light," Science Spectra, Number **9**, 18-24 (1997).
43. P. R. Villeneuve and M. Piché, "Photonic Band Gaps in Periodic Dielectric Structures," Prog. Quant. Electr. **18**, 153-200 (1994).
44. D. Wiersma, P. Bartolini, A. Lagendijk and R. Righini, "Localization of Light in a Disordered Medium," Nature **390**, 671-673 (1997).
45. D. Wiersma, in this volume.
46. E. Yablonovitch, T. Gmitter, R. Meade, D. Brommer, A. Rappe and J. Joannopoulos, "Donor and Acceptor Modes in Photonic Periodic Structure," Phys. Rev. **B 44**, 13772-13774 (1991).

PHOTONIC BAND GAP MATERIALS

C. M. Soukoulis
Ames Laboratory and Department of Physics and Astronomy,
Iowa State University, Ames, IA 50011

Abstract. An overview of the theoretical and experimental efforts in obtaining a photonic band gap, a frequency band in three-dimensional dielectric structures in which electromagnetic waves are forbidden, is presented.

1. INTRODUCTION AND HISTORY

Electron waves traveling in the periodic potential of a crystal are arranged into energy bands separated by gaps in which propagating states are prohibited[1]. It is interesting to see if analogous band gaps exist when electromagnetic (EM) waves propagate in a periodic dielectric structure (e.g., a periodic lattice of dielectric spheres of dielectric constant ϵ_a embedded in a uniform dielectric background ϵ_b). If such a band gap or frequency gap exists, EM waves with frequencies inside the gap cannot propagate in any direction inside the material. These frequency gaps are referred to as "photonic band gaps."

Photonic band gaps can have a profound impact on many areas in pure and applied physics[2,3]. Due to the absence of optical modes in the gap, spontaneous emission is suppressed for photons with frequencies in the forbidden region[4,5]. It has been suggested that, by tuning the photonic band gap to overlap with the electronic band edge, the electron-hole recombination process can be controlled in a photonic band gap material, leading to enhanced efficiency and reduced noise in the operation of semiconductor lasers and other solid state devices[3,5]. The suppression of spontaneous emission can also be used to prolong the lifetime of selected chemical species in catalytic processes[6]. Photonic band gap materials can also find applications in frequency-selective mirrors, band-pass filters, and resonators. Besides technical applications in various areas, scientists are interested in the possibility of observing the localization of EM waves by the introduction of

J.-P. Fouque (ed.), Diffuse Waves in Complex Media, 93–107.

defects and disorder in a photonic band gap material[7-9]. This will be an ideal realization of the phenomenon of localization uncomplicated by many-body effects present in the case of electron localization. Another interesting effect is that, zero-point fluctuations, which are present even in vacuum, are absent for frequencies inside a photonic gap. Electromagnetic interaction governs many properties of atoms, molecules, and solids. The absence of EM modes and zero point fluctuations inside the photonic gap can lead to unusual physical phenomena[7-12]. For example, atoms or molecules embedded in such a material can be locked in excited states if the photons emitted to release the excess energy have frequency within the forbidden gap. All the aforementioned ideas[2,3] about new physics and new technology hinge upon the assumption of the existence of material with photonic gaps.

To search for the appropriate structures, scientists at Bellcore employed a "cut-and-try" approach in which various periodic dielectric structures were fabricated in the microwave regime and the dispersion of EM waves were measured to see if a frequency gap existed[13]. The process was time consuming and not very successful. After attempting dozens of structures over a period of two years, Yablonovitch and Gmitter identified[13] only one structure with a photonic band gap. This structure consists of a periodic array of overlapping spherical holes inside a dielectric block. The centers of the holes are arranged in a face-centered-cubic (fcc) lattice and the holes occupy 86% of the volume of the block.

Stimulated by the experimental work, theorists became interested in the solution of the photonic band problem and in the search for structures with photonic band gaps. Early work in this area employed the "scalar wave approximation" which assumed the two polarizations of the EM waves can be treated separately, thus decoupling the problem into the solution of two scalar wave equations. When we first became involved with the photon band problem, calculations had already been completed for the experimental structure in the scalar wave approximation[14,15]. The results showed the existence of a gap but the position and size of the gap were not in quantitative agreement with the experiment, indicating the need for a full vector wave treatment. It turned out from subsequent calculations that the errors made in neglecting the vector nature of the EM wave were more serious than initially anticipated, and the scalar wave calculations actually gave qualitatively wrong results.

The vector wave solution of Maxwell's equations for a periodic dielectric system was carried out independently by several groups shortly after the appearance of the scalar wave results[16-18]. All of the methods employ a plane wave expansion of the electromagnetic fields and use Bloch's theorem to reduce the problem to the solution of a set of linear equations.

When the photon band structure for the experimental fcc structure[13] of 86% air spheres in a dielectric matrix, was calculated[18], it showed that the experimental fcc structure does not have a complete photonic band gap for the lowest-lying bands. A very large depletion of DOS is found, called a "pseudo-gap." Actually, this result was also obtained earlier by two other groups[16,17], although at that time we were not aware of their results. At this point, the existence of photonic gap materials was seriously doubted[19]. However, since we found that the plane wave expansion method[16-18] can solve the photon band problem efficiently and much faster than the experimental "cut-and-try" method, we used it to investigate whether other structures could succeed where the fcc air sphere structure failed.

2. PHOTONIC BAND GAP STRUCTURES WITH THE DIAMOND LATTICE SYMMETRY

Ho, Chan, and Soukoulis were the first to give a prescription for a periodic dielectric structure[18] that possesses a full photonic band gap rather than a pseudogap. This proposed structure is a periodic arrangement of dielectric spheres in a diamond-like structure. A systematic examination[18] of the photonic band structures for dielectric spheres and air spheres on a diamond lattice, as a function of the refractive index contrasts and filling ratios, was made. It was found that photonic band gaps exist over a wide region of filling ratios for both dielectric spheres and air spheres for refractive-index contracts as low as 2. However, this diamond dielectric structure is not easy to fabricate, especially in the micron and submicron length scales for infrared or optical devices. However, after we communicated our findings about the diamond structure, Yablonovitch very quickly devised[20] an ingenious way of constructing a diamond lattice. He noted that the diamond lattice is a very open structure characterized by open channels along the[110] directions. Thus, by drilling cylindrical holes through a dielectric block, a structure with the symmetry of the diamond structure can be created. Since there are 6 sets of equivalent[110] directions in the lattice, there are 6 sets of holes drilled. If the crystal is oriented such that the[111] surface is exposed, then three sets of these holes will be slanted at angles of 35.26° with respect to the normal[111] direction. The remaining three sets of holes have their axes parallel to the[111] surface and are harder to make on a thin film oriented in the[111] direction. Thus, in the end, the experimentalists decided to abandon the second three sets of holes and construct a structure with only the first three sets of holes (see Fig. 15, in the article by Yablonovitch in Ref. 3) which became the first experimental structure that demonstrates the existence of a photonic band gap, in agreement with the predictions[21] of the theoretical calculations. This is a successful example where the theory was used to design dielectric structures with desired properties.

We repeated our calculations for several variations on the diamond lattice[21]. One calculation uses the diamond lattice generated by 6 sets of air cylinders or dielectric cylinders in the six[110] directions. The other calculation uses a diamond rod lattice in which, instead of putting spheres at the lattice sites, we joined them together by nearest-neighbor rods. We also tested the effects on the photon band gap when 3 sets of cylinders are omitted in the 6-cylinder diamond structure. All of these structures exhibit photonic band gaps, with the best performance coming from a diamond rod lattice, which achieves a maximum gap of 30% for a refractive index contrast of 3.6.

Very narrow photonic band gaps have also been found[22] in a simple cubic geometry. For 2D systems, theoretical studies have shown[23-26] that a triangular lattice of air columns in a dielectric background[23-25] and a graphite lattice[26] of dielectric rods in air background are the best overall 2D structure, which gives the largest photonic gap with the smallest retroactive index contrast. In addition, it was demonstrated[27-30] that lattice imperfections in a 2D and/or 3D periodic arrays of a dielectric material can give rise to fully localized EM wave functions. Experimental investigations of the photonic band gaps in either 2D or 3D have been mostly done[20,27,29,30] at microwave frequencies because of the difficulty in fabricating ordered dielectric structures at optical length scales. In fact, the main challenge in the photonic band gap field is the discovery of a 3D dielectric structure that exhibits a photonic gap but, in addition, can be built by microfabrication techniques on the scale of optical wavelengths.

3. LAYER-BY-LAYER PHOTONIC BAND GAP STRUCTURES

The search for simplifying the structure and reducing the dimensionality of the structural building blocks continued. The Iowa State group has designed[31] a novel three-dimensional layer-by-layer structure that has a full three-dimensional photonic band gap over a wide range of structural parameters. The new structure (Fig. 1) consists of layers of one-dimensional rods with a stacking sequence that repeats every fourth layer with a repeat distance of c. Within each layer the rods are arranged in a simple one-dimensional pattern and are separated by a distance a, a significant simplification from the two-dimensional grid found earlier. The rods in the next layer are rotated by an angle θ has the value of 90° but in general could vary from 90° to 60° but still have a full three-dimensional photonic band gap. The rods in the second neighbor plane are shifted by half the

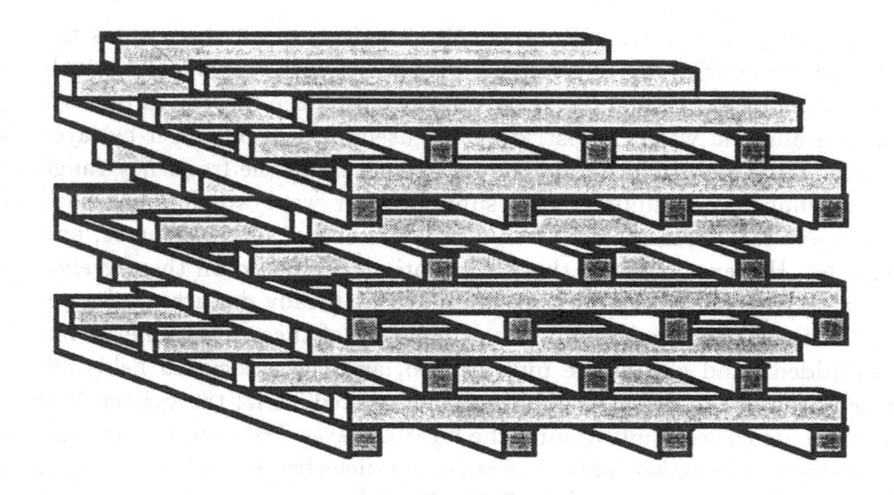

Figure 1. The new layer-by-layer structure producing full three-dimensional photonic band gaps. The structure is constructed by an orderly stacking of dielectric rods, with a simple one-dimensional pattern of rods in each layer. Although rods of rectangular cross-section are shown here, the rods may also have cylindrical or elliptical cross sections.

spacing, a, relative to rods in the first plane in a direction perpendicular to the rods. The rods in every alternate layer are parallel (Fig. 1). This structure has the symmetry of a face centered tetragonal (fct) lattice. For the special case of $c/a = \sqrt{2}$, the lattice can be derived from a fcc unit cell with a 10

basis of two rods. This layered structure can be derived from diamond by replacing the 110 chains of atoms with the rods.

This structure was first fabricated[32] in the microwave regime by stacking alumina cylinders and demonstrated to have a full three-dimensional photonic band gap at microwave frequencies (12-14 GHz). A similar structure was also fabricated with alumina rods that had a band gap between 18 and 24 GHz. We have also fabricated[33-35] the layer-by-layer structure with rectangular rods of silicon by micromachining silicon[110] wafers, using anisotropic etching properties of silicons and an orderly stacking procedure. The structure with rectangular Si-rods have been fabricated for three different length scales producing midgap frequencies of 95 GHz, 140 GHz, and

450 GHz using progressively thinner silicon wafers. In all three cases the band edge frequencies are in excellent agreement with the calculated values. The structure with midgap at 94 GHz has also been fabricated by laser machining alumina wafers, illustrating the usefulness of our layer-by-layer structure. This performance puts the new structure in the frequency range where a number of millimeter and submillimeter wave applications have been proposed, including efficient mm wave antennas, filters, sources, and waveguides. However, most of these applications are based on the presence of defect or cavity modes, which are obtained by locally disturbing the periodicity of the photonic crystal. The frequency of these modes lie within the forbidden band gap of the pure crystal, and the associated fields are localized around the defect. We have demonstrated[36,37] the existence of such cavity structures built around the layer-by-layer PBG crystal. The defects are formed by either adding or removing dielectric material to or from the crystal. We have observed[36,37] localized defect modes with peak and high Q values. The measurements are in good agreement with theoretical calculations.

An interesting class of photonic crystals is the A7-family of structures[38]. These structures have rhombohedral symmetry and can be generated by connecting lattice points of the A7 structure by cylinders. The A7 class of structures can be described a two structural parameters - an internal displacement u and a shear angle α- that can be varied to optimize the gap. For special values of the parameters the structure reduces to simple cubic, diamond, and the Yablonovitch 3-cylinder structure. Gaps as large as 50% are found[38] in the A7 class of structures for well optimized values of the structural parameters and fabrication of these structures would be most interesting. It is worth noting that the fcc structure *does* have[39,40] a true photonic band gap between the eight and the ninth bands (see Fig. 2). The fcc lattice does *not* have[16-18] a PBG between the lowest bands (bands 2 and 3).

4. FABRICATION OF PHOTONIC BAND GAP STRUCTURES

There have been intensive efforts to build and test photonic band gap structures, dating back to the original efforts of Yablonovitch[13] shortly after his first proposal for PBG crystals. Fabrication can be either easy or extremely difficult, depending upon the desired wavelength of the band gap and the level of dimensionality. Since the wavelength of the band gap scales directly with the lattice constant of the photonic crystal, lower frequency structures that require larger dimensions will be easier to fabricate. At microwave frequencies, where the wavelength is on the order of one centimeter, the photonic crystals are decidedly macroscopic, and simple machining techniques

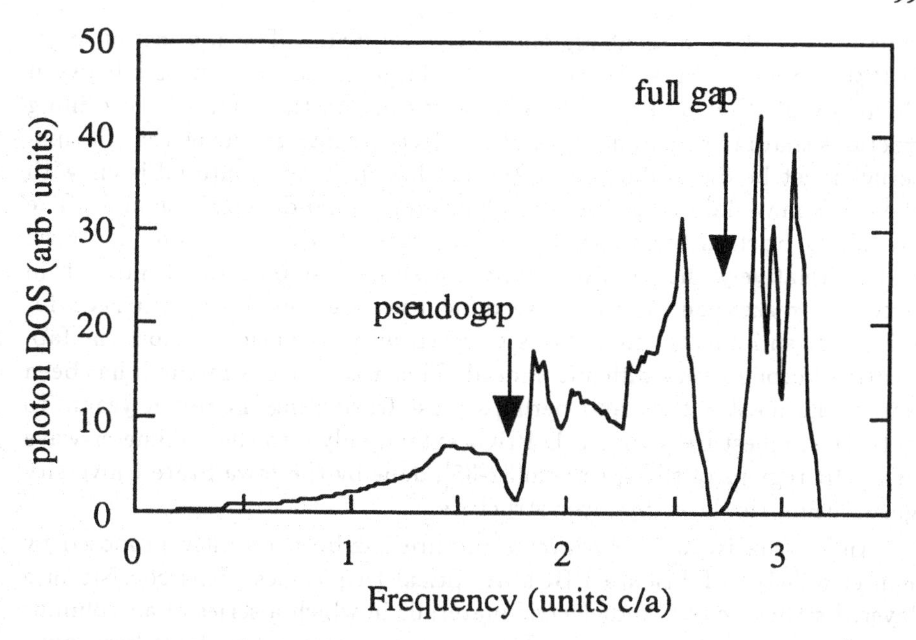

Figure 2. Density of states for the fcc system of low dielectric spheres (filling ratio 0.74) in a high dielectric background, with a refractive index contrast of 3.1, displaying the full gap between the 8 and 9 bands and the weaker pseudogap between bands 2 and 3. Frequencies are in dimensionless units where c is the speed of light (in the dielectric background) and a the lattice constant.

or rapid prototyping methods can be employed in building the crystals. At the other extreme, optical wavelength PBGs require crystal lattice constants less than one micron. Building PBGs in the optical regime requires methods that push current state-of-the art micro- or nano-fabrication techniques. In a similar manner, the dimensionality of the PBG has a big impact on the ease or difficulty of fabrication. Since one-dimensional PBGs require periodic variation of the dielectric constant in only one direction, they are relatively easy to build at all length scales. One-dimensional PBG mirrors (more commonly known as distributed Bragg reflectors) have used in building optical and near-infrared photonic devices for many years. Two common examples of devices using 1-D PBGs are distributed feedback lasers and vertical-cavity surface-emitting lasers. Two-dimensional PBGs require somewhat more fabrication, but relatively main-stream fabrication techniques can be employed to achieve such structures. There are several examples of 2-D PBGs operating at mid- and near-IR wavelengths[26]. Clearly, the most challenging PBG structures are fully 3-D structures with band gaps in the IR or optical regions of the spectrum. The fabrication of 3-D PBGs is complicated by the need for large dielectric contrasts between the materials that make up the PBG crystal, and the relatively low filling fractions that are required. The large dielectric contrast means that the materials must be dissimilar, and often the low-dielectric material is air with the other material being a semiconductor or a high-dielectric ceramic. The low filling fraction means that the PBG crystal with air as one dielectric will be relatively the high dielectric material must be formed into a thin network or skeleton. When these difficulties are combined with need for micron or sub-micron dimensions to reach into the optical region, the fabrication becomes very difficult, indeed. This area of PBG research has been one of the most active, and perhaps most frustrating, in recent years. To date, the highest frequency 3-D PBGs extend only into the millimeter-wave or far-IR regions of the spectrum[32-35], done by the Iowa State University group using the layer-by-layer structure.

An alternative layer-by-layer structure has been recently proposed by Fan et al.[41] to fabricate PBGs at optical frequencies. This consists of a layered structure of two dielectric materials in which a series of air columns is drilled into the top surface. The structural parameters have been optimized to yield 3D photonic gap to midgap ratios of 14% using Si, SiO_2 and air, to 23% using Si and air (i.e., the SiO_2 layers are replaced by air.

As we have discussed above, the first successful PBG crystal was fabricated by Yablonovitch[20] at the millimeter-wave region of the spectrum. Since 1991, both Yablonovitch and Scherer have been working towards reducing the size of the structure to micrometer length scales[42].

Another approach was undertaken by a group at the Institute of Microtechnology in Mainz, Germany in collaboration with the Research Center of Crete and Iowa State University. They fabricated PBG structures using deep x-ray lithography[43]. PMMA resist layers with thickness of 500 microns were irradiated to form a "three-cylinder" structure. Since the dielectric constant of the PMMA is not large enough for the formation of a PBG, the holes in the PMMA structure were filled with a ceramic material. After the evaporation of the solvent, the samples were heat treated at 1100°C, and a lattice of ceramic rods corresponding to the holes in the PMMA structure remained. A few layers of this structure were fabricated with a measured band gap centered at 2.5 THz. Recent experiments are currently trying to fill the PMMA holes with metal.

The layer-by-layer structure shown in Fig. 1 were fabricated[44] by laser rapid prototyping using laser-induced direct-write deposition from the gas phase. The PBG structure consisted of oxide rods and the measured photonic band gap was centered at 2 THz. However, our calculations were unable to confirm those measurements. More experiments on this promising direction are needed.

Very recent work in Sandia Labs by Lin[45] have been able to grow up to five layers, of the layer-by-layer structure shown in Fig. 1. Their measured transmittance shows a band gap centered at 25 THz. They were able to overcome very difficult technological challenges, in planarization, orientation, and 3-D growth at micrometer length scales.

Finally, colloidal suspensions[46-48] have the ability to spontaneously form bulk 3D crystals with lattice parameters on the order of 1-1000 nm. Also, 3D dielectric lattices have been developed[49] from a solution of artificially grown monodisperse spherical SiO_2 particles. However, both these procedures give structures with quite small dielectric contrast ratio (less than 2), which is not enough to give a full band gap. A lot of effort is going into finding new methods in increasing the dielectric contrast ratio. Researchers at Iowa State University[50] and the University of California at Santa Barbara[51] are trying to produce ordered macroporous materials of titania, silica, and zirconia by using the emulsion droplets as templates around which material is deposited through a sol-gel process. Subsequent drying and heat treatment yields solid materials with spherical pores left behind the emulsion droplets.

5. THEORETICAL TECHNIQUES AND TRANSFER MATRIX RESULTS

All of the theoretical results discussed above were obtained with the plane-wave expansion technique[16-18], which is now very well developed. However, most of the theoretical techniques concentrate on the calculation of the dispersion of the photon bands in the infinite periodic structure, while experimental investigations focus mainly on the transmission of EM waves through a finite slab of the photonic band gap patterned in the required periodic structure. Even with the knowledge of the photon band structure, it is still a non-trivial task to obtain the transmission coefficient for comparison with experiment. Another important quantity for the photonic band gap experiments and devices is the attenuation length for incident EM waves inside the photonic band gap. Another topic of interest is the behavior of impurity modes associated with the introduction of defects into the photonic band gap structure. While this problem can be tackled within a plane wave approach using the supercell method[27,28] in which a simple defect is placed within each supercell of an artificially periodic system, the calculations require a lot of computer time and memory. Recently, Pendry and MacKinnon[52] introduced a complimentary technique for studying photonic band gap structures which is called the transfer-matrix method. In the transfer matrix method (TMM), the total volume of the system is divided into small cells and the fields in each cell are coupled to those in the neighboring cells. Then the transfer matrix is defined by relating the incident fields on one side of the PBG structure with the outgoing fields on the other side. Using the TMM, the band structure of an infinite periodic system can be calculated. The main advantage of this method is the calculation of the transmission and reflection coefficients for EM waves of various frequencies incident on a finite thickness slab of the PBG material. In this case, the material is assumed to be periodic in the directions parallel to the interfaces.

We want to stress that this technique can also be applied to cases where the plane-wave method fails or becomes too time consuming. For example, when the dielectric constant is frequency dependent, or has a non-zero imaginary part, and when defects are present in an otherwise periodic system, this technique works well. The TMM has previously been applied in studies of defects in 2D PBG structures[53], of PBG materials in which the dielectric constants are complex and frequency dependent[54], of 3D layer-by-layer PBG materials[34], of 2D metallic PBG structures[55,56] and 3D metallic structures[56]. In all these examples, the agreement between theoretical predictions and experimental measurements is very good, as can be seen in Fig. 3.

In particular, for 2D systems consisting of metallic cylinders[56], there is considerable difference between the two polarizations. For p-polarized waves, the results are qualitatively similar to the dielectric PBG systems. Propagating modes are interrupted by band gaps appearing close to the edges of the Brillouin zone. On the other hand, for s-polarized waves, there is a cut-off frequency ν_c. There are no propagating modes for frequencies between zero and ν_c, so the transmission has a very sharp drop in this frequency range. Above ν_c, there is the usual behavior of bands interrupted by gaps.

For 3D metallic PBG structures[56], the results are very sensitive on the topology of the structure. Systems with isolated metallic scatterers (cermet topology) exhibit similar behavior to the dielectric PBG materials. But, for metallic scatterers forming a continuous network (network topology), there are no propagating modes for frequencies smaller than a cut-off frequency for both polarizations and for any incident angle. Note that for dielectric PBG materials, there is no cut-off frequency for both types of the topology. We have shown this behavior, in both 2D and 3D cases, can be explained using a simple waveguide model where the ν_c is predicted with good accuracy. This cut-off frequency is well below the plasma frequency and is related to the structure of the system.

In all the periodic cases studied, the absorption can be largely neglected for metallic PBG structures with lattice constants, a less than about 100

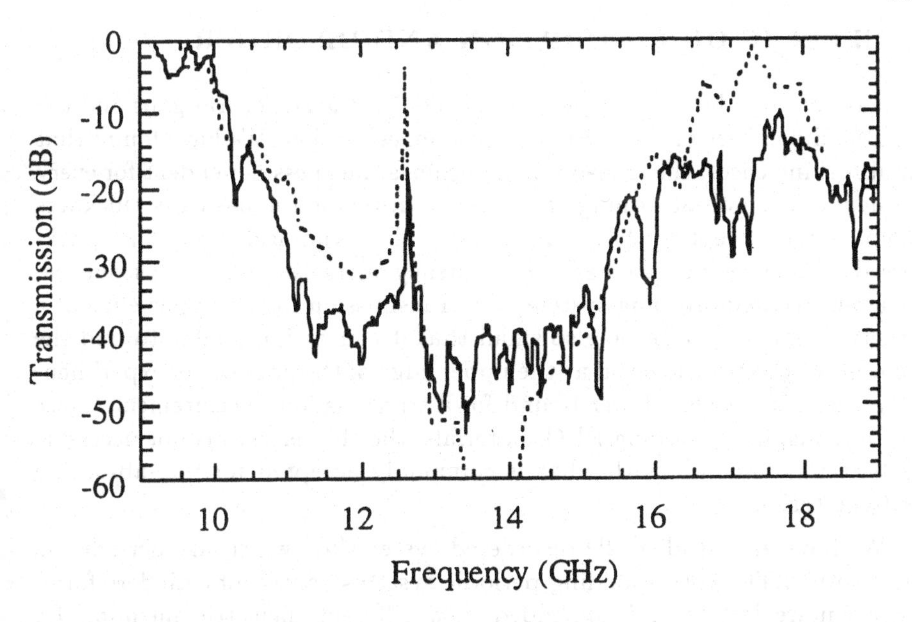

Figure 3. Theoretical (dashed line) and experimental (solid line) transmission characteristics of a defect structure.

μm which correspond to frequencies below about 1 THz. Therefore, for frequencies less than about 1 THz, wide stop-band filters constructed from periodic metallic PBG materials can be used as alternatives to similar filters constructed from dielectric PBG.

By breaking the connections in the 3D metallic networks, defect states appear below the cut-off frequency, resulting in a peak in the transmission. The smaller the volume of the removed metal, the smaller the frequency where the defect peak appears. This is a very interesting feature of the metallic PBG which, in connection with the fact that the filling ratio of the metal can be less than 0.01, can be used in the construction of narrow band-pass filters smaller in size than those constructed from dielectric PBG. By increasing the lattice constant, the Q factor and the transmission at the defect peak increase by order of magnitudes, while the dimensionless defect frequency remains almost constant. The absorption at the frequency where the defect peak appears increases as the lattice constant increases, an effect which may create problems in some of the possible applications. An important advantage of metallic PBG structures is they could be smaller in size and lighter than the corresponding dielectric PBG materials.

6. EFFECTS OF ABSORPTION AND DISORDER

We have studied how the absorption affects[54] the structural gaps and the possible difficulties of their experimental investigation. We have found that for absorbing cases, the transmission becomes thickness dependent for every frequency. For non-absorbing cases, the transmission is basically thickness independent, except for frequencies inside the structural gaps. This thickness dependence increases as the frequency increases. As a consequence, for a very thick absorbing system, the transmission in the upper edge of a structural gap could become so small that it can be impossible to find the recovery of the transmission at the upper edge of the gap, since, experimentally, there is always a lower bound for a transmission measurement (noise level). Thus, for absorbing PBG materials, the thicker slab is not necessarily better in contrast with what is commonly accepted for non-absorbing PBG materials.

We have also studied 2D disordered systems[57] which are periodic on the average. The corresponding periodic systems consist of cylinders forming a square lattice and embedded in a different dielectric medium. By introducing disorder in these periodic systems, the higher gaps, which are narrow, disappear quickly and the logarithmic average of the transmission, $< ln(T) >$, or the localization length, ℓ, becomes almost constant at relatively high frequencies (ω higher than about c/a). These high frequency values of the localization length depend upon the filling ratio and can be as small as 5.2a (a is the lattice constant of the unperturbed periodic system) for the cases that we have studied. On the other hand, for low frequencies, $< ln(T) >$ is not affected by the disorder and it is close to zero which corresponds to very high localization lengths. At intermediate frequencies, there are large drops in the $< ln(T) >$ which correspond to the lowest gaps of the periodic cases. The wider the gaps of the periodic cases, the higher the amount of disorder needed to close these gaps. The gaps of the s-polarized waves are generally wider and survive a high amount of disorder, in contrast with the gaps of the p-polarized waves which are destroyed easily by the disorder. A systematic study of the optimum conditions for the appearance of the gaps has shown these conditions are fulfilled for cylinders of high dielectric material with filling ratio around 0.25 for the s-polarized waves. In these cases, the gaps are wider and they survive even a high amount of disorder, resulting to localization lengths smaller than 5a.

Finally, we have also studied 3D disordered systems[58] which are periodic on the average. Preliminary results show that for structures with network topology, the gaps disappear easily by breaking the network. In contrast, introducing positional disorder but keeping the network topology, the gaps survive a high amount of disorder and the corresponding localization length at frequencies close to the gap can be small.

Yan et al.[59] have also considered many types of fabrication related disorders, from the variation of the layer thickness and channel depth, to the misalignment of channels between successive layers. They also found that the band gap suffered little change. They argued that the apparent insensitivity of the photonic band gap arises from the fact that the wavelength of the light is much larger than the length scale of the disorder.

7. CONCLUSIONS

In summary, we have reviewed the theoretical and experimental efforts in obtaining 2D and 3D dielectric structures that possess a full photonic band gap. The plane-wave method results of Ho, Chan, and Soukoulis suggested the first structure to exhibit a true photonic band gap, and the Yablonovitch "3-cylinder" structure of diamond symmetry was the first experimental structure with a photonic band gap. We have demonstrated that a systematic search for the structures that possess optimal photonic gaps can be conducted via theoretical calculations. We find that the photonic band gap depends crucially on i) the local connectivity or geometry, ii) the refractive index contrast and iii) the filling ratio of a structure. Multiply-connected geometries produce larger gaps than simply connected structures.

We have designed a new layer-by-layer structure that has a full three-dimensional photonic band gap. Each layer consists of a set of one-dimensional pattern of parallel dielectric rods. The rods were rotated by 90° between neighboring layers and shifted by half the distance a between second neighbor layers. This stacking procedure led to a unit cell of four layers. This structure has been fabricated by stacking alumina rods producing full 3-dimensional photonic band gaps between 12 and 14 GHz. The structure has been fabricated by micromachining silicon wafers and stacking the wafers in an orderly fashion producing millimeter wave photonic band gap structures at progressively smaller length scales. We have achieved these photonic band gap structures with midgap frequencies of 100 and 500 GHz. A number of applications of the microwave and millimeter wave PBG crystals may be realized with the structures we have already fabricated. This layer-by-layer structure is very promising for the extension of photonic band crystals into the infrared and optical regimes an area that will surely lead to new areas in basic physics together with novel applications. We are excited about the future applications of photonic band gaps and the prospects of using our calculational techniques to design and help the fabrication of these photonic band gap materials.

8. ACKNOWLEDGMENTS

This work was done in collaboration K.-M. Ho, C. T. Chan, M. Sigalas, and R. Biswas. Ames Laboratory is operated by the U.S. Department of Energy by Iowa State University under Contract No. W-7405-Eng-82. This work was supported by the Director for Energy Research, Office of Basic Energy Sciences and Advanced Energy Projects and by NATO Grant No. 940647.

References

1. See e.g., C. Kittel, *Introduction of Solid State Physics*. (5th Edition, Wiley, 1976) Ch. 7.
2. See the proceedings of the NATO ARW, *Photonic Band Gaps and Localization*, ed. by C. M. Soukoulis, (Plenum. N.Y., 1993); *Photonic Band Gap Materials*, ed. by C. M. Soukoulis, NATO ASI, Series E, vol. 315.
3. For a recent review see the special issue of J. Opt. Soc. Am. B **10**, 280-408 (1993) on the Development of Applications of Materials Exhibiting Photonic Band Gaps: and J. D. Joannopoulos, R. D. Mead, and J. N. Winn *Photonic Crystals*, (Princeton, 1995).
4. E. M. Purcell. *Phys. Rev.* **69**, 681 (1946).
5. E. Yablonovitch, *Phys. Rev. Lett.* **58**, 2059 (1987).
6. N. Lawandy, in *Photonic Band Gaps and Localization*, ed. by C. M. Soukoulis (Plenum Publ., N.Y., 1993). p. 355
7. S. John, *Phys. Rev. Lett.* **58**, 2486 (1987); S. John, *Comments Cond. Matt. Phys.* **14**, 193 (1988); S. John, *Physics Today* **32**, 33 (1991).
8. *Scattering and Localization of Classical Waves in Random Media*, ed. by P. Sheng (World Scientific, Singapore, 1990).
9. J. M. Drake and A. Z. Genack, *Phys. Rev. Lett.* **63**, 259 (1989).
10. C. A. Condat and T. R. Kirkpatrick, *Phys. Rev. B* **36**, 6783 (1987).
11. J. Martorell and N. M. Lawandy,*Phys. Rev. Lett.* **65**, 1877 (1990).
12. G. Kurizki and A. Z. Genack, *Phys. Rev. Lett.* **66**, 1850 (1991).
13. E. Yablonovitch and T. J. Gmitter, *Phys. Rev. Lett.* **63**, 1950 (1989).
14. S. Satpathy, Z. Zhang, and M. R. Salehpour, *Phys. Rev. Lett.* **64**, 1239 (1990).
15. K. M. Leung and Y. F. Liu, *Phys. Rev. B* **41**, 10188 (1990).
16. K. M. Leung and Y. F. Liu, *Phys. Rev. Lett.* **65**, 2646 (1990).
17. Z. Zhang and S. Satpathy, *Phys. Rev. Lett.* **65**, 2650 (1990).
18. K. M. Ho, C. T. Chan, and C. M. Soukoulis, *Phys. Rev. Lett.* **65**, 3152 (1990).
19. J. Maddox, *Nature* **348**, 481 (1990).
20. E. Yablonovitch, T. J. Gmitter, and K. M. Leung, *Phys. Rev. Lett.* **67**, 2295 (1991); E. Yablonovitch and K. M. Leung, *Nature* **351**, 278 (1991).
21. C. T. Chan, K. M. Ho, and C. M. Soukoulis, *Europhys. Lett.* **16**, 563 (1991).
22. H. S. Sözuer and J. W. Haus, *J. Opt. Soc. Am. B* **10**, 296 (1993) and references therein.
23. P. R. Villeneuve and M. Piche, *Phys. Rev. B* **46**, 4964 (1992); ibid **46**, 4973 (1992).
24. R. D. Meade, K. D. Brommer, A. M. Rappe, and J. D. Joannopoulos, *Appl. Phys. Lett.* **61**, 495 (1992).
25. M. Plihal, A. Shambrook, A. A. Maradudin, and P. Sheng, *Opt. Commun.* **80**, 199 (1991); M. Plihal and A. A. Maradudin, *Phys. Rev. B* **44**, 8565 (1991).
26. A. Barra, D. Cassagne, and C. Uonanin, *Appl. Phys. Lett.* **72**, 627 (1998) and references therein.
27. E. Yablonovitch, T. J. Gmitter, R. D. Meade, A. M. Rappe, K. D. Brommer, and J. D. Joannopoulos, *Phys. Rev. Lett.* **67**, 3380 (1991).

28. R. D. Meade, K. D. Brommer, A. M. Rappe, and J. D. Joannopoulos, *Phys. Rev. B* **44**, 13772 (1991).
29. S. L. McCall, P. M. Platzman, R. Dalichaouch, D. Smith and S. Schultz, *Phys. Rev. Lett.* **67**, 2017 (1991).
30. W. Robertson, G. Arjavalingan, R. D. Meade, K. D. Brommer, A. M. Rappe and J. D. Joannopoulos,*Phys. Rev. Lett.* **68**, 2023 (1992).
31. K. M. Ho, C. T. Chan, C. M. Soukoulis, R. Biswas, and M. Sigalas, *Solid State Comm.* **89**, 413 (1994).
32. E. Ozbay, A. Abeyta, G. Tuttle, M. C. Tringides, R. Biswas, M. Sigalas, C. M. Soukoulis, C. T. Chan, and K. M. Ho,*Phys. Rev. B* **50**, 1945 (1994).
33. E. Ozbay, G. Tuttle, R. Biswas, M. Sigalas, and K. M. Ho, *Appl. Phys. Lett.* **64**, 2059 (1994).
34. E. Ozbay, E. Michel, G. Tuttle, R. Biswas, K. M. Ho, J. Bostak, and D. M. Bloom, *Optics Lett.* **19**, 1155 (1994).
35. E. Ozbay, G. Tuttle, R. Biswas, K. M. Ho, J. Bostak, and D. M. Bloom, *Appl. Phys. Lett.* **65**, 1617 (1994).
36. E. Ozbay, G. Tuttle, J. S. McCalmont, M. Sigalas, R. Biswas, C. M. Soukoulis, and K. M. Ho, *Appl. Phys. Lett.* **67**, 1969 (1995).
37. Ozbay, G. Tuttle, M. Sigalas, C. M. Soukoulis, and K. M. Ho, *Phys. Rev. B* **51**, 13961 (1995).
38. C. T. Chan, S. Datta, K. M. Ho, and C. M. Soukoulis, *Phys. Rev. B* **49**, 1988 (1994).
39. H. S. Sozuer, J. W. Haus, and R. Inguva, *Phys. Rev. B* **45**, 13962 (1992).
40. T. Suzuki and P. Yu, *J. Opt. Soc. of Am. B* **12**, 571 (1995).
41. S. Fan, P. Villeneuve, P. Meade, and J. Joannopoulos, *Appl. Phys. Lett.* **65**, 1466 (1994).
42. C. Cheng and A. Scherer, *J. Vac. Sci. Tech. B* **13**, 2696 (1995).
43. G. Feiertag et al. in *Photonic Band Gap Materials* ed. by C. M. Soukoulis (Kluwer, Dordrecht, 1996), p. 63; G. Feiertag et al. *Appl. Phys. Lett.* **71**, 1441 (1997).
44. M. C. Wanke, O. Lehmann, K. Muller, Q. Wen, and M. Stuke, *Science* **275**, 1284 (1997).
45. S. Y. Lin, private communication.
46. R. J. Hunter, *Foundations of Colloidal Science* (Clarendon, Oxford, 1993).
47. I. Tarhan and G. H. Watson, *Phys. Rev. Lett.* **76**, 315 (1996); in *Photonic Band Gap Materials* ed. by C. M. Soukoulis (Kluwer, Dordrecht, 1996), p. 93.
48. W. L. Vos, R. Sprik, A. van Blaaderen, A. Imhof, A. Lagendijk, G. H. G. H. Wegdam, *Phys. Rev. B* **53**, 16231 (1996).
49. Yu A. Vlasov et al., *Appl. Phys. Lett.* **71**, 1616 (1997) and references therein.
50. G. Subramania, private communication.
51. A. Imhof and D. J. Pine, *Nature* (1998).
52. J. B. Pendry and A. MacKinnon, *Phys. Rev. Lett.* **69**, 2772 (1992).
53. M. M. Sigalas, C. M. Soukoulis, E. N. Economou, C. T. Chan, and K. M. Ho, *Phys. Rev. B* **48**, 14121 (1993).
54. M. M. Sigalas, C. M. Soukoulis, C. T. Chan, and K. M. Ho, *Phys. Rev. B* **49**, 11080 (1994).
55. D. R. Smith, S. Shultz, N. Kroll, M. M. Sigalas, K. M. Ho, and C. M. Soukoulis, *Appl. Phys. Lett.* **65**, 645 (1994).
56. M. M. Sigalas, C. T. Chan, K. M. Ho, and C. M. Soukoulis, *Phys. Rev. B* **52**, 11744 (1995).
57. M. M. Sigalas, C. M. Soukoulis, C. T. Chan, and D. Turner, *Phys. Rev. B* **53**, 8340 (1996).
58. M. Sigalas, C. M. Soukoulis, C. T. Chan, and K. M. Ho in *Photonic Band Gap Materials*, ed. by C. M. Soukoulis (Kluwer, Dordrecht, 1996), p. 173.
59. S. Fan, P. R. Villeneuve and J. D. Joannopoulos, *Appl. Phys.* **78**, 1415 (1995).

HIGH-CONTRAST PHOTONIC CRYSTALS

ALEXANDER FIGOTIN
Department of Mathematics
University of California, Irvine
Irvine, CA 92697-3875
afigotin@math.uci.edu

Abstract. Problems related to the spectral theory of two- and three-dimensional photonic crystals exhibiting appreciable gaps are considered. The asymptotic models of high-contrast photonic crystals which can be analyzed rigorously are discussed. The asymptotic models suggest geometries of photonic crystals which favor appreciable spectral gaps. The properties of two-dimensional tunable photonic crystals are considered.

1. Introduction

There is growing interest to lossless periodic dielectric media, called photonic crystals . The interest is motivated by potential novel optical devices and the theoretical significance of the phenomena involved (see [21], [6], [40], [41], [18], [37] and references therein). One of the most important properties of photonic crystals as in the case of solids is the band gap structure of the spectrum which is due to periodicity of the media, [1]. In particular, the existence of spectral gaps, as well as conditions which provide it, are of theoretical and practical importance. The physical reason for the raise of spectral gaps is the multiple scattering and destructive wave interference. Another important phenomenon closely related to the raise of spectral gaps and of the same physical origin is wave localization. The localization phenomenon of electromagnetic (EM) and other classical waves is discussed in detail in [10] and [24].

A variety of methods have been used to compute the band gap structure numerically, [19], [28], [30], [31], [8] (see also surveys [40] and [18] and references therein), and for some photonic crystals spectral gaps were observed experimentally (see survey [18]). In [12]-[14] we have introduced and devel-

J.-P. Fouque (ed.), Diffuse Waves in Complex Media, 109–136.

oped some analytic methods for the spectral analysis of $2D$ high-contrast dielectric structures of square geometry as in Fig. 1 (a). For these photonic crystals we have rigorously established the existence of spectral gaps and found their locations. In [15] we introduced some asymptotic models for $2D$ and $3D$ high-contrast photonic crystals, for which we proved rigorously the existence of spectral gaps and estimated their locations.

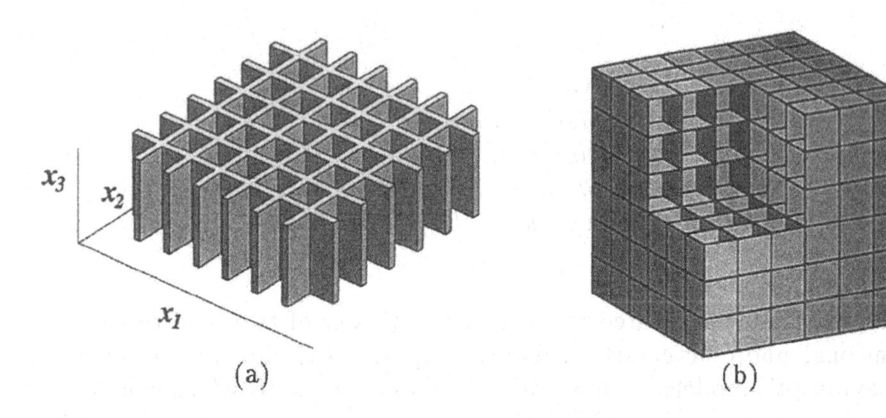

Figure 1. (a) A slab of tetragonal $2D$ photonic crystal composed of parallel air rods of square cross section imbedded in a lossless dielectric material. (b) A slab of a $3D$ photonic crystal composed of a periodic array of cubic air cavities imbedded in a lossless dielectric material.

The theory of photonic crystals deals with a variety of issues: (i) the optimal design of periodic dielectric structure providing prescribed spectral properties, for instance, the largest second spectral gap centered at a given frequency; (ii) selection of a dielectric substance with minimal losses in a prescribed frequency range; (iii) anisotropy of EM wave propagation; (iv) the propagation constant describing the penetration depth of EM wave of a frequency in a spectral gap; (iv) properties of a finite slab of a photonic crystal; (v) defect states generated by an impurity in a photonic crystal and more. Our focus here will be primarily on analytic approaches in the spectral theory of photonic crystals aiming at the following problems:

1. The geometries of $2D$ and $3D$ photonic crystals for which the spectral analysis can be carried out analytically.
2. High-contrast $2D$ and $3D$ asymptotic models of photonic crystals whose spectra can be found exactly.
3. Computational methods based on analytic models with accuracy control.

One of the important problems in the design of photonic crystals is to find within giving constraints the values of its characteristic parameters which yield appreciable spectral gaps. Keeping that in mind, we will call some periodic dielectric structures "better" if they have larger spectral gaps. Discussing that problem we address the following issues:

1. The effect of the space arrangement of the dielectric component on the spectrum. What arrangement is better: air domains imbedded in the optically dense component or its inverse structure, when the dielectric domains are suspended in the air? Is there any simple relation between the spectra of the two mutually inverse structures?
2. The dependence of the spectrum of a photonic crystal on general features of its geometry. Apparently polyhedral air domains separated by a thin film of a dielectric substance produce appreciable spectral gaps.
3. The relationship between the dielectric contrast and the thickness of the dielectric film. The value of the dielectric constant of the optically dense component must be rather tightly related to the geometry of the photonic crystals to achieve appreciable spectral gaps.

The paper is organized as follows. In Section 2 we consider the basics of the theory, introduce and describe photonic crystals that should exhibit appreciable spectral gaps, introduce and briefly discuss the properties of high-contrast photonic crystals and their asymptotic models. We also provide in this section some general arguments for narrowing down the class of photonic crystals that can have appreciable spectral gaps. In Section 3 we discuss primarily $2D$ photonic crystals of square geometry, their spectrum and properties of the eigenmodes. In Section 4 we introduce and discuss the properties of $2D$ tunable photonic crystals. Finally, in Section 5 we introduce and discuss the asymptotic models of high-contrast $3D$ and $2D$ photonic crystals.

2. General properties

Our treatment of the EM waves is based on the classical Maxwell equations

$$\nabla \cdot \mathbf{D} = 0, \ \nabla \times \mathbf{E} = -\frac{1}{c}\frac{\partial \mathbf{B}}{\partial t}, \ \mathbf{D} = \varepsilon \mathbf{E}, \tag{1}$$

$$\nabla \cdot \mathbf{B} = 0, \ \nabla \times \mathbf{H} = \frac{1}{c}\frac{\partial \mathbf{D}}{\partial t}, \ \mathbf{B} = \mu \mathbf{H}, \tag{2}$$

where $\mathbf{E}$ and $\mathbf{D}$ are respectively the electric field and electric induction, $\mathbf{H}$ and $\mathbf{B}$ are the magnetic field and magnetic induction, and c is the velocity of light. The medium parameters $\varepsilon = \varepsilon(\mathbf{x})$ and $\mu = \mu(\mathbf{x})$ are assumed to be position dependent. We neglect frequency dependence of ε and μ. We will assume that $\mu \equiv 1$ since this is true for many dielectric materials and simplifies the spectral analysis.

Remark. *The general spectral theory of dielectric, acoustic, and elastic media when all the media coefficients are inhomogeneous is considered in [24], [25]. The combined effect of all the coefficients of an elastic periodic medium on spectral gaps is discussed in [23].*

Let us introduce now the vector potential $\mathbf{\Phi} = \mathbf{\Phi}(\mathbf{x}, t)$ such that

$$\mathbf{D}(\mathbf{x}, t) = \nabla \times \mathbf{\Phi}(\mathbf{x}, t), \ \mathbf{H}(\mathbf{x}, t) = \frac{1}{c}\frac{\partial}{\partial t}\mathbf{\Phi}(x, t), \ \nabla \cdot \mathbf{\Phi}(\mathbf{x}, t) = 0. \quad (3)$$

Then the Maxwell equations (1), (2) take the form

$$\nabla \times \frac{1}{\varepsilon(\mathbf{x})}\nabla \times \mathbf{\Phi}(\mathbf{x}, t) = -\frac{1}{c^2}\frac{\partial^2}{\partial t^2}\mathbf{\Phi}(\mathbf{x}, t), \ \nabla \cdot \mathbf{\Phi}(\mathbf{x}, t) = 0 \quad (4)$$

with the EM energy density defined by

$$\mathcal{E}(\mathbf{x}, t) = \frac{1}{8\pi}\left[\frac{1}{\varepsilon(\mathbf{x})}|\nabla \times \mathbf{\Phi}(\mathbf{x}, t)|^2 + \frac{1}{c^2}\left|\frac{\partial}{\partial t}\mathbf{\Phi}(x, t)\right|^2\right]. \quad (5)$$

Hence, the spectral properties of the dielectric medium are reduced to the spectral theory of the Maxwell operator

$$M\mathbf{\Phi}(\mathbf{x}) = \nabla \times \frac{1}{\varepsilon(\mathbf{x})}\nabla \times \mathbf{\Phi}(\mathbf{x}), \quad (6)$$

where in the case of lossless photonic crystal $\varepsilon(\mathbf{x})$ is a real-valued periodic function.

The analytic results from [12]-[15] suggest that if a photonic crystal consists of a periodic array of low optical density domains, for instance air cavities, separated by a thin optically dense film, then it can have appreciable spectral gaps. Latter on we will give some arguments showing that *this way to design photonic crystals with appreciable spectral gaps may be the only one.* Assume for simplicity that the optically light component is just air with dielectric constant $\varepsilon_a = 1$ (we will often use the symbol ε_a to keep the track of its physical dimensions). The photonic crystal we are interested in is obtained by partitioning the space by dielectric sheets of thickness α (see Fig. 2 (a)) arranged periodically. For appreciable spectral gaps the dielectric constant ε_d of the dielectric sheets must be chosen appropriately depending on α.

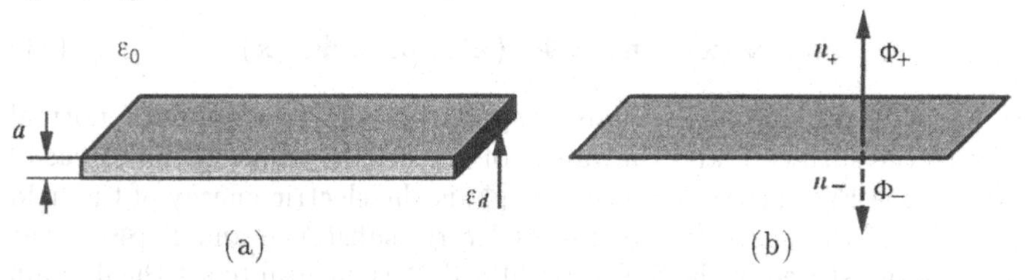

Figure 2. (a) A dielectric sheet of the thickness α suspended in air. (b) An ideal zero thickness dielectric sheet between two adjacent air cavities.

For the clarity of analytic consideration we consider the situation of asymptotically high-contrast $\varepsilon_d/\varepsilon_a$ and send ε_d to infinity. In this case, it turns out, that to have appreciable spectral gaps we also have to send the thickness α to zero in such a way that the product $\varepsilon_d \alpha$ is kept constant. In three-dimensional case for small α it is natural to introduce the "surface dielectric constant" ε_σ by

$$\varepsilon_\sigma = \varepsilon_d \alpha. \tag{7}$$

The dimension $[\varepsilon_\sigma]$ of the surface dielectric constant ε_σ is evidently $[\varepsilon_\sigma] = [\varepsilon][L]$.

Let us write now the asymptotic expression for the *bilinear form of the electric energy* as $\alpha \to 0$ which is fundamental for the spectral analysis. To do that, we introduce the set Γ^α of the points constituting the dielectric film. Assume that Γ^α converge to a limit surface Γ as α approaches 0 and that the dielectric surface Γ partitions the space into disconnected domains (*air cavities*) Q_m. In the photonic crystal case the domains form a periodic array and tile the space. As a simple example, one can think of the cubes tiling the space with Γ being the union of all the faces of the cubes (see Fig. 1 (b)).

Following [15], we get the expression for the electric energy in terms of the vector potential $\mathbf{\Phi}$ from (3) :

$$\mathcal{E}[\mathbf{\Phi}] = \mathcal{E}_a[\mathbf{\Phi}] + \mathcal{E}_d[\mathbf{\Phi}], \text{ where} \tag{8}$$

$$\mathcal{E}_a[\mathbf{\Phi}] = \sum_m \int_{Q_m} \frac{1}{\varepsilon_a} |\nabla \times \mathbf{\Phi}(\mathbf{x})|^2 \, d\mathbf{x}, \tag{9}$$

$$\mathcal{E}_d[\mathbf{\Phi}] = \int_\Gamma \frac{1}{\varepsilon_\sigma} |\nabla_\sigma \times \mathbf{\Phi}(\mathbf{x})|^2 \, d_\sigma\mathbf{x}, \tag{10}$$

where d_σ is the element of area and ∇_σ is the "surface curl" defined by the formula [39]

$$\nabla_\sigma \times \mathbf{\Phi}(\mathbf{x}) = \mathbf{n}_+ \times \mathbf{\Phi}_+(\mathbf{x}) + \mathbf{n}_- \times \mathbf{\Phi}_-(\mathbf{x}). \tag{11}$$

Here $\mathbf{n}_+ = -\mathbf{n}_-$ and $\mathbf{\Phi}_\pm(\mathbf{x})$ are respectively the two opposite normal vectors to the surface Γ and the values of the field in adjacent to Γ areas of the space (see Fig. 2 (b)). Note that $\mathcal{E}_a[\mathbf{\Phi}]$ is the electric energy of the field in the part of the space free of the dielectric substance and $\mathcal{E}_d[\mathbf{\Phi}]$ is the electric energy stored in the dielectric film Γ. It is natural to set the domain of the energy form $\mathcal{E}[\mathbf{\Phi}]$ to include all the fields $\mathbf{\Phi}$ with finite energy $\mathcal{E}[\mathbf{\Phi}]$. Observe also that $\mathbf{\Phi}(\mathbf{x})$ *can be discontinuous at* Γ. The discontinuity, in view of (10), results in the stored electric energy

$$\begin{aligned}
\mathcal{E}_d[\mathbf{\Phi}] &= \int_\Gamma \frac{1}{\varepsilon_\sigma} |\mathbf{n}_+(\mathbf{x}) \times \mathbf{\Phi}_+(\mathbf{x}) + \mathbf{n}_- \times \mathbf{\Phi}_-(\mathbf{x})|^2 \, d_\sigma \mathbf{x} \qquad (12)\\
&= \int_\Gamma \frac{1}{\varepsilon_\sigma} |\mathbf{n}_\pm(\mathbf{x}) \times (\mathbf{\Phi}_+(\mathbf{x}) - \mathbf{\Phi}_-(\mathbf{x}))|^2 \, d_\sigma \mathbf{x}\\
&= \int_\Gamma \frac{1}{\varepsilon_\sigma} |\mathbf{\Phi}_{\tau;+}(\mathbf{x}) - \mathbf{\Phi}_{\tau;-}(\mathbf{x})|^2 \, d_\sigma \mathbf{x},
\end{aligned}$$

where $\mathbf{\Phi}_{\tau;\pm}$ are the components of $\mathbf{\Phi}_\pm$ tangent to the dielectric surface Γ . Note also that the field $\mathbf{\Phi}(\mathbf{x})$ must be divergent free and, hence,

$$\begin{aligned}
\nabla \cdot \mathbf{\Phi}(\mathbf{x}) &= 0 \text{ for } \mathbf{x} \text{ in } Q_m, \qquad (13)\\
\nabla_\sigma \cdot \mathbf{\Phi}(\mathbf{x}) &= 0 \text{ for } \mathbf{x} \text{ in } \Gamma, \text{ where} \qquad (14)\\
\nabla_\sigma \cdot \mathbf{\Phi}(\mathbf{x}) &= \mathbf{n}_+(\mathbf{x}) \cdot \mathbf{\Phi}_+(\mathbf{x}) + \mathbf{n}_-(\mathbf{x}) \cdot \mathbf{\Phi}_-(\mathbf{x}). \qquad (15)
\end{aligned}$$

According to the variational principles, the frequency spectrum of the dielectric medium can be found from the ratio of two quadratic forms associated with the electric and magnetic energy, [17],

$$\frac{\mathcal{E}[\mathbf{\Phi}]}{\mathcal{E}_M[\mathbf{\Phi}]} = \frac{\mathcal{E}_a[\mathbf{\Phi}] + \mathcal{E}_d[\mathbf{\Phi}]}{\int_{\mathbb{R}^3} |\mathbf{\Phi}(\mathbf{x})|^2 \, d\mathbf{x}}, \tag{16}$$

where $\mathbb{R}^3$ denotes three-dimensional space. The extreme values of the ratio (16) and the corresponding fields $\mathbf{\Phi}(\mathbf{x})$ constitute the eigenvalues and eigenmodes. In particular, for an eigenmode $\mathbf{\Phi}_\omega(\mathbf{x})$ we have

$$\frac{\mathcal{E}[\mathbf{\Phi}]}{\mathcal{E}_M[\mathbf{\Phi}]} = \frac{\omega^2}{c^2} = k_\omega^2, \tag{17}$$

where ω is the corresponding eigenfrequency and k_ω can be interpreted as the quasiwavenumber. Note that since $\mathbf{\Phi}_\omega$ *is not a plane wave* the meaning of k_ω as the wavenumber is not as clear and fundamental as the wave time frequency ω. Then for the potential $\mathbf{\Phi}(\mathbf{x}) = \mathbf{\Phi}_\omega(\mathbf{x})$ we have

$$\mathbf{D}(\mathbf{x},t) = e^{i\omega t}\nabla \times \mathbf{\Phi}_\omega(\mathbf{x}), \ \mathbf{H}(\mathbf{x},t) = \frac{i\omega}{c}e^{i\omega t}\mathbf{\Phi}_\omega(\mathbf{x}). \tag{18}$$

In addition to that, for monochromatic waves $\mathbf{\Phi}(\mathbf{x},t) = e^{i\omega t}\mathbf{\Phi}(\mathbf{x})$ we get from (4) the following eigenvalue problems

$$\nabla \times \frac{1}{\varepsilon(\mathbf{x})}\nabla \times \mathbf{\Phi}(\mathbf{x}) = \lambda \mathbf{\Phi}(\mathbf{x}), \ \lambda = \frac{\omega^2}{c^2}, \ \mathbf{x} \in \mathbb{R}^3. \tag{19}$$

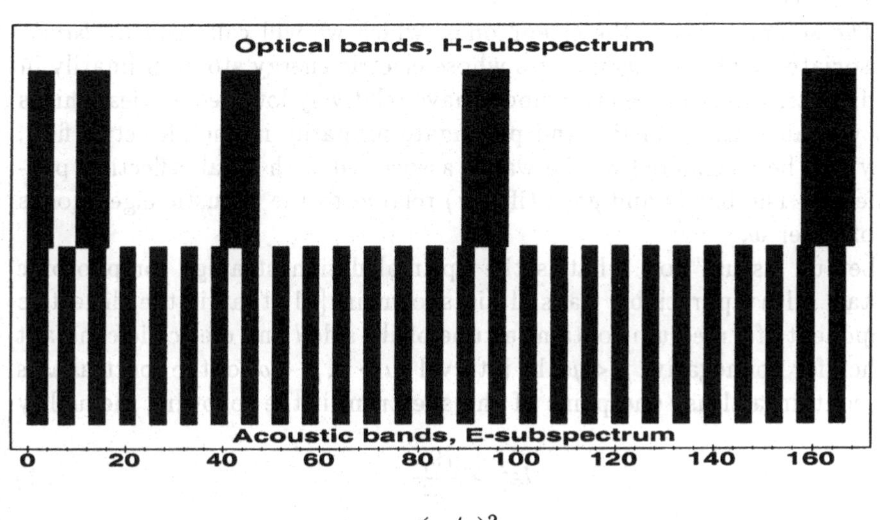

$(\omega/c)^2$

Figure 3. The qualitative picture of the optical and acoustic bands of the spectrum of $2D$ photonic crystals as in Fig. 1 (a). The horizontal sides of shaded rectangles indicate the spectral bands positions. The optical bands are chiefly due to H-polarized EM waves (TE modes) and the the acoustic bands are chiefly due to E-polarized EM waves (TM modes).

As it follows from [12]-[15] , *all the eigenmodes can be divided into two classes. The first class*, which we will call *optical band eigenmodes*, consists of the eigenmodes whose electric energy stored primarily in the light dielectric component (air). These eigenmodes have relatively high frequencies (this is why we call them optical) which are approximately equal to $\lambda_s/\varepsilon_a, s = 0, 1, \ldots$, where λ_s are the resonant frequencies of the cavity Q_0 with perfectly conducting boundary. More precisely, in the regime of optical band eigenmodes we can view the medium as a set of identical cavity resonators associated with the air domains Q_m which are coupled by the dielectric film. The coupling is described by the term $\mathcal{E}_d[\Phi]$ and is proportional to the *coupling constant*

$$w = 1/\varepsilon_\sigma = 1/(\varepsilon_d \alpha). \tag{20}$$

This implies that: (i) the optical bands are the intervals centered approximately at the resonant frequencies λ_s/ε_a of the air cavities Q_m; (ii) the band lengths are proportional to the coupling constant $w = 1/(\alpha\varepsilon_d)$. For small w (or large $\varepsilon_\sigma = \varepsilon_d\alpha$) there must be spectral gaps with the lengths approximately equal to to the distances between adjacent frequencies λ_s/ε_a (see Fig. 3).

The second class of the eigenmodes, which we will call *acoustic bands*, is associated with the eigenmodes whose electric energy stored primarily in the dielectric film. These eigenmodes have relatively low frequencies (this is why we call them acoustic) and propagate primarily in the dielectric film. They can be considered as the waves associated with total reflection phenomenon. The bands and gaps (if any) related to the acoustic eigenmodes are of order w.

Let us discuss now what is the optimal design strategy for photonic crystals with appreciable gaps. It is shown in [11] that if the dielectric component of a medium contains a cube of the side l and dielectric constant ε_1 then for nonnegative $\rho < \mu$ the interval $(\mu - \rho, \mu + \rho)$ of the spectral axis will contain at least one point of the spectrum if the following inequality holds:

$$l\varepsilon_1^2 \geq \frac{79\mu}{\rho^2}. \tag{21}$$

This inequality is based on the following rather simple statement (see the details in [11]). Let A be a self-adjoint operator, μ is real and ρ is nonnegative. Then if there exists a wave function φ, which we call testing, such that

$$\|A\varphi - \mu\varphi\| \leq \rho \|\varphi\| \tag{22}$$

then the distance from the point μ to the spectrum of the operator A is not greater than ρ. In our case the problem of finding the testing function can be efficiently reduced to similar problem for Laplacian operator $-\varepsilon_1^{-1}\Delta$. The inequality (21) immediately implies that if we want to have a spectral gap larger then 2ρ in a spectral interval $(0, N)$ we must satisfy the inequality

$$l\varepsilon_1^2 < \frac{79N}{\rho^2}. \tag{23}$$

This inequality, in turn, shows that *to have a spectral gap of a given width in a given spectral interval, the dielectric component must not contain any cube of the side larger than the following critical length*

$$l_{\mathrm{cr}}(\varepsilon_1) = \frac{79N}{\varepsilon_1^2 \rho^2}. \tag{24}$$

In the case of not too peculiar geometric arrangement, this condition basically says that *to provide desirable gaps, the dielectric component must form a film of the thickness at most* $l_{\mathrm{cr}}(\varepsilon_1)$. *This argument also suggests that periodic arrays of air domains of polyhedral shape are better for appreciable gaps than arrays of balls or cylinders. Indeed, in the case of polyhedra the same amount of the dielectric material is distributed more evenly forming a thinner film and, hence, allowing larger values of* ρ *for the spectral gaps.*

Let us turn now to the issue of the existence of a simple *relation between the spectra of a dielectric medium and its inverse counterpart*, i.e. the medium for which the location of the dielectric component and the air are interchanged. Note that the *total reflection phenomenon assigns substantially different roles to the optically light and dense components.* In particular, the geometry of the dielectric component and connectedness, for instance, can have a decisive impact on the spectral structure. *These arguments do not support the expectation of a simple relationships between the spectra of mutually inverse dielectric structures. Based on the above arguments we suggest that photonic crystals formed by periodic arrays of polyhedral air domains separated by a film of a dielectric substance are better for appreciable spectral gaps. The thickness of the film must be chosen appropriately based on the value of its dielectric constant.*

From now on we will consider only polyhedral (polygonal for $2D$ case) two-component photonic crystal photonic crystal. We give now a more accurate description of the geometry of polyhedral photonic crystals. Assume that the optically light component, for instance the air, has the dielectric constant $\varepsilon_a = 1$. The dielectric constant of the optically dense dielectric component is denoted by $\varepsilon_d > 1$. It is clear that the geometry of a two-component photonic crystal is completely described by the location of its dielectric component. Let us consider the case when the dielectric component forms a film of the thickness α which partitions the space into separated polyhedral air cavities (see Fig. 1 (a), (b)). In more formal terms, suppose that the space $\mathbb{R}^d$, $d = 2, 3$, is tiled periodically by polyhedra (polygons in two-dimensional case) Q_m without overlapping. The periodic structure then can be introduced through the lattice $\mathbb{L}$ which is a discrete group of translations in $\mathbb{R}^d$. The simplest example is a cubic structure of linear dimension ℓ, where $\mathbb{L} = \ell\mathbb{Z}^d$ and

$$\begin{aligned} \mathbb{Z}^d &= \{\mathbf{n} = (n_1, ..., n_d) : \ n_j \text{ in } \mathbb{Z},\ j = 1, ..., d\}, \\ &\quad \mathbb{Z} \text{ is the set of integers,} \end{aligned} \tag{25}$$

and the fundamental polyhedron is the cube

$$Q_0 = \ell \{\mathbf{x} = (x_1, ..., x_d) : 0 \leq x_j \leq 1, \; j = 1, ..., d\}. \tag{26}$$

In this case all the polyhedra are obtained by $\ell\mathbb{Z}^d$-translations of Q_0, namely

$$Q_m = m + Q_0, \; m \text{ in } \ell\mathbb{Z}^d. \tag{27}$$

Let us denote by ∂Q the boundary of the domain Q. We introduce then in $3D$ case the surface Γ which is the union of the boundaries of all the polyhedra Q_m, i.e.

$$\Gamma = \text{union of the boundaries } \partial Q_m \text{ when } m \text{ runs over } \mathbb{L}. \tag{28}$$

Clearly Γ separates the polyhedra Q_m. In $2D$ case Γ will be a periodic graph in the plane. In addition to that, the interface Γ is invariant under the shifts $\Gamma + m = \Gamma$ for all m in $\mathbb{L}$.

Now we can define the film of thickness α on the surface Γ as follows. For a given α we introduce the α-neighborhood Γ^α of Γ, i.e. the set of points $\mathbf{x}$ such that dist $(x, \Gamma) \leq \alpha$. We assign the localication of the dielectric film to be the set Γ^α and hence

$$\varepsilon(\mathbf{x}) = \begin{cases} \varepsilon_d & \text{if} \quad \mathbf{x} \text{ in } \Gamma^\alpha, \\ \varepsilon_a & \quad\text{otherwise.} \end{cases} \tag{29}$$

For future considerations we also introduce dimensionless parameters

$$\hat{\varepsilon} = \varepsilon_d / \varepsilon_a, \; \hat{\alpha} = \alpha / \ell. \tag{30}$$

3. 2D photonic crystals

$2D$ photonic crystals are three-dimensional structures that are periodic in two directions (see Fig. 1 (a)). To describe them, we denote by x_1, x_2, x_3 and e_1, e_2, e_3 the space coordinates and the standard basis vectors and assume that the two special directions are x_1 and x_2. Then the dielectric constant ε of the $2D$ photonic crystals will depend only on x_1 and x_2, i.e. $\varepsilon = \varepsilon(x_1, x_2)$. In the simplest case of the square lattice of periods we have

$$\varepsilon(x_1 + \ell n_1, x_2 + \ell n_2) = \varepsilon(x_1, x_2), \tag{31}$$

where ℓ is the linear dimension of the square primitive cell $\ell[0,1]^2$ and n_j are integers.

Various theoretical methods for $2D$ photonic crystals were developed in [40] for sinusoidally and rectangularly modulated dielectric constants, and in [30], [34] for a periodic array of parallel dielectric rods of circular cross-section whose intersections with perpendicular planes form a triangular or square lattice. Similar structures were studied theoretically and experimentally in [29] and [31]. We have carried out rigorous mathematical studies of the spectra of some $2D$ photonic crystals in [12]-[14]. The results from these references form the base of our considerations here. Though some of our results hold for rather general periodic structures (as, for instance, in Fig. 4 (a), (b)), we will limit our consideration for simplicity to the square structures as in Fig. 1 (a).

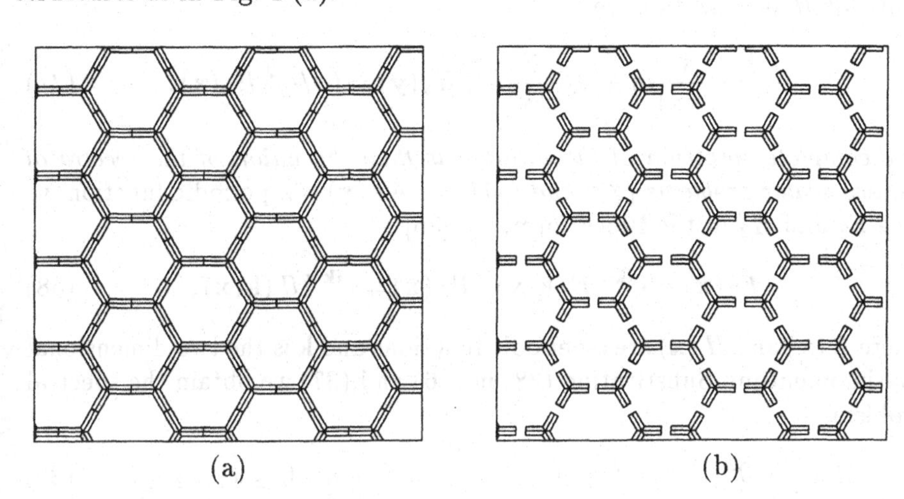

Figure 4. (a) Cross section of a $2D$ photonic crystal where the air rods are completely separated by a dielectric materials. (b) cross section of a $2D$ photonic crystal where the walls of the dielectric are perforated and the air rods form a connected structure.

For $2D$ photonic crystal we consider only the waves propagating parallel to the plane x_1, x_2 and, hence, the magnetic and electric fields $\mathbf{H}$ and $\mathbf{E}$ will depend only on (x_1, x_2). Abbreviating $\mathbf{x} = (x_1, x_2)$, we then get

$$\mathbf{H} = \mathbf{H}(\mathbf{x}),\ \nabla \cdot \mathbf{H} = 0; \quad \mathbf{E} = \mathbf{E}(\mathbf{x}),\ \nabla \cdot \varepsilon \mathbf{E} = 0. \tag{32}$$

Plugging in the time harmonic fields

$$\mathbf{H}(\mathbf{x}, t) = \mathbf{H}(\mathbf{x}) e^{-i\omega t}, \mathbf{E}(\mathbf{x}, t) = \mathbf{E}(\mathbf{x}) e^{-i\omega t} \tag{33}$$

into Maxwell equations (1)–(2) we readily get the eigenvalue problem

$$\nabla \times \mathbf{E}(\mathbf{x}) = \frac{i\omega}{c} \mathbf{H}(\mathbf{x}),\ \nabla \times \mathbf{H}(\mathbf{x}) = -\frac{i\omega}{c} \varepsilon(\mathbf{x}); \tag{34}$$

$$\nabla \cdot \varepsilon(\mathbf{x}) \mathbf{E}(\mathbf{x}) = 0,\ \nabla \cdot \mathbf{H}(\mathbf{x}) = 0. \tag{35}$$

It is well known, [20], that the spectral analysis of the problem (34)-(35) can be reduced to two scalar equations associated with two basic polarizations: (i) E-polarized fields (or TM modes) when $H_3 = 0$ and $E_1 = E_2 = 0$ and (ii) H-polarized field (or TE modes) when $E_3 = 0$ and $H_1 = H_2 = 0$. For E-polarized fields the following equation holds

$$-\left(\partial_1^2 + \partial_2^2\right) E_3(\mathbf{x}) = (\omega/c)^2 \varepsilon(\mathbf{x}) E_3(\mathbf{x}), \ \partial_j = \frac{\partial}{\partial x_j}, \qquad (36)$$

while for H polarization we have

$$-\left[\partial_1 \frac{1}{\varepsilon(\mathbf{x})} \partial_1 + \partial_2 \frac{1}{\varepsilon(\mathbf{x})} \partial_2\right] H_3(\mathbf{y}) = (\omega/c)^2 H_3(\mathbf{x}). \qquad (37)$$

The complete spectrum of the medium is then the union of the spectra of the two scalar problems (36) and (37). Since $\varepsilon(\mathbf{x})$ is a periodic function we take E_3 and H_3 in the Bloch form, [1], [35]

$$E_3(\mathbf{x}) = e^{i\mathbf{k}\cdot\mathbf{x}} E(\mathbf{k}, \mathbf{x}), \ H_3(\mathbf{x}) = e^{i\mathbf{k}\cdot\mathbf{x}} H(\mathbf{k}, \mathbf{x}), \qquad (38)$$

where $E(\mathbf{x})$ and $H(\mathbf{x})$ are ℓ-periodic functions and $\mathbf{k}$ is the two-dimensional quasimomentum. Substituting (38) in (36) and (37) we obtain the spectral problems

$$-\left(\partial_{1,k_1}^2 + \partial_{2,k_2}^2\right) E = (\omega/c)^2 \bar{\varepsilon} E, \ \partial_{j,\kappa} = \partial_j + i\kappa, \ 0 \le x_1, x_2 \le \ell, \qquad (39)$$

$$-\left[\partial_{1,k_1} \frac{1}{\bar{\varepsilon}} \partial_{1,k_1} + \partial_{2,k_2} \frac{1}{\bar{\varepsilon}} \partial_{2,k_2}\right] H = (\omega/c)^2 H, \ 0 \le x_1, x_2 \le \ell. \qquad (40)$$

The properties of the eigenvalue problems (39) and (40) were thoroughly analyzed in [12]–[13] by analytic methods. The qualitative picture of the complete spectrum is displayed in Fig. 3. Under the asymptotic conditions

$$\ell = \text{const}, \ \varepsilon_a = \text{const}, \ \varepsilon_d \alpha = \varepsilon_\sigma = \text{const}, \ \varepsilon_d \to \infty \qquad (41)$$

all the bands for the spectral parameter $(\omega/c)^2$ in a fixed bounded spectral interval are of order $(\ell \varepsilon_d \alpha)^{-1}$. In particular, under asymptotic condition (41) we call the medium high-contrast if

$$\frac{\varepsilon_d \alpha}{\varepsilon_a \ell} \gg 1 \qquad (42)$$

which is equivalent to $\varepsilon_d \alpha \gg \varepsilon_a \ell$ or $\hat{\varepsilon} \hat{\alpha} \gg 1$. The rigorous analysis shows that entire spectrum splits into two subspectra, namely, optical bands (of higher frequencies) and acoustic bands (of lower frequencies) (see Fig. 3) confirming the informal arguments of the previous sections.

Optical bands. The optical bands are chiefly due to H-polarized fields and have the form

$$\left[\frac{\pi^2\mathbf{n}^2}{\varepsilon_a\ell^2}-\frac{\hat{b}_{\mathbf{n}}^-}{\varepsilon_d\alpha\ell},\frac{\pi^2\mathbf{n}^2}{\varepsilon_a\ell^2}+\frac{\hat{b}_{\mathbf{n}}^+}{\varepsilon_d\alpha\ell}\right]=\frac{\pi^2}{\varepsilon_a\ell^2}\left[\mathbf{n}^2-\frac{\hat{b}_{\mathbf{n}}^-}{\hat{\varepsilon}\hat{\alpha}},\mathbf{n}^2+\frac{\hat{b}_{\mathbf{n}}^+}{\hat{\varepsilon}\hat{\alpha}}\right], \tag{43}$$

where $\mathbf{n}=(n_1,n_2)\neq\mathbf{0}$ runs over the lattice $\mathbb{Z}^2$ of pairs of integers and $\hat{b}_{\mathbf{n}}^{\pm}$ are positive constants. Note that the typical gaps between the optical bands are of order $\left(\varepsilon_a\ell^2\right)^{-1}$. Observe also that the gap-to-band ratio has the order

$$\text{gap/band} \sim (\varepsilon_d\alpha)/(\varepsilon_a\ell)=\hat{\varepsilon}\hat{\alpha}. \tag{44}$$

Hence, in the high-contrast case (42) we have for the optical bands

$$\text{gap/band} \gg 1. \tag{45}$$

The electric energy of the corresponding eigenmodes is stored primarily in the air domains.

Acoustic bands. The acoustic bands are entirely due to E-polarized fields and have the form

$$(\varepsilon_d\alpha\ell)^{-1}[0,4],\ \pi(\varepsilon_d\alpha\ell)^{-1}[2n,2n+1] \text{ for } n=1,2,\ldots. \tag{46}$$

The electric energy of the corresponding eigenmodes is stored primarily in the dielectric film. We remind that the eigenmodes in acoustic bands are due to total refection phenomenon. One can easily recognize both spectra (43) and (46) in Fig. 3.

Remark. *For more general geometries of 2D photonic crystals, for instance such as in Fig. 4 (a), (b), in the case of high-contrast media under conditions (41) the spectra of the scalar equations (36) and (37) are similar to (46) and (43) respectively, [14], [15], and [27].*

3.1. COMPUTATION OF THE SPECTRA

The computation of bands and gaps and other spectral attributes for $2D$ photonic crystals, not to mention $3D$ periodic dielectrics, is a challenging numerical problem because of the high dimensions of the matrices involved. The common approach to the computations of bands and gaps is based on the decomposition of the fields into plane waves, i.e. standard Fourier series, with the consequent series truncation. The clear advantage of this approach is that theoretically it works for any periodic dielectric structure. However, in practice, truncation severely limits the accuracy of the plane-wave method [40], [38], [8]. To maintain a reasonable accuracy, one has to deal with matrices of enormous dimensions, [40], [38]. Another approach to the computation of the bandgap structure is based on a finite-difference approximation of Maxwell's equations. In the T-matrix method [32], a transfer matrix is calculated by integration of the wavefield to find the change of the electric and magnetic fields in adjacent layers of the dielectric. Unlike the T-matrix approach, the R-matrix method relates the electric field of adjacent layers to the magnetic field on both sides. A comparison of the two methods is discussed in [7].

An alternative to the plane-wave method was developed in [8] for the case of $2D$ photonic crystals of square geometry. It is based on analytic methods from [12], [13]. In [8] instead of plane waves we use a set of orthogonal functions $\overset{\circ}{W}_{\mathbf{n}}$ associated with an exactly solvable model well approximating the medium with given $\varepsilon(x)$. The construction of the basis waves $\overset{\circ}{W}_{\mathbf{n}}$ employs eigenfunctions of auxiliary $1D$ Schrödinger equations that are similar to $1D$ equations for classical waves. For those $1D$ eigenproblems we develop an efficient analytic approach using a phase function allowing fast and precise computation of a large number of $1D$ eigenfunctions and eigenvalues (see [8] for details). Computation of the spectra of some $2D$ photonic crystals based on [12]–[13] was carried out also in [26] by a somewhat different approach. The computation of the density of states carried out in [8] shown in Fig. 5 is completely consistent with the qualitative picture of the spectrum in Fig. 3.

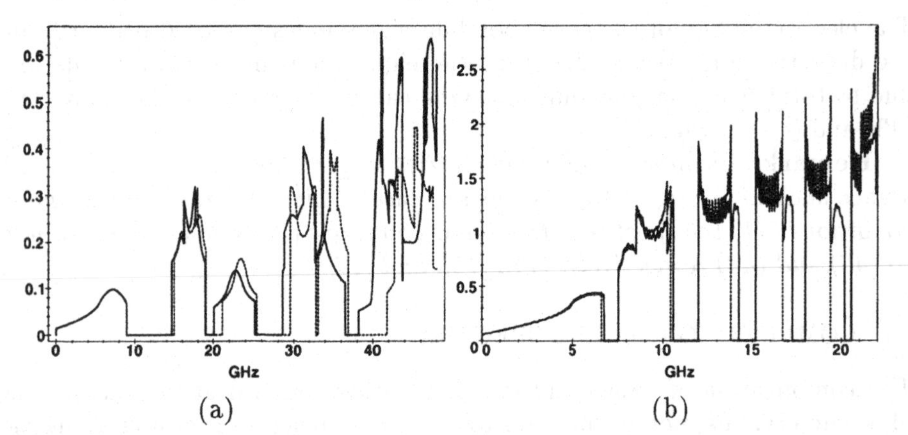

(a) (b)

Figure 5. Density of states for (a) E polarization and (b) H polarization for $\varepsilon_d = 50$, $\hat{\alpha} = 0.04$, $\ell = 1$cm. The solid lines correspond to the main problems (39)–(40). The dashed curves refer to the auxiliary exactly solvable models, [8].

As to the accuracy, the results of computations in [8] were checked in different ways. The spectra of E and H-polarized waves completely agree with those in the asymptotic case for which exact estimates were obtained in

[12]-[13]. Truncation analysis shows that for $\varepsilon = 20$ and $\hat{\alpha} = 0.1$ it requires 146 basis functions in the Rayleigh-Ritz method to get 10 first spectral bands of E and H-polarized waves with an accuracy of four significant digits.

4. 2D tunable photonic crystals

In this section we consider tunable photonic crystals whose characteristics can be controlled by a moderate external magnetic or electric field, [9]. For a photonic crystal to be tunable at least one of its constituting components must display a nonlinearity in either the electric or the magnetic susceptibility. If the amplitudes $E(t)$ and $H(t)$ of the electric and magnetic fields of the propagating EM wave are sufficiently small then the wave can be treated within the linear approximation. In addition to that, if the external uniform field $\mathbf{H}_0$ or $\mathbf{E}_0$ is strong enough, it may substantially alter the material tensors

$$\varepsilon = \varepsilon(\mathbf{E}_0, \mathbf{H}_0) \quad \text{or} \quad \mu = \mu(\mathbf{E}_0, \mathbf{H}_0) \tag{47}$$

and consequently alter the entire spectrum of the medium. We will use the subscript "0" to refer to the external (controlling) field which is assumed to be strong enough to cause a nonlinear response as in (47).

With few exceptions, the magnitude of an applied external field must be much greater compared to the amplitude of the corresponding components of the propagating electromagnetic wave, i.e.

$$E_0 \gg E(t) \quad \text{or} \quad H_0 \gg H(t). \tag{48}$$

In most cases condition (48) must be imposed to get a linear problem. *In some cases though, condition (48) is not required.* For instance, nearly static external field may cause substantial nonlinear response. At the same time *a propagating EM wave with sufficiently high frequency can be treated within the linear approximation even if its amplitude is comparable with the amplitude of the external field.* This is especially likely if the main effect caused by an external quasistationary field $\mathbf{E}_0$ (or $\mathbf{H}_0$) reduces to a rearrangement of the domain structure in a thermodynamic equilibrium state. A similar effect may occur if the external field $\mathbf{E}_0$ (or $\mathbf{H}_0$) alternates, thereby causing a resonant response of the medium. Indeed, in the resonant case there can be a pronounced nonlinear behavior even for a relatively small amplitude of the controlling field. If the frequency of the propagating wave is not resonant, it still can be treated within the linear theory.

In general, the controlling field may be time- and space-dependent and must be treated as an inseparable part of the electrodynamic problem. But in view of the previous considerations on the relationships between the external fields and the propagating EM wave, one may consider much simpler linear problem where $\mathbf{E}_0$ (or $\mathbf{H}_0$) is just a stationary or quasi-stationary parameter that alters parametrically the material tensors as in (47).

A controllable alteration of the photonic band structure may have numerous physical and practical aspects. We will focus primarily on a single and basic question: how does the external field affect the propagation of EM waves for a given fixed frequency Ω? At first glance, the only remarkable *effect of the tunability is the possibility of switching between transparent and opaque states*, depending on whether the frequency Ω falls in a transmittance band or a photonic band gap. This appears to be true for the case of one-dimensional periodical structures. But for two- and three-dimensional periodicity more subtle consideration shows that in addition to that there are other interesting phenomena. For instance, if a fixed frequency Ω was originally situated within a photonic band gap, then the gradual alteration of the photonic band structure caused by the external controlling field will result in at least two distinctive transitions accompanied by a dramatic modification of the character of EM wave propagation. The corresponding transitions are rather similar to those well-known in the theory of electronic topological phase transitions (see [4] and references therein). *In particular, for selected frequencies the EM wave propagation can be extremely anisotropic* (see Fig. 8 (b)).

4.1. MATERIALS FOR TUNABLE PHOTONIC CRYSTALS

There exist many dielectric materials with pronounced nonlinearity in the electric or magnetic properties. In particular, most of ferroelectrics display substantial dependence $\varepsilon = \varepsilon(\mathbf{E}_0)$. On the other hand, magnetically ordered crystals, especially ferromagnets and ferrimagnets, are likely to manifest a magnetic nonlinearity $\mu = \mu(\mathbf{H}_0)$ even in a relatively low external magnetic field $\mathbf{H}_0$. We have chosen to consider here the case of the electric-field-dependent tensor $\varepsilon = \varepsilon(\mathbf{E}_0)$, having in mind that the entire consideration holds for the case of a controlling magnetic field as well. Note though, that in spite of the formal mathematical similarity, the electric and magnetic cases may differ significantly. For the known lossless dielectric materials with strong dependence $\varepsilon(\mathbf{E}_0)$, the electromagnetic properties are substantially different from those of the media with "magnetic-type" nonlinearity like $\mu(\mathbf{H}_0)$, especially so if the appropriate frequency range is concerned.

As already mentioned, one of the two constitutive components of the tunable photonic crystal must be made of a material with substantial nonlinearity of the electric permittivity $\varepsilon = \varepsilon(\mathbf{E}_0)$. The overwhelming majority of ferroelectric materials do meet this requirement. Hence, the main problem is to find those *dielectrics* which, first, will be practically lossless in the given frequency range and, second, will display sufficiently high electric permittivity at that frequency range. For some materials the above restrictions may be critical for infrared or optical frequencies, but *in the microwave range up to* $10^{11} sec^{-1}$ *there exist hundreds of dielectrics with satisfactory physical characteristics.* Particularly, the most attractive would be a situation when a small shift in the impressed controlling field would lead to a significant alteration of ε in one of the two constitutive components (see [9] for details).

As to possible applications of *magnetic materials* as the active elements of tunable photonic devices we note that the most of the so-called soft ferromagnets and ferrimagnets display high magnetic permeability with the tensor $\mu = \mu(\mathbf{H}_0)$ being strongly dependent on $\mathbf{H}_0$, and, from this point of view, they can be ideal materials for tunable photonic crystals. The problem though is that magnetic susceptibility $\chi(\omega)$ of the common ferromagnets and ferrimagnets at high frequencies becomes very small. The frequency at which $\mu(\omega)$ drops significantly is usually much lower than that of $\varepsilon(\omega)$ and lies somewhere within the radio-frequency range. Another problem is that in the presence of the magnetic field $\mathbf{H}_0$ the temporal dispersion of magnetic susceptibility tensor may involve a substantial increase of the imaginary antisymmetric components like those responsible for Faraday rotation. This fact may significantly complicate the entire electromagnetic band structure.

4.2. TWO-DIMENSIONAL TETRAGONAL PHOTONIC CRYSTAL

Let us consider a $2D$ photonic crystal with tetragonal symmetry as shown in Fig. 1. As before we assume that the magnetic permeability μ is just the identity tensor $\mathbf{I}$, i.e. $\mu = \mathbf{I}$. The electric permittivity tensor $\varepsilon = \varepsilon(\mathbf{r})$ is assumed to be real and position-dependent. It takes on two different values ε_1 and ε_2, since there are just two constitutive components. In the absence of the external field $\mathbf{E}_0$, both tensors ε_1 and ε_2 are assumed to be isotropic

$$\text{for } \mathbf{E}_0 = 0: \quad \varepsilon_1 = \varepsilon\mathbf{I}, \ \varepsilon_2 = \mathbf{I}. \tag{49}$$

For simplicity, the second constitutive component of the tetragonal structure in Fig. 1 (a) is assumed to be the air, therefore $\varepsilon_2 = \mathbf{I}$ regardless of the external field. Space symmetry of this $2D$ periodic structure belongs to the tetragonal point group $4/mmm$, [1].

The uniform electric field $\mathbf{E}_0$ applied along the x_3-direction affects the tensor $\varepsilon_1 = \varepsilon_1(\mathbf{E_0})$ as follows

$$\varepsilon_1 = \begin{bmatrix} \varepsilon_\perp & 0 & 0 \\ 0 & \varepsilon_\perp & 0 \\ 0 & 0 & \varepsilon_\| \end{bmatrix}; \quad \varepsilon_2 = \mathbf{I}, \tag{50}$$

where

$$\varepsilon_\perp = \varepsilon_\perp(\mathbf{E_0}),\ \varepsilon_\| = \varepsilon_\|(\mathbf{E_0}); \quad \mathbf{E_0} \| \mathbf{e_3}. \tag{51}$$

Since in out problems the medium depends on the external field $\mathbf{E_0}$ parametrically, there is no need to present the explicit dependence of $\varepsilon_\perp$ and $\varepsilon_\|$ on $\mathbf{E_0}$. Instead, in further considerations we deal only with the dependence of different spectral characteristic on the quantities $\varepsilon_\perp$ and $\varepsilon_\|$ (see Fig. 6).

We will call the axis x_3 the principal axis of the photonic crystal. The imposed external electric field $\mathbf{E_0}$ aligned along the principal axis x_3 of the photonic crystal, i.e.

$$\mathbf{E_0} = \begin{bmatrix} 0 \\ 0 \\ h \end{bmatrix} \tag{52}$$

alters the tensor $\varepsilon = \varepsilon\left(\mathbf{x}; \mathbf{h}\right) : \varepsilon_\perp = \varepsilon_\perp\left(\mathbf{x}; h\right)$ and $\varepsilon_\| = \varepsilon_\|\left(\mathbf{x}; h\right)$.

As before, we consider only EM waves propagating perpendicularly to the principal axis x_3. Proceeding as in the case of simple $2D$ photonic crystals and using the standard symmetry arguments we find that the vector spectral problem can be reduced to the analysis of two kind of modes: (i) E-polarized fields (or TM modes) when $H_3 = 0$ and $E_1 = E_2 = 0$ and (ii) H-polarized fields (or TE modes) when $E_3 = 0$ and $H_1 = H_2 = 0$. In addition to that, as in the case of simple $2D$ photonic crystals we get the following two scalar eigenvalue problems

$$-\Delta E_3\left(\mathbf{x}\right) = \left(\omega/c\right)^2 \varepsilon_\|\left(\mathbf{x}\right) E_3\left(\mathbf{x}\right), \tag{53}$$

$$-\nabla \varepsilon_\perp^{-1}\left(\mathbf{x}\right) \nabla H_3\left(\mathbf{x}\right) = \left(\omega/c\right)^2 H_3\left(\mathbf{x}\right). \tag{54}$$

Since $\varepsilon_\perp\left(\mathbf{x}\right)$ and $\varepsilon_\|\left(\mathbf{x}\right)$ are ℓ-periodic functions, we seek the eigenfunctions of the spectral problems (53) and (54) in the Bloch form (38) and then end up with the eigenvalue problems

$$-\left[\partial_{1,k_1}^2 + \partial_{2,k_2}^2\right] E_{\mathbf{k}}\left(\mathbf{x}\right) = \left(\omega/c\right)^2 \varepsilon_\|\left(\mathbf{x}\right) E_{\mathbf{k}}\left(\mathbf{x}\right), \tag{55}$$

$$-\left[\partial_{1,k_1} \varepsilon_\perp^{-1}\left(\mathbf{x}\right) \partial_{1,k_1} + \partial_{2,k_2} \varepsilon_\perp^{-1}\left(\mathbf{x}\right) \partial_{2,k_2}\right] H_{\mathbf{k}}\left(\mathbf{x}\right) = \left(\omega/c\right)^2 H_{\mathbf{k}}\left(\mathbf{x}\right), \tag{56}$$

where $\mathbf{x}$ is in the primitive cell of the two-dimensional lattice $\mathbb{L}$ and $\mathbf{k}$ runs the primitive cell of the lattice $\mathbb{L}'$ dual to $\mathbb{L}$.

The mathematical properties of the eigenvalue problems (55) and (56) for square periodic geometries were thoroughly analyzed in [12] and [13] by analytic methods and then in [8] numerically.

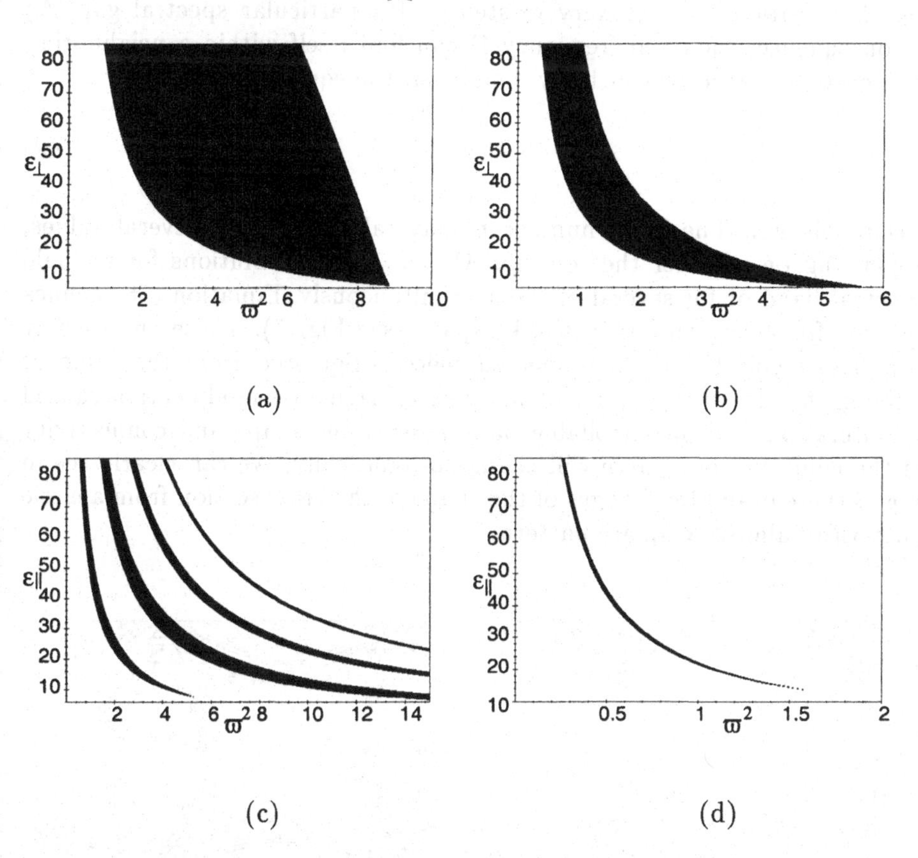

Figure 6. Dependence of photonic gaps (shaded areas) on the dielectric constant: (a) for H-mode, $\hat{\alpha} = 0.1$; (b) $\hat{\alpha} = 0.3$; (c) for E -mode $\hat{\alpha} = 0.1$; (d) $\hat{\alpha} = 0.3$. $\varpi^2 = (\omega\ell/c)^2\varepsilon_\perp^{(0)}$, where $\varepsilon_\perp^{(0)}$ the value of the dielectric constant for vanishing controlling field.

4.3. GROUP VELOCITY ANOMALIES

Most of the qualitative results concerning different aspects of tunability are equally applicable to E- and H-modes. Let us pick the frequency Ω in such a way that in the absence of the external field it falls in a gap of the electromagnetic spectrum $\omega_n(\mathbf{k})$. This implies that equation $\omega_n(\mathbf{k}) = \Omega$ has no solutions for any $\mathbf{k}$ lying in the x_1x_2-plane, i.e.,

$$\Omega \neq \omega_n(\mathbf{k}) \quad \text{for any } n \text{ and } \mathbf{k}. \tag{57}$$

The external field can alter the entire electromagnetic spectrum including the location and the very existence of a particular spectral gap. As a consequence, the fixed frequency Ω can find itself within a neighboring *transmittance band* in which, by definition, the equation

$$\omega_n(\mathbf{k}) = \Omega \tag{58}$$

has a solution. The band number n may take on one or several values, depending on whether the equation $\Omega = \omega_n(\mathbf{k})$ has solutions for a single spectral band or for several of them simultaneously. Equation (58) defines the equifrequency curves in the $\mathbf{k}$- space (see Fig. 7). These curves may comprise connected or disconnected pieces originated from the same or different bands. In the course of the spectral structure modification caused by alteration of the controllable parameter ε, the shape and connectivity of the equifrequency curve will change dramatically. We can clearly see in Fig. 7 these dramatic changes of the shape of the cross section from a circle pattern to almost a square pattern.

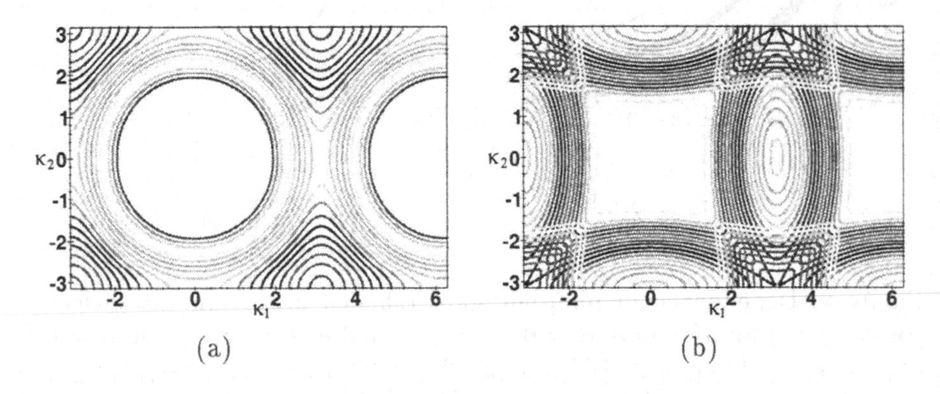

Figure 7. The equifrequency curves of equation (58) for E-polarized modes: (a) $\varepsilon = 5 \ldots 22$; (b) $\varepsilon = 28 \ldots 45$.

To translate these changes into EM wave propagation we consider the group velocity

$$\mathbf{v}_n(\mathbf{k}) = \nabla_{\mathbf{k}} \omega_n(\mathbf{k}) \tag{59}$$

characterizing the speed and direction of the wave propagation.

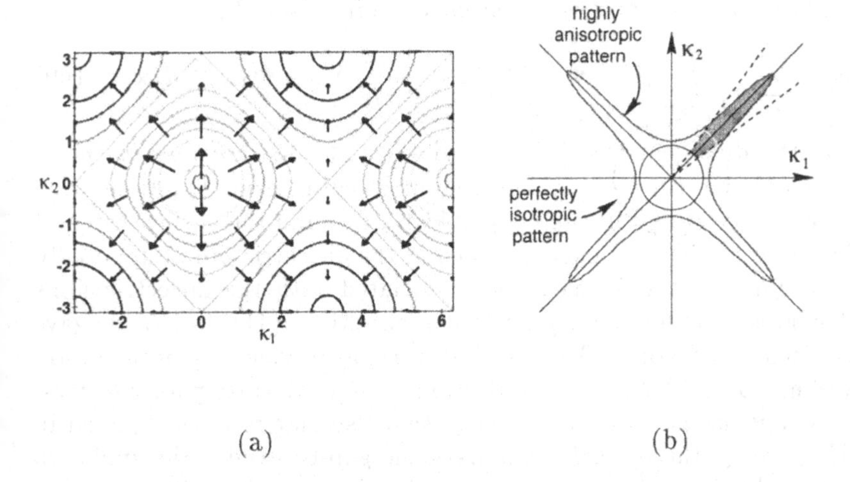

Figure 8. (a) The equifrequency curves of equation (58) together with the group velocity displayed as arrows. (b) The qualitative pircture of the mode distribution over the directions for the wave propagation indicated by the group velocity.

The group velocity vector at any point **k** is perpendicular to the corresponding equifrequency curve (see Fig. 8 (a)). Observe that *the circle pattern of an equifrequency curve indicates* that at the given frequency the distribution of all the modes over the directions is even and, hence, *the propagation of the EM waves at this frequency is isotropic.* In contrast, in the case of *the square pattern there are just two special directions, perpendicular the sides of the square, and nearly all the eigenmodes can propagate only is the two directions (see Fig. 8 (b)). This is a clear manifistation of the extreme anisotropy of EM wave propagation at the given frequency.*

Since the analysis of the tunability effects was based on the precise band-structure calculations, the predicted anomalies, we believe, can be unambiguously identified experimentally. Most of the qualitative aspects of the tunability effects, we believe, can be extended to the case of $3D$ photonic crystals.

5. Asymptotic models

As we discussed in Section 2, a photonic crystal formed by a thin dielectric film Γ which partitions the space into polyhedral air domains, for instance cubes, apparently is the best for appreciable spectral gaps. In $3D$, *the asymptotic bilinear form of the electric energy* (8)-(10)

$$\mathcal{E}[\mathbf{\Phi}] = \sum_m \frac{1}{\varepsilon_a} \int_{Q_m} |\nabla \times \mathbf{\Phi}(\mathbf{x})|^2 \, d\mathbf{x} + \frac{1}{\varepsilon_\sigma} \int_\Gamma |\nabla_\sigma \times \mathbf{\Phi}(\mathbf{x})|^2 \, d_\sigma \mathbf{x} \qquad (60)$$

defined on the divergent free fields satisfying (13)-(15) *becomes the main subject of mathematical studies.* The volume integral terms $\mathcal{E}_a[\mathbf{\Phi}]$ describe the electric energy stored in the air polyhedra Q_m and the surface integral $\mathcal{E}_d[\mathbf{\Phi}]$ describes the electric energy stored in the dielectric film. Supposedly $\mathcal{E}[\mathbf{\Phi}]$ is the limit of the forms $\mathcal{E}_\alpha[\mathbf{\Phi}]$ associated with the dielectric films of the thickness α under the asymptotic assumptions (41). In [15] we gave some arguments in favor of that fact but did not provide a rigorous proof.

According to our discussions in Section 2, high-contrast photonic crystals possess *optical and acoustic spectral bands* similar to ones depicted in Fig. 3. If $\varepsilon_\sigma \gg \varepsilon_a \ell$ the *optical bands* have the gap-to-band ratio similar to one in (44) and (45) for $2D$ photonic crystals. *The situation with the acoustic bands is substantially more complicated.* In particular, (i) the eigenmodes of acoustic bands are confined to a neighborhood of the dielectric surface Γ by the total refection phenomena; (ii) both the bands and the spectral gaps, if any, are of the order $(\varepsilon_d \alpha \ell)^{-1}$. The existence of spectral gaps for acoustic bands is a subject of future studies.

Remark. *In the case of scalar waves, the acoustic bands associated with spectral problem (53) in any dimension have been studied in [14] and [27].*

We can conclude from previous discussions that if the acoustic bands are somehow eliminated then only the optical bands with nice gaps remain (see Fig. 3). A way to eliminate the acoustic band waves is to impose instead of zero divergence conditions (13)-(15) the following more restrictive condition on the fields on the surface Γ

$$\nabla \cdot \mathbf{\Phi}(\mathbf{x}) = 0 \text{ for } \mathbf{x} \text{ in } Q_m, \qquad (61)$$

$$\mathbf{n}_+(\mathbf{x}) \cdot \mathbf{\Phi}_+(\mathbf{x}) = \mathbf{n}_-(\mathbf{x}) \cdot \mathbf{\Phi}_-(\mathbf{x}) = 0 \text{ for } \mathbf{x} \text{ in } \Gamma, \qquad (62)$$

which basically allows only magnetic fields of zero normal component on both sides of the surface Γ. *Note that conditions (61) and (62) are clearly in full compliance with zero divergence conditions (13)-(15) and, hence, with Maxwell equations.* The question of how practically enforce these conditions is an issue of another study. Our guess on this is that to eliminate the normal component of the magnetic field one can imbed in already thin lossless dielectric sheet a thinner sheet of perfectly conducting metal. The thickness of the conducting sheet should be chosen appropriately depending on the thickness of the dielectric sheet.

Thus, from now on we will consider the dielectric medium with the electric energy defined by (60) subject to the interface conditions (61), (62). In view of the previous discussions we expect the medium to have the spectrum similar to one for the optical bands in Fig. 3.

Let us start our spectral analysis from the simplest possible case of the extremely high-contrast when $\varepsilon_\sigma = \infty$ and, hence, $\mathcal{E}_d[\mathbf{\Phi}] = 0$. In this case we have $\mathcal{E}[\mathbf{\Phi}] = \mathcal{E}_a[\mathbf{\Phi}]$. The bilinear form $\mathcal{E}_a[\mathbf{\Phi}]$ describes the array of identical and completely decoupled oscillators with the energy form

$$\mathcal{E}_{Q_0}[\mathbf{\Phi}] = \frac{1}{\varepsilon_a}\widehat{\mathcal{E}}_{Q_0}[\mathbf{\Phi}], \; \widehat{\mathcal{E}}_{Q_0}[\mathbf{\Phi}] = \int_{Q_0} |\nabla \times \mathbf{\Phi}(\mathbf{x})|^2 \, d\mathbf{x} \tag{63}$$

subject to the boundary conditions

$$\begin{aligned} \nabla \cdot \mathbf{\Phi}(\mathbf{x}) &= 0 \text{ for } \mathbf{x} \text{ in } Q_0, & (64)\\ \mathbf{n}_+(\mathbf{x}) \cdot \mathbf{\Phi}_+(\mathbf{x}) &= \mathbf{n}_-(\mathbf{x}) \cdot \mathbf{\Phi}_-(\mathbf{x}) = 0 \text{ for } \mathbf{x} \text{ in } \partial Q_0, & (65) \end{aligned}$$

where ∂Q_0 is the boundary of Q_0. Note that $\widehat{\mathcal{E}}_{Q_0}[\mathbf{\Phi}]$ corresponds, in fact, to the cavity Q_0 with perfectly conducting boundary. The form $\widehat{\mathcal{E}}_{Q_0}[\mathbf{\Phi}]$, subject to the boundary conditions (64) and (65), has a discrete spectrum $\{\lambda_s\}$ of the eigenvalues (see [2], [3], [5], [39])) such that

$$\begin{aligned} 0 &< \lambda_1 \le \lambda_2 \le \ldots, \text{ where } \lambda_s \to \infty \text{ as } s \to \infty, & (66)\\ \lambda_s &= \lambda_s(Q_0). \end{aligned}$$

The normalized eigenmodes of $\widehat{\mathcal{E}}_{Q_0}[\mathbf{\Phi}]$ corresponding to the eigenvalues λ_s are denoted by $\mathbf{\Xi}_s$. In the extreme case of $\varepsilon_\sigma = \infty$, the spectrum of the form $\mathcal{E}[\mathbf{\Phi}] = \mathcal{E}_a[\mathbf{\Phi}]$ is a discrete set $\{\lambda_s/\varepsilon_a\}$ of the eigenvalues λ_s/ε_a of infinite multiplicity since every of identical cavities Q_m will produce the same spectrum $\{\lambda_s/\varepsilon_a\}$. Observe that in the case when Q_0 is a cube of the side ℓ the cavity eigenvalues and eigenmodes can be explicitly found, [17], and, in particular,

$$\{\lambda_s\} = \{\pi^2 \mathbf{m}^2/\ell^2, \; \mathbf{m} \text{ in } \mathbb{Z}^3\} \tag{67}$$

where $\mathbb{Z}^3$ is the set of three-dimensional vectors with integer-valued coordinates.

Let us turn now to more interesting case of a large but finite ε_σ. First, after singling out the geometric factor in $\mathcal{E}_d[\mathbf{\Phi}]$

$$\mathcal{E}_d[\mathbf{\Phi}] = \frac{1}{\varepsilon_\sigma}\widehat{\mathcal{E}}_d[\mathbf{\Phi}], \; \widehat{\mathcal{E}}_d[\mathbf{\Phi}] = \int_\Gamma |\nabla_\sigma \times \mathbf{\Phi}(\mathbf{x})|^2 \, d_\sigma \mathbf{x} \tag{68}$$

we treat the surface term $\mathcal{E}_d[\mathbf{\Phi}]$ as a small, of order $1/\varepsilon_\sigma$, analytic perturbation of the term $\mathcal{E}_a[\mathbf{\Phi}]$. The analytic perturbation theory is a well developed tool (see for instance [35]), and the only question is of its applicability here since both forms $\mathcal{E}_a[\mathbf{\Phi}]$ and $\mathcal{E}_d[\mathbf{\Phi}]$ are unbounded. The standard analytic perturbation theory requires the perturbation term $\mathcal{E}_d[\mathbf{\Phi}]$ to be subordinated to the form $\mathcal{E}_a[\mathbf{\Phi}]$ in the following sense: for every positive constant ξ there should exist another positive constant $C(\xi)$ such that for any $\mathbf{\Phi}$ the following inequality holds

$$\widehat{\mathcal{E}}_d[\mathbf{\Phi}] \leq \xi \mathcal{E}_a[\mathbf{\Phi}] + C(\xi)\|\mathbf{\Phi}\|^2, \; \|\mathbf{\Phi}\|^2 = \int_{\mathbb{R}^3} |\mathbf{\Phi}(\mathbf{x})|^2 \, dx. \tag{69}$$

Note that for $\xi \to 0$ we should have $C(\xi) \to \infty$. *The physical significance of the inequality (69) lies in the fact that it imposes constraints on allowed singularities for the potential $\mathbf{\Phi}$ and consequently electromagnetic field near the dielectric surface* Γ. In our situation the validity of the inequality follows from the following two inequalities. The first one holds for any number $0 < \zeta < 1$, [16],

$$\int_{\partial Q_0} |u(\mathbf{x})|^2 \, d\mathbf{x} \leq C_1 \left[\zeta \int_{Q_0} \sum_{j,i=1}^{3} |\partial_j u(\mathbf{x})|^2 \, d\mathbf{x} + \zeta^{-1} \int_{Q_0} |u(\mathbf{x})|^2 \, d\mathbf{x} \right], \tag{70}$$

where constant C_1 depends only on the polyhedron Q_0. The second inequality is a nontrivial estimate of the full derivative of the vector field $\mathbf{\Phi}(\mathbf{x})$ through its curl

$$\int_{Q_0} \sum_{j,i=1}^{3} |\partial_j \Phi_i(\mathbf{x})|^2 \, d\mathbf{x} \;\leq\; C_2 \int_{Q_0} |\nabla \times \mathbf{\Phi}(\mathbf{x})|^2 \, d\mathbf{x}, \tag{71}$$

$$\nabla \cdot \mathbf{\Phi}(\mathbf{x}) \;=\; 0, \; \mathbf{n}(\mathbf{x}) \cdot \mathbf{\Phi}(\mathbf{x}) = 0, \;\; \text{for } \mathbf{x} \text{ in } \partial Q_0, \tag{72}$$

where $\mathbf{n}(\mathbf{x})$ is the normal vector to the boundary ∂Q_0. *It holds for any convex polyhedron* Q_0, [2]. *Observe that the condition of convexity of the polyhedron Q_0 is essential and if dropped the inequality (71) can be violated*, [2]. Note also that we have used the zero normal component condition (65) and (72) without which we will not have (71) and (69). If (69) does not hold then the standard perturbation arguments can not be used. More detailed arguments on the perturbation theory and the inequalities (69)-(71) are given in [15].

In our case, since the inequality (69) holds, we can apply the standard perturbation theory to the sum $(1/\varepsilon_a)\widehat{\mathcal{E}}_a[\mathbf{\Phi}] + (1/\varepsilon_\sigma)\widehat{\mathcal{E}}_d[\mathbf{\Phi}]$ for small $1/\varepsilon_\sigma$. Before that, we use the fact of periodicity of our medium and apply Floquet-Bloch theory to simplify the spectral analysis, [35], [1]. Recall that if $\mathbf{b}_j$, $j = 1, \ldots, d$, is the basis dual to the basis $\mathbf{a}_j$ of the lattice of periods $\mathbb{L}$ then

$$\mathbf{b}_j \cdot \mathbf{a}_m = \delta_{jm}, \; j, m = 1, \ldots, d,$$

where δ_{jm} is the Kronecker symbol. Let BZ be the Brillouin zone (fundamental domain) of the lattice $\mathbb{L}'$ dual to $\mathbb{L}$, namely

$$BZ = \left\{ \mathbf{k} = (k_1, \ldots, k_d) : \mathbf{k} = \sum_{j=1}^{d} k_j \mathbf{b}_j,\ 0 \le k_j \le 2\pi,\ 1 \le j \le d \right\}. \quad (73)$$

Evidently, $\Gamma + \mathbb{L} = \Gamma$ and it is easy to see that for any point x on the boundary ∂Q_0 of the fundamental polyhedron Q_0 there exists a unique cite $\varsigma = \varsigma(x)$ belonging to the lattice $\mathbb{L} =$ such that $x + \varsigma(x)$ is in ∂Q_0. In other words, every point x on the boundary Γ_0 has a unique "$\mathbb{L}$-opposite" point in Γ_0. It is also can be easily verified that for the polyhedral domain Q_0 the quantity $\varsigma(x)$ is constant on every face of Q_0.

According to Floquet-Bloch theory, [35], [1], a periodic bilinear form $\mathcal{E}$ can be decomposed into the so-called *fibers* $\mathcal{E}(\mathbf{k})$, where quasimomentum $\mathbf{k}$ runs the *Brillouin zone* BZ. In our case $\mathcal{E}(\mathbf{k})$ takes the form, [15],

$$\begin{aligned} \mathcal{E}(\mathbf{k})[\mathbf{\Phi}] &= \frac{1}{\varepsilon_a}\mathcal{E}_{Q_0}[\mathbf{\Phi}] + \frac{1}{\varepsilon_\sigma}\widehat{\mathcal{E}}_{\partial Q_0}(\mathbf{k})[\mathbf{\Phi}], \quad (74)\\ \mathcal{E}_a[\mathbf{\Phi}] &= \int_{Q_0} |\nabla \times \mathbf{\Psi}(\mathbf{x})|^2\, d\mathbf{x},\\ \widehat{\mathcal{E}}_{\partial Q_0}(\mathbf{k})[\mathbf{\Phi}] &= \frac{1}{2}\int_{\partial Q_0} \left| \mathbf{n}(\mathbf{x}) \times \left[\mathbf{\Phi}(\mathbf{x}) - e^{i(\mathbf{k}\varsigma(\mathbf{x}))}\mathbf{\Phi}(\mathbf{x} + \varsigma(\mathbf{x})) \right] \right|^2 d_\sigma \mathbf{x},\\ \nabla \cdot \mathbf{\Phi}(\mathbf{x}) &= 0,\ \mathbf{x} \in Q_0,\ \mathbf{n}(\mathbf{x}) \cdot \mathbf{\Phi}(\mathbf{x}) = 0,\ \mathbf{x} \in \partial Q_0. \end{aligned}$$

If we denote now the eigenvalues of the form $\mathcal{E}(\mathbf{k})[\mathbf{\Phi}]$ by $\Lambda_s(\mathbf{k})$ then the analytic perturbation theory, [35], gives

$$\Lambda_s(\mathbf{k}) = \frac{\lambda_s}{\varepsilon_a} + \frac{1}{\varepsilon_\sigma}\widehat{\mathcal{E}}_{\partial Q_0}(\mathbf{k})[\Xi_s] + O\left(\frac{1}{\varepsilon_\sigma^2}\right),\quad s = 1, 2, \ldots, \quad (75)$$

where λ_s and Ξ_s are respectively the eigenvalues and normalized eigenmodes of the cavity form $\widehat{\mathcal{E}}_{Q_0}[\mathbf{\Phi}]$. Since the spectrum of the entire form $\mathcal{E}[\mathbf{\Phi}]$ is the union of the spectra of its fibers $\mathcal{E}(\mathbf{k})[\mathbf{\Phi}]$ as $\mathbf{k}$ runs the Brillouin zone BZ, the spectrum of $\mathcal{E}[\mathbf{\Phi}]$ consists of the following bands (intervals)

$$\left\{ \left[\frac{\lambda_s}{\varepsilon_a} - \frac{b_s^-}{\varepsilon_\sigma} + O\left(\frac{1}{\varepsilon_\sigma^2}\right), \frac{\lambda_s}{\varepsilon_a} - \frac{b_s^+}{\varepsilon_\sigma} + O\left(\frac{1}{\varepsilon_\sigma^2}\right) \right],\ s = 1, 2, \ldots \right\}, \quad (76)$$

where

$$b_s^+ = \max_{\mathbf{k} \text{ in } BZ} \widehat{\mathcal{E}}_{\partial Q_0}(\mathbf{k})[\Xi_s],\ b_s^- = \min_{\mathbf{k} \text{ in } BZ} \widehat{\mathcal{E}}_{\partial Q_0}(\mathbf{k})[\Xi_s]. \quad (77)$$

These bands are evidently similar in form to the $2D$ optical bands in (43) and the similarity becomes even more pronounced in the case of the cubic structure in Fig. 1 (b) when we have (67). The spectrum defined by (76) and (77) clearly exhibits spectral gaps for small $(1/\varepsilon_\sigma)$.

The geometry of the spectrum (76) is consistent with the simple physical interpretation of the high-contrast photonic crystal as an array of identical cavity resonators Q_m coupled by the thin dielectric film with the coupling constant $\varepsilon_\sigma = \varepsilon_d \alpha$. We remind that high-contrast photonic crystal is associated with the electric energy form defined by (60), subject to the interface conditions (61), (62).

Scalar case. Similar considerations give the following asymptotic bilinear form for the acoustic waves with inhomogeneous mass density and $2D$ photonic crystals (see the equation (54))

$$A[\psi] = \frac{1}{\varepsilon_a} \sum_m \int_{Q_m} |\nabla \psi(\mathbf{x})|^2 \, d\mathbf{x} + \frac{1}{\varepsilon_\sigma} \int_\Gamma |\psi_+(\mathbf{x}) - \psi_-(\mathbf{x})|^2 \, d_\sigma \mathbf{x}. \quad (78)$$

Acknowledgment and Disclaimer: Effort of A. Figotin is sponsored by the Air Force Office of Scientific Research, Air Force Materials Command, USAF, under grant number F49620-97-1-0019. The US Government is authorized to reproduce and distribute reprints for governmental purposes notwithstanding any copyright notation thereon. The views and conclusions contained herein are those of the author and should not be interpreted as necessarily representing the official policies or endorsements, either expressed or implied, of the Air Force Office of Scientific Research or the US Government.

I want to thank Dr. Yu. Godin, who coauthored some the cited results, for generating the most of the figures and many useful suggestions which have improved the manuscript.

References

1. Ashcroft N. and Mermin N. (1976) *Solid State Physics* (Harcourt Brace College Publishers, Philadelphia).
2. Birman M. (1992) Three Problems of the Theory of Continuous Media in Polyhedra, *Zapiski nauchnykh seminarov POMI*, vol. **200**, pp. 27-37.
3. Birman M. and Solomyak M. (1987) L_2-Theory of the Maxwell operator in arbitrary domains, *Russian Math. Surveys* **42**, no. 6: 75-96.
4. Blunter Ya. et al. (1994) *Physics Reports* **245**, 159.
5. Dautray R. and Lions J. L. (1990) *Mathematical Analysis and Numerical Methods for Science and Technology*, Springer-Verlag, New York.
6. (1993) Development and Applications of Materials Exhibiting Photonic Band Gaps, *Journal of the Optical Society of America* **B 10**, pp. 280-413.
7. Elson J. M. and Tran P. (1996) *Phys. Rev. Lett.* **63**, 259.
8. A. Figotin and Godin Yu. (1997) The Computation of Spectra of some 2D photonic crystals, *Journal of Computational Physics*, **136**, pp. 585-598.

9. Figotin A., Godin Yu. and Vitebsky I. (1998) Two-dimensional Tunable Photonic Crystals, *Phys. Rev.* **B 57**, pp. 2841-2848.
10. A. Figotin and Klein A. (1998) Localization of Light in Lossless Inhomogeneous Dielectrics, *J. Opt. Soc. Am. A*, **15**, pp. 1423-1435.
11. Figotin A. and Klein A. (1997) Localized Classical Waves Created by Defects, *J. Stat. Phys.* **86**, pp. 165-177.
12. Figotin A. and Kuchment P. (1996) Band-Gap Structure of the Spectrum of Periodic Dielectric and Acoustic Media. I. Scalar model, *SIAM J. Appl. Math.* **58**(1), pp. 68-88.
13. Figotin A. and Kuchment P. (1996) Band-Gap Structure of the Spectrum of Periodic Dielectric and Acoustic Media. II. 2D Photonic Crystals, *SIAM J. Appl. Math.* **56**, pp. 1561-1620.
14. Figotin A. and Kuchment P. (1998) Spectral properties of classical waves in high-contrast periodic periodic media, *SIAM J. Appl. Math.* **58**, No. 2, pp. 683-702.
15. Figotin A. and Kuchment P. (1998) Asymptotic models of high-contrast periodic dielectric and acoustic media, *preprint*.
16. Grisvard P. (1985) *Elliptic Problems in Nonsmooth Domains*, Pitman, New York.
17. Harrington R. F. (1961) *Time-Harmonic Electromagnetic Fields*, McGraw-Hill, New York.
18. Hui P. M. and Johnson N. F. (1995) *Photonic Band-Gap Materials*, in"Solid State Physics, vol. **49**", H. Ehrenreich and F. Spaepen, eds., Academic Press, pp. 151-203.
19. Ho K. M. , Chan C. T. and Soukoulis C. M. (1990) Existence of a Photonic Gap in Periodic Dielectric Structures, *Phys. Rev. Lett.* **65**, pp. 3152-3156).
20. J. D. Jackson (1975) *Classical Electrodynamics*, Wiley, New York.
21. John S. (1993) *The Localization of Light*, in "Photonic Band Gaps and Localization", NATO ASI Series B: Physical **308**.
22. Joannopolous J. D., Meade R. D., and Winn J. N., *Photonic Crystals. Molding the Flow of Light* (1995), Princeton University Press, Princeton, NJ.
23. Kafesaki M., Economou E. and Sigalas M. (1996) Elastic Waves in Periodic Media, in NATO ASI Series E, Vol. **315**, *Photonic Band Gap Materials*, edited by C. M. Soukoulis, Kluwer Academic Publishers, p. 143-164.
24. Klein A. Localization of Light in Randomized Periodic Media, in this volume.
25. Klein A. and Koines A. A General Framework for Localization of Classical Waves, in preparation.
26. Kuchment P. and Ponomaryov I. (1997) Separation of Variables in the Computation of Spectra in 2D Photonic Crystals, *preprint*.
27. Kuchment P. and Kunyansky L. (1997), Spectral Properties of High-Contrast Band-Gap Materials and Operators on Graphs, to appear in *Experimental Mathematics.*
28. Leung K. M. and Liu Y. F. (1990) Full Vector Wave Calculation of Photonic Band Structures in Face-Centered-Cubic Dielectric Media, *Phys. Rev. Lett.* **65**, pp. 2646-2650.
29. McCall S. L., Platzman P. M., Dalichaouch R., Smith D. and Schultz S. (1991) *Phys. Rev. Lett.* **67**, 2017.
30. Maradudin A. A., McGurn A. R. (1993) Photonic band gaps of a truncated, two-dimensional periodic dielectric media, *Journal of the Optical Society of America* **B**, **10**, pp. 307-313.
31. Meade R. D., Brommer K. D., A. Rapper A. M. and Joannopoulos J. D. (1992) Existence of Photonic Band Gap in Two Dimensions, *Appl. Phys. Lett.* **61**, pp. 495-497.
32. Pendry J. B. and Bell P. M. (1996) Transfer Matrix Techniques for Electromagnetic Waves, in *Photonic Band Gap Materials*, edited by C. M. Soukoulis, Kluwer, Dordrecht, [NATO ASI Series E, Vol. **318**], p. 203.
33. Prokhorov A. and Kuz'minov Yu. (1990) *Ferroelectric Crystals for Laser Radiation Control*, Adam Hilger.

34. Plihal M. and Maradudin A. A. (1991) Photonic band structure of two-dimensional systems: The triangular lattice, *Phys. Rev.*, 44, 8565.
35. Reed M., Simon B. (1978) *Methods of Modern Mathematical Physics*, Vol.IV, Academic Press.
36. Sheng P. (1990) *Scattering and Localization of Classical Waves*, World Scientific, Singapor.
37. Soukoulis C. M. *Photonic Band Gap Materials*, in this volume.
38. Sözüer H. S. , Haus J. W. and Inguva R. (1992) *Phys. Rev.* **B 45**, p. 13962.
39. Van Bladel J. (1995) *Singular Electromagnetic Fields and Sources*, IEEE Press, New York.
40. Villeneure P. R. and Piché M. (1991)Photonic band gaps of transverse-electric models in two-dimensionally periodic Media, *Journal of the Optical Society of America* **A**, **8**, pp. 1296-1305.
41. Wendt J. R., Vawter G. A. , Gourley P. L., Brennan T. M. and Hammons B. E. (1993) Nanofabrication of photonic lattice structures in GaAs/AlGaAs, *J. Vac. Sci. Technol.*, **B 11**(6), pp. 2637-2640.

PHOTON STATISTICS OF A RANDOM LASER

C.W.J. BEENAKKER
Instituut-Lorentz, Leiden University
P.O. Box 9506, 2300 RA Leiden, The Netherlands

Abstract. A general relationship is presented between the statistics of thermal radiation from a random medium and its scattering matrix S. Familiar results for black-body radiation are recovered in the limit $S \to 0$. The mean photocount $\bar{n}$ is proportional to the trace of $\mathbb{1} - S \cdot S^{\dagger}$, in accordance with Kirchhoff's law relating emissivity and absorptivity. Higher moments of the photocount distribution are related to traces of powers of $\mathbb{1} - S \cdot S^{\dagger}$, a generalization of Kirchhoff's law. The theory can be applied to a random amplifying medium (or "random laser") below the laser threshold, by evaluating the Bose-Einstein function at a negative temperature. Anomalously large fluctuations are predicted in the photocount upon approaching the laser threshold, as a consequence of overlapping cavity modes with a broad distribution of spectral widths.

1. Introduction

The name "random laser" made its appearance a few years ago [1], in connection with experiments on amplifying random media [2]. The concept goes back to Letokhov's 1967 proposal to use a mirrorless laser as an optical frequency standard [3]. Laser action requires gain and feedback. In any laser, gain results from stimulated emission of radiation by atoms in a non-equilibrium state. The random laser differs from a conventional laser in that the feedback is provided by multiple scattering from disorder rather than by confinement from mirrors. Because of the randomness, there is no geometry-dependent shift of the laser line with respect to the atomic transition frequency (hence the potential as a frequency standard). Stellar atmospheres may form a naturally occuring realization of a random laser [4].

J.-P. Fouque (ed.), Diffuse Waves in Complex Media, 137–164.

Possible applications as "paint-on lasers" [5] have sparked an intensive experimental and theoretical investigation of the interplay of multiple scattering and stimulated emission. The topic has been reviewed by Wiersma and Lagendijk (see Ref. [6] and these Proceedings). A particularly instructive experiment [7] was the demonstration of the narrowing of the coherent backscattering cone as a result of stimulated emission below the laser threshold. This experiment can be explained within the framework of *classical wave* optics. *Wave* optics, as opposed to ray optics, because coherent backscattering is an interference effect. *Classical* optics, as opposed to quantum optics, because stimulated emission can be described by a classical wave equation. (What is needed is a dielectric constant with a negative imaginary part.)

In a recent work [8] we went beyond classical optics by studying the photodetection statistics of amplified spontaneous emission from a random medium. Spontaneous emission, as opposed to stimulated emission, is a quantum optical phenomenon that can not be described by a classical wave equation. In this contribution we review our theory, with several extensions (notably in Secs. 3.4, 3.5, 4.5, and 4.6).

We start out in Sec. 2 with a discussion of the quantization of the electromagnetic field in absorbing or amplifying media. There exists a variety of approaches to this problem [9–17]. We will use the method of input–output relations developed by Gruner and Welsch [15, 16], and by Loudon and coworkers [12, 13, 14, 17]. The central formula of this section is a fluctuation-dissipation relation, that relates the commutator of the operators describing the quantum fluctuations in the electromagnetic field to the deviation $\mathbb{1} - S \cdot S^{\dagger}$ from unitarity of the scattering matrix S of the system. The relation holds both for absorbing and amplifying media. The absorbing medium is in thermal equilibrium at temperature T, and expectation values can be computed in terms of the Bose-Einstein function

$$f(\omega, T) = [\exp(\hbar\omega/k_{\rm B}T) - 1]^{-1}. \tag{1}$$

The amplifying medium is not in thermal equilibrium, but the expectation values can be obtained from those in the absorbing medium by evaluating the Bose-Einstein function at a negative temperature [12, 17].

In Sec. 3 we apply this general framework to a photodetection measurement. Our central result is a relationship between the probability distribution $P(n)$ to count n photons in a long time t (long compared to the coherence time of the radiation) and the eigenvalues $\sigma_1, \sigma_2, \ldots \sigma_N$ of the matrix product $S \cdot S^{\dagger}$. We call these eigenvalues "scattering strengths". They are between 0 and 1 for an absorbing medium and greater than 1 for an amplifying medium. The mean photocount $\bar{n}$ is proportional to the spectral average $N^{-1} \sum_n (1 - \sigma_n)$ of the scattering strengths. This spectral average is the absorptivity of the medium, being the fraction of the radiation incident in N modes that is absorbed. (An amplifying medium has a negative absorptivity.) The relation between mean photocount and absorptivity constitutes Kirchhoff's law of thermal radiation. We generalize Kirchhoff's law to higher moments of the counting distribution by relating the p-th factorial cumulant of n to $N^{-1}\sum_n(1-\sigma_n)^p$. While the absorptivity ($p = 1$) can be obtained from the radiative transfer equation, the spectral averages with $p > 1$ can not. Fortunately, random-matrix theory provides a set of powerful tools to compute such spectral averages [18].

We continue in Sec. 4 with the application of our formula for the photodetection distribution to specific random media. We focus on two types of geometries: An open-ended waveguide and a cavity containing a small opening. Randomness is introduced by disorder or (in the case of the cavity) by an irregular shape of the boundaries. Radiation is emitted into N propagating modes, which we assume to be a large number. (In the case of the cavity, N is the number of transverse modes in the opening.) It is unusual, but essential, that all the emitted radiation is incident on the photodetector. We show that if only a single mode is detected, the counting distribution contains solely information on the absorptivity, while all information on higher spectral moments of the scattering strengths is lost. To characterize the fluctuations in the photocount we compute the variance $\mathrm{Var}\, n = \overline{n^2} - \bar{n}^2$. The variance can be directly measured from the auto-correlator of the photocurrent $I(t) = \bar{I} + \delta I(t)$, according to

$$\int_{-\infty}^{\infty} dt\, \overline{\delta I(0)\delta I(t)} = \lim_{t\to\infty} \frac{1}{t} \mathrm{Var}\, n. \tag{2}$$

The bar $\overline{\cdots}$ indicates an average over many measurements on the same sample. The mean photocount $\bar{n}$ (and hence the mean current $\bar{I} = \bar{n}/t$) contains information on the absorptivity. The new information contained in the variance of n (or the auto-correlator of I) is the effective number of degrees of freedom ν_{eff}, defined by [19] $\mathrm{Var}\, n = \bar{n}(1+\bar{n}/\nu_{\mathrm{eff}})$. For black-body radiation in a narrow frequency interval $\delta\omega$, one has $\nu_{\mathrm{eff}} = Nt\delta\omega/2\pi \equiv \nu$. The counting distribution is then a negative-binomial distribution with ν degrees of freedom,

$$P(n) \propto \binom{n+\nu-1}{n} \exp(-n\hbar\omega/k_{\mathrm{B}}T). \tag{3}$$

The quantity ν_{eff} generalizes the notion of degrees of freedom to radiation from systems that are not black bodies.

A black body has scattering matrix $S = 0$. (Any incident radiation is fully absorbed.) In other words, the scattering strengths σ_n of a black

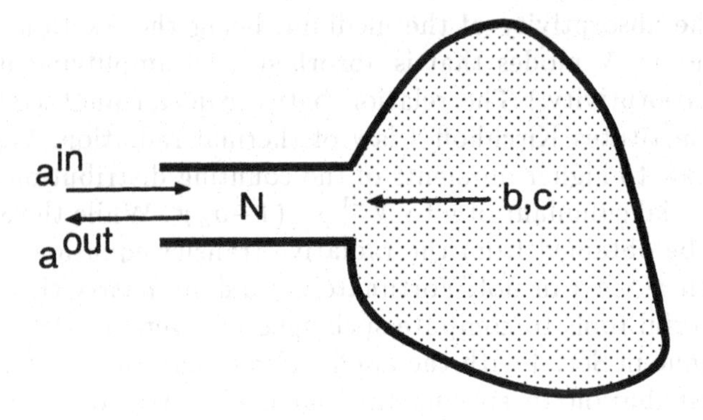

Figure 1. Scattering geometry consisting of a random medium (dotted) coupled to free space via an N-mode waveguide. The N-component vector of outgoing-mode operators a^{out} is linearly related to the incoming-mode operators a^{in} and the spontaneous-emission operators b, c.

body are all equal to zero. A random medium, in contrast, has in general a broad (typically bimodal) density of scattering strengths. We show that this results in a substantial reduction of ν_{eff} below ν. In other words, the noise in the photocount is anomalously large in a random medium. The reduction in ν_{eff} holds both for absorbing and amplifying media. The only requirement is a broad distribution of scattering strengths. For the random laser, we predict that the ratio ν_{eff}/ν vanishes on approaching the laser threshold. No such reduction is expected in a conventional laser. We discuss the origin of this difference in the concluding Sec. 5, together with a discussion of the relationship between ν_{eff} and the Thouless number of mesoscopic physics.

2. Quantization of the electromagnetic field

2.1. INPUT–OUTPUT RELATIONS

We consider a dielectric medium coupled to free space via a waveguide with $N(\omega)$ propagating modes (counting polarizations) at frequency ω (see Fig. 1). The incoming and outgoing modes in the waveguide are represented by two N-component vectors of annihilation operators $a^{\text{in}}(\omega)$, $a^{\text{out}}(\omega)$. They satisfy the canonical commutation relations

$$[a_n^{\text{in}}(\omega), a_m^{\text{in}\dagger}(\omega')] = \delta_{nm}\delta(\omega-\omega'), \quad [a_n^{\text{in}}(\omega), a_m^{\text{in}}(\omega')] = 0, \tag{4}$$
$$[a_n^{\text{out}}(\omega), a_m^{\text{out}\dagger}(\omega')] = \delta_{nm}\delta(\omega-\omega'), \quad [a_n^{\text{out}}(\omega), a_m^{\text{out}}(\omega')] = 0. \tag{5}$$

The input–output relations take the form

$$a^{\text{out}} = S\cdot a^{\text{in}} + U\cdot b + V\cdot c^{\dagger}. \tag{6}$$

The two sets of operators $b, b^\dagger$ and $c, c^\dagger$ commute with each other and with the set of input operators $a^{\text{in}}, a^{\text{in}\dagger}$. They satisfy the canonical commutation relations

$$[b_n(\omega), b_m^\dagger(\omega')] = \delta_{nm}\delta(\omega - \omega'), \quad [b_n(\omega), b_m(\omega')] = 0, \tag{7}$$
$$[c_n(\omega), c_m^\dagger(\omega')] = \delta_{nm}\delta(\omega - \omega'), \quad [c_n(\omega), c_m(\omega')] = 0, \tag{8}$$

provided the $N \times N$ matrices $U(\omega)$ and $V(\omega)$ are related to the scattering matrix $S(\omega)$ by

$$U \cdot U^\dagger - V \cdot V^\dagger = \mathbb{1} - S \cdot S^\dagger \tag{9}$$

($\mathbb{1}$ denoting the $N \times N$ unit matrix). Equation (9) can be understood as a fluctuation-dissipation relation: The left-hand side accounts for quantum fluctuations in the electromagnetic field due to spontaneous emission or absorption of photons, the right-hand side accounts for dissipation due to absorption (or stimulated emission in the case of an amplifying medium). Equation (9) also represents a link between classical optics (the scattering matrix S) and quantum optics (the quantum fluctuation matrices U, V).

The matrix $\mathbb{1} - S \cdot S^\dagger$ is positive definite in an absorbing medium, so we can put $V = 0$ and write

$$a^{\text{out}} = S \cdot a^{\text{in}} + U \cdot b, \quad U \cdot U^\dagger = \mathbb{1} - S \cdot S^\dagger. \tag{10}$$

(These are the input–output relations of Ref. [16].) Conversely, in an amplifying medium $\mathbb{1} - S \cdot S^\dagger$ is negative definite, so we can put $U = 0$ and write

$$a^{\text{out}} = S \cdot a^{\text{in}} + V \cdot c^\dagger, \quad V \cdot V^\dagger = S \cdot S^\dagger - \mathbb{1}. \tag{11}$$

(The operator c represents the inverted oscillator of Ref. [12].) Both matrices U, V and operators b, c are needed if $\mathbb{1} - S \cdot S^\dagger$ is neither positive nor negative definite, which might happen if the medium contains both absorbing and amplifying regions. In what follows we will not consider that case any further.

2.2. EXPECTATION VALUES

We assume that the absorbing medium is in thermal equilibrium at temperature T. Thermal emission is described by the operator b with expectation value

$$\langle b_n^\dagger(\omega) b_m(\omega') \rangle = \delta_{nm}\delta(\omega - \omega') f(\omega, T), \tag{12}$$

where f is the Bose-Einstein function (1).

The inverted oscillator c accounts for spontaneous emission in an amplifying medium. We consider the regime of linear amplification, below the laser threshold. Formally, this regime can be described by a thermal distribution at an effective *negative* temperature $-T$ [12, 17]. For a two-level atomic system, with level spacing $\hbar\omega_0$ and an average occupation $N_{\rm upper} > N_{\rm lower}$ of the two levels, the effective temperature is given by $N_{\rm upper}/N_{\rm lower} = \exp(\hbar\omega_0/k_{\rm B}T)$. The zero-temperature limit corresponds to a complete population inversion. The expectation value is given by

$$\langle c_n(\omega)c_m^\dagger(\omega')\rangle = -\delta_{nm}\delta(\omega-\omega')f(\omega,-T), \tag{13}$$

or equivalently by

$$\langle c_n^\dagger(\omega)c_m(\omega')\rangle = \delta_{nm}\delta(\omega-\omega')f(\omega,T). \tag{14}$$

(We have used that $f(\omega,T)+f(\omega,-T)=-1$.)

Higher order expectation values are obtained by pairwise averaging, as one would do for Gaussian variables, after having brought the operators into normal order (all creation operators to the left of the annihilation operators). This procedure is an example of the "optical equivalence theorem" [19, 20]. To do the Gaussian averages it is convenient to discretize the frequency as $\omega_p = p\Delta$, $p = 1, 2, \ldots$, and send Δ to zero at the end. The expectation value of an arbitrary functional $\mathcal{F}$ of the operators $b, b^\dagger$ (or $c, c^\dagger$) can then be written as a multiple integral over an array of complex numbers z_{np} $[1 \le n \le N(\omega_p)]$,

$$\langle : \mathcal{F}[\{b_n(\omega_p)\},\{b_n^\dagger(\omega_p)\}] : \rangle = Z^{-1}\int dz\, \mathrm{e}^{-\Phi}\mathcal{F}[\{z_{np}\},\{z_{np}^*\}], \tag{15}$$

with the definitions

$$\Phi = \sum_{n,p} \frac{|z_{np}|^2\Delta}{f(\omega_p,T)}, \tag{16}$$

$$Z = \int dz\, \mathrm{e}^{-\Phi} = \prod_{n,p} \frac{\pi f(\omega_p,T)}{\Delta}. \tag{17}$$

The colons in Eq. (15) indicate normal ordering, and $\int dz$ indicates the integration over the real and imaginary parts of all the z_{np}'s.

3. Photodetection statistics

3.1. GENERAL FORMULAS

We consider the case that the incoming radiation is in the vacuum state, while the outgoing radiation is collected by a photodetector. We assume a mode and frequency independent detection efficiency of α photoelectrons per photon. The probability that n photons are counted in a time t is given by the Glauber-Kelley-Kleiner formula [21, 22]

$$P(n) = \frac{1}{n!}\langle : I^n \mathrm{e}^{-I} : \rangle, \tag{18}$$

$$I = \alpha \int_0^t dt' \, a^{\mathrm{out}\dagger}(t') \cdot a^{\mathrm{out}}(t'), \tag{19}$$

where we have defined the Fourier transform

$$a^{\mathrm{out}}(t) = (2\pi)^{-1/2} \int_0^\infty d\omega \, \mathrm{e}^{-\mathrm{i}\omega t} a^{\mathrm{out}}(\omega). \tag{20}$$

The factorial cumulants κ_p of $P(n)$ are the cumulants of the factorial moments $\overline{n(n-1)\cdots(n-p+1)}$. For example, $\kappa_1 = \bar{n}$ and $\kappa_2 = \overline{n(n-1)} - \bar{n}^2 = \mathrm{Var}\, n - \bar{n}$. The factorial cumulants have the generating function

$$F(\xi) = \sum_{p=1}^{\infty} \frac{\kappa_p \xi^p}{p!} = \ln\left(\sum_{n=0}^{\infty} (1+\xi)^n P(n)\right). \tag{21}$$

Once $F(\xi)$ is known, the distribution $P(n)$ can be recovered from

$$P(n) = \frac{1}{2\pi\mathrm{i}} \oint_{|z|=1} dz \, z^{-n-1} \mathrm{e}^{F(z-1)} = \lim_{\xi\to -1} \frac{1}{n!} \frac{d^n}{d\xi^n} \mathrm{e}^{F(\xi)}. \tag{22}$$

From Eq. (18) one finds the expression

$$\mathrm{e}^{F(\xi)} = \langle : \mathrm{e}^{\xi I} : \rangle. \tag{23}$$

To evaluate Eq. (23) for the case of an absorbing medium, we combine Eq. (10) with Eqs. (19) and (20), and then compute the expectation value with the help of Eq. (15),

$$\mathrm{e}^{F(\xi)} = Z^{-1} \int dz \, \exp\left(-\Delta \sum_{n,p} \sum_{n',p'} z^*_{np} M_{np,n'p'} z_{n'p'}\right), \tag{24}$$

$$M_{np,n'p'} = \frac{\delta_{nn'}\delta_{pp'}}{f(\omega_p, T)} - \frac{\xi\alpha\Delta}{2\pi} \int_0^t dt' \, \mathrm{e}^{\mathrm{i}(\omega_p - \omega_{p'})t'} \sum_m U^\dagger_{nm}(\omega_p) U_{mn'}(\omega_{p'}). \tag{25}$$

Evaluation of the Gaussian integrals results in the compact expression

$$F(\xi) = \mathrm{constant} - \ln \|M\|, \tag{26}$$

where $\|\cdots\|$ indicates the determinant. (The ξ-independent constant can be found from the normalization requirement that $F(0) = 0$.) The matrix

U is related to the scattering matrix S by $U \cdot U^\dagger = \mathbb{1} - S \cdot S^\dagger$ [Eq. (10)]. This relation determines U up to a transformation $U \to U \cdot A$, with $A(\omega)$ an arbitrary unitary matrix. Since the determinant $\|M\|$ is invariant under this transformation, we can say that knowledge of the scattering matrix suffices to determine the counting distribution.

The result for an amplifying medium is also given by Eqs. (25) and (26), with the replacement of U by V and $f(\omega_p, T)$ by $-f(\omega_p, -T)$ [in accordance with Eqs. (11) and (13)].

The determinant $\|M\|$ can be simplified in the limit of large and small counting times t. These two regimes will be discussed separately in the next two subsections. A simple expression valid for all t exists for the mean photocount,

$$\bar{n} = t \int_0^\infty \frac{d\omega}{2\pi} \alpha f \,\mathrm{Tr}\,(\mathbb{1} - S \cdot S^\dagger). \tag{27}$$

The quantity $N^{-1}\mathrm{Tr}\,(\mathbb{1} - S \cdot S^\dagger)$ is the absorptivity, defined as the fraction of the incident power that is absorbed at a certain frequency, averaged over all incoming modes. The relation (27) between thermal emission and absorption is *Kirchhoff's law.* It holds also for an amplifying medium, upon replacement of $f(\omega, T)$ by $f(\omega, -T)$.[1]

3.2. LONG-TIME REGIME

The long-time regime is reached when $\omega_c t \gg 1$, with ω_c the frequency interval within which $S \cdot S^\dagger$ does not vary appreciably. In this regime we may choose the discretization $\omega_p = p\Delta$, $\Delta = 2\pi/t$, satisfying

$$\int_0^t dt'\, \mathrm{e}^{\mathrm{i}(\omega_p - \omega_{p'})t'} = t\delta_{pp'}. \tag{28}$$

The matrix (25) then becomes diagonal in the indices p, p',

$$M_{np,n'p'} = \frac{\delta_{nn'}\delta_{pp'}}{f(\omega_p, T)} - \xi\alpha\delta_{pp'}\Big(U^\dagger(\omega_p)\cdot U(\omega_p)\Big)_{nn'}, \tag{29}$$

so that the generating function (26) takes the form

$$\begin{aligned} F(\xi) &= -t\int_0^\infty \frac{d\omega}{2\pi} \ln\|\mathbb{1} - (\mathbb{1} - S \cdot S^\dagger)\xi\alpha f\| \\ &= -t\int_0^\infty \frac{d\omega}{2\pi} \sum_{n=1}^{N(\omega)} \ln[1 - [1 - \sigma_n(\omega)]\xi\alpha f(\omega, T)]. \end{aligned} \tag{30}$$

[1] Eq. (26) for an amplifying system is obtained by the replacement of U by V and $f(\omega, T)$ by *minus* $f(\omega, -T)$. Since $V \cdot V^\dagger$ equals *minus* $\mathbb{1} - S \cdot S^\dagger$ [Eq. (11)], the two minus signs cancel and the net result for Eq. (27) is that we should replace $f(\omega, T)$ by *plus* $f(\omega, -T)$.

We have introduced the *scattering strengths* $\sigma_1, \sigma_2, \ldots \sigma_N$, being the eigenvalues of the scattering-matrix product $S \cdot S^\dagger$. The result (30) holds also for an amplifying system, if we replace $f(\omega, T)$ by $f(\omega, -T)$.

Expansion of the logarithm in powers of ξ yields the factorial cumulants [cf. Eq. (21)]

$$\kappa_p = (p-1)!\, t \int_0^\infty \frac{d\omega}{2\pi} (\alpha f)^p \sum_{n=1}^{N} (1-\sigma_n)^p. \tag{31}$$

The p-th factorial cumulant of $P(n)$ is proportional to the p-th spectral moment of the scattering strengths. It is a special property of the long-time regime that the counting distribution is determined entirely by the eigenvalues of $S \cdot S^\dagger$, independently of the eigenfunctions. Eq. (31) can be interpreted as a generalization of Kirchhoff's law (27) to higher moments of the counting distribution.

3.3. SHORT-TIME REGIME

The short-time regime is reached when $\Omega_c t \ll 1$, with Ω_c the frequency range over which $S \cdot S^\dagger$ differs appreciably from the unit matrix. (The reciprocal of Ω_c is the coherence time of the thermal emissions.) In this regime we may replace $\exp[i(\omega_p - \omega_{p'})t']$ in Eq. (25) by 1, so that M simplifies to

$$M_{np,n'p'} = \frac{\delta_{nn'}\delta_{pp'}}{f(\omega_p, T)} - \frac{t\xi\alpha\Delta}{2\pi} \sum_m U^\dagger_{nm}(\omega_p) U_{mn'}(\omega_{p'}). \tag{32}$$

If we suppress the mode indices n, n', Eq. (32) can be written as

$$M_{p,p'} = \frac{\delta_{pp'}}{f(\omega_p, T)} \mathbb{1} - \frac{t\xi\alpha\Delta}{2\pi} U^\dagger(\omega_p) \cdot U(\omega_{p'}). \tag{33}$$

The determinant $\|M_{pp'}\|$ can be evaluated with the help of the formula[2]

$$\|\delta_{pp'}\mathbb{1} + A_p \cdot B_{p'}\| = \|\mathbb{1} + \sum_q B_q \cdot A_q\| \tag{34}$$

(with $\{A_p\}$, $\{B_p\}$ two arbitrary sets of matrices). The resulting generating function is[3]

$$F(\xi) = -\ln\|\mathbb{1} - t \int_0^\infty \frac{d\omega}{2\pi} (\mathbb{1} - S \cdot S^\dagger)\xi\alpha f\|. \tag{35}$$

[2] To verify Eq. (34), take the logarithm of each side and use $\ln \|M\| = \mathrm{Tr} \ln M$. Then expand each logarithm in powers of A and equate term by term. I am indebted to J.M.J. van Leeuwen for helping me with this determinant.

[3] We adopt the convention that the matrix $\mathbb{1} - S \cdot S^\dagger$ is embedded in an infinite-dimensional matrix by adding zeroes. The matrix $\mathbb{1}$ outside the integral over ω in Eq. (35) is then interpreted as an infinite-dimensional unit matrix.

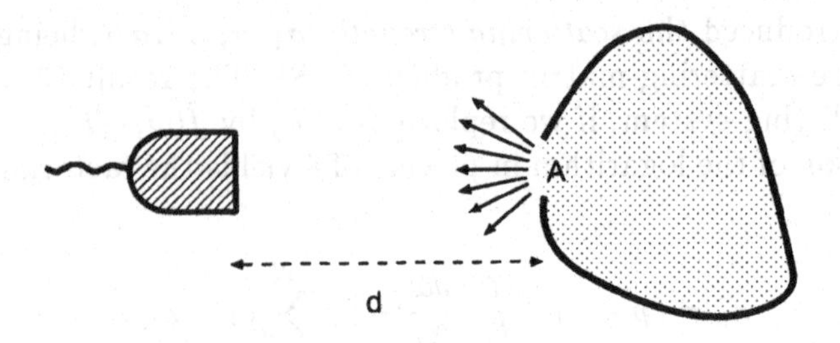

Figure 2. A cavity radiates through a hole with an area $\mathcal{A}$. The number N of radiating modes at wavelength λ is $2\pi\mathcal{A}/\lambda^2$ (counting polarizations). All N modes are detected if the photocathode covers the hole. Upon increasing the separation d between hole and photocathode, fewer and fewer modes are detected. Finally, single-mode detection is reached when the area of the photocathode becomes less than the coherence area $\simeq d^2/N$ of the radiation.

Again, to apply Eq. (35) to an amplifying system we need to replace $f(\omega, T)$ by $f(\omega, -T)$.

The short-time limit (35) is more complicated than the long-time limit (30), in the sense that the former depends on both the eigenvalues and eigenvectors of $S \cdot S^\dagger$. There is therefore no direct relation between factorial cumulants of $P(n)$ and spectral moments of scattering strengths in the short-time regime.

3.4. SINGLE-MODE DETECTION

We have assumed that each of the N radiating modes is detected with equal efficiency α. At the opposite extreme, we could assume that only a single mode is detected. This would apply if the photocathode had an area smaller than the coherence area of the emitted radiation (see Fig. 2). Single-mode detection is less informative than multi-mode detection, for the following reason.

Suppose that only a single mode is detected. The counting distribution $P(n)$ is still given by Eq. (18), but now I contains only a single element (say, number 1) of the vector of operators a^{out},

$$I = \alpha \int_0^t dt'\, a_1^{\text{out}\dagger}(t') a_1^{\text{out}}(t'). \tag{36}$$

This amounts to the replacement of the matrix U in Eq. (25) by the projection $\mathcal{P} \cdot U$, with $\mathcal{P} = \delta_{nm}\delta_{n1}$. Instead of Eq. (27) we have the mean photocount

$$\bar{n} = \int_0^\infty d\omega\, \frac{d\bar{n}}{d\omega}; \quad \frac{d\bar{n}}{d\omega} = \frac{t\alpha f}{2\pi}[1 - (S \cdot S^\dagger)_{11}]. \tag{37}$$

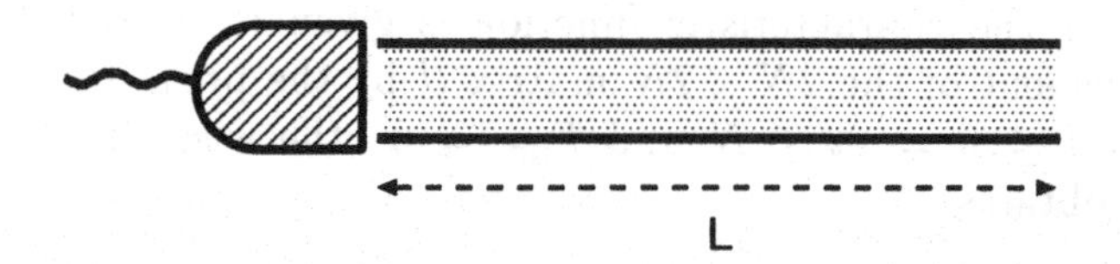

Figure 3. Disordered waveguide (length L) connected at one end to a photodetector.

The generating function now takes the form

$$F(\xi) = -t \int_0^\infty \frac{d\omega}{2\pi} \ln \left(1 - \frac{2\pi\xi}{t} \frac{d\bar{n}}{d\omega}\right) \tag{38}$$

in the long-time regime, and

$$F(\xi) = -\ln(1 - \xi\bar{n}) \tag{39}$$

in the short-time regime. We see that the entire counting distribution is determined by the mean photocount, hence by the absorptivity of the single detected mode. This strong version of Kirchhoff's law is due to Bekenstein and Schiffer [23]. It holds only for the case of single-mode detection. Multimode detection is determined not just by $\bar{n}$, being the first spectral moment of the scattering strengths, but also by higher moments.

3.5. WAVEGUIDE GEOMETRY

Figs. 1 and 2 show a cavity geometry. Alternatively, one can consider the waveguide geometry of Fig. 3. The waveguide has cross-section $\mathcal{A}$, corresponding to $N = 2\pi\mathcal{A}/\lambda^2$ modes at wavelength λ. The $2N \times 2N$ scattering matrix S consists of four $N \times N$ blocks,

$$S = \begin{pmatrix} r & t \\ t' & r' \end{pmatrix}, \tag{40}$$

namely two reflection matrices r, r' (reflection from the left and from the right) and two transmission matrices t, t' (transmission from right to left and from left to right). Reciprocity relates t and t' (they are each others transpose).

A photodetector detects the radiation emitted at one end of the waveguide, while the radiation emitted at the other end remains undetected. If the radiation from *both* ends would be detected (by two photodetectors), then the eigenvalues of $S \cdot S^\dagger$ would determine the counting distribution in the long-time limit, as in the cavity geometry. But for detection at the left end only, one needs instead the eigenvalues of the matrix $r \cdot r^\dagger + t \cdot t^\dagger$ (or $r' \cdot r'^\dagger + t' \cdot t'^\dagger$ for detection at the right end). More precisely, the general expression for the characteristic function is given by Eqs. (24) and (25) upon replacement of the $2N \times 2N$ matrix U by the projection $\mathcal{P} \cdot U$, with $\mathcal{P}_{nm} = 1$ if $1 \le n = m \le N$ and $\mathcal{P}_{nm} = 0$ otherwise. In the long-time regime one obtains

$$F(\xi) = -t \int_0^\infty \frac{d\omega}{2\pi} \ln \|\mathbb{1} - (\mathbb{1} - r \cdot r^\dagger - t \cdot t^\dagger)\xi\alpha f\|, \tag{41}$$

and in the short-time regime

$$F(\xi) = -\ln \|\mathbb{1} - t \int_0^\infty \frac{d\omega}{2\pi} (\mathbb{1} - r \cdot r^\dagger - t \cdot t^\dagger)\xi\alpha f\|. \tag{42}$$

The eigenvalues of $r \cdot r^\dagger + t \cdot t^\dagger$ differ from the sum $R_n + T_n$ of the reflection and transmission eigenvalues (eigenvalues of $r \cdot r^\dagger$ and $t \cdot t^\dagger$, respectively), because the two matrices $r \cdot r^\dagger$ and $t \cdot t^\dagger$ do not commute. The absorptivity

$$N^{-1} \mathrm{Tr}\,(\mathbb{1} - r \cdot r^\dagger - t \cdot t^\dagger) = N^{-1} \sum_{n=1}^{N} (1 - R_n - T_n)$$

does depend only on the reflection and transmission eigenvalues. It determines the mean photocount

$$\bar{n} = t \int_0^\infty \frac{d\omega}{2\pi} \alpha f \, \mathrm{Tr}\,(\mathbb{1} - r \cdot r^\dagger - t \cdot t^\dagger), \tag{43}$$

in accordance with Kirchhoff's law. Higher moments of the counting distribution can not be obtained from the reflection and transmission eigenvalues, but require knowledge of the eigenvalues of $r \cdot r^\dagger + t \cdot t^\dagger$. A substantial simplification occurs, in the case of absorption, if the waveguide is sufficiently long that there is no transmission through it. Then $t \cdot t^\dagger$ can be neglected and the counting distribution depends entirely on the reflection eigenvalues.

4. Applications

4.1. BLACK-BODY RADIATION

Let us first check that we recover the familiar results for black-body radiation [19, 20]. The simplest "step-function model" of a black body has

$$S(\omega) = \begin{cases} 0 & \text{for } |\omega - \omega_0| < \frac{1}{2}\Omega_c, \\ \mathbb{1} & \text{for } |\omega - \omega_0| > \frac{1}{2}\Omega_c. \end{cases} \tag{44}$$

Incident radiation is fully absorbed within the frequency interval Ω_c around ω_0 and fully reflected outside this interval. Typically, $\Omega_c \ll \omega_0$ so that we may neglect the frequency dependence of $N(\omega)$ and $f(\omega,T)$, replacing these quantities by their values at $\omega=\omega_0$.

The generating function (30) in the long-time regime then becomes

$$F(\xi)=-\frac{Nt\Omega_{\rm c}}{2\pi}\ln(1-\xi\alpha f). \tag{45}$$

The inversion formula (22) yields the counting distribution

$$P(n)=\frac{\Gamma(n+\nu)}{n!\Gamma(\nu)}\frac{(\bar{n}/\nu)^n}{(1+\bar{n}/\nu)^{n+\nu}}. \tag{46}$$

This is the negative-binomial distribution with $\nu=Nt\Omega_{\rm c}/2\pi$ degrees of freedom. [For integer ν, the ratio of Gamma functions forms the binomial coefficient $\binom{n+\nu-1}{n}$ that counts the number of partitions of n bosons among ν states, cf. Eq. (3).] Note that $\nu\gg 1$ in the long-time regime. The mean photocount is $\bar{n}=\nu\alpha f$. In the limit $\bar{n}/\nu\to 0$, the negative-binomial distribution tends to the Poisson distribution

$$P(n)=\frac{1}{n!}\bar{n}^n{\rm e}^{-\bar{n}} \tag{47}$$

of independent photocounts. The negative-binomial distribution describes photocounts that occur in "bunches". Its variance

$$\operatorname{Var} n=\bar{n}(1+\bar{n}/\nu) \tag{48}$$

is larger than the Poisson value by a factor $1+\bar{n}/\nu$.

Similarly, the short-time limit (35) becomes

$$F(\xi)=-N\ln\left(1-\frac{t\Omega_{\rm c}}{2\pi}\xi\alpha f\right), \tag{49}$$

corresponding to a negative-binomial distribution with N degrees of freedom.

In the step-function model (44) S changes abruptly from 0 to $\mathbb{1}$ at $|\omega-\omega_0|=\frac{1}{2}\Omega_{\rm c}$. A more realistic model would have a gradual transition. A Lorentzian frequency profile is commonly used in the literature [24], for the case of single-mode detection. Substitution of $1-(S\cdot S^\dagger)_{11}=[1+4(\omega-\omega_0)^2/\Omega_{\rm c}^2]^{-1}$ into Eq. (37) and integration over ω in Eq. (38) (neglecting the frequency dependence of f) yields the generating function in the long-time regime,

$$F(\xi)=\tfrac{1}{2}t\Omega_{\rm c}\left(1-\sqrt{1-\xi\alpha f}\right). \tag{50}$$

The corresponding counting distribution is

$$P(n)=\frac{C}{n!}\left(\frac{\bar{n}}{\sqrt{1+\alpha f}}\right)^n K_{n-1/2}\left(\tfrac{1}{2}t\Omega_{\rm c}\sqrt{1+\alpha f}\right), \tag{51}$$

with $\bar{n} = \frac{1}{4}t\Omega_c \alpha f$ and K a Bessel function. [The normalization constant is $C = \exp(\frac{1}{2}t\Omega_c)(t\Omega_c/\pi)^{1/2}(1+\alpha f)^{1/4}$.] This distribution was first obtained by Glauber [21]. It is closely related to the socalled K-distribution in the theory of scattering from turbulent media [25, 26]. The counting distribution (39) in the short-time regime remains negative-binomial.

In most realizations of black-body radiation the value of the Bose-Einstein function $f(\omega_0, T)$ is $\ll 1$. The difference between the two distributions (46) and (51) is then quite small, both being close to the Poisson distribution (47).

4.2. REDUCTION OF DEGREES OF FREEDOM

We now turn to applications of our general formulas to specific random media. We concentrate on the long-time regime and assume a frequency-resolved measurement, in which photons are only counted within a frequency interval $\delta\omega$ around ω_0. (For a black body, this would correspond to the step-function model with Ω_c replaced by $\delta\omega$.) We take $\delta\omega$ smaller than any of the characteristic frequencies ω_c, Ω_c, but necessarily greater than $1/t$. The factorial cumulants are then given by

$$\kappa_p = (p-1)!\,\nu(\alpha f)^p \frac{1}{N}\sum_{n=1}^{N}(1-\sigma_n)^p, \tag{52}$$

where $\nu = Nt\delta\omega/2\pi$. For comparison with black-body radiation we parameterize the variance in terms of the effective number ν_{eff} of degrees of freedom [19],

$$\operatorname{Var} n = \bar{n}(1 + \bar{n}/\nu_{\text{eff}}), \tag{53}$$

with $\nu_{\text{eff}} = \nu$ for a black body [cf. Eq. (48)]. Eq. (52) implies

$$\frac{\nu_{\text{eff}}}{\nu} = \frac{[\sum_n (1-\sigma_n)]^2}{N\sum_n (1-\sigma_n)^2} \leq 1. \tag{54}$$

We conclude that the super-Poissonian noise of a random medium corresponds to a black body with a *reduced* number of degrees of freedom. The reduction occurs only for multi-mode emission. (Eq. (54) with $N = 1$ gives $\nu_{\text{eff}} = \nu$.) In addition, it requires multi-mode detection to observe the reduction, because single-mode detection contains no other information than the absorptivity (cf. Sec. 3.4).

An ensemble of random media has a certain scattering-strength density

$$\rho(\sigma) = \left\langle \sum_{n=1}^{N} \delta(\sigma - \sigma_n) \right\rangle, \tag{55}$$

where the brackets $\langle \cdots \rangle$ denote the ensemble average. In the large-N regime sample-to-sample fluctuations are small, so the ensemble average is representative for a single system.[4] We may therefore replace $\sum_n$ by $\int d\sigma\, \rho(\sigma)$ in Eqs. (52) and (54). In the applications that follow we will restrict ourselves to the large-N regime, so that we can ignore sample-to-sample fluctuations. All that we need in this case is the function $\rho(\sigma)$. Random-matrix theory [18] provides a method to compute this function for a variety of random media.

4.3. DISORDERED WAVEGUIDE

As a first example we consider the thermal radiation from a disordered absorbing waveguide (Fig. 3). The length of the waveguide is L, the transport mean free path in the medium is l, the velocity of light is c, and $\tau_s = l/c$ is the scattering time. The absorption time τ_a is related to the imaginary part $\varepsilon'' > 0$ of the (relative) dielectric constant by $1/\tau_a = \omega_0 \varepsilon''$. We assume that τ_s and τ_a are both $\gg 1/\omega_0$, so that scattering as well as absorption occur on length scales large compared to the wavelength. We define the normalized absorption rate[5]

$$\gamma = \frac{16}{3}\frac{\tau_s}{\tau_a}. \tag{56}$$

We call the system weakly absorbing if $\gamma \ll 1$, meaning that the absorption is weak on the scale of the mean free path. If $\gamma \gg 1$ we call the system strongly absorbing, $\gamma \to \infty$ being the black-body limit. For simplicity we restrict ourselves to the case of an infinitely long waveguide (more precisely, $L \gg l/\sqrt{\gamma}$), so that transmission through it can be neglected.[6]

In the absence of transmission the scattering matrix S coincides with the reflection matrix r, and the scattering strengths σ_n coincide with the reflection eigenvalues R_n (eigenvalues of $r \cdot r^\dagger$). The density $\rho(\sigma)$ for this system is known for any value of N [27, 28]. The general expression is a series of Laguerre polynomials, which in the large-N regime of present

[4] This statement is strictly speaking not correct for amplifying systems. The reason is that the ensemble average is dominated by a small fraction of members of the ensemble that are above the laser threshold, and this fraction is non-zero for any non-zero amplification rate. This is a non-perturbative finite-N effect that does not appear if the ensemble average is computed using the large-N perturbation theory employed here.

[5] The coefficient 16/3 in Eq. (56) is chosen to facilitate the comparison between waveguide and cavity in the next subsection, and refers to three-dimensional scattering. In the case of two-dimensional scattering the coefficient is $\pi^2/2$. The present definition of γ differs from that used in Ref. [27] by a factor of two.

[6] Results for an absorbing waveguide of finite length follow from Eqs. (73)–(75) upon changing the sign of γ.

interest simplifies to

$$\rho(\sigma) = \frac{N\sqrt{\gamma}}{\pi}\frac{(\sigma^{-1} - 1 - \frac{1}{4}\gamma)^{1/2}}{(1-\sigma)^2}, \quad 0 < \sigma < \frac{1}{1+\frac{1}{4}\gamma}. \tag{57}$$

(The large-N regime requires $N \gg 1$, but in weakly absorbing systems the condition is stronger: $N \gg 1/\sqrt{\gamma}$.) This leads to the effective number of degrees of freedom

$$\frac{\nu_{\text{eff}}}{\nu} = \frac{[\int d\sigma\, \rho(\sigma)(1-\sigma)]^2}{N\int d\sigma\, \rho(\sigma)(1-\sigma)^2} = 4[(1+4/\gamma)^{1/4} + (1+4/\gamma)^{-1/4}]^{-2}, \tag{58}$$

plotted in Fig. 4, with a mean photocount of

$$\bar{n} = \tfrac{1}{2}\nu\alpha f\gamma\left(\sqrt{1+4/\gamma} - 1\right). \tag{59}$$

For strong absorption, $\gamma \gg 1$, we recover the black-body result $\nu_{\text{eff}} = \nu$, as expected. For weak absorption, $\gamma \ll 1$, we find $\nu_{\text{eff}} = 2\nu\sqrt{\gamma}$.

The characteristic function in the long-time frequency-resolved regime follows from

$$F(\xi) = -\frac{\nu}{N}\int d\sigma\, \rho(\sigma)\ln[1-(1-\sigma)\xi\alpha f]. \tag{60}$$

Substitution of Eq. (57) into Eq. (60) yields a hypergeometric function, which in the limit $\gamma \ll 1$ of weak absorption simplifies to

$$F(\xi) = \nu_{\text{eff}}\left(1 - \sqrt{1-\xi\alpha f}\right), \quad \nu_{\text{eff}} = 2\nu\sqrt{\gamma}. \tag{61}$$

The counting distribution corresponding to Eq. (61),

$$P(n) \propto \frac{1}{n!}\left(\frac{\bar{n}}{\sqrt{1+\alpha f}}\right)^n K_{n-1/2}\left(\nu_{\text{eff}}\sqrt{1+\alpha f}\right), \tag{62}$$

is Glauber's distribution (51) with an effective number of degrees of freedom. Note that Eq. (51) resulted from single-mode detection over a broad frequency range, whereas Eq. (62) results from multi-mode detection over a narrow frequency range. It appears as a coincidence that the two distributions have the same functional form (with different parameters).

4.4. CHAOTIC CAVITY

Our second example is an optical cavity radiating through a small hole covered by a photodetector (Fig. 2). The area $\mathcal{A}$ of the hole should be

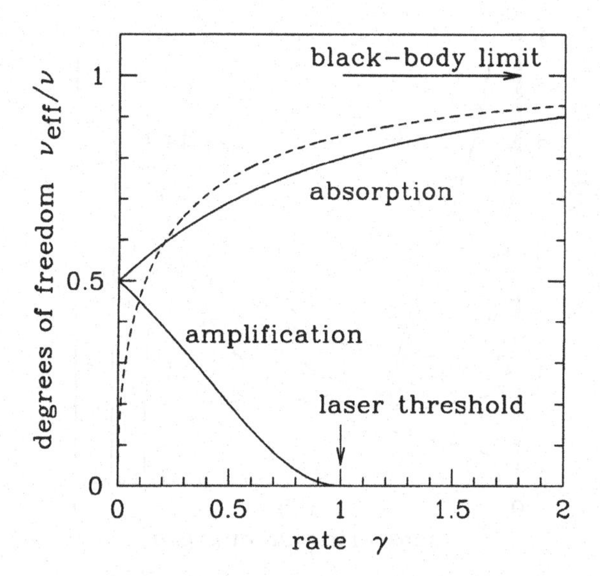

Figure 4. Effective number of degrees of freedom as a function of normalized absorption or amplification rate. The dashed curve is Eq. (58) for an absorbing, infinitely long disordered waveguide, the solid curves are Eqs. (67) and (72) for the chaotic cavity. For the cavity both the cases of absorption and amplification are shown. (See Fig. 7 for the amplifying waveguide.) The black-body limit for absorbing systems and the laser threshold for amplifying systems are indicated by arrows.

small compared to the surface area of the cavity. The cavity should have an irregular shape, or it should contain random scatterers — to ensure chaotic scattering of the radiation inside the cavity. It should be large enough that the spacing $\Delta\omega$ of the cavity modes near frequency ω_0 is $\ll \omega_0$. For this system we define the normalized absorption rate as

$$\gamma = \frac{\tau_{\text{dwell}}}{\tau_{\text{a}}}; \quad \tau_{\text{dwell}} \equiv \frac{2\pi}{N\Delta\omega}. \tag{63}$$

The time τ_{dwell} is the mean dwell time of a photon in the cavity without absorption. The frequency $1/\tau_{\text{dwell}}$ represents the broadening of the cavity modes due to the coupling to the $N = 2\pi\mathcal{A}/\lambda^2$ modes propagating through the hole. The broadening is much greater than the spacing $\Delta\omega$ for $N \gg 1$. The large-N regime requires in addition $N \gg 1/\gamma$. The scattering-strength density in the large-N regime can be calculated using the perturbation theory of Ref. [29].

The result is a rather complicated algebraic function, see the Appendix.

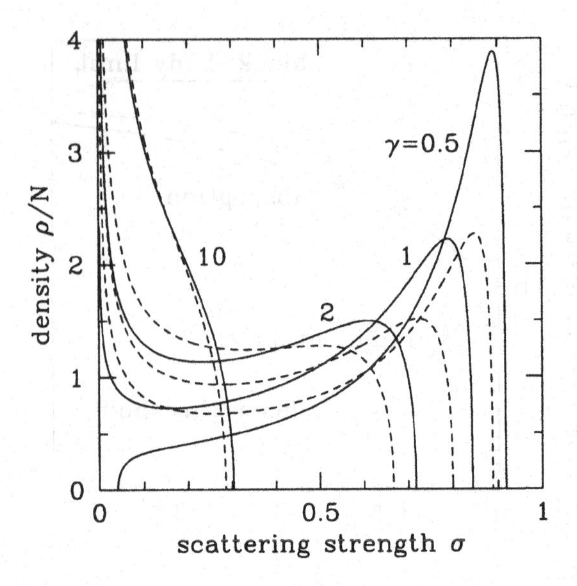

Figure 5. Solid curves: Scattering-strength density of an absorbing chaotic cavity in the large-N regime, calculated from Eq. (91) for four values of the dimensionless absorption rate γ. The density (57) for an absorbing, infinitely long disordered waveguide is included for comparison (dashed). For $\gamma \gg 1$ the results for cavity and waveguide coincide.

It has a simple form in the limit $\gamma \ll 1$ of weak absorption,

$$\rho(\sigma) = \frac{N}{2\pi} \frac{(\sigma - \sigma_-)^{1/2}(\sigma_+ - \sigma)^{1/2}}{(1-\sigma)^2}, \quad \sigma_- < \sigma < \sigma_+, \tag{64}$$

with $\sigma_\pm = 1 - 3\gamma \pm 2\gamma\sqrt{2}$. In the opposite limit $\gamma \gg 1$ of strong absorption, $\rho(\sigma)$ is given by the same Eq. (57) as for the infinitely long disordered waveguide. The crossover from weak to strong absorption is shown in Fig. 5. The value $\gamma = 1$ is special in the sense that $\rho(\sigma)$ goes to zero or infinity as $\sigma \to 0$, depending on whether γ is smaller or greater than 1.

We find the mean and variance of the photocount

$$\bar{n} = \frac{\nu\alpha f\gamma}{1+\gamma}, \tag{65}$$

$$\mathrm{Var}\, n = \bar{n} + \nu(\alpha f)^2\gamma^2 \frac{\gamma^2 + 2\gamma + 2}{(1+\gamma)^4}, \tag{66}$$

corresponding to the effective number of degrees of freedom

$$\frac{\nu_{\text{eff}}}{\nu} = \frac{(1+\gamma)^2}{\gamma^2 + 2\gamma + 2}. \tag{67}$$

Again, $\nu_{\rm eff} = \nu$ for $\gamma \gg 1$. For $\gamma \ll 1$ we find $\nu_{\rm eff} = \frac{1}{2}\nu$. This factor-of-two reduction of the number of degrees of freedom is a "universal" result, independent of any parameters of the system. The chaotic cavity is compared with the disordered waveguide in Fig. 4. The ratio $\nu_{\rm eff}/\nu$ for the chaotic cavity remains finite no matter how weak the absorption, while this ratio goes to zero when $\gamma \to 0$ in the case of the infinitely long disordered waveguide.[7]

4.5. RANDOM LASER

The examples of the previous subsections concern thermal emission from absorbing systems. As we discussed in Sec. 3.2, our general formulas can also be applied to amplified spontaneous emission, by evaluating the Bose-Einstein function f at a negative temperature [12, 17]. Complete population inversion corresponds to $f = -1$. The amplification rate $1/\tau_{\rm a} = \omega_0|\varepsilon''|$ should be so small that we are well below the laser threshold, in order to stay in the regime of linear amplification. The laser threshold occurs when the normalized amplification rate γ reaches a critical value $\gamma_{\rm c}$. (Sample-to-sample fluctuations in the laser threshold [30] are small in the large-N regime.) For the cavity $\gamma_{\rm c} = 1$. For the disordered waveguide one has

$$\gamma_{\rm c} = \left(\frac{4\pi l}{3L}\right)^2 \quad \text{if } L \gg l. \tag{68}$$

Since $\gamma_{\rm c} \to 0$ in the limit $L \to \infty$, the infinitely long waveguide is above the laser threshold no matter how weak the amplification.

A duality relation [31] between absorbing and amplifying systems greatly simplifies the calculation of the scattering strengths. Dual systems differ only in the sign of the imaginary part ε'' of the dielectric constant (positive for the absorbing system, negative for the amplifying system). Therefore, dual systems have the same value of $\tau_{\rm a}$ and γ. The scattering matrices of dual systems are related by $S_-^\dagger = S_+^{-1}$, hence $S_- \cdot S_-^\dagger = (S_+ \cdot S_+^\dagger)^{-1}$. (The subscript $+$ denotes the absorbing system, the subscript $-$ the dual amplifying system.) We conclude that the scattering strengths $\sigma_1, \sigma_2, \ldots \sigma_N$ of an amplifying system are the reciprocal of those of the dual absorbing system. The densities $\rho_\pm(\sigma)$ are related by

$$\sigma^2 \rho_-(\sigma) = \rho_+(1/\sigma). \tag{69}$$

In Fig. 6 we show the result of the application of the transformation (69) to the densities of Fig. 5 for the case of a cavity. The critical value $\gamma_{\rm c} = 1$

[7] For a waveguide of finite length L ($\gg l$) one has instead $\nu_{\rm eff}/\nu \to 5\,l/L$ in the limit $\gamma \to 0$ (cf. Sec. 4.5).

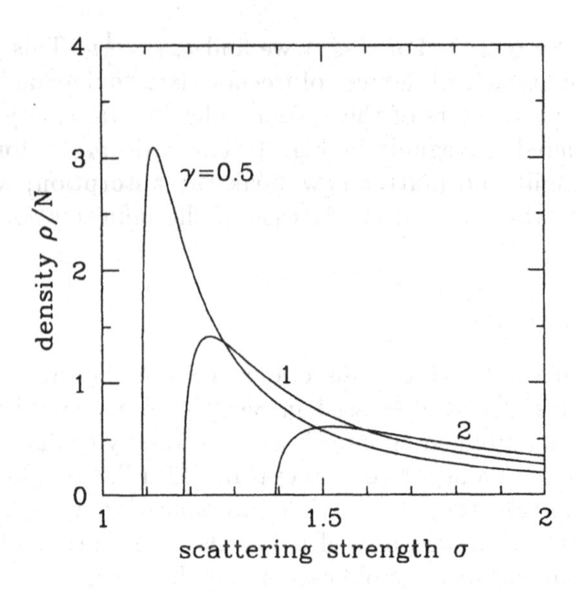

Figure 6. Scattering-strength density of an amplifying chaotic cavity, calculated from the results in Fig. 5 by means of the duality relation (69). First and higher moments diverge for $\gamma \geq 1$.

is such that the first and higher moments are finite for $\gamma < \gamma_c$ and infinite for $\gamma \geq \gamma_c$.

We find that the expressions for $\bar{n}$, $\mathrm{Var}\, n$, and ν_{eff}/ν in the amplifying chaotic cavity differ from those in the dual absorbing cavity by the substitution of γ by $-\gamma$:

$$\bar{n} = -\frac{\nu \alpha f \gamma}{1-\gamma}, \tag{70}$$

$$\mathrm{Var}\, n = \bar{n} + \nu(\alpha f)^2 \gamma^2 \frac{\gamma^2 - 2\gamma + 2}{(1-\gamma)^4}, \tag{71}$$

$$\frac{\nu_{\mathrm{eff}}}{\nu} = \frac{(1-\gamma)^2}{\gamma^2 - 2\gamma + 2}. \tag{72}$$

The Bose-Einstein function f is now to be evaluated at a negative temperature, so that $f < 0$. In Fig. 4 we compare ν_{eff}/ν for amplifying and absorbing cavities. In the limit $\gamma \to 0$ the two results coincide, but the γ-dependence is strikingly different: While the ratio ν_{eff}/ν increases with γ in the case of absorption, it decreases in the case of amplification — vanishing at the laser threshold. Of course, close to the laser threshold [when $\gamma \gtrsim 1 - (\Omega_c \tau_{\mathrm{dwell}})^{-1/2}$] the approximation of a linear amplifier breaks down and a non-linear treatment (along the lines of Ref. [32]) is required.

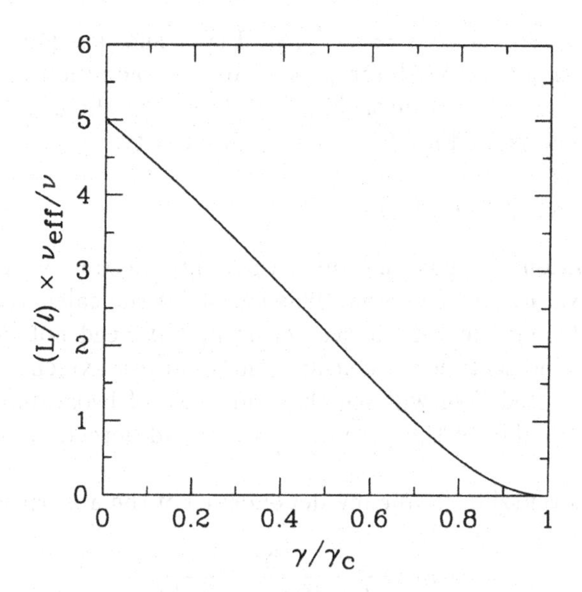

Figure 7. Effective number of degrees of freedom for an amplifying disordered waveguide of finite length L (much greater than the transport mean free path l), computed from Eq. (75). The laser threshold occurs at $\gamma = \gamma_c \equiv (4\pi l/3L)^2$.

For the amplifying disordered waveguide we can not use the infinite-length formulas of Sec. 4.3, because then we would be above threshold for arbitrarily small γ. The counting distribution $P(n)$ at a finite L requires the density of eigenvalues of the matrix $r \cdot r^\dagger + t \cdot t^\dagger$, as explained in Sec. 3.5. The density itself is not known, but its first few moments have been calculated recently by Brouwer [33]. This is sufficient to compute ν_{eff}, since we only need the first two moments of $P(n)$. The results for $\gamma, \gamma_c \ll 1$ are[8]

$$\bar{n} = -\nu\alpha f \frac{\sqrt{\gamma}}{\sin s}(1 - \cos s), \tag{73}$$

$$\operatorname{Var} n = \bar{n} + \nu(\alpha f)^2 \frac{\sqrt{\gamma}}{2\sin^4 s}(s - s\cos s + s\sin^2 s + \sin s - 3\sin^3 s - \cos^3 s \sin s), \tag{74}$$

$$\frac{\nu_{\text{eff}}}{\nu} = \frac{2\sqrt{\gamma}(1-\cos s)^2 \sin^2 s}{s - s\cos s + s\sin^2 s + \sin s - 3\sin^3 s - \cos^3 s \sin s}, \tag{75}$$

[8] Ref. [33] considers an absorbing waveguide. The amplifying case follows by changing the sign of the parameter γ. Eq. (13c) in Ref. [33] contains a misprint: The second and third term between brackets should have, respectively, signs minus and plus instead of plus and minus.

where we have abbreviated $s = \pi\sqrt{\gamma/\gamma_c}$. Fig. 7 shows a plot of Eq. (75). Notice the limit $\nu_{\rm eff}/\nu = 5\,l/L$ for $\gamma/\gamma_c \to 0$. The reduction of the number of degrees of freedom on approaching the laser threshold is qualitatively similar to that shown in Fig. 4 for the chaotic cavity.

4.6. BROAD-BAND DETECTION

In these applications we have assumed that only photons within a narrow frequency interval $\delta\omega$ are detected. This simplifies the calculations because the frequency dependence of the scattering matrix need not be taken into account. In this subsection we consider the opposite extreme that all frequencies are detected. We will see that this case of broad-band detection is qualitatively similar to the case of narrow-band detection considered so far.

We take a Lorentzian frequency dependence of the absorption or amplification rate,

$$\gamma(\omega) = \frac{\gamma_0}{1 + 4(\omega - \omega_0)^2/\Gamma^2}. \tag{76}$$

The characteristic frequency Ω_c for the scattering strengths is defined by

$$\Omega_c = \Gamma\sqrt{1 + \gamma_0}. \tag{77}$$

The two frequencies Ω_c and Γ are essentially the same for $\gamma_0 \lesssim 1$, but for $\gamma_0 \gg 1$ the former is much bigger than the latter. The reason is that what matters for the deviation of the scattering strengths from zero is the relative magnitude of $\gamma(\omega)$ with respect to 1, not with respect to γ_0. As in Sec. 4.1, we assume that $\Omega_c \ll \omega_0$, so that we may neglect the frequency dependence of N and f. The mean and variance of the photocount in the long-time regime are given by

$$\bar{n} = t\alpha f \int \frac{d\omega}{2\pi} \int d\sigma\, \rho(\sigma, \omega)(1 - \sigma), \tag{78}$$

$$\operatorname{Var} n = \bar{n} + t(\alpha f)^2 \int \frac{d\omega}{2\pi} \int d\sigma\, \rho(\sigma, \omega)(1 - \sigma)^2. \tag{79}$$

Again, we have assumed that N is sufficiently large that sample-to-sample fluctuations can be neglected and we may replace $\sum_n$ by $\int d\sigma$. The scattering-strength density ρ depends on ω through the rate $\gamma(\omega)$.

In the absorbing, infinitely long disordered waveguide $\rho \propto \sqrt{\gamma}$ for $\gamma \ll 1$, hence the integrands in Eqs. (78) and (79) decay $\propto 1/|\omega - \omega_0|$ and the integrals over ω diverge. A cutoff is provided by the finite length L of the waveguide. When γ drops below $(l/L)^2$, radiation can be transmitted through the waveguide with little absorption. Only the frequency range $|\omega - \omega_0| \lesssim \Gamma(L/l)\sqrt{\gamma_0}$, therefore, contributes effectively to the integrals (78)

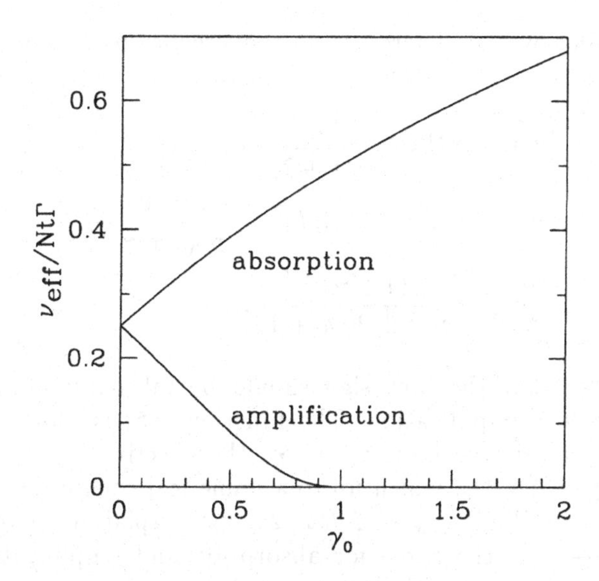

Figure 8. Plot of Eq. (85) for the effective number of degrees of freedom of an absorbing or amplifying chaotic cavity, in the case of broad-band detection with a Lorentzian frequency dependence $\gamma(\omega) = \gamma_0[1+4(\omega-\omega_0)^2/\Gamma^2]^{-1}$ of the absorption or amplification rate. For $\gamma_0 \lesssim 1$ the curves are qualitatively similar those plotted in Fig. 4 for the case of narrow-band detection. In the absorbing cavity ν_{eff} increases $\propto \sqrt{\gamma_0}$ as $\gamma_0 \to \infty$, instead of saturating as in Fig. 4, because the characteristic frequency Ω_c increases $\propto \sqrt{\gamma_0}$ in that limit [cf. Eq. (77)].

and (79). To leading order in $(L/l)\sqrt{\gamma_0}$ we can take the infinite-L result for ρ with the cutoff we mentioned. The result (for $(l/L)^2 \ll \gamma_0 \ll 1$) is

$$\bar{n} = \frac{Nt\Gamma}{2\pi}\alpha f\sqrt{\gamma_0}\left[\ln\left(\frac{L}{l}\sqrt{\gamma_0}\right) + \mathcal{O}(1)\right]. \tag{80}$$

$$\operatorname{Var} n = \bar{n} + \frac{Nt\Gamma}{2\pi}(\alpha f)^2\tfrac{1}{2}\sqrt{\gamma_0}\left[\ln\left(\frac{L}{l}\sqrt{\gamma_0}\right) + \mathcal{O}(1)\right]. \tag{81}$$

If we write $\operatorname{Var} n = \bar{n}(1+\bar{n}/\nu_{\text{eff}})$, as before, then

$$\frac{\nu_{\text{eff}}}{Nt\Gamma} = \pi^{-1}\sqrt{\gamma_0}\left[\ln\left(\frac{L}{l}\sqrt{\gamma_0}\right) + \mathcal{O}(1)\right]. \tag{82}$$

In the case of narrow-band detection considered in Sec. 4.3 we had $\nu_{\text{eff}}/Nt\delta\omega = \pi^{-1}\sqrt{\gamma_0}$ for $(l/L)^2 \ll \gamma_0 \ll 1$. The difference with Eq. (82) (apart from the replacement of $\delta\omega$ by $\Gamma \approx \Omega_c$) is the logarithmic enhancement factor, but still $\nu_{\text{eff}} \ll Nt\Gamma$ for $\gamma_0 \ll 1$.

For the chaotic cavity, we can compute $\bar{n}$ and $\operatorname{Var} n$ directly from the narrow-band results (65), (66), (70), and (71), by substituting Eq. (76) for

γ and integrating over ω. There are no convergence problems in this case. The results are

$$\bar{n} = \pm Nt\Gamma\alpha f \frac{\gamma_0}{4\sqrt{1 \pm \gamma_0}}, \tag{83}$$

$$\mathrm{Var}\, n = \bar{n} + Nt\Gamma(\alpha f)^2 \gamma_0^2 \frac{9\gamma_0^2 \pm 20\gamma_0 + 16}{64(1 \pm \gamma_0)^{7/2}}, \tag{84}$$

$$\frac{\nu_{\text{eff}}}{Nt\Gamma} = \frac{4(1 \pm \gamma_0)^{5/2}}{9\gamma_0^2 \pm 20\gamma_0 + 16}. \tag{85}$$

The $\pm$ indicates that the plus sign should be taken for absorption and the minus sign for amplification. The function (85) is plotted in Fig. 8. In the strongly absorbing limit $\gamma_0 \to \infty$, the effective number of degrees of freedom $\nu_{\text{eff}} \to \frac{4}{9} Nt\Omega_c$, which up to a numerical coefficient corresponds to the narrow-band limit $\nu_{\text{eff}} \to Nt\delta\omega/2\pi$ upon replacement of $\delta\omega$ by Ω_c. The limit $\gamma_0 \to 0$ is the same for absorption and amplification, $\nu_{\text{eff}} \to \frac{1}{4} Nt\Omega_c$, again corresponding to the narrow-band result $\nu_{\text{eff}} \to Nt\delta\omega/4\pi$ up to a numerical coefficient. Finally, the ratio $\nu_{\text{eff}}/Nt\Gamma$ tends to zero upon approaching the laser threshold $\gamma_0 \to 1$ in an amplifying system, similarly to the narrow-band case. The qualitative behavior of ν_{eff}/ν is therefore the same for broad-band and narrow-band detection.

5. Conclusion

5.1. SUMMARY

In conclusion, we have shown that the photodetection statistics contains substantially more information on the scattering properties of a medium than its absorptivity. The mean photocount $\bar{n}$ is determined just by the absorptivity, as dictated by Kirchhoff's law. Higher order moments of the counting distribution, however, contain information on higher spectral moments of the scattering strengths σ_n (being the N eigenvalues of the scattering matrix product $S \cdot S^\dagger$). These higher moments are independent of the absorptivity, which is determined by the first moment. While the absorptivity follows from the radiative transfer equation, higher spectral moments are outside of the range of that approach. We have used random-matrix theory for their evaluation.

To measure these higher spectral moments, it is necessary that the counting time t is greater than the coherence time $1/\Omega_c$ of the thermal radiation (being the inverse of the absorption or amplification line width). It is also necessary that the area of the photocathode is greater than the coherence area (being the area corresponding to one mode emitted by the medium). Single-mode detection yields solely information on the absorptivity. Multi-mode detection is unusual in photodetection experiments, but required if one wants to go beyond Kirchhoff's law.

We have shown that the variance $\mathrm{Var}\, n$ of the photocount contains information on the width of the density $\rho(\sigma)$ of scattering strengths. We have computed this density for a disordered waveguide and for a chaotic cavity, and find that it is very wide and strongly non-Gaussian. In an absorbing medium, the deviations from Poisson statistics of independent photocounts are small because the Bose-Einstein function is $\ll 1$ for all practical frequencies and temperatures. Since the Poisson distribution contains the mean photocount as the only parameter, one needs to be able to measure the super-Poissonian fluctuations in order to obtain information beyond the absorptivity. The deviations from Poisson statistics are easier to detect in an amplifying medium, where the role of the Bose-Einstein function is played by the relative population inversion of the atomic states.

We have shown that the super-Poissonian fluctuations in a linearly amplifying random medium are much greater than would be expected from the mean photocount. If we write $\mathrm{Var}\, n - \bar{n} = \bar{n}^2/\nu_{\mathrm{eff}}$, then ν_{eff} would equal $Nt\delta\omega/2\pi \equiv \nu$ if all N modes reaching the photodetector would have the same scattering strength. (We assume for simplicity that only a narrow band $\delta\omega$ is detected, in a time t; For broad-band detection $\delta\omega$ should be replaced by Ω_{c}.) The effective number ν_{eff} of degrees of freedom is much smaller than ν for a broad $\rho(\sigma)$, hence the anomalously large fluctuations. On approaching the laser threshold, the ratio ν_{eff}/ν goes to zero. In a conventional laser the noise itself increases with increasing amplification rate because $\bar{n}$ increases, but ν_{eff} does not change below the laser threshold. Typically, light is emitted in a single mode, hence ν_{eff} equals $t\delta\omega/2\pi$ independent of the amplification rate. In a random laser a large number N of cavity modes contribute to the radiation, no matter how small the frequency window $\delta\omega$, because the cavity modes overlap. The overlap is the consequence of the much weaker confinement created by disorder in comparison to that created by a mirror. The reduction of the number of degrees of freedom is a quantum optical effect of overlapping cavity modes that should be observable experimentally.

5.2. RELATION TO THOULESS NUMBER

The Thouless number N_{T} plays a central role in mesoscopic physics [34]. It is a dimensionless measure of the coupling strength of a closed system to the outside world,

$$N_{\mathrm{T}} \simeq \frac{1}{\tau_{\mathrm{dwell}}\Delta\omega}. \tag{86}$$

(We use $\simeq$ instead of $=$ because we are ignoring numerical coefficients of order unity.) As before, $\Delta\omega$ is the spacing of the eigenfrequencies of the closed system and $\tau_{\rm dwell}$ is the mean time a particle (electron or photon) entering the system stays inside. In a conducting metal, $N_{\rm T}$ is the conductance in units of the conductance quantum e^2/h. The metal-insulator transition occurs when $N_{\rm T}$ becomes of order unity. It is assumed that there is no absorption or amplification, as is appropriate for electrons.

For the two types of systems considered in this work, one has $N_{\rm T} \simeq Nl/L$ for the disordered waveguide and $N_{\rm T} \simeq N$ for the chaotic cavity. We notice that $N_{\rm T}$ is related to the effective number of degrees of freedom in the limit of zero absorption and amplification,

$$\lim_{\gamma\to 0} \frac{\nu_{\rm eff}}{\nu} \simeq \frac{N_{\rm T}}{N}. \tag{87}$$

The ratio of $\nu_{\rm eff}$ to the black-body value ν is the same as that of $N_{\rm T}$ to the number of propagating modes N. We believe that the relation (87) holds for all random media, not just for those considered here.

Acknowledgments

I have benefitted from discussions with A. Lagendijk, R. Loudon, M. Patra, D. S. Wiersma, and J. P. Woerdman. This research was supported by the "Nederlandse organisatie voor Wetenschappelijk Onderzoek" (NWO) and by the "Stichting voor Fundamenteel Onderzoek der Materie" (FOM).

Appendix. Scattering-strength density of a chaotic cavity

As derived in Ref. [35], absorption in a chaotic cavity (with rate $1/\tau_{\rm a}$) is statistically equivalent to the loss induced by a fictitious waveguide that is weakly coupled to the cavity. The coupling has transmission probability Γ' for each of the N' modes in the fictitious waveguide. The equivalence requires the limit $N' \to \infty$, $\Gamma' \to 0$, at fixed $N'\Gamma' = 2\pi/\tau_{\rm a}\Delta\omega$ (with $\Delta\omega$ the spacing of the cavity modes).

We are therefore led to consider the system illustrated in Fig. 9: A chaotic cavity without absorption containing two openings. One opening is coupled to an N-mode waveguide with transmission probability 1 per mode, the other opening is coupled to the fictitious N'-mode waveguide with transmission probability Γ' per mode. The scattering strength σ_n equals $1 - T_n$, with T_n an eigenvalue of the transmission-matrix product $t \cdot t^\dagger$ (t being the $N \times N'$ transmission matrix from one waveguide to the other). The density of transmission eigenvalues $\rho(T) = \langle \sum_n \delta(T - T_n) \rangle$ can be calculated in the large-N regime using the perturbation theory of Ref. [29]. The scattering-strength density $\rho(\sigma)$ then follows from $\sigma = 1 - T$.

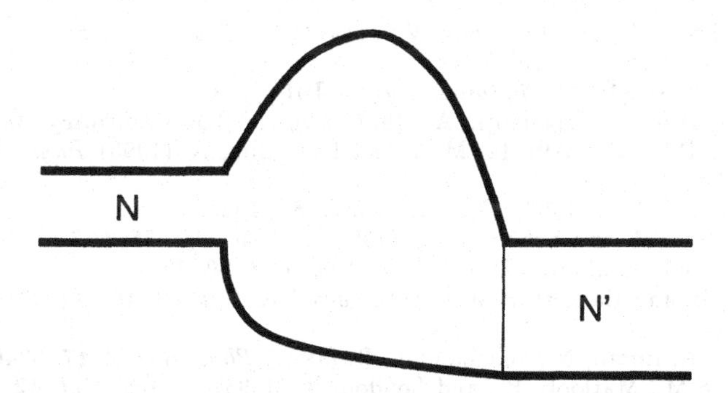

Figure 9. An absorbing cavity with one opening is statistically equivalent to the non-absorbing cavity with two openings shown here. The thin line in the second opening indicates a barrier with transmission probability Γ' for each of the N' modes in the waveguide attached to the opening. (The limit $\Gamma' \to 0$, $N' \to \infty$ at fixed $N'\Gamma'$ is required for the equivalence with absorption.)

The result is non-zero for $\sigma_{\rm min} < \sigma < \sigma_{\rm max}$, with the definitions

$$\sigma_{\rm min} = \begin{cases} \sigma_- & \text{if } \gamma < 1, \\ 0 & \text{if } \gamma > 1, \end{cases} \tag{88}$$

$$\sigma_{\rm max} = \sigma_+, \tag{89}$$

$$\sigma_\pm = \frac{8 + 20\gamma^2 - \gamma^4 \pm \gamma(8+\gamma^2)^{3/2}}{8(1+\gamma)^3}. \tag{90}$$

[We use the same dimensionless absorption rate $\gamma = N'\Gamma'/N = 2\pi/N\tau_{\rm a}\Delta\omega$ as in Eq. (63).] Inside this interval the scattering-strength density is given by

$$\begin{aligned} \rho(\sigma) &= \frac{6N\sqrt{3}}{\pi}\Big((a+b)^{1/3} - (a-b)^{1/3}\Big)\bigg(\Big[(a+b)^{1/3} + (a-b)^{1/3} \\ &\quad - 2\gamma + 2 - 6\sigma\Big]^2 + 3\Big[(a+b)^{1/3} - (a-b)^{1/3}\Big]^2\bigg)^{-1}, \end{aligned} \tag{91}$$

$$a = (\gamma - 1)^3 + 9(1 + \tfrac{1}{2}\gamma^2)\sigma, \tag{92}$$

$$b = (3 + 3\gamma)^{3/2}[\sigma(\sigma - \sigma_-)(\sigma_+ - \sigma)]^{1/2}. \tag{93}$$

Eq. (91) is plotted in Fig. 5 for several values of γ.

References

1. Wiersma, D.S., Van Albada, M.P., and Lagendijk, A. (1995) *Nature* **373**, 203.
2. Lawandy, N.M., Balachandran, R.M., Gomes, A.S.L., and Sauvain, E. (1994) *Nature* **368**, 436.
3. Letokhov, V.S. (1967) *Zh. Eksp. Teor. Fiz.* **53**, 1442 [(1968) *Sov. Phys. JETP* **26**, 835].

4. Lavrinovich, N.N. and Letokhov, V.S. (1974) *Zh. Eksp. Teor. Fiz.* **67**, 1609 [(1975) *Sov. Phys. JETP* **40**, 800].
5. Lawandy, N.M. (1994) *Photonics Spectra* **July**, 119.
6. Wiersma, D.S. and Lagendijk, A. (1997) *Physics World* **January**, 33.
7. Wiersma, D.S., Van Albada, M.P., and Lagendijk, A. (1995) *Phys. Rev. Lett.* **75**, 1739.
8. Beenakker, C.W.J. (1998) *Phys. Rev. Lett.* **81**, 1829.
9. Fleischhauer, M. and Schubert, M. (1991) *J. Mod. Opt.* **38**, 677.
10. Knöll, L. and Leonhardt, U. (1992) *J. Mod. Opt.* **39**, 1253.
11. Huttner, B. and Barnett, S.M. (1992) *Europhys. Lett.* **18**, 487; (1992) *Phys. Rev. A* **46**, 4306.
12. Jeffers, J.R., Imoto, N., and Loudon, R. (1993) *Phys. Rev. A* **47**, 3346.
13. Barnett, S.M., Matloob, R., and Loudon, R. (1995) *J. Mod. Opt.* **42**, 1165.
14. Matloob, R., Loudon, R., Barnett, S.M., and Jeffers, J. (1995) *Phys. Rev. A* **52**, 4823.
15. Gruner, T. and Welsch, D.-G. (1996) *Phys. Rev. A* **53**, 1818.
16. Gruner, T. and Welsch, D.-G. (1996) *Phys. Rev. A* **54**, 1661.
17. Matloob, R., Loudon, R., Artoni, M., Barnett, S.M., and Jeffers, J. (1997) *Phys. Rev. A* **55**, 1623.
18. Beenakker, C.W.J. (1997) *Rev. Mod. Phys.* **69**, 731.
19. Mandel, L. and Wolf, E. (1995) *Optical Coherence and Quantum Optics*, Cambridge University, Cambridge.
20. Loudon, R. (1983) *The Quantum Theory of Light*, Clarendon, Oxford.
21. Glauber, R.J. (1963) *Phys. Rev. Lett.* **10**, 84; (1965) in C. DeWitt, A. Blandin, and C. Cohen-Tannoudji (eds.), *Quantum Optics and Electronics*, Gordon and Breach, New York.
22. Kelley, P.L. and Kleiner, W.H. (1964) *Phys. Rev.* **136**, A316.
23. Bekenstein, J.D. and Schiffer, M. (1994) *Phys. Rev. Lett.* **72**, 2512.
24. Mehta, C.L. (1970) in E. Wolf (ed.), *Progress in Optics*, Vol. VIII, North-Holland, Amsterdam.
25. Jakeman, E. and Pusey, P.N. (1978) *Phys. Rev. Lett.* **40**, 546.
26. Jakeman, E. (1980) *J. Phys. A* **13**, 31.
27. Beenakker, C.W.J., Paasschens, J.C.J., and Brouwer, P.W. (1996) *Phys. Rev. Lett.* **76**, 1368.
28. Bruce, N.A. and Chalker, J.T. (1996) *J. Phys. A* **29**, 3761.
29. Brouwer, P.W. and Beenakker, C.W.J. (1996) *J. Math. Phys.* **37**, 4904.
30. Zyuzin, A.Yu. (1995) *Phys. Rev. E* **51**, 5274.
31. Paasschens, J.C.J., Misirpashaev, T.Sh., and Beenakker, C.W.J. (1996) *Phys. Rev. B* **54**, 11887.
32. Brunner, W. and Paul, H. (1969) *Ann. Physik* **23**, 152; **23**, 384; **24**, 38.
33. Brouwer, P.W. (1998) *Phys. Rev. B* **57**, 10526.
34. Imry, Y. (1997) *Introduction to Mesoscopic Physics*, Oxford University, Oxford.
35. Brouwer, P.W. and Beenakker, C.W.J. (1997) *Phys. Rev. B* **55**, 4695.

SPECTRA OF LARGE RANDOM MATRICES: A METHOD OF STUDY

E. KANZIEPER
The Abdus Salam International Centre for Theoretical Physics
P.O.B. 586, 34100 Trieste, Italy

AND

V. FREILIKHER
The Jack and Pearl Resnick Institute for Advanced Technology
Department of Physics, Bar–Ilan University
52900 Ramat–Gan, Israel

Abstract. A formalism for study of spectral correlations in non–Gaussian, unitary invariant ensembles of large random matrices with strong level confinement is reviewed. It is based on the Shohat method in the theory of orthogonal polynomials. The approach presented is equally suitable for description of both local and global spectral characteristics, thereby providing an overall look at the phenomenon of spectral universality in Random Matrix Theory.

1. Introduction: Motivation and Basic Results

1.1. UBIQUITY OF INVARIANT RANDOM MATRIX MODELS

Random matrices [1, 2] have been introduced in a physical context since the pioneering works by Wigner [3] and Dyson [4, 5]. Initially proposed as an effective phenomenological model for description of the higher excitations in nuclei [6] they found numerous applications in very diverse fields of physics such as two dimensional quantum gravity [7], quantum chromodynamics [8], quantum chaos [9], and mesoscopic physics [10, 11]. One can state that from the standpoint of the mathematical formalism all these fields are pooled by the Random Matrix Theory (RMT). Such an ubiquity of random matrices owes its origin to the exclusive role played by symmetry. The most amazing evidence to this fact comes from *invariant random*

J.-P. Fouque (ed.), Diffuse Waves in Complex Media, 165–211.

matrix models which will be the focus of this review. The main feature of these models lies in that they discard (irrelevant) microscopic details of the physical system in question, but they do properly take into account its underlying fundamental symmetries. In accordance with the very idea of the construction of invariant matrix models, they do *not* relate to any dynamical properties of the physical object under study: general symmetry requirements alone lead to appearance of knowledge about the system. As far as the symmetry constraints follow from the first principles, the Random Matrix Theory turns out to be a general and powerful field–theoretical approach leading to a unified mathematical description of the quite different physical problems mentioned above.

A great variety of invariant random matrix ensembles can be assigned to three irreducible symmetry classes [5]. To specify them, we consider the typical line of arguing used in applications of the Random Matrix Theory to the disordered quantum mechanical systems, where it was first invented by Gor'kov and Eliashberg [12]. Since in this situation the microscopic Hamiltonian $\mathcal{H}$ is rather intricate, the integration of exact equations is impossible. It is therefore useful to appeal to statistical description by conjecturing that the operator $\mathcal{H}$ can be modelled by an $N \times N$ random matrix $\mathbf{H}$ whose eigenvalues and eigenvectors reproduce statistically the eigenlevels and eigenfunctions of the real microscopic Hamiltonian in the thermodynamic limit which corresponds to the matrix dimension N going to infinity. With this conjecture accepted, an ensemble of large random matrices $\mathbf{H}$, characterized by the joint distribution function $P[\mathbf{H}]$ of the matrix elements $\mathbf{H}_{ij}$ of the corresponding Hamiltonian $\mathcal{H}$, becomes the main object of study. Once the primary role of symmetry is postulated in the RMT–approach, the matrix $\mathbf{H}$ must adequately reflect the symmetry properties of the physical system under study. The matrix $\mathbf{H}$ is chosen to be real symmetric if the underlying physical system possesses time–reversal and rotational invariance. Systems with broken time–reversal symmetry are characterized by a Hermitean matrix $\mathbf{H}$, while systems with conserved time–reversal symmetry but with broken rotational invariance are described by a self–dual Hermitean matrix. These three symmetry classes referred to as the orthogonal, unitary and symplectic symmetry classes, respectively, can be characterized by a symmetry parameter β equals the number of independent elements in the off–diagonal entries of the matrix $\mathbf{H}$. The parameter $\beta = 1$ for a real symmetric matrix (orthogonal symmetry), $\beta = 2$ for a Hermitean matrix (unitary symmetry), and $\beta = 4$ for a self–dual Hermitean matrix (symplectic symmetry).

So far we did not yet specify the form of the joint distribution function $P[\mathbf{H}]$ of the matrix elements $\mathbf{H}_{ij}$. By definition, the invariant random matrix ensembles are characterized by

$$P[\mathbf{H}] = \frac{1}{\mathcal{Z}_N} \exp\{-\beta \mathrm{Tr} V[\mathbf{H}]\}. \tag{1}$$

Here the function $V[\mathbf{H}]$ should ensure the existence of the partition function $\mathcal{Z}_N$, which is defined by the normalization condition $\int P[\mathbf{H}]\, d[\mathbf{H}] = 1$ with the elementary volume $d[\mathbf{H}]$ depending on the symmetry of the matrix $\mathbf{H}$. For $\beta = 1$ the volume element $d[\mathbf{H}] = \prod_{i\leq j} d\mathbf{H}_{ij}$, for $\beta = 2$ the volume element $d[\mathbf{H}] = \prod_{i\leq j} d\,\mathrm{Re}\,\mathbf{H}_{ij} \prod_{i<j} d\mathrm{Im}\mathbf{H}_{ij}$, while for $\beta = 4$ it equals $d[\mathbf{H}] = \prod_{i\leq j} d\mathbf{H}_{ij}^{(0)} \prod_{\sigma=1}^{3} \prod_{i<j} d\mathbf{H}_{ij}^{(\sigma)}$. The presence of the trace in Eq. (1) leads to the invariance of the probability density $P[\mathbf{H}]\, d[\mathbf{H}]$ under the similarity transformation $\mathbf{H} \to \mathcal{R}_\beta^{-1}\mathbf{H}\mathcal{R}_\beta$ with $\mathcal{R}_\beta$ being an orthogonal, unitary or symplectic $N \times N$ matrix for $\beta = 1, 2$ or 4, respectively. In turns, the invariance built in the probability density $P[\mathbf{H}]\, d[\mathbf{H}]$ implies that *there is no preferential basis in the space of matrix elements.* From the physical point of view, this means that invariant matrix models given by Eq. (1) are applicable to particular regimes of a physical system where (i) all the normalized linear combinations of the eigenstates have similar properties, and where (ii) the dimensionality is irrelevant. In disordered systems this is just a metallic state where the typical electron states are extended and hence structureless.

Notice that up to this point we have no constraints allowing us to uniquely choose the function $V[\mathbf{H}]$ (referred to as "confinement potential"). However, if we impose the additional requirement that the entries of the random matrix $\mathbf{H}$ be statistically independent of each other, we immediately arrive at the Gaussian Orthogonal, Unitary and Symplectic Ensembles (GOE, GUE, GSE) which are characterized by the quadratic confinement potential $V[\mathbf{H}] \propto \mathbf{H}^2$. This particular form of confinement potential leads to significant mathematical simplifications which allowed the complete treatment of these three ensembles many years ago [13].

It is remarkable that (even for non–Gaussian distributions of $P[\mathbf{H}]$) the invariant random matrix ensembles possess a great degree of mathematical tractability, and, what is more important, they have a high physical relevance, being much more than just a mathematical construction. In this respect we mention that in the physics of disordered systems the applicability of Gaussian invariant random matrix ensembles to description of weakly disordered systems has been proven by Efetov [14] by using the nonlinear σ–model. In that study [14] the statistical properties of energy levels for metallic particles with volume imperfections were considered by solving the Schrödinger equation with nonperturbative averaging over the random potential configurations within the framework of the supersymmetry method. Random Matrix Theory appears there as a zero–dimensional version of a more general microscopic nonlinear σ–model, thereby proving the validity of the basic principles used for an RMT phenomenological description of energy levels of noninteracting electrons confined in a restricted volume. This connection takes place at times much larger than the ergodic time needed for diffusive particle to completely and homogeneously fill the available volume of the sample provided it is in the metallic regime characterized by a dimensionless Thouless conductance $g \gg 1$.

Let us point out that the choice of quadratic confinement potential $V[\mathbf{H}]$ can hardly be justified. Indeed, it was understood from the very beginning [13] that the requirement of statistical independence of the matrix elements $\mathbf{H}_{ij}$ is not motivated by the first principles and, therefore, the important problem of elucidating the influence of a particular form of confinement potential on the predictions of the Random Matrix Theory developed for Gaussian Invariant Ensembles had been posed already in the sixties. Despite this fact, considerable progress in study of spectral properties of *non–Gaussian* random matrix ensembles was achieved almost thirty years later when RMT experienced a great renaissance due to new ideas in the physics of disordered/chaotic systems which had led the birth of mesoscopic physics, as well as due to a penetration of Random Matrix Theory to quantum chromodynamics (QCD). In the latter field, the Random Matrix Theory turned out to be a useful tool for understanding the spectral properties of low–lying eigenvalues of the Dirac operator. The idea of introducing the RMT–approach in QCD is very similar to that in the physics of disordered systems and is based on the conjecture [8, 15] that the spectral density of the Dirac operator very close to the spectrum origin should depend only on the symmetries in question. One startling consequence of this conjecture is that the spectral density of the Dirac operator near the origin need not be computed within the framework of the gauge theories at all, but it can be extracted from much simpler random matrix ensembles reflecting the symmetry of the problem. The matrix models appearing in the context of QCD are manifestly non–Gaussian possessing an additional chiral structure[1] [16]. Another motivation for studying the spectral properties of non–Gaussian random matrix ensembles comes from the theories of $2D$ quantum gravity [7].

1.2. NON–GAUSSIAN RANDOM MATRIX ENSEMBLES AND

[1]Throughout the paper we consider non–chiral non–Gaussian random matrix ensembles. It can be shown that an arbitrary chiral matrix model can be reduced to an auxiliary one without chirality; see, for instance, Ref. [17].

The examples above clearly demonstrate an important role the non–Gaussian random matrix ensembles play in different physics theories, and serve as a compelling evidence of the necessity to have a powerful method for study of their spectral properties which could be equally applicable to rather different probability measures $P[\mathbf{H}]$. During (mostly) the recent decade a number of methods were developed in order to treat non–Gaussian random matrix ensembles. All of them can schematically be related to two groups.

1.2.1. *Global Universality*

The first group includes different approximate methods useful to explore *global spectral characteristics* which manifest themselves on the scale of $n \gg 1$ eigenlevels. Among these methods there are (i) the mean–field approximation proposed by Dyson [4, 18] which allows one to compute the density of levels in random matrix ensemble; (ii) the Schwinger–Dyson loop equations' technique [19, 20, 21] that had led to discovery of the phenomenon of the *global spectral universality*; (iii) the method of functional derivative of Beenakker [22, 23], and (iv) the diagrammatic approach of Brézin and Zee [24] whose development enabled their authors to study the phenomenon of global universality [19] in more detail as well as to generalize it in the context of mesoscopic physics.

It was found that contrary to the one–point spectral characteristics (such as one–point Green's function or level density) which essentially depend on the measure $P[\mathbf{H}]$, i.e. on the explicit form of the confinement potential, the *functional form* of (connected) two–point correlators becomes insensitive to the details of confinement potential upon smoothing over the scale which is much larger than the mean level spacing Δ_N but much smaller than the scale of the entire spectrum support. This is the essence of the phenomenon of global spectral universality. For example, the smoothed connected density–density correlator computed for the matrix model Eq. (1) is given by the universal function

$$\rho_c^{(N)}(\lambda, \lambda') = -\frac{1}{\pi^2 \beta (\lambda - \lambda')^2} \frac{\mathcal{D}_N^2 - \lambda\lambda'}{(\mathcal{D}_N^2 - \lambda^2)^{1/2} (\mathcal{D}_N^2 - \lambda'^2)^{1/2}}, \tag{2}$$

where $\lambda \neq \lambda'$. This form of $\rho_c^{(N)}$ is valid for random matrices whose spectrum is supported on a single, symmetric interval $(-\mathcal{D}_N, +\mathcal{D}_N)$. Here, the universality implies that all the information about the particular form of $V[\mathbf{H}]$ is encoded into $\rho_c^{(N)}$ only through the end point $\mathcal{D}_N$ of the spectrum support.

The methods mentioned above, being applicable to study of the spectral correlations in the long–range regime for all three symmetry classes, leave aside the fine structure of eigenvalue correlations manifested on the scale of the mean eigenvalue spacing. In this sense, these approaches are less informative as compared to the method of orthogonal polynomials [1] which, along with the supersymmetry approach [25, 26], enters the second group.

1.2.2. *Local Universality*

At present, the orthogonal polynomial technique, originally developed by Gaudin and Mehta [27], seems to be the most powerful one furnishing us the possibility to probe both the global spectral characteristics (which are of importance in computing the integral spectral properties) and the local spectral correlations (which describe a dynamics of underlying physical system) for arbitrary symmetry classes.

(i) The early attempts [28, 29, 30] to go beyond the Gaussian distribution of $P[\mathbf{H}]$ were concentrated on unitary invariant, $\mathrm{U}(N)$, matrix ensembles ($\beta = 2$) associated with classical orthogonal polynomials. It was found in Refs. [28, 29, 30, 31] that (in the thermodynamic limit $N \to \infty$) the scalar two-point kernel (in terms of which all n-point correlation functions are expressible, see Sec. 2.2) computed in the bulk of the spectrum, i.e. far from the end points of the eigenvalue support, follows the *sine law*

$$K_{\text{bulk}}(s, s') = \frac{\sin[\pi(s - s')]}{\pi(s - s')}, \tag{3}$$

that inevitably leads to the famous Wigner–Dyson level statistics [32] inherent in nuclear physics, weakly disordered and chaotic systems. This form of the scalar kernel corresponds to the *bulk scaling limit*, when the initial spectral variables λ, λ' are measured in the units of the mean eigenvalue spacing Δ_N, so that the scaled variable $s = \lambda/\Delta_N$. The same form of the scalar kernel Eq. (3) obtained for different ensembles associated with classical orthogonal polynomials was the first evidence to the phenomenon currently known as the *local spectral universality.* Later, it was shown in Ref. [31] that spectral properties of orthogonal $\mathrm{O}(N)$ and symplectic $\mathrm{Sp}(N)$ ($\beta = 1$ and $\beta = 4$) invariant ensembles associated with the weights of classical orthogonal polynomials[2] follow the predictions of GOE and GSE, respectively.

A further significant progress in the field came with the works [35, 36] whose authors considered spectral properties of $\mathrm{U}(N)$ invariant random matrix ensembles associated with strong symmetric confinement potentials of the form $V[\mathbf{H}] = \mathbf{H}^2 + \gamma\mathbf{H}^4$ ($\gamma > 0$) and $V[\mathbf{H}] = \sum_{k=1}^{p} a_k \mathbf{H}^{2k}$ ($a_p > 0$),

[2]In this situation the spectral correlations are expressible through the 2×2 matrix kernel, that can be computed by using so-called skew orthogonal polynomials [33]. In fact, as was recently shown in Ref. [34], the ensembles with $\beta = 1$ and $\beta = 4$ can be also treated without appealing to the skew polynomials.

respectively. Both works, based on different conjectures about the functional form of asymptotics of polynomials orthogonal with respect to a non–Gaussian measure, restricted their attention to the spectrum bulk, where the two–point kernel was shown, once again, to follow the sine law, Eq. (3). A rigorous treatment of a richer class of $\mathrm{U}(N)$ invariant random matrix ensembles related to the Freud and Erdös–type orthogonal polynomials was given in Refs. [37, 38]. It was proven there that the universal sine law for the two–point kernel holds for a wide class of monotonic (not necessarily of polynomial form) non–singular confinement potentials $V(\lambda)$ which increase at least as $|\lambda|$ at infinity, and can grow as or even faster than any polynomial at infinity. Confinement potentials satisfying these properties are referred to as *strong confinement potentials.* This definition takes its origin in the limits [37] of spectral universality and is non–accidently connected to the problem of determinate and indeterminate moments [39]. Also, it was demonstrated in Ref. [38] that an intimate connection exists between the structure of Szegö functional [40] entering the strong pointwise asymptotics of orthogonal polynomials and the mean–field equation by Dyson for mean level density that has been derived in Ref. [38] within the framework of orthogonal polynomial technique.

(ii) All the random matrix ensembles treated in Refs. [28, 29, 30, 31, 35, 36, 37, 38] were characterized by strong confinement potential with no singularities. However, (logarithmic) singularities do appear when one considers chiral matrix ensembles, arising in the context of QCD [8], in the theory of mesoscopic electron transport [41] and in description of electron level statistics in normalconducting–superconducting hybrid structures [42]. An example of a rather general (though non–chiral) random matrix ensemble possessing a log–singular level confinement is given by the distribution

$$P[\mathbf{H}] = \frac{1}{\mathcal{Z}_N} |\det \mathbf{H}|^{\alpha\beta} \exp\{-\beta \mathrm{Tr} V[\mathbf{H}]\}, \tag{4}$$

where the function $V[\mathbf{H}]$ is a well behaved function which has not singular points. Ensemble Eq. (4), being a natural generalization of the matrix ensemble proposed by Bronk [30], was first considered in the situations where associated orthogonal polynomials were classical [30, 43, 44, 45]. For quadratic confinement potential and $\beta = 2$ one obtains that in the vicinity of the singularity, $\lambda = 0$, the scalar kernel satisfies the *Bessel law*

$$K_{\mathrm{orig}}(s,s') = \frac{\pi}{2}(ss')^{1/2} \frac{J_{\alpha+1/2}(\pi s) J_{\alpha-1/2}(\pi s') - J_{\alpha-1/2}(\pi s) J_{\alpha+1/2}(\pi s')}{s - s'}, \tag{5}$$

where s and s' are scaled by the level spacing $\Delta_N(0)$ near the spectrum origin, $s = \lambda/\Delta_N(0)$, and $\alpha > -1/2$. This scaling procedure is referred to as the *origin scaling limit.* Extensions to two other symmetry classes, as well as to the chiral matrix ensembles, can be found in Refs. [43, 44, 45, 46]. An important breakthrough in understanding the universal character of this kernel was given in Refs. [47, 17]. These authors, guided by QCD applications, have shown that the Bessel kernel Eq. (5) is again universal, being independent of the details of strong confinement potential $V[\mathbf{H}]$. An alternative proof of universality that holds more generally was presented in Ref. [48].

(iii) The third type of universal correlations takes place near the soft edge of the spectrum support, which is of special interest in the models of two-dimensional quantum gravity. The first study of the level density near the end point of the spectrum support is due to Wigner [49]. More comprehensive description of the tails of the density of states was done in Ref. [50]. It was shown there that at $\beta = 2$ a universal crossover occurs from a nonzero density of states to a vanishing one that is independent of the confining potential in the *soft-edge scaling limit.* Later it was demonstrated [51] that eigenvalue *correlations* in the $\mathrm{U}(N)$ invariant matrix ensembles with quartic and sextic confinement potentials are determined by the scalar kernel obeying the *Airy law* [52]

$$K_{\mathrm{soft}}(s, s') = \frac{\mathrm{Ai}(s)\,\mathrm{Ai}'(s') - \mathrm{Ai}(s')\,\mathrm{Ai}'(s)}{s - s'}, \tag{6}$$

suggesting that the Airy kernel should be universal as well. (Here the rescaling $s \propto N^{2/3}(\lambda/\mathcal{D}_N - 1)$ determines the soft-edge scaling limit). This conjecture has been proven in Ref. [53], where it was also shown that the Airy correlations, being universal for a class of matrix models with monotonic confinement potential or with that having light local extrema, are indeed a particular case of more general universal multicritical correlations [53].

1.2.3. *Universality in a Broader Context*

Are the universal scalar kernels in unitary invariant random matrix models with strong level confinement exhausted by the universal sine, Bessel and Airy laws given by Eqs. (3), (5) and (6) above? The present state of the art leads us to the negative answer. Indeed, by adding a singular component to the strong level confinement we may either accumulate a finite number of eigenvalues near the singular point λ_{sing} or repel them from it. Such a rearrangement of eigenlevels will lead to emergence of a new scalar two-point kernel in the vicinity of the singular point of the spectrum. (In particular case of the logarithmic singularity this kernel will follow the Bessel law Eq. (5)). Generically, it is natural to expect that the functional form of the kernel near λ_{sing} will be sensitive to the particular type of the singular deformation. However, the new two-point kernel will be insensitive to the details of the background component of the confinement potential, as it takes place in the case of the Bessel kernel. In this sense, one can still say about a phenomenon of universality. One of the latest evidences to this fact can be found in Ref. [54] where deformed ensembles of large random matrices associated with massive Dirac operators were considered.

By the same token, the universal global spectral correlator expressed by Eq. (2) is not the only one arising in Random Matrix Theory. As it was already stressed, the universal function Eq. (2) is inherent in matrix models with spectra possessing a single connected support. Correspondingly, ensembles of large random matrices with eigenvalue densities having more than one–cut support will give rise to the novel global spectral correlators whose universality classes (for a given symmetry parameter β) are characterized entirely by the number of cuts in the support of spectral density. This was explicitly demonstrated in the recent studies [20, 55] by means of the loop equation technique.

1.3. THE AIM

At this point, it is appropriate to notice that all the mentioned above (universal) results for both global and local eigenvalue correlations have been obtained by using different methods, each of them was only suitable for a particular problem under consideration. Any deformations of $P[\mathbf{H}]$ (which will preserve its invariance) would cause principal difficulties in elucidating the influence of these deformations on spectral properties of corresponding random matrix ensembles. The goal of this paper is to represent a general method (recently developed in a series of publications [53, 48, 56]), which is equally suitable for study of both local and global eigenvalue correlations, and easily leads to generalizations. The approach we introduce is based on a simple and elegant idea [57] by J. Shohat (which goes back to 1930), providing a detailed description of the spectral properties of non–Gaussian $\mathrm{U}(N)$ invariant random matrix ensembles through the analysis of the three–term recurrence equation for associated orthogonal polynomials. We show that for the most situations of interest, the knowledge of the large–N behavior of the coefficients in the recurrence equation is sufficient to directly reconstruct the local eigenvalue correlations of arbitrary order, as well as to explore the global spectral statistics. In the case of a non–singular, well behaved confinement potential, the knowledge of such a large–N behavior of the recurrence coefficients is equivalent, in fact, to a knowledge of the Dyson density of states for the corresponding random matrix ensemble. The latter assertion leads to a rather unexpected conclusion: Once the Dyson density of states (which is a rather crude one–point spectral characteristics) is available, the scalar kernel (and hence the n–point spectral correlators) can immediately be recovered through the solution of a certain second-order differential equation (See Sec. 4.3).

We believe that this method offers not only new computational potentialities, but also provides a different, overall look at the problem of eigenvalue correlations in unitary invariant random matrix ensembles in arbitrary spectrum range and in arbitrary scaling limits. It seems that together with the very recent works [58, 59] establishing a precise connection of the scalar kernel for random matrix ensembles with $\mathrm{U}(N)$ symmetry with the 2×2 matrix kernels in ensembles with $\mathrm{O}(N)$ and $\mathrm{Sp}(N)$ symmetries, the formalism to be reviewed below gives a rather complete solution of the problem of eigenvalue correlations in invariant matrix models with strong level confinement.

The review is organized as follows. Section 2 contains a brief description of the Gaudin–Mehta calculational scheme, that introduces the orthogonal polynomials as a tool for exact evaluation of n-point correlation functions in $\mathrm{U}(N)$ invariant random matrix ensembles. In Section 3 the Shohat method in the theory of orthogonal polynomials is presented. Section 4 is devoted to a detailed study of spectral properties of large Hermitean random matrices with a single connected eigenvalue support. In Section 5 we extend this analysis to random matrices with eigenvalue gap. Section 6 contains conclusions. The most lengthy calculations are collected in three Appendices.

2. Elements of the Gaudin–Mehta formalism

2.1. INVARIANT RANDOM MATRIX MODEL IN EIGENVALUE REPRESENTATION: TWO INTERPRETATIONS

The invariance of the distribution function $P[\mathbf{H}]$ implies that different matrices with the same eigenvalues have the same probability of occurring. To study spectral characteristics of an invariant random matrix model it is convenient to integrate out "auxiliary" angular variables in the construction $P[\mathbf{H}]\, d[\mathbf{H}]$ in order to get the matrix model in the eigenvalue representation. To proceed with this, we have to pass from the integration over independent elements $\mathbf{H}_{ij}$ of the matrix $\mathbf{H}$ to the integration over the smaller space of its N eigenvalues $\{\lambda\}$, calculating the corresponding Jacobian J.

Let us introduce the matrix $\mathcal{R}_\beta$ that diagonalizes the random matrix $\mathbf{H}$, so that $\mathbf{H} = \mathcal{R}_\beta^{-1} \Lambda \mathcal{R}_\beta$ and $\Lambda = \mathrm{diag}(\lambda_1, ..., \lambda_N)$, and consider the infinitesimal variation of $\mathbf{H}$,

$$\delta\mathbf{H} = \mathcal{R}_\beta^{-1} \left(\mathcal{R}_\beta \delta \mathcal{R}_\beta^{-1} \Lambda + \delta\Lambda + \Lambda \delta \mathcal{R}_\beta \mathcal{R}_\beta^{-1} \right) \mathcal{R}_\beta = \mathcal{R}_\beta^{-1} \left(\delta\Lambda - i\,[\Lambda, \delta s] \right) \mathcal{R}_\beta. \tag{7}$$

Here we have denoted $\delta s = i \delta \mathcal{R}_\beta \mathcal{R}_\beta^{-1}$. The norm, $\|\delta\mathbf{H}\|^2 = \mathrm{Tr}\,(\delta\mathbf{H})^2$, of the infinitesimal variation of $\mathbf{H}$ is

$$\|\delta\mathbf{H}\|^2 = \mathrm{Tr}\,(\delta\Lambda)^2 - 2i\,\mathrm{Tr}\,([\delta\Lambda, \Lambda]\,\delta s) + 2\,\mathrm{Tr}\left(-\delta s \Lambda \delta s \Lambda + (\delta s)^2 \Lambda^2 \right). \tag{8}$$

The second term in the last expression vanishes as the matrices $\delta\Lambda$ and Λ are diagonal. We then find

$$||\delta\mathbf{H}||^2 = \sum_i (\delta\lambda_i)^2 + \sum_{i,j} (\lambda_i - \lambda_j)^2 |\delta s_{ij}|^2 . \quad (9)$$

The independent variables are the variations of the eigenvalues $\delta\lambda_i$ and δs_{ij} for $\beta = 1$, $\mathrm{Re}\,\delta s_{ij}$, $\mathrm{Im}\,\delta s_{ij}$ for $\beta = 2$, or $\delta s_{ij}^{(\sigma)}$ with $\sigma = 0,1,2,3$ for $\beta = 4$, $i < j$. Then, from Eq. (9) we obtain the Jacobian J of the transformation $\mathbf{H} \mapsto \{\lambda_i, \mathcal{R}_\beta\}$ that is equal to $\sqrt{\det G}$, where G is the metric tensor with $\det G = \prod_{i\neq j} |\lambda_i - \lambda_j|^\beta$. Hence

$$J = \prod_{i<j} |\lambda_i - \lambda_j|^\beta = |\Delta(\lambda)|^\beta \quad (10)$$

with $\Delta(\lambda)$ being the Vandermonde determinant. The construction Eq. (10) is also known as the Jastrow factor.

Combining Eqs. (1) and (10) we arrive at the famous expression for the joint probability density function of the eigenvalues $\{\lambda\}$ of the matrix $\mathbf{H}$:

$$\begin{aligned} \rho(\lambda_1, \ldots, \lambda_N) &= \mathcal{Z}_N^{-1} \exp\{-\beta \sum_{i=1}^{N} V(\lambda_i)\} |\Delta(\lambda)|^\beta & (11) \\ &= \mathcal{Z}_N^{-1} \exp\{-\beta[\sum_i V(\lambda_i) - \sum_{i<j} \ln|\lambda_i - \lambda_j|]\}. & (12) \end{aligned}$$

The probability distribution given by Eq. (12) has the form of a Gibbs distribution for a classical one–dimensional system of N "particles" at "positions" λ_i confined by the external one–body potential $V(\lambda)$ and *interacting* with each other through the pairwise logarithmic law originated from the Jacobian J of the transformation $\mathbf{H} \mapsto \{\lambda_i, \mathcal{R}_\beta\}$. The symmetry parameter β plays the role of the equilibrium temperature. Such an interpretation of Eq. (12) gives rise to approximate mean–field methods in the Random Matrix Theory.

For invariant matrix ensembles with unitary symmetry ($\beta = 2$), the probability distribution in the form of Eq. (11) can alternatively be related to a system of fictitious *non-interacting fermions*, that can be described by effective one–particle Schrödinger equation [53]. This equation is a cornerstone of the method under review. Although such a simple, transparent interpretation cannot be ascribed to $\rho(\lambda_1, ..., \lambda_N)$ for two other symmetry classes, $\beta = 1$ and $\beta = 4$, a recently discovered deep connection [58, 59] between orthogonal, unitary and symplectic ensembles makes the unitary invariant ensembles of random matrices to be the central and most important.

2.2. ORTHOGONAL POLYNOMIALS' TECHNIQUE: $\beta = 2$

For $\beta = 2$ the structure of Eq. (11) enables us to exactly represent all the statistical characteristics of the spectrum in the terms of polynomials orthogonal with respect to a non–Gaussian measure $d\alpha$. To demonstrate this, it is useful to write down the joint probability density function of the eigenvalues $\{\lambda\}$ in the form

$$\rho(\lambda_1, \ldots, \lambda_N) = \Psi_0^2(\lambda_1, \ldots, \lambda_N), \tag{13}$$

$$\Psi_0(\lambda_1, \ldots, \lambda_N) = \mathcal{Z}_N^{-1/2} \exp\{-\sum_{i=1}^{N} V(\lambda_i)\} \Delta(\lambda). \tag{14}$$

One can see that Ψ_0 can be thought of as a wave function of N fictitious non–interacting fermions. Namely, noting that

$$\Delta(\lambda) = \prod_{i>j=1}^{N} (\lambda_i - \lambda_j) = \det \left\| \lambda_i^{j-1} \right\|, \tag{15}$$

and taking the linear combinations of the columns of the initial matrix with entries λ_i^{j-1}, one can reduce this matrix to the matrix whose entries are arbitrary polynomials $P_{j-1}(\lambda_i)$ of degrees $j - 1 = 0, 1, \ldots, N - 1$,

$$\Delta(\lambda) = \det \| P_{j-1}(\lambda_i) \| . \tag{16}$$

Further, choosing these polynomials to be orthogonal with respect to the measure $d\alpha(\lambda) = \exp\{-2V(\lambda)\} d\lambda$,

$$\int_{\lambda \in (-\infty, +\infty)} d\alpha(\lambda) P_n(\lambda) P_m(\lambda) = \delta_{nm}, \tag{17}$$

we arrive at the conclusion that the function Ψ_0 in Eq. (14) can be represented as a Slater determinant

$$\Psi_0(\lambda_1, \ldots, \lambda_N) = \frac{1}{\sqrt{N!}} \det \| \varphi_{j-1}(\lambda_i) \| \tag{18}$$

that formally describes the system of N fictitious non–interacting fermions located at "spatial" points λ_i and characterized by the set of orthonormal "eigenfunctions"

$$\varphi_n(\lambda) = P_n(\lambda) \exp\{-V(\lambda)\}, \tag{19}$$

$$\int_{-\infty}^{+\infty} d\lambda \varphi_n(\lambda) \varphi_m(\lambda) = \delta_{nm}. \tag{20}$$

Bearing in mind the representation Eq. (13), and taking into account Eqs. (17) - (19), we readily get that

$$\rho(\lambda_1,\ldots,\lambda_N) = \frac{1}{N!}\det\|K_N(\lambda_i,\lambda_j)\|_{i,j=1\ldots N}, \tag{21}$$

where

$$K_N(\lambda,\lambda') = \sum_{k=0}^{N-1}\varphi_k(\lambda)\varphi_k(\lambda') \tag{22}$$

is the *scalar two-point kernel.* Making use of the Christoffel–Darboux theorem [60] (see Eq. (32) below), the two-point kernel can be reduced to the form

$$K_N(\lambda,\lambda') = c_N\frac{\varphi_N(\lambda')\varphi_{N-1}(\lambda) - \varphi_N(\lambda)\varphi_{N-1}(\lambda')}{\lambda'-\lambda}, \tag{23}$$

convenient for the further analysis. In Eq. (23) c_N is the coefficient in the three-term recurrence equation (see Eq. (28) below) for the set P_n of the polynomials orthogonal with respect to the measure $d\alpha$. This formula simplifies significantly RMT calculations, since the "eigenfunctions" φ_N with large "quantum numbers" $N \gg 1$ entering Eq. (23) can be replaced by their large-N asymptotics.

The n-point correlation function is determined in the Random Matrix Theory by the formula [1]

$$R_n(\lambda_1,\ldots,\lambda_n) = \frac{N!}{(N-n)!}\prod_{k=n+1}^{N}\int_{-\infty}^{+\infty}d\lambda_k\rho(\lambda_1,\ldots,\lambda_N). \tag{24}$$

It describes the probability to find n levels around each of the points $\lambda_1,\ldots,\lambda_n$ when the positions of the remaining levels are unobserved. The multiple integrals in the last equation can exactly be calculated by using the representation Eq. (21). The result of the integration reads [1]

$$R_n(\lambda_1,\ldots,\lambda_n) = \det\|K_N(\lambda_i,\lambda_j)\|_{i,j=1\ldots n}. \tag{25}$$

Equation (25) implies that the knowledge of the scalar two-point kernel $K_N(\lambda,\lambda')$ is sufficient to calculate the energy level correlation function of any order. In particular, the averaged density of states is expressed as

$$\langle\nu_N(\lambda)\rangle = R_1(\lambda) = K_N(\lambda,\lambda). \tag{26}$$

Analogously, Eq. (25) leads to the following expression for connected "density-density" correlation function $\rho_c^{(N)} = \langle\nu_N(\lambda)\nu_N(\lambda')\rangle - \langle\nu_N(\lambda)\rangle\langle\nu_N(\lambda')\rangle$,

$$\begin{aligned}\rho_c^{(N)} &= \delta(\lambda-\lambda')R_1(\lambda) + R_2(\lambda,\lambda') \\ &= \delta(\lambda-\lambda')K_N(\lambda,\lambda) - K_N^2(\lambda,\lambda').\end{aligned} \tag{27}$$

Thus, all the nontrivial information about eigenlevel correlations is contained in the squared two–point kernel $-K_N^2(\lambda, \lambda')$.

3. The Shohat Method: General Relations

Equations (13) and (18) suggest that the joint distribution function of N eigenvalues of a U(N) invariant random matrix ensemble can be interpreted as a probability of finding N fictitious non–interacting fermions to be confined in a one–dimensional space. The effective one–particle Schrödinger equation for the wave functions $\varphi_n(\lambda)$, Eq. (19), of these fictitious fermions can be derived by mapping a three–term recurrence equation for orthogonal polynomials, Eq. (17), onto a second–order differential equation. The method of reducing a recurrence equation to a differential equation is essentially due to J. Shohat who proved in 1939 that orthogonal polynomials associated with exponential weights satisfy a second–order differential equation [57]. Much later the Shohat method has got a further development in the work by Bonan and Clark [61]. By now, rather extended mathematical literature exists on this subject [62, 63, 64].

Let as consider a set of polynomials $P_n(\lambda)$ orthogonal on the entire real axis with respect to the measure $d\alpha(\lambda) = \exp\{-2V(\lambda)\}\, d\lambda$. If $V(\lambda)$ is an even function[3], $V(-\lambda) = V(\lambda)$, this set of orthogonal polynomials can be defined by the recurrence equation

$$\lambda P_{n-1}(\lambda) = c_n P_n(\lambda) + c_{n-1} P_{n-2}(\lambda), \tag{28}$$

where the coefficients c_n are uniquely determined by the measure $d\alpha$.

In order to derive the differential equation for the wave functions $\varphi_n(\lambda) = P_n(\lambda) \exp\{-V(\lambda)\}$, we note that the following identity takes place,

$$\frac{dP_n(\lambda)}{d\lambda} = A_n(\lambda) P_{n-1}(\lambda) - B_n(\lambda) P_n(\lambda), \tag{29}$$

with functions $A_n(\lambda)$ and $B_n(\lambda)$ to be determined from the following consideration. Since $dP_n(\lambda)/d\lambda$ is a polynomial of the degree $n-1$, it can be represented [60] through the Fourier expansion in the terms of the kernel $Q_n(t, \lambda) = \sum_{k=0}^{n-1} P_k(\lambda) P_k(t)$ as

$$\frac{dP_n(\lambda)}{d\lambda} = \int d\alpha(t) \frac{dP_n(t)}{dt} Q_n(t, \lambda). \tag{30}$$

Integrating by parts in the last equation we get that

$$\frac{dP_n(\lambda)}{d\lambda} = 2 \int d\alpha(t)\, Q_n(t, \lambda) \left(\frac{dV}{dt} - \frac{dV}{d\lambda} \right) P_n(t). \tag{31}$$

[3]For asymmetric confinement potentials the recurrence equation takes the form $(\lambda - b_n) P_{n-1}(\lambda) = c_n P_n(\lambda) + c_{n-1} P_{n-2}(\lambda)$. The additional parameter b_n can easily be incorporated into the calculational scheme.

Now, making use of the Christoffel–Darboux identity [60]

$$Q_n(t,\lambda) = \sum_{k=0}^{n-1} P_k(\lambda) P_k(t) = c_n \frac{P_n(t) P_{n-1}(\lambda) - P_n(\lambda) P_{n-1}(t)}{t-\lambda}, \quad (32)$$

we are led to

$$\begin{aligned} \frac{dP_n(\lambda)}{d\lambda} &= 2c_n \int d\alpha(t) \frac{V'(t) - V'(\lambda)}{t-\lambda} P_n(t) \\ &\times \left[P_n(t) P_{n-1}(\lambda) - P_n(\lambda) P_{n-1}(t)\right]. \end{aligned} \quad (33)$$

Comparison of this expression with Eq. (29) yields

$$A_n(\lambda) = 2c_n \int d\alpha(t) \frac{V'(t) - V'(\lambda)}{t-\lambda} P_n^2(t), \quad (34)$$

$$B_n(\lambda) = 2c_n \int d\alpha(t) \frac{V'(t) - V'(\lambda)}{t-\lambda} P_n(t) P_{n-1}(t). \quad (35)$$

Now one can obtain the *exact* differential equation for the eigenfunctions φ_n. Differentiating Eq. (29), consequently applying Eqs. (29) and (28), and taking into account Eq. (19) we derive after somewhat lengthy calculations

$$\frac{d^2\varphi_n(\lambda)}{d\lambda^2} - \mathcal{F}_n(\lambda) \frac{d\varphi_n(\lambda)}{d\lambda} + \mathcal{G}_n(\lambda) \varphi_n(\lambda) = 0, \quad (36)$$

where

$$\mathcal{F}_n(\lambda) = \frac{1}{A_n} \frac{dA_n}{d\lambda}, \quad (37)$$

$$\begin{aligned} \mathcal{G}_n(\lambda) &= \frac{dB_n}{d\lambda} + \frac{c_n}{c_{n-1}} A_n A_{n-1} - B_n \left(B_n + 2\frac{dV}{d\lambda} + \frac{1}{A_n}\frac{dA_n}{d\lambda}\right) \\ &+ \frac{d^2V}{d\lambda^2} - \left(\frac{dV}{d\lambda}\right)^2 - \frac{1}{A_n}\frac{dA_n}{d\lambda}\frac{dV}{d\lambda}. \end{aligned} \quad (38)$$

When deriving Eqs. (36), (37) and (38) we made use of the sum rule

$$B_n + B_{n-1} - \frac{\lambda}{c_{n-1}} A_{n-1} = -2\frac{dV}{d\lambda}, \quad (39)$$

directly following from Eqs. (34), (35), (28) and from oddness of $dV/d\lambda$.

Equation (36) is valid for arbitrary n. Previously, an equation of this type was known in the context of the Random Matrix Theory only for GUE, where $V(\lambda) = \lambda^2/2$. For such a confinement potential both functions $A_n(\lambda)$ and $B_n(\lambda)$ can readily be obtained from Eqs. (34) and (35), and are given by $A_n(\lambda) = 2c_n$ and $B_n(\lambda) = 0$. Taking into account that for GUE

$c_n = (n/2)^{1/2}$, we end up with $\mathcal{F}_n(\lambda) = 0$ and $\mathcal{G}_n(\lambda) = 2n + 1 - \lambda^2$. This allows us to interpret $\varphi_n(\lambda)$ as a wave function of the fermion confined by a parabolic potential,

$$\frac{d^2 \varphi_n^{\mathrm{GUE}}(\lambda)}{d\lambda^2} + \left(2n + 1 - \lambda^2\right) \varphi_n^{\mathrm{GUE}}(\lambda) = 0. \tag{40}$$

The effective Schrödinger equation (36) applies to general non–Gaussian random matrix ensembles as well, although the explicit calculation of $\mathcal{F}_n(\lambda)$ and $\mathcal{G}_n(\lambda)$ in this situation is a rather complicated task. In two cases of relatively simple measures with $V(\lambda) = \lambda^4/8 + q_3\lambda^3/6 + q_2\lambda^2/4 + q_1\lambda/2$ and $V(\lambda) = \lambda^6/12$ the functions $\mathcal{F}_n(\lambda)$ and $\mathcal{G}_n(\lambda)$ are known explicitly [62]. Significant simplifications, however, arise in the limit $n = N \gg 1$, which is just a thermodynamic limit of the Random Matrix Theory representing for us the most interest.

4. Random Matrices with Single Eigenvalue Support

In this Section we will be interested in the study of eigenvalue statistics for unitary invariant non–Gaussian large random matrices characterized by a distribution function $P[\mathbf{H}]$ given by Eq. (4). The confinement potential associated with this model is

$$V_\alpha(\lambda) = v(\lambda) - \alpha \log|\lambda|. \tag{41}$$

Here $v(\lambda)$ is the regular part of level confinement

$$v(\lambda) = \sum_{k=1}^{p} \frac{d_k}{2k} \lambda^{2k}, \tag{42}$$

with a positive leading coefficient, $d_p > 0$; the signs of the rest of the d_k can be arbitrary but they should lead to an eigenvalue density supported on a single connected interval, $\{\lambda\} \in (-\mathcal{D}_N, +\mathcal{D}_N)$. The parameter $\alpha > -1/2$ is the strength of the logarithmic singularity.

In Subsection 4.1 below, we demonstrate how the one–point spectral characteristics (density of states) can be obtained by making use of the recurrence equation (28). In Subsection 4.2, we turn to the study of the smoothed connected "density–density" correlator, also starting with recurrence equation (28). Finally, in Subsection 4.3, we obtain the universal scalar kernels in the origin, bulk and soft–edge scaling limits by solving the effective Schrödinger equation for fictitious fermions.

4.1. MACROSCOPIC LEVEL DENSITY FROM RECURRENCE EQUATION

We start with an explanation of the main idea of the derivation to make clear all the subsequent calculations. Here we mainly follow Ref. [65]. Our basic observation is that in the large–N limit the density of states[4] $\nu_N(\lambda)$ consists of two parts,

$$\nu_N(\lambda) = \nu_N^{\text{smooth}}(\lambda) + \nu_N^{\text{osc}}(\lambda). \tag{43}$$

The smooth part $\nu_N^{\text{smooth}}(\lambda)$ contributes to different integral characteristics determined by the density of states, while the oscillating part does not, because any integration will level the oscillating features. Then, for some smooth, well behaved, even function $f(\lambda)$ we have

$$\int_{-\infty}^{+\infty} d\lambda f(\lambda)\, \nu_N(\lambda) \stackrel{N\to\infty}{\rightrightarrows} \int_{\lambda\in\text{support}} d\lambda f(\lambda)\, \nu_N^{\text{smooth}}(\lambda). \tag{44}$$

Let us implement this scheme by choosing (without any loss of generality) $f(\lambda) = \lambda^{2s}$, with s being a positive integer. This choice is possible due to the evenness of $\nu_N(\lambda)$. By definition Eq. (26), we obtain from Eq. (23)

$$\nu_N(\lambda) = c_N \exp\{-2V_\alpha(\lambda)\} \left(P_N^{(\alpha)\prime}(\lambda)\, P_{N-1}^{(\alpha)}(\lambda) - P_{N-1}^{(\alpha)\prime}(\lambda)\, P_N^{(\alpha)}(\lambda)\right). \tag{45}$$

Having in mind the relation Eq. (29), the sum rule Eq. (39) and the definition Eq. (19) we come down to

$$\begin{aligned}\nu_N(\lambda) \; = \; & c_N \left[A_N^{(\alpha)}(\lambda)\left(\varphi_{N-1}^{(\alpha)}(\lambda)\right)^2 + \frac{c_N}{c_{N-1}} A_{N-1}^{(\alpha)}(\lambda)\left(\varphi_N^{(\alpha)}(\lambda)\right)^2 \right. \\ & \left. -\varphi_N^{(\alpha)}(\lambda)\,\varphi_{N-1}^{(\alpha)}(\lambda)\left(\frac{\lambda}{c_{N-1}} A_{N-1}^{(\alpha)}(\lambda) + B_N^{(\alpha)}(\lambda) - B_{N-1}^{(\alpha)}(\lambda)\right)\right].\end{aligned} \tag{46}$$

Here the upper index (α) is used to reflect the presence of a log–singular component in confinement potential $V_\alpha(\lambda)$, Eq. (41). Thus, the level density is expressed through the wave functions $\varphi_N^{(\alpha)}(\lambda)$, and through the functions $A_N^{(\alpha)}(\lambda)$ and $B_N^{(\alpha)}(\lambda)$, given by Eqs. (34) and (35).

For further convenience we introduce two quantities,

$$\Lambda_{2\sigma}^{(N)} = \int d\alpha(t)\left(P_N^{(\alpha)}(t)\right)^2 t^{2\sigma} = \int_{-\infty}^{+\infty} dt \left(\varphi_N^{(\alpha)}(t)\right)^2 t^{2\sigma}, \tag{47}$$

$$\Gamma_{2\sigma+1}^{(N)} = \int d\alpha(t)\, P_N^{(\alpha)}(t)\, P_{N-1}^{(\alpha)}(t)\, t^{2\sigma+1} = \int_{-\infty}^{+\infty} dt \varphi_N^{(\alpha)}(t)\, \varphi_{N-1}^{(\alpha)}(t)\, t^{2\sigma+1}, \tag{48}$$

for which the alternative explicit integral representations are shown to exist in the Appendix A. Without going into details of those calculations, we

[4]Hereafter we use the notation $\nu_N(\lambda)$ for averaged density of states, Eq. (26).

only stress an extremely important role played by both the large–N limit, Eq. (124), of the recurrence equation for associated orthogonal polynomials and the asymptotic expansion, Eq. (125), deduced from Eq. (124). We also notice that owing to the existence of these explicit representations, derived by using a large–N version of the recurrence equation (28), all further calculations became possible.

The functions $A_N^{(\alpha)}$ and $B_N^{(\alpha)}$ entering Eq. (46) can be expressed in terms of $\Lambda_\sigma^{(N)}$ and $\Gamma_\sigma^{(N)}$. Namely, bearing in mind the definitions given by Eqs. (34) and (35) we obtain for $A_N^{(\alpha)}(\lambda) = A_{\text{reg}}^{(N)}(\lambda) + \alpha A_{\text{sing}}^{(N)}(\lambda)$,

$$A_{\text{reg}}^{(N)}(\lambda) = 2c_N \sum_{k=1}^{p} d_k \sum_{\sigma=1}^{k} \Lambda_{2(k-\sigma)}^{(N)} \lambda^{2\sigma-2}, \tag{49}$$

$$A_{\text{sing}}^{(N)}(\lambda) = 2c_N \int \frac{d\alpha(t)}{t} \left(P_N^{(\alpha)}(t)\right)^2. \tag{50}$$

In the same way, the function $B_N^{(\alpha)}(\lambda) = B_{\text{reg}}^{(N)}(\lambda) + \alpha B_{\text{sing}}^{(N)}(\lambda)$ is given by

$$B_{\text{reg}}^{(N)}(\lambda) = 2c_N \sum_{k=1}^{p} d_k \sum_{\sigma=1}^{k} \Gamma_{2(k-\sigma)-1}^{(N)} \lambda^{2\sigma-1}, \tag{51}$$

$$B_{\text{sing}}^{(N)}(\lambda) = \frac{2c_N}{\lambda} \int \frac{d\alpha(t)}{t} P_N^{(\alpha)}(t)\, P_{N-1}^{(\alpha)}(t). \tag{52}$$

When deriving these formulas we have used the fact of evenness of the measure $d\alpha(t)/dt$ and of $\left(P_N^{(\alpha)}(t)\right)^2$, as well as the expansion

$$\frac{t^k - \lambda^k}{t - \lambda} = \sum_{m=1}^{k} t^{k-m} \lambda^{m-1}. \tag{53}$$

The "singular" components, $A_{\text{sing}}^{(N)}$ and $B_{\text{sing}}^{(N)}$, can easily be determined. First, due to oddness of the integrand in Eq. (50), we have $A_{\text{sing}}^{(N)}(\lambda) \equiv 0$. Second, in order to find $B_{\text{sing}}^{(N)}$, we notice that the quantity

$$\gamma_n^{(\alpha)} = c_n \int \frac{d\alpha(t)}{t} P_n^{(\alpha)}(t)\, P_{n-1}^{(\alpha)}(t), \tag{54}$$

where n is not necessarily large, obeys the identity $\gamma_n^{(\alpha)} + \gamma_{n-1}^{(\alpha)} = 1$, which is a direct consequence of the recurrence equation (28). As far as $\gamma_{2n}^{(\alpha)} =$

$\gamma_{2n-2}^{(\alpha)} = \ldots = \gamma_2^{(\alpha)} \equiv 0$, we conclude that $\gamma_n^{(\alpha)} = [1-(-1)^n]/2$ and therefore, $B_{\text{sing}}^{(N)}(\lambda) = 2\gamma_N^{(\alpha)}/\lambda$. Hence we are led to the following representations,

$$A_N^{(\alpha)}(\lambda) = 2c_N \sum_{k=1}^{p} d_k \sum_{\sigma=1}^{k} \Lambda_{2(k-\sigma)}^{(N)} \lambda^{2\sigma-2}, \tag{55}$$

$$B_N^{(\alpha)}(\lambda) = 2c_N \sum_{k=1}^{p} d_k \sum_{\sigma=1}^{k} \Gamma_{2(k-\sigma)-1}^{(N)} \lambda^{2\sigma-1} + \alpha \frac{1-(-1)^N}{\lambda}. \tag{56}$$

With these preliminarily calculations in hand we are ready to implement the idea of recovering the Dyson density of states from the recurrence equation. In accordance with Eqs. (44) and (46) there are five contributions to the integral in the r.h.s. of Eq. (44) corresponding to five terms in Eq. (46). Substituting Eqs. (55) and (56) into Eq. (46) and performing a formal integration with the help of Eqs. (47) and (48), these contributions are found to be

$$\rho_1 = 2c_N^2 \sum_{k=1}^{p} d_k \sum_{\sigma=1}^{k} \Lambda_{2(k-\sigma)}^{(N)} \Lambda_{2(\sigma+s-1)}^{(N-1)}, \tag{57}$$

$$\rho_2 = 2c_N^2 \sum_{k=1}^{p} d_k \sum_{\sigma=1}^{k} \Lambda_{2(k-\sigma)}^{(N-1)} \Lambda_{2(\sigma+s-1)}^{(N)}, \tag{58}$$

$$\rho_3 = -2c_N \sum_{k=1}^{p} d_k \sum_{\sigma=1}^{k} \Lambda_{2(k-\sigma)}^{(N-1)} \Gamma_{2(\sigma+s)-1}^{(N)}, \tag{59}$$

$$\begin{aligned} \rho_4 &= -2c_N^2 \sum_{k=1}^{p} d_k \sum_{\sigma=1}^{k-1} \Gamma_{2(k-\sigma)-1}^{(N)} \Gamma_{2(\sigma+s)-1}^{(N)} \\ &\quad -\alpha c_N \left[1-(-1)^N\right] \Gamma_{2s-1}^{(N)}, \end{aligned} \tag{60}$$

$$\begin{aligned} \rho_5 &= 2c_N c_{N-1} \sum_{k=1}^{p} d_k \sum_{\sigma=1}^{k-1} \Gamma_{2(k-\sigma)-1}^{(N-1)} \Gamma_{2(\sigma+s)-1}^{(N)} \\ &\quad +\alpha c_N \left[1-(-1)^{N-1}\right] \Gamma_{2s-1}^{(N)}. \end{aligned} \tag{61}$$

For large–N matrix models with a single spectrum support there are asymptotic identities $c_N \approx c_{N-1}$, $\Lambda_\sigma^{(N)} \approx \Lambda_\sigma^{(N-1)}$ and $\Gamma_\sigma^{(N)} \approx \Gamma_\sigma^{(N-1)}$ (see Appendix A for details) which simplify matters greatly. Collecting Eqs. (57) – (61) we come down to

$$\int_{-\infty}^{+\infty} d\lambda \lambda^{2s} \nu_N(\lambda) \overset{N\to\infty}{=} -2\alpha c_N (-1)^N \Gamma_{2s-1}^{(N)}$$

$$+ \quad 2c_N \sum_{k=1}^{p} d_k \sum_{\sigma=1}^{k} \Lambda^{(N)}_{2(k-\sigma)} \left(\Lambda^{(N)}_{2(\sigma+s-1)} - 2c_N \Gamma^{(N)}_{2(\sigma+s)-1} \right). \tag{62}$$

Further double summation over indices k and σ is performed by making use of the integral representations for $\Lambda_\sigma^{(N)}$ and $\Gamma_\sigma^{(N)}$ given by Eqs. (129) and (131) in Appendix A. Straightforward calculations lead to

$$\int_{-\infty}^{+\infty} d\lambda \lambda^{2s} \nu_N(\lambda) \overset{N\to\infty}{\to} \int_{-\mathcal{D}_N}^{\mathcal{D}_N} d\lambda \lambda^{2s} \rho_\Sigma(\lambda), \tag{63}$$

with $\mathcal{D}_N = 2c_N$ and

$$\begin{aligned} \rho_\Sigma(\lambda) &= \frac{2}{\pi^2}\left(\mathcal{D}_N^2 - \lambda^2\right)^{1/2} \mathcal{P} \int_0^{\mathcal{D}_N} \frac{dt}{\left(\mathcal{D}_N^2 - t^2\right)^{1/2}} \frac{tv'(t) - \lambda v'(\lambda)}{t^2 - \lambda^2} \\ &- \frac{\alpha}{\pi} \frac{(-1)^N}{\left(\mathcal{D}_N^2 - \lambda^2\right)^{1/2}}. \end{aligned} \tag{64}$$

For $\lambda = \mathcal{D}_N z$ with $|z| < 1$ the term proportional to $\lambda dv/d\lambda$ vanishes identically due to the principal value $\mathcal{P}$ of the integral over variable t, while the term proportional to α is a subleading one in the large–N limit. Then, in accordance with our concept Eq. (44), we end up with

$$\nu_N^{\text{smooth}}(\lambda) = \frac{2}{\pi^2}\left(\mathcal{D}_N^2 - \lambda^2\right)^{1/2} \mathcal{P} \int_0^{\mathcal{D}_N} \frac{tdt}{\left(\mathcal{D}_N^2 - t^2\right)^{1/2}} \frac{dv/dt}{t^2 - \lambda^2}. \tag{65}$$

Equation (65) is exactly the Dyson density $\nu_D(\lambda)$ of states with $\mathcal{D}_N$ being the end point of the eigenvalue support. We reconstructed the macroscopic level density Eq. (65) directly from the recurrence equation (28), alternatively to the traditional mean–field–theory derivation [1]. Notice that the spectrum end point $\mathcal{D}_N$ is the positive root to the integral equation

$$\frac{\pi N}{2} = \int_0^{\mathcal{D}_N} \frac{tdt}{\left(\mathcal{D}_N^2 - t^2\right)^{1/2}} \frac{dv}{dt} \tag{66}$$

following from the normalization of the level density.

4.2. GLOBAL CONNECTED "DENSITY–DENSITY" CORRELATION FUNCTION

The same technology is applicable to the study of the smoothed connected "density–density" correlator $\rho_c^{(N)}$. It is defined in terms of the scalar kernel by Eq. (27), so that

$$\rho_c^{(N)}(\lambda,\lambda') = -\overline{K_N^2(\lambda,\lambda')} = -\frac{c_N^2}{(\lambda-\lambda')^2} \times \left[\overline{\left(\varphi_N^{(\alpha)}(\lambda)\right)^2}\,\overline{\left(\varphi_{N-1}^{(\alpha)}(\lambda')\right)^2} + \overline{\left(\varphi_{N-1}^{(\alpha)}(\lambda)\right)^2}\,\overline{\left(\varphi_N^{(\alpha)}(\lambda')\right)^2} \right. \left. -2\overline{\varphi_N^{(\alpha)}(\lambda)\varphi_{N-1}^{(\alpha)}(\lambda)}\,\overline{\varphi_N^{(\alpha)}(\lambda')\varphi_{N-1}^{(\alpha)}(\lambda')} \right]. \tag{67}$$

We remind that here $\varphi_N^{(\alpha)}(\lambda) = \exp\{-V_\alpha(\lambda)\} P_N^{(\alpha)}(\lambda)$ are fictitious wave functions, $\lambda \neq \lambda'$, and $\overline{(\ldots)}$ denotes averaging over rapid oscillations manifested on the characteristic scale of the mean level spacing. The averaging in Eq. (67) can be done along the lines of the previous Subsection with two modifications. First, as far as $\lambda \neq \lambda'$ we can run averaging over λ and λ' independently (this is already reflected in Eq. (67)). Second, we have to take into account the evenness of $\left(\varphi_N^{(\alpha)}(\lambda)\right)^2$ and the oddness of $\varphi_N^{(\alpha)}(\lambda)\varphi_{N-1}^{(\alpha)}(\lambda)$.

There are two integrals

$$I_1^{(N)} = \int_{-\infty}^{+\infty} d\lambda \lambda^{2s} \exp\{-2V_\alpha(\lambda)\} \left(P_{N-1}^{(\alpha)}(\lambda)\right)^2, \tag{68}$$

$$I_2^{(N)} = \int_{-\infty}^{+\infty} d\lambda \lambda^{2s+1} \exp\{-2V_\alpha(\lambda)\} P_N^{(\alpha)}(\lambda) P_{N-1}^{(\alpha)}(\lambda) \tag{69}$$

to be evaluated in the large-N limit. With the help of Eqs. (47) and (48) one immediately recognizes them as the objects $\Lambda_{2s}^{(N)}$ and $\Gamma_{2s+1}^{(N)}$, respectively, calculated in Appendix A. By comparing of Eqs. (68) and (69) with Eqs. (129) and (131) we deduce that

$$\overline{\left(\varphi_N^{(\alpha)}(\lambda)\right)^2} = \frac{1}{\pi}\left(\mathcal{D}_N^2 - \lambda^2\right)^{-1/2}, \tag{70}$$

$$\overline{\varphi_N^{(\alpha)}(\lambda)\varphi_{N-1}^{(\alpha)}(\lambda)} = \frac{\lambda}{\pi \mathcal{D}_N}\left(\mathcal{D}_N^2 - \lambda^2\right)^{-1/2}, \tag{71}$$

for $|\lambda| < \mathcal{D}_N$. Substituting Eqs. (70) and (71) into Eq. (67) is a final step leading us to the smoothed "density–density" correlator

$$\rho_c^{(N)}(\lambda,\lambda') = -\frac{1}{2\pi^2(\lambda-\lambda')^2} \frac{\mathcal{D}_N^2 - \lambda\lambda'}{\left(\mathcal{D}_N^2-\lambda^2\right)^{1/2}\left(\mathcal{D}_N^2-\lambda'^2\right)^{1/2}} \tag{72}$$

announced by Eq. (2) with $\beta = 2$. We stress, that it has been obtained here from the recurrence equation for orthogonal polynomials associated with the random matrix ensemble in question.

4.3. LOCAL EIGENVALUE CORRELATIONS BY SOLVING EFFECTIVE SCHRÖDINGER EQUATION

4.3.1. *Effective Schrödinger Equation*

Local eigenvalue correlations in the matrix ensemble Eq. (4) can be studied by using the asymptotic version of the effective Schrödinger equation (36) obtained in Section 3. For the confinement potential V_α introduced by Eq. (41) we obtain in the large–N limit the following expressions for the functions $\mathcal{F}_n^{(\alpha)}(\lambda)$ and $\mathcal{G}_n^{(\alpha)}(\lambda)$ (see Eqs. (37) and (38)),

$$\mathcal{F}_N^{(\alpha)}(\lambda) = \frac{d}{d\lambda}\log A_N^{(\alpha)}(\lambda), \tag{73}$$

$$\begin{aligned}\mathcal{G}_N^{(\alpha)}(\lambda) &= \left(A_N^{(\alpha)}(\lambda)\right)^2\left[1-\left(\frac{\lambda}{\mathcal{D}_N}\right)^2\right] \\ &\quad +(-1)^N\frac{\alpha}{\lambda}\left(\frac{d}{d\lambda}\log A_N^{(\alpha)}(\lambda)\right)+\frac{\alpha(-1)^N-\alpha^2}{\lambda^2}.\end{aligned} \tag{74}$$

Here we have used the sum rule Eq. (39) to eliminate $B_N^{(\alpha)}$. For confinement potential with a smooth regular part v, the second term in Eq. (74) is a subleading and hence it must be discarded[5]. Note, that it is rather interesting that the confinement potential V_α does not appear in both equations above in an explicit way. It is even more exciting that in the considered approximation the function $A_N^{(\alpha)}$ can be solely expressed through the Dyson density, Eq. (65). Indeed, taking into account the representation Eq. (132) and the fact that $A_N^{(\alpha)}(\lambda) = A_{\rm reg}^{(N)}(\lambda)$, we are led to the asymptotic relation

$$A_N^{(\alpha)}(\lambda) = \frac{\pi\nu_D(\lambda)}{\left[1-(\lambda/\mathcal{D}_N)^2\right]^{1/2}}, \tag{75}$$

where, in accordance with Eq. (65),

$$\nu_D(\lambda) = \frac{2}{\pi^2}\mathcal{P}\int_0^{\mathcal{D}_N}\frac{tdt}{t^2-\lambda^2}\frac{dv}{dt}\left(\frac{\mathcal{D}_N^2-\lambda^2}{\mathcal{D}_N^2-t^2}\right)^{1/2}. \tag{76}$$

This allows us to arrive at the following remarkable effective one–particle Schrödinger equation for the wave functions

$$\varphi_N^{(\alpha)}(\lambda) = |\lambda|^\alpha P_N^{(\alpha)}(\lambda)\exp\{-v(\lambda)\} \tag{77}$$

[5]This is not the case for the *multicritical* correlations near the origin $\lambda = 0$. A detailed discussion of this important situation can be found in the very recent paper of Ref. [66].

of fictitious non-interacting fermions in the large-N limit [48, 53]

$$\frac{d^2\varphi_N^{(\alpha)}}{d\lambda^2} - \left[\frac{d}{d\lambda}\log\left(\frac{\pi\nu_D(\lambda)}{\left[1-(\lambda/\mathcal{D}_N)^2\right]^{1/2}}\right)\right]\frac{d\varphi_N^{(\alpha)}}{d\lambda} + \left(\pi^2\nu_D^2(\lambda) + \frac{(-1)^N\alpha - \alpha^2}{\lambda^2}\right)\varphi_N^{(\alpha)}(\lambda) = 0. \tag{78}$$

Also, due to Eq. (29), one can verify that the wave functions $\varphi_{N-1}^{(\alpha)}(\lambda)$ and $\varphi_N^{(\alpha)}(\lambda)$ of two successive quantum states are connected by the relationship

$$\frac{d\varphi_N^{(\alpha)}}{d\lambda} = \frac{\pi\nu_D(\lambda)}{\left[1-(\lambda/\mathcal{D}_N)^2\right]^{1/2}}\left(\varphi_{N-1}^{(\alpha)}(\lambda) - \frac{\lambda}{\mathcal{D}_N}\varphi_N^{(\alpha)}(\lambda)\right) + (-1)^N\frac{\alpha}{\lambda}\varphi_N^{(\alpha)}(\lambda). \tag{79}$$

Equations (78) and (79) serve as a general basis for the study of eigenvalue correlations in non-Gaussian random matrix ensembles in an *arbitrary spectral range.*

It is instructive to analyze them in the particular case of GUE, where the Dyson density of states is the celebrated semicircle,

$$\nu_D^{\mathrm{GUE}}(\lambda) = \pi^{-1}\left(\mathcal{D}_N^2 - \lambda^2\right)^{1/2} \tag{80}$$

with $\mathcal{D}_N = (2N)^{1/2}$. The square-root law for $\nu_D^{\mathrm{GUE}}(\lambda)$ immediately removes the first derivative $d\varphi_N^{(\alpha)}/d\lambda$ in Eq. (78), providing us with the possibility to interpret the fictitious fermions as those confined by a quadratic potential ($\alpha = 0$). As far as the semicircle is a distinctive feature of density of states in GUE only, one will always obtain a first derivative in the effective Schrödinger equation for non-Gaussian unitary ensembles of random matrices. Therefore, fictitious non-interacting fermions associated with non-Gaussian ensembles of random matrices occur in a non-Hermitean quantum mechanics.

An interesting property of these equations is that they do not contain the regular part of confinement potential explicitly, but only involve the Dyson density ν_D (analytically continued on the entire real axis) and the spectrum end point $\mathcal{D}_N$. In contrast, the logarithmic singularity (that does not affect the Dyson density) introduces additional singular terms into Eqs. (78) and (79), changing significantly the behavior of the wave function $\varphi_N^{(\alpha)}$ near the origin $\lambda = 0$. The influence of the singularity decreases rather rapidly outward from the origin.

Structure of the effective Schrödinger equation leads us to the following statements [48] valid in the thermodynamic limit:

• *Eigenvalue correlations are stable with respect to non–singular deformations of the confinement potential.*

• In the random matrix ensembles with well behaved confinement potential *the knowledge of Dyson density* (that is rather crude one–point characteristics coinciding with the real density of states only in the spectrum bulk) *is sufficient to determine the genuine density of states, as well as the n–point correlation function, everywhere.*

The latter conclusion is rather unexpected since it considerably reduces the knowledge required for computing n–point correlators.

Effective Schrödinger equation obtained above enables us to examine in a unified way the local eigenvalue correlations in non–Gaussian ensembles with $\mathrm{U}(N)$ symmetry in different scaling limits. As we show below, it inevitably leads to the universal Bessel correlations in the origin scaling limit [47, 17, 53], to the universal sine correlations in the bulk scaling limit [36, 25, 38], and to the universal Airy correlations in the soft–edge scaling limit [53]. Corresponding scalar kernels are given by Eqs. (85), (87) and (94), respectively.

4.3.2. *Origin scaling limit and the universal Bessel law*

Origin scaling limit deals with the region of the spectrum close to $\lambda = 0$ where confinement potential displays the logarithmic singularity. In the vicinity of the origin the Dyson density can be taken as being approximately a constant, $\nu_D(0) = 1/\Delta_N(0)$, where $\Delta_N(0)$ is the mean level spacing at the origin in the absence of the logarithmic deformation of potential v. Within the framework of this approximation, Eq. (78) reads

$$\frac{d^2\varphi_N^{(\alpha)}}{d\lambda^2} + \left(\frac{\pi^2}{\Delta_N^2(0)} + \frac{(-1)^N\alpha - \alpha^2}{\lambda^2}\right)\varphi_N^{(\alpha)}(\lambda) = 0. \tag{81}$$

Solution to this equation that remains finite at $\lambda = 0$ can be expressed by means of Bessel functions

$$\varphi_{2N}^{(\alpha)}(\lambda) = a\sqrt{\lambda}J_{\alpha-1/2}\left(\frac{\pi\lambda}{\Delta(0)}\right), \tag{82}$$

$$\varphi_{2N+1}^{(\alpha)}(\lambda) = b\sqrt{\lambda}J_{\alpha+1/2}\left(\frac{\pi\lambda}{\Delta(0)}\right), \tag{83}$$

where a and b are constants to be determined later, and $\Delta(0) = \Delta_{2N}(0) \approx \Delta_{2N+1}(0)$. Inserting these solutions into Eq. (23) we find that the scalar kernel can be written down as

$$K_{2N}^{(\alpha)}(\lambda,\lambda') = c\frac{\sqrt{\lambda\lambda'}}{\lambda'-\lambda}\left[J_{\alpha+1/2}\left(\frac{\pi\lambda}{\Delta(0)}\right)J_{\alpha-1/2}\left(\frac{\pi\lambda'}{\Delta(0)}\right)\right.$$

$$-J_{\alpha+1/2}\left(\frac{\pi\lambda'}{\Delta(0)}\right) J_{\alpha-1/2}\left(\frac{\pi\lambda}{\Delta(0)}\right)\Bigg], \tag{84}$$

where the unknown factor c can be found from the requirement $K_{2N}^{(\alpha=0)}(\lambda,\lambda) = 1/\Delta(0)$. This immediately yields us the value $c = -\pi/\Delta(0)$. Defining now the scaled variable $s = \lambda_s/\Delta(0)$, we obtain that in the origin scaling limit the scalar kernel $K_{\text{orig}}(s,s') = \lim_{N\to\infty} \lambda'_s K_{2N}^{(\alpha)}(\lambda_s,\lambda_{s'})$ takes the universal Bessel law,

$$K_{\text{orig}}(s,s') = \frac{\pi}{2}(ss')^{1/2} \frac{J_{\alpha+1/2}(\pi s) J_{\alpha-1/2}(\pi s') - J_{\alpha-1/2}(\pi s) J_{\alpha+1/2}(\pi s')}{s-s'}. \tag{85}$$

Equation (85) is valid for arbitrary $\alpha > -1/2$, thus extending a recent proof [17] of universality of the Bessel kernel.

4.3.3. *Bulk scaling limit and the universal sine law*

Bulk scaling limit is associated with a spectrum range where the confinement potential is well behaved (that is far from the logarithmic singularity $\lambda = 0$), and where the density of states can be taken as being approximately a constant on the scale of a few levels. In accordance with this definition one has

$$K_{\text{bulk}}(s,s') = \lim_{s,s'\to\infty} K_{\text{orig}}(s,s'), \tag{86}$$

where s and s' should, nevertheless, remain far enough from the end point $\mathcal{D}_N$ of the spectrum support.

Taking this limit in Eq. (86), we arrive at the universal sine law

$$K_{\text{bulk}}(s,s') = \frac{\sin[\pi(s-s')]}{\pi(s-s')} \tag{87}$$

deeply connected to the Wigner–Dyson level statistics [32].

4.3.4. *Soft–edge scaling limit and the universal Airy law*

Soft–edge scaling limit is relevant to the tail of eigenvalue support where crossover occurs from a nonzero density of states to a vanishing one. It is known [50, 67] that by tuning coefficients d_k which enter the regular part v of confinement potential (see Eq. (42)), one can obtain a macroscopic (Dyson) density of states which possesses a singularity of the type $\nu_D(\lambda) \propto (1-\lambda^2/\mathcal{D}_N^2)^{m+1/2}$ with the multicritical index $m = 0, 2, 4$, etc. (Odd indices m are inconsistent with our choice that the leading coefficient d_p, entering the regular component $v(\lambda)$ of confinement potential, be positive in order to keep a convergence of integral for partition function $\mathcal{Z}_N$ in Eq. (4)). It was shown in Ref. [53] within the Shohat method that as long as the multicriticality of the order m is reached, the eigenvalue correlations in the vicinity of the soft edge become universal, and are independent of the particular potential chosen. The order m of the multicriticality is the only parameter which governs spectral correlations in the soft–edge scaling limit. Here, however, we restrict ourselves to a general confinement potential without tuning to the multicritical point, that corresponds to $m = 0$.

Let us move the spectrum origin to its end point $\mathcal{D}_N$, making the replacement

$$\lambda_s = \mathcal{D}_N \left[1 + (s/2) \left(2(\pi \mathcal{D}_N \mathcal{R}_N(1))^{-1} \right)^{2/3} \right], \tag{88}$$

that defines the *soft–edge scaling limit* provided $s \ll (\mathcal{D}_N \mathcal{R}_N(1))^{2/3} \propto N^{2/3}$. It is straightforward to show from Eqs. (78) and (79) that in the vicinity of the end point $\mathcal{D}_N$ the function $\widehat{\varphi}_N(s) = \varphi_N^{(\alpha)}(\lambda_s - \mathcal{D}_N)$ obeys the universal differential equation

$$\widehat{\varphi}_N''(s) - s\widehat{\varphi}_N(s) = 0, \tag{89}$$

and that the following relation takes place,

$$\widehat{\varphi}_{N-1}(s) = \widehat{\varphi}_N(s) + \left(\frac{2}{\pi \mathcal{D}_N \mathcal{R}_N(1)} \right)^{1/3} \widehat{\varphi}_N'(s). \tag{90}$$

Solution to Eq. (89) which decreases at $s \to +\infty$ (that is at far tails of the density of states) can be represented through the Airy function

$$\mathrm{Ai}(s) = \frac{1}{3} \begin{cases} s^{1/2} \left[I_{-1/3}\left(\frac{2}{3}s^{3/2}\right) - I_{1/3}\left(\frac{2}{3}s^{3/2}\right) \right], & s > 0, \\ |s|^{1/2} \left[J_{-1/3}\left(\frac{2}{3}|s|^{3/2}\right) + J_{1/3}\left(\frac{2}{3}|s|^{3/2}\right) \right], & s < 0. \end{cases} \tag{91}$$

as follows

$$\widehat{\varphi}_N(s) = a\mathrm{Ai}(s), \tag{92}$$

with a being an unknown constant. Making use of Eq. (90), we obtain that in the vicinity of the soft edge the scalar kernel is

$$K_N(\lambda_s, \lambda_{s'}) = b \frac{\mathrm{Ai}(s)\,\mathrm{Ai}'(s') - \mathrm{Ai}(s')\,\mathrm{Ai}'(s)}{s - s'}, \tag{93}$$

where b is an unknown constant again. It can be found by fitting [51] the density of states $K_N(\lambda_s, \lambda_s)$, Eq. (93), to the Dyson density of states $\nu_D(\lambda_s)$, Eq. (76), near the soft edge provided $1 \ll s \ll N^{2/3}$. This yields us the value $b = c_N^{-1}(\pi c_N \mathcal{R}_N(1))^{2/3}$. Thus, we obtain that in the soft–edge scaling limit, Eq. (88), the scalar kernel $K_{\text{soft}}(s, s') = \lim_{N\to\infty} \lambda_s' K_N(\lambda_s, \lambda_{s'})$ satisfies the universal Airy law

$$K_{\text{soft}}(s, s') = \frac{\mathrm{Ai}(s)\,\mathrm{Ai}'(s') - \mathrm{Ai}(s')\,\mathrm{Ai}'(s)}{s - s'} \tag{94}$$

which does not depend on the details of the confinement potential. In fact, the Airy law is a particular case ($m = 0$) of more general multicritical correlations characterized by the index m of the multicriticality. For more details we refer the reader to Refs. [53, 48].

It follows from Eq. (94) that the density of states in the same scaling limit

$$\nu_{\text{soft}}(s) = \left(\frac{d}{ds}\mathrm{Ai}(s)\right)^2 - s\,[\mathrm{Ai}(s)]^2 \tag{95}$$

is also universal. The large–$|s|$ behavior of ν_{soft} can be deduced from the known asymptotic expansions [68] of the Bessel functions,

$$\nu_{\text{soft}} = \begin{cases} \frac{|s|^{1/2}}{\pi} - \frac{1}{4\pi|s|}\cos\left(\frac{4}{3}|s|^{3/2}\right), & s \to -\infty, \\ \frac{1}{8\pi s}\exp\left(-\frac{4}{3}s^{3/2}\right), & s \to +\infty. \end{cases} \tag{96}$$

Note that the leading order behavior as $s \to -\infty$ is consistent with the $|s|^{1/2}$ singularity of the bulk density of states.

4.4. DISCUSSION

Looking back at the formalism developed we should reiterate that the crucial point in the derivations above is the large–N limit, Eq. (124), of the recurrence equation for associated orthogonal polynomials. It was precisely this limit that led us to the important relation Eq. (75) and to the effective Schrödinger equation in the form of Eq. (78) which is a nonuniversal in general. However, it takes locally universal forms in the spectrum bulk (where ν_D is approximatelly a constant on the scale of a few eigenlevels), near the spectrum origin (where all the nontrivial information is contained in λ^{-2} term in front of $\varphi_N^{(\alpha)}$ in Eq. (78)), and near the soft edge of the spectrum (where universality shows up in the universal square–root singularity of ν_D). These three locally universal features of Eq. (78) have led us to the universal sine, Bessel and Airy kernels in corresponding scaling limits.

We stress that Eq. (124) is, in fact, the leading–order–limit as $N \to \infty$. How accurate is this approximation, and do situations exist where the next–order terms in the recurrence equation should be taken into account? A partial answer to this question was given in Ref. [66] whose authors, remaining within the framework of the Shohat method, convincingly demonstrated that corrections to Eq. (124) are of importance in a problem of multicritical spectral correlations near the spectrum origin. The effective Schrödinger equation obtained there for two particular matrix ensembles with fine–tuned confinement potentials was shown not only to depend on the macroscopic spectral density $\nu_D(\lambda)$ but, in addition, to contain contributions from subdominant terms in $1/N$ expansion for the recurrence coefficients. It is important however, that in the situation in question the resulting differential equation contained the universal functions $\mathcal{F}_N$ and $\mathcal{G}_N$ involving certain universal combinations of recurrence coefficients and coupling constants responsible for the fine tuning of the multicritical confinement potential. Having a complicated form, the effective Schrödinger equation could not be solved analytically, but, remarkably, the authors of Ref. [66] succeded in identifying a certain "mesoscopic" limit, in which the numerical solution of the exact differential equation and the analytical solution of the approximate differential equation obtained by making use of the relation Eq. (75) were shown to have quite similar qualitative features. With increasing of the order of the multicriticality near the spectrum origin, the approximate (analytical) and exact (numerical) solutions were shown to approach each other even quantitatively, demonstrating thus the potentialities of the Shohat method even in its simplest formulation presented above.

5. Two–Band Random Matrices

5.1. MULTI–BAND SPECTRAL REGIMES

Ensembles of large random matrices $\mathbf{H}$ generated by the joint distribution function $P[\mathbf{H}]$, Eq. (1), may display phase transitions under non–monotonic deformation of the confinement potential $V[\mathbf{H}]$. Different phases are characterized by topologically different arrangements of eigenvalues in random matrix spectra that may have multiple–band structure. Random matrices, whose spectra undergo phase transitions, appear in quantizing two–dimensional gravity [69, 70, 71], in the context of quantum chromodynamics [72, 73], as well as in some models of particles interacting in high dimensions [74]. Transition regimes realized in invariant random matrix ensembles have implications for a certain class of Calogero–Sutherland–Moser models [75]. These matrix models may also be applicable to chaotic systems having a forbidden gap in the energy spectrum.

It is convenient to parametrize the confinement potential $V(\lambda)$ entering Eq. (1) by a set of coupling constants $\{d\} = \{d_1, ..., d_p\}$,

$$V(\lambda) = \sum_{k=1}^{p} \frac{d_k}{2k} \lambda^{2k}, \; d_p > 0, \tag{97}$$

so that we may consider the phase transitions as occurring in $\{d\}$–space. Because the confinement potential is an even function, the associated random matrix model possesses so–called $Z2$–symmetry.

Variations of the coupling constants affect the Dyson density ν_D, that can be found by minimizing the free energy $F_N = -\log \mathcal{Z}_N$, Eq. (1), subject to a normalization constraint $\int \nu_D(\lambda)\, d\lambda = N$,

$$\frac{dV}{d\lambda} - \mathcal{P} \int dt \frac{\nu_D(t)}{\lambda - t} = 0, \tag{98}$$

where $\mathcal{P}$ indicates a principal value of the integral. When all d_k are positive, so that confinement potential is monotonic, the spectral density ν_D has a single–band support, $\mathcal{N}_{\mathrm{b}} = 1$. Non–monotonic deformation of the confinement potential can be carried out by changing the signs of some of d_k ($k \neq p$). Such a *continuous* variation of coupling constants may lead, under certain conditions, to a *discontinuous* change of the topological structure of spectral density ν_D, when the eigenvalues $\{\lambda\}$ are arranged in $\mathcal{N}_{\mathrm{b}} > 1$ "allowed" bands separated by "forbidden" gaps.

The phase structure of Hermitean ($\beta = 2$) one–matrix model Eq. (1) has been studied in a number of works [76, 77, 78, 79], where the simplest examples of non–monotonic quartic and sextic confinement potentials have been examined. It has been found that there are domains in the phase space of coupling constants where only a particular solution for ν_D exists, and it has a fixed number $\mathcal{N}_{\mathrm{b}}$ of allowed bands. However, in some regions of the phase space, one can have more than one kind of solution of the saddle–point equation Eq. (98). In this situation, solutions with different number of bands $\mathcal{N}_{\mathrm{b}}^{(1)}, \mathcal{N}_{\mathrm{b}}^{(2)}, \ldots$ are present simultaneously. When such an overlap appears, one of the solutions, say $\mathcal{N}_{\mathrm{b}}^{(k)}$, has the lowest free energy $F_N^{(k)}$, and this solution is dominant, while the others are subdominant. Moreover, numerical calculations [78] showed that some special regimes exist in which the *bulk* spectral density obtained as a solution to the saddle–point equation Eq. (98) differs significantly from the genuine level density computed numerically within the framework of the orthogonal polynomial technique. It was then argued that such a genuine density of levels cannot be interpreted as a multi–band solution with an integer number of bands. A full understanding of this phenomenon is still absent.

Recently, interest was renewed in multi–band regimes in invariant random matrix ensembles. An analysis based on a loop equation technique [20, 55] showed that fingerprints of phase transitions appear not only in the Dyson density but also in the (universal) wide–range eigenvalue correlators, which in the multi–band phases differ from those known in the single–band phase [19, 36, 23]. A renormalization group approach developed in Ref. [80] supported the results found in Refs. [20, 55] for the particular case of two allowed bands, referring a new type of universal wide–range eigenlevel correlators to an additional attractive fixed point of a renormalization group transformation.

As it was already stressed in the Introduction, the method of loop equations [20, 55], used for a treatment of non–Gaussian, unitary invariant, random matrix ensembles fallen in a multi-band phase, is only suitable for computing the global characteristics of spectrum. Therefore, an appropriate approach is needed capable of analyzing local characteristics of spectrum (manifested on the scale of a few eigenlevels). A possibility to probe the local properties of eigenspectrum is offered by the method of orthogonal polynomials. A step in this direction was taken in the paper [81], where an ansatz was proposed for large-N asymptotes of orthogonal polynomials associated with a random matrix ensemble having two allowed bands in its spectrum. Because the asymptotic formula proposed there is of the Plancherel-Rotach type [60], it is only applicable for studying eigenvalue correlations in the spectrum bulk and cannot be used for studying local correlations in an arbitrary spectrum range (for example, near the edges of two-band eigenvalue support).

Below we demonstrate that the Shohat method needs minimal modifications to allow a unified treatment of eigenlevel correlations in the unitary invariant $\mathrm{U}(N)$ matrix model ($\beta = 2$) with a forbidden gap. In particular, we will be able to study both the fine structure of local characteristics of the spectrum in different scaling limits and smoothed global spectral correlations. As is the single-band phase, the treatment presented below is based on the direct reconstruction of spectral correlations from the recurrence equation for the corresponding orthogonal polynomials.

5.2. EFFECTIVE SCHRÖDINGER EQUATION IN THE TWO-BAND PHASE AND LOCAL EIGENVALUE CORRELATIONS

Let us consider the situation when the confinement potential has two deep wells leading to the Dyson density supported on two disjoint intervals located symmetrically about the origin, $\mathcal{D}_N^- < |\lambda| < \mathcal{D}_N^+$. In this situation, the recurrence coefficients c_n entering Eq. (28) are known to be double-valued functions of the number n [69, 79]. This means that for $n = N \gg 1$ and in contrast with a single-band phase, one must distinguish between coefficients $c_{N\pm 2q} \approx c_N$ and coefficients $c_{N-1\pm 2q} \approx c_{N-1}$, belonging to two different, smooth (in index) sub-sequences; here, integer $q \sim \mathcal{O}(N^0)$. Bearing this in mind, the large-N version of recurrence equation Eq. (28) can be rewritten as

$$\left[\lambda^2 - \left(c_N^2 + c_{N-1}^2\right)\right] P_N(\lambda) = c_N c_{N-1} \left[P_{N-1}(\lambda) + P_{N+1}(\lambda)\right], \qquad (99)$$

whence the two analogues of the asymptotic expansion Eq. (125) can be obtained. They are given by Eqs. (135) and (136) of Appendix B. We notice that this is the only point crucial for extending the Shohat method to the double-well matrix models considered in this Section. These new expansions make it possible to compute the required functions $\mathcal{F}_N$ and $\mathcal{G}_N$

entering the differential equation Eq. (36) for fictitious wave functions in the limit $N \gg 1$.

In accordance with the general framework of the Shohat method, we have to compute two functions (compare with Eqs. (55) and (56))

$$A_N(\lambda) = 2c_N \sum_{k=1}^{p} d_k \sum_{\sigma=1}^{k} \Lambda_{2(k-\sigma)}^{(N)} \lambda^{2\sigma-2}, \tag{100}$$

$$B_N(\lambda) = 2c_N \sum_{k=1}^{p} d_k \sum_{\sigma=1}^{k} \Gamma_{2(k-\sigma)-1}^{(N)} \lambda^{2\sigma-1}, \tag{101}$$

involving the objects $\Lambda_{2\sigma}^{(N)}$ and $\Gamma_{2\sigma+1}^{(N)}$, for which there exist the useful integral representations given by Eqs. (140) and (148) in Appendix B. Substituting them into Eqs. (100) and (101) one is able to perform the double summation over indices k and σ. Omiting details of straightforward calculations we present the final answer given by the formulas

$$A_N(\lambda) = \frac{2}{\pi}\left(\mathcal{D}_N^+ - (-1)^N \mathcal{D}_N^-\right) \mathcal{P} \int_{\mathcal{D}_N^-}^{\mathcal{D}_N^+} \frac{dV}{dt} \frac{t^2}{t^2-\lambda^2} \times \frac{dt}{\left[\left(\mathcal{D}_N^+\right)^2 - t^2\right]^{1/2} \left[t^2 - \left(\mathcal{D}_N^-\right)^2\right]^{1/2}}, \tag{102}$$

$$B_N(\lambda) = \frac{2\lambda}{\pi} \mathcal{P} \int_{\mathcal{D}_N^-}^{\mathcal{D}_N^+} \frac{dV}{dt} \frac{t^2 - (-1)^N \mathcal{D}_N^- \mathcal{D}_N^+}{\left[\left(\mathcal{D}_N^+\right)^2 - t^2\right]^{1/2} \left[t^2 - \left(\mathcal{D}_N^-\right)^2\right]^{1/2}} \frac{dt}{t^2-\lambda^2} - \frac{dV}{d\lambda}. \tag{103}$$

Note, that along with a different (compared to the single–band phase) functional form of the functions $A_N(\lambda)$ and $B_N(\lambda)$, these functions are, in fact, double–valued in index N, and behave in a different way for odd and even N. This is a direct consequence of the "period–two" behavior [69, 79] of the recurrence coefficients c_n.

Having obtained the explicit expressions for functions A_N and B_N, it is easy to verify that coefficients $\mathcal{F}_n(\lambda)$ and $\mathcal{G}_n(\lambda)$ entering the differential equation Eq. (36) for the fictitious wave function $\varphi_n(\lambda)$ may be expressed in terms of the Dyson density $\nu_D^{(\mathrm{II})}$ in the two–cut phase supported on two disconnected intervals $\lambda \in \left(-\mathcal{D}_N^+, -\mathcal{D}_N^-\right) \cup \left(\mathcal{D}_N^-, \mathcal{D}_N^+\right)$,

$$\nu_D^{(\mathrm{II})}(\lambda) = \frac{2}{\pi^2} |\lambda| \mathcal{P} \int_{\mathcal{D}_N^-}^{\mathcal{D}_N^+} dt \frac{dV/dt}{t^2-\lambda^2} \left(\frac{\left(\mathcal{D}_N^+\right)^2 - \lambda^2}{\left(\mathcal{D}_N^+\right)^2 - t^2}\right)^{1/2} \left(\frac{\lambda^2 - \left(\mathcal{D}_N^-\right)^2}{t^2 - \left(\mathcal{D}_N^-\right)^2}\right)^{1/2} \tag{104}$$

when $N \gg 1$. This formula can be obtained either via the procedure of the Sec. 4.1 or within the mean–field approach, Eq. (98). Here $\mathcal{D}_N^-$ and $\mathcal{D}_N^+$ are the end points of the eigenvalue support that obey the two integral equations

$$\int_{\mathcal{D}_N^-}^{\mathcal{D}_N^+} \frac{dV}{dt} \frac{t^2 dt}{\left[\left(\mathcal{D}_N^+\right)^2 - t^2\right]^{1/2} \left[t^2 - \left(\mathcal{D}_N^-\right)^2\right]^{1/2}} = \frac{\pi N}{2}, \tag{105}$$

$$\int_{\mathcal{D}_N^-}^{\mathcal{D}_N^+} \frac{dV}{dt} \frac{dt}{\left[\left(\mathcal{D}_N^+\right)^2 - t^2\right]^{1/2} \left[t^2 - \left(\mathcal{D}_N^-\right)^2\right]^{1/2}} = 0, \tag{106}$$

derived in Appendix C.

By making use of the Eqs. (37), (38) and (102) – (104) we obtain in the large–N limit

$$\mathcal{F}_N(\lambda) = \frac{d}{d\lambda} \log \left(\frac{\pi |\lambda| \nu_D^{(\mathrm{II})}(\lambda)}{\left[\left(\mathcal{D}_N^+\right)^2 - \lambda^2\right]^{1/2} \left[\lambda^2 - \left(\mathcal{D}_N^-\right)^2\right]^{1/2}} \right), \tag{107}$$

$$\mathcal{G}_N(\lambda) = \left(\pi \nu_D^{(\mathrm{II})}(\lambda)\right)^2, \tag{108}$$

so that for $N \gg 1$ the effective Schrödinger equation in the two–cut phase reads [56]

$$\frac{d^2 \varphi_N(\lambda)}{d\lambda^2} - \left[\frac{d}{d\lambda} \log \left(\frac{\pi |\lambda| \nu_D^{(\mathrm{II})}(\lambda)}{\left[\left(\mathcal{D}_N^+\right)^2 - \lambda^2\right]^{1/2} \left[\lambda^2 - \left(\mathcal{D}_N^-\right)^2\right]^{1/2}} \right) \right] \frac{d\varphi_N(\lambda)}{d\lambda} + \left(\pi \nu_D^{(\mathrm{II})}(\lambda)\right)^2 \varphi_N(\lambda) = 0. \tag{109}$$

As $\mathcal{D}_N^-$ tends to zero, we reproduce the equation (78) with $\alpha = 0$ valid in the single–band regime.

Local eigenvalue correlations in the spectra of two–band random matrices are completely determined by the Dyson density of states entering the effective Schrödinger equation Eq. (109).

(i) In the spectrum bulk, the Dyson density is a well behaved function that can be taken approximately as being a constant on the scale of a few eigenlevels. Then, in the vicinity of some λ_0 that is chosen to be far enough from the spectrum end points $\pm\mathcal{D}_N^{\pm}$, Eq. (109) takes the form

$$\frac{d^2\varphi_N(\lambda)}{d\lambda^2} + [\pi/\Delta(\lambda_0)]^2 \varphi_N(\lambda) = 0, \tag{110}$$

with $\Delta(\lambda_0) = 1/\nu_D^{(\mathrm{II})}(\lambda_0)$ being the mean level spacing in the vicinity of λ_0. Clearly, the universal sine law, Eq. (87), for the two–point kernel follows immediately.

(ii) To study the eigenvalue correlations near the end points of an eigenvalue support we notice that in the absence of the fine tunning of confinement potential, the Dyson density has a universal square–root singularity in the vicinity of $|\lambda| = \mathcal{D}_N^{\pm}$, that is $\nu_D^{(\mathrm{II})}(\lambda) \propto \left(1 - \left(\lambda/\mathcal{D}_N^{\pm}\right)^2\right)^{1/2}$. We then readily recover the universal Airy correlations, Eq. (94), previously found in the soft–edge scaling limit for $\mathrm{U}(N)$ invariant matrix model in the single–band phase.

5.3. GLOBAL CONNECTED "DENSITY–DENSITY" CORRELATOR

Let us turn to the study of the smoothed connected "density–density" correlator that is expressed in terms of the scalar kernel as follows (see Eq. (27)),

$$\begin{aligned} \rho_{\mathrm{cII}}^{(N)}(\lambda, \lambda') &= -\frac{c_N^2}{(\lambda - \lambda')^2} \left\{ \overline{\varphi_N^2(\lambda)}\,\overline{\varphi_{N-1}^2(\lambda')} + \overline{\varphi_N^2(\lambda')}\,\overline{\varphi_{N-1}^2(\lambda)} \right. \\ &\left. -2\overline{\varphi_N(\lambda)\varphi_{N-1}(\lambda)}\,\overline{\varphi_N(\lambda')\varphi_{N-1}(\lambda')} \right\}. \end{aligned} \tag{111}$$

Here $\lambda \neq \lambda'$. Equation (111) contains (before averaging) rapid oscillations on the scale of the mean level spacing. These oscillations are due to presence of oscillating wave functions φ_N and φ_{N-1}.

To average over the rapid oscillations, we integrate, over the entire real axis, rapidly varying wave functions in Eq. (111) multiplied by an arbitrary, smooth, slowly varying function, which without any loss of generality can be choosen to be λ^{2s} for $\varphi_N^2(\lambda)$ and λ^{2s+1} for $\varphi_N(\lambda)\varphi_{N-1}(\lambda)$ (s is an arbitrary positive integer, $s > 0$). Consider, first, the integral

$$I_1^{(N)} = \int_{-\infty}^{+\infty} d\lambda \lambda^{2s} \varphi_N^2(\lambda) = \Lambda_{2s}^{(N)}. \tag{112}$$

With the help of Eq. (140), and bearing in mind that $\varphi_N^2(\lambda)$ is an even function, we conclude that

$$I_1^{(N)} = \frac{1}{\pi} \int_{\mathcal{D}_N^- < |\lambda| < \mathcal{D}_N^+} \frac{|\lambda| \, \lambda^{2s} d\lambda}{\left[\left(\mathcal{D}_N^+\right)^2 - \lambda^2\right]^{1/2} \left[\lambda^2 - \left(\mathcal{D}_N^-\right)^2\right]^{1/2}}, \tag{113}$$

whence, in the large–N limit,

$$\overline{\varphi_N^2(\lambda)} = \frac{\Omega_\lambda}{\pi} \frac{|\lambda|}{\left[\left(\mathcal{D}_N^+\right)^2 - \lambda^2\right]^{1/2} \left[\lambda^2 - \left(\mathcal{D}_N^-\right)^2\right]^{1/2}}. \tag{114}$$

Here

$$\Omega_\lambda = \Theta\left(\mathcal{D}_N^+ - |\lambda|\right) \Theta\left(|\lambda| - \mathcal{D}_N^-\right) \tag{115}$$

with Θ being a step function. The same procedure should be carried out with the expression $\varphi_N(\lambda)\varphi_{N-1}(\lambda)$ in Eq. (111). Since this construction is an odd function of λ, we have to consider the integral

$$I_2^{(N)} = \int_{-\infty}^{+\infty} d\lambda \lambda^{2s+1} \varphi_N(\lambda) \varphi_{N-1}(\lambda) = \Gamma_{2s+1}^{(N)}. \tag{116}$$

With the help of Eq. (148), and exploiting the oddness of $\varphi_N(\lambda)\varphi_{N-1}(\lambda)$, we rewrite Eq. (116) in the form

$$\begin{aligned} I_2^{(N)} &= \frac{1}{\pi\left[\mathcal{D}_N^+ - (-1)^N \mathcal{D}_N^-\right]} \\ &\times \int_{\mathcal{D}_N^- < |\lambda| < \mathcal{D}_N^+} \frac{\left[\lambda^2 - (-1)^N \mathcal{D}_N^- \mathcal{D}_N^+\right] \mathrm{sgn}(\lambda) \, d\lambda}{\left[\left(\mathcal{D}_N^+\right)^2 - \lambda^2\right]^{1/2} \left[\lambda^2 - \left(\mathcal{D}_N^-\right)^2\right]^{1/2}}, \end{aligned} \tag{117}$$

whence

$$\begin{aligned} \overline{\varphi_N(\lambda)\varphi_{N-1}(\lambda)} &= \frac{\Omega_\lambda \, \mathrm{sgn}(\lambda)}{\pi\left[\mathcal{D}_N^+ - (-1)^N \mathcal{D}_N^-\right]} \\ &\times \frac{\lambda^2 - (-1)^N \mathcal{D}_N^- \mathcal{D}_N^+}{\left[\left(\mathcal{D}_N^+\right)^2 - \lambda^2\right]^{1/2} \left[\lambda^2 - \left(\mathcal{D}_N^-\right)^2\right]^{1/2}}. \end{aligned} \tag{118}$$

Combining Eqs. (111), (114), (118) and (147), we finally arrive at the following formula for smoothed "density–density" correlator [56]

$$\begin{aligned} \rho_{\mathrm{cII}}^{(N)}(\lambda, \lambda') &= -\frac{\mathrm{sgn}(\lambda\lambda')}{2\pi^2} \frac{\Omega_\lambda \Omega_{\lambda'}}{Q_N(\lambda) Q_N(\lambda')} \left\{(-1)^N \mathcal{D}_N^- \mathcal{D}_N^+ \right. \\ &\left. + \frac{1}{(\lambda - \lambda')^2} \left[\lambda\lambda' - (\mathcal{D}_N^-)^2\right] \left[(\mathcal{D}_N^+)^2 - \lambda\lambda'\right]\right\}, \end{aligned} \tag{119}$$

where

$$Q_N(\lambda) = \left[(\mathcal{D}_N^+)^2 - \lambda^2\right]^{1/2} \left[\lambda^2 - (\mathcal{D}_N^-)^2\right]^{1/2}. \tag{120}$$

It is seen from Eq. (120) that smoothed "density–density" correlator in the two–band phase differs from that in the single–band phase, Eq. (72). However, it is still universal in the sense that the information of the distribution Eq. (1) is encoded into the "density–density" correlator only through the end points $\mathcal{D}_N^\pm$ of the eigenvalue support. The striking parity effect in the new universal function Eq. (120), that is the *sharp* dependence of correlations on the oddness/evenness of the dimension N of the random matrices, is the main *qualitative* difference as compared to the global corelations in random matrices fallen in the single–band phase. This effect is most pronounced in the case of unbounded spectrum. The origin of this unusual large–N behavior will be discussed later on.

Finally, let us speculate about the universal correlator Eq. (120) in the limit of *unbounded* spectrum, $\mathcal{D}_N^+ \to \infty$, with a gap. Inasmuch as it describes correlations between the eigenlevels which are repelled from each other in accordance with the logarithmic law, that is known to be realized [82, 83] in weakly disordered systems on the energy scale $|\lambda - \lambda'| \ll E_c$ (E_c is the Thouless energy), we may conjecture that the corresponding limiting universal expression

$$\begin{aligned} \lim_{\mathcal{D}_N^+ \to +\infty} \rho_{\mathrm{cII}}^{(N)}(\lambda, \lambda') &= -\frac{\operatorname{sgn}(\lambda\lambda')}{2\pi^2 (\lambda - \lambda')^2} \Theta(|\lambda| - \Delta)\, \Theta(|\lambda'| - \Delta) \\ &\quad \times \frac{\lambda\lambda' - \Delta^2}{[\lambda^2 - \Delta^2]^{1/2} [\lambda'^2 - \Delta^2]^{1/2}}, \end{aligned} \tag{121}$$

reflects the universal properties of real chaotic systems with a forbidden gap $\Delta = \mathcal{D}_N^-$ and broken time reversal symmetry, provided $|\lambda - \lambda'| \ll E_c$. In two limiting situations (i) of gapless spectrum, $\Delta = 0$, and (ii) far from the gap, $|\lambda|, |\lambda'| \gg \Delta$, the correlator Eq. (121) coincides with that known in the Random Matrix Theory of gapless ensembles [36, 23] and derived in Ref. [82] within the framework of diagrammatic technique for spectrum of electron in a random impurity potential.

5.4. DISCUSSION

In this Section we have demonstrated how the Shohat method should be transformed in order to study both global and local spectral characteristics of $\mathrm{U}(N)$ invariant ensembles of large random matrices possessing $Z2$–symmetry, and deformed in such a way that their spectra contain a forbidden gap. We proved that in the pure two–band phase, the local eigenvalue correlations are insensitive to this deformation both in the bulk and soft–edge scaling limits. In contrast, global smoothed eigenvalue correlations in the two–band phase differ drastically from those in the single–band phase, and generically satisfy a universal law, Eq. (120), which is unusually sensitive to the oddness/evenness of the random matrix dimension provided the spectrum support is bounded. On the formal level, this sensitivity is a direct consequence of the "period–two" behavior [69, 79] of the recurrence coefficients c_n that is characteristic of two–band phase of reduced Hermitean matrix model. To see this, consider the simplest connected correlator $\langle \mathrm{Tr}\,\mathbf{H}\,\mathrm{Tr}\,\mathbf{H}\rangle_c$ that can be *exactly* represented in terms of recurrence coefficients for any n,

$$\langle \mathrm{Tr}\,\mathbf{H}\,\mathrm{Tr}\,\mathbf{H}\rangle_c = c_n^2. \tag{122}$$

Since in the two–band phase c_n is a double–valued function of index n, alternating between two different functions as n goes from odd to even, the large–N limit of the correlator $\langle \mathrm{Tr}\,\mathbf{H}\,\mathrm{Tr}\,\mathbf{H}\rangle_c$ strongly depends on whether infinity is approached through odd or even N. Then, an implementation of a double–valued behavior of c_n into the higher order correlators of the form $\left\langle \mathrm{Tr}\,\mathbf{H}^k\,\mathrm{Tr}\,\mathbf{H}^l\right\rangle_c$ contributing to the connected "density–density" correlator gives rise to the universal expression Eq. (120), which is valid for the two–band random matrix model with pure Z2–symmetry.

Let us, however, point out that no such sensitivity has been detected in a number of previous studies [20, 55] exploiting a loop equation technique. One possible explanation comes from the following reasons. In the method of loop equations, used for a treatment of non–Gaussian random matrix ensembles fallen in a multi–band phase, one has to keep the most general (non–symmetric) confinement potential $V(\lambda) = \sum_{k=1}^{2p} \tilde{d}_k \lambda^k/k$ until very end of the calculations, and to take the thermodynamic limit $N \to \infty$ prior to any others. Therefore, $Z2$–symmetry in this calculational scheme can only be implemented by restoring $Z2$–symmetry at the final stage of the calculations, setting all the extra coupling constants $\tilde{d}_{2k+1}$ to zero. Doing so, one arrives at the results reported in Refs. [20, 55]. From this point of view, our treatment corresponds to the *opposite sequence* of thermodynamic and $Z2$–symmetry limits, since we have considered the random matrix model that possesses $Z2$–symmetry from the beginning. Qualitatively different large–N behavior of the smoothed connected "density–density" correlator, Eq. (120), and of the smoothed connected two–point Green's function given by Eq. (15) of Ref. [55] provides a direct evidence that the order of thermodynamic and $Z2$–symmetry limits is indeed important when studying global spectral characteristics of multi–band random matrices.

The parity effect manifested in global spectral correlators of double–well matrix models was the focus of the discussion in the recent study [84].

The authors of Ref. [84] noted that, contrary to the method of orthogonal polynomials, the standard large-N limit techniques of analyzing matrix models like the loop equation method [19, 20] and the renormalization group approach [85] assume a smooth behavior with respect to N in the thermodynamic limit. The result Eq. (120) obtained by the authors of Ref. [84] in a different way [81] led them to conclusion that these methods need to be revisited when one deals with matrix models posessing eigenvalue gaps.

6. Conclusion

In this review we presented a formalism for statistical description of spectra of $\mathrm{U}(N)$ invariant ensembles of large random matrices. It lies within the general framework of the orthogonal polynomials' technique, and consists of the direct reconstruction of spectral densities and spectral correlations from the recurrence equation for orthogonal polynomials associated with a given random matrix ensemble. We have demonstrated the potentialities of this method, considering in a unified way both global and local spectral characteristics in matrix models with and without an eigenvalue gap. Although we directed our main attention to the most known bulk, origin and soft–edge scaling limits characterized by the universal sine, Bessel and Airy kernels, respectively, there are examples in the recent literature signaling about applicability of the described formalism to more refined situations – such as multicritical correlations near the soft edge of the spectrum support [53] and near the spectrum origin [66].

Attaching special significance to the study of the large-N limit of the recurrence equation for associated orthogonal polynomials, this method turns the recurrence equation into a kind of laboratory allowing the construction of matrix models with nonstandard properties – for example with eigenvalue gaps – by guessing a particular ansatz for the behavior of the recurrence coefficients c_N in the thermodynamic limit. Just this feature of the formalism presented forces us, finally, to mention a crucial difference between the random matrix ensembles with strong level confinement considered here and the random matrix ensembles with extremely soft level confinement. While the former ensembles (with confinement potentials of the Freud and Erdös type [38]) are characterized by a powerlike large-N limit of recurrence coefficients, $c_N \propto N^\rho$ $(\rho > 0)$, the latter (representing a class of q-deformed potentials [86, 87]) exhibit a qualitatively different, exponential rate of growth, $c_N \propto q^N$ $(q > 1)$. This results in a different large-N limit of the recurrence equation that will not already be as simple as stated in Eq. (124), and the emergence of different nontrivial classes of spectral statistics is inevitable. We consider a treatment of q-deformed random matrix ensembles as a challenge to the Shohat method which, going back to 1930, had to wait so long to find its application in Random Matrix Theory.

Acknowledgments

One of the authors (E. K.) thanks G. Akemann for useful discussions at the Abdus Salam International Centre for Theoretical Physics and for collaboration. Financial support through the Rothschild Postdoctoral Fellowship (E. K.) is gratefully acknowledged.

A. Integral Representation of $\Lambda_{2\sigma}^{(N)}$ and $\Gamma_{2\sigma+1}^{(N)}$: Single–Band Phase

Consider the integral

$$\Lambda_{2\sigma}^{(N)} = \int d\alpha\,(t) \left(P_N^{(\alpha)}(t)\right)^2 t^{2\sigma} \tag{123}$$

with integer $\sigma \geq 0$. In the large–N limit an alternative explicit integral representation can be found for $\Lambda_{2\sigma}^{(N)}$. This is achieved by making use of the large–N version of the recurrence equation (28). It is known that in the single–band phase of the matrix model the recurrence coefficients approach [69, 79] a smooth (in index N) single–valued function, so that $c_{N+q} \approx c_N$ for q being of order $\mathcal{O}(N^0)$. Within this approximation one obtains from Eq. (28)

$$\lambda P_N^{(\alpha)}(\lambda) = c_N \left(P_{N-1}^{(\alpha)}(\lambda) + P_{N+1}^{(\alpha)}(\lambda)\right), \tag{124}$$

whence it follows that

$$\lambda^m P_N^{(\alpha)}(\lambda) = c_N^m \sum_{j=0}^{m} \binom{m}{j} P_{N+2j-m}^{(\alpha)}(\lambda),\; m \geq 1. \tag{125}$$

The identity (125) can be proven by the mathematical induction. The advantage of the asymptotic expansion Eq. (125) is that being substituted (for $m = 2\sigma$) into Eq. (123), it immediately allows us to explicitly perform the integration due to the orthogonality property Eq. (17). This yields

$$\Lambda_{2\sigma}^{(N)} = c_N^{2\sigma} \sum_{j=0}^{2\sigma} \binom{2\sigma}{j} \delta_{2j}^{2\sigma}, \tag{126}$$

with δ_j^k being the Kronecker symbol. To evaluate the sum in Eq. (126) we make use of the integral representation of the Kronecker symbol,

$$\delta_j^k = \mathrm{Re} \int_0^{2\pi} \frac{d\theta}{2\pi} \exp\{i\theta\,(j-k)\}. \tag{127}$$

We then find that

$$\Lambda_{2\sigma}^{(N)} = c_N^{2\sigma} \operatorname{Re} \int_0^{2\pi} \frac{d\theta}{2\pi} \sum_{j=0}^{2\sigma} \binom{2\sigma}{j} \exp\{2i\theta(j-\sigma)\}, \tag{128}$$

and hence, after summation over j and some rearrangements,

$$\Lambda_{2\sigma}^{(N)} = \frac{2}{\pi} \int_0^{\mathcal{D}_N} \frac{dt\, t^{2\sigma}}{(\mathcal{D}_N^2 - t^2)^{1/2}}. \tag{129}$$

Here $\mathcal{D}_N = 2c_N$.

The integral

$$\Gamma_{2\sigma+1}^{(N)} = \int d\alpha(t)\, P_N^{(\alpha)}(t)\, P_{N-1}^{(\alpha)}(t)\, t^{2\sigma+1} \tag{130}$$

with integer $\sigma \geq 0$ is computable in the same manner, with an answer given by

$$\Gamma_{2\sigma+1}^{(N)} = \frac{2}{\pi \mathcal{D}_N} \int_0^{\mathcal{D}_N} \frac{dt\, t^{2\sigma+2}}{(\mathcal{D}_N^2 - t^2)^{1/2}}. \tag{131}$$

Notice that due to the asymptotic property $c_{N+q} \approx c_N$ mentioned above, $\Lambda_{2\sigma}^{(N+q)} \approx \Lambda_{2\sigma}^{(N)}$ and $\Gamma_{2\sigma+1}^{(N+q)} \approx \Gamma_{2\sigma+1}^{(N)}$ for $q \sim \mathcal{O}(N^0)$.

Finally, we demonstrate the usefulness of the formulas (129) and (131) by finding the explicit expressions for the functions $A_{\text{reg}}^{(N)}(\lambda)$ and $B_{\text{reg}}^{(N)}(\lambda)$, defined by Eqs. (49) and (51). Substitution of Eq. (129) into Eq. (49) followed by summation over σ yields

$$\begin{aligned} A_{\text{reg}}^{(N)}(\lambda) &= \frac{2\mathcal{D}_N}{\pi} \mathcal{P} \int_0^{\mathcal{D}_N} \frac{dt}{(\mathcal{D}_N^2 - t^2)^{1/2}} \sum_{k=1}^{p} d_k \frac{t^{2k} - \lambda^{2k}}{t^2 - \lambda^2} \\ &= \frac{2\mathcal{D}_N}{\pi} \mathcal{P} \int_0^{\mathcal{D}_N} \frac{dt}{(\mathcal{D}_N^2 - t^2)^{1/2}} \frac{t dv/dt}{t^2 - \lambda^2}. \end{aligned} \tag{132}$$

Analogously, we obtain

$$\begin{aligned} B_{\text{reg}}^{(N)}(\lambda) &= \frac{2}{\pi} \mathcal{P} \int_0^{\mathcal{D}_N} \frac{t dt}{(\mathcal{D}_N^2 - t^2)^{1/2}} \frac{\lambda dv/dt - t dv/d\lambda}{t^2 - \lambda^2} \\ &= \frac{2\lambda}{\pi} \mathcal{P} \int_0^{\mathcal{D}_N} \frac{dt}{(\mathcal{D}_N^2 - t^2)^{1/2}} \frac{t dv/dt}{t^2 - \lambda^2} - \frac{dv}{d\lambda}. \end{aligned} \tag{133}$$

One can convince himself that Eqs. (132) and (133) obey the sum rule Eq. (39) for the confinement potential V_α, Eq. (41), and for the functions $A_N^{(\alpha)}$

and $B_N^{(\alpha)}$, given by Eqs. (55) and (56). Notice that this is the recurrence equation (28) which enabled us to obtain closed analytic expressions (132) and (133) relating the functions $A_{\mathrm{reg}}^{(N)}$ and $B_{\mathrm{reg}}^{(N)}$ to the regular part v of the confinement potential.

B. Integral Representation of $\Lambda_{2\sigma}^{(N)}$ and $\Gamma_{2\sigma+1}^{(N)}$: Two–Band Phase

Consider the integral

$$\Lambda_{2\sigma}^{(N)} = \int d\alpha\,(t)\, P_N^2\,(t)\, t^{2\sigma} \tag{134}$$

with integer $\sigma \geq 0$. It follows from Eq. (99) that in the two–band phase the following asymptotic identities exist,

$$\begin{aligned} \lambda^{2m} P_N\,(\lambda) &= \left(c_N^2 + c_{N-1}^2\right)^m \sum_{k=0}^{m} \binom{m}{k} \left(\frac{c_N c_{N-1}}{c_N^2 + c_{N-1}^2}\right)^k \\ &\times \sum_{j=0}^{k} \binom{k}{j} P_{N+4j-2k}\,(\lambda)\,, \end{aligned} \tag{135}$$

and

$$\begin{aligned} \lambda^{2m+1} P_N\,(\lambda) &= \left(c_N^2 + c_{N-1}^2\right)^m \sum_{k=0}^{m} \binom{m}{k} \left(\frac{c_N c_{N-1}}{c_N^2 + c_{N-1}^2}\right)^k \sum_{j=0}^{k} \binom{k}{j} \\ &\times \left[c_{N-1} P_{N+4j-2k+1}\,(\lambda) + c_N P_{N+4j-2k-1}\,(\lambda)\right] \end{aligned} \tag{136}$$

with integer $m \geq 0$. Both Eqs. (135) and (136) can be proven by the mathematical induction. Making use of Eq. (135) we rewrite $\Lambda_{2\sigma}^{(N)}$ in the form

$$\begin{aligned} \Lambda_{2\sigma}^{(N)} &= \left(c_N^2 + c_{N-1}^2\right)^\sigma \sum_{k=0}^{\sigma} \binom{\sigma}{k} \left(\frac{c_N c_{N-1}}{c_N^2 + c_{N-1}^2}\right)^k \\ &\times \sum_{j=0}^{k} \binom{k}{j} \int d\alpha\,(t)\, P_N\,(t)\, P_{N+4j-2k}\,(t)\,. \end{aligned} \tag{137}$$

Orthogonality of P_n allows us to integrate over the measure $d\alpha$, thus simplifying Eq. (137) to

$$\Lambda_{2\sigma}^{(N)} = \left(c_N^2 + c_{N-1}^2\right)^\sigma \sum_{k=0}^{\sigma} \binom{\sigma}{k} \left(\frac{c_N c_{N-1}}{c_N^2 + c_{N-1}^2}\right)^k \sum_{j=0}^{k} \binom{k}{j} \delta_{2j}^k\,. \tag{138}$$

Substituting the integral representation (127) for the Kronecker symbol, and performing the double summation over indices j and k, we obtain

$$\Lambda_{2\sigma}^{(N)} = \int_0^{2\pi} \frac{d\theta}{2\pi} \left(c_N^2 + c_{N-1}^2 + 2c_N c_{N-1} \cos\theta \right)^\sigma . \tag{139}$$

Introducing a new integration variable $t^2 = c_N^2 + c_{N-1}^2 + 2c_N c_{N-1} \cos\theta$, we derive the final formula

$$\Lambda_{2\sigma}^{(N)} = \frac{2}{\pi} \int_{\mathcal{D}_N^-}^{\mathcal{D}_N^+} \frac{t^{2\sigma+1} dt}{\left[\left(\mathcal{D}_N^+ \right)^2 - t^2 \right]^{1/2} \left[t^2 - \left(\mathcal{D}_N^- \right)^2 \right]^{1/2}} \tag{140}$$

with

$$\mathcal{D}_N^\pm = |c_N \pm c_{N-1}| . \tag{141}$$

The integral

$$\Gamma_{2\sigma+1}^{(N)} = \int d\alpha(t)\, P_N(t)\, P_{N-1}(t)\, t^{2\sigma+1} \tag{142}$$

with integer $\sigma \geq 0$ is evaluated in the same way. Making use of expansion Eq. (136), we rewrite Eq. (142) in the form that allows us to perform the integration over the measure $d\alpha$,

$$\begin{aligned} \Gamma_{2\sigma+1}^{(N)} &= \frac{1}{2} \left(c_N^2 + c_{N-1}^2 \right)^\sigma \int d\alpha(t)\, P_{N-1}(t) \sum_{k=0}^{\sigma} \binom{\sigma}{k} \left(\frac{c_N c_{N-1}}{c_N^2 + c_{N-1}^2} \right)^k \\ &\times \sum_{j=0}^{k} \binom{k}{j} \left[c_{N-1} P_{N+4j-2k+1}(t) + c_N P_{N+4j-2k-1}(t) \right] . \end{aligned} \tag{143}$$

After integration, we get

$$\begin{aligned} \Gamma_{2\sigma+1}^{(N)} &= \frac{1}{2} \left(c_N^2 + c_{N-1}^2 \right)^\sigma \sum_{k=0}^{\sigma} \binom{\sigma}{k} \left(\frac{c_N c_{N-1}}{c_N^2 + c_{N-1}^2} \right)^k \\ &\times \sum_{j=0}^{k} \binom{k}{j} \left[c_{N-1} \delta_{2j+1}^k + c_N \delta_{2j}^k \right] . \end{aligned} \tag{144}$$

The double summation in Eq. (144) can be performed by using the integral representation for the Kronecker symbol given by Eq. (127),

$$\Gamma_{2\sigma+1}^{(N)} = \frac{1}{2} \int_0^{2\pi} \frac{d\theta}{2\pi} \left(c_N^2 + c_{N-1}^2 + 2c_N c_{N-1} \cos\theta \right)^\sigma \left[c_N + c_{N-1} \cos\theta \right] . \tag{145}$$

Introducing a new integration variable $t^2 = c_N^2 + c_{N-1}^2 + 2c_N c_{N-1}\cos\theta$, we get

$$\Gamma_{2\sigma+1}^{(N)} = \frac{1}{\pi c_N}\int_{\mathcal{D}_N^-}^{\mathcal{D}_N^+} \frac{t^{2\sigma+1}\,dt}{\left[\left(\mathcal{D}_N^+\right)^2 - t^2\right]^{1/2}\left[t^2 - \left(\mathcal{D}_N^-\right)^2\right]^{1/2}}\left[t^2 + c_N^2 - c_{N-1}^2\right]. \tag{146}$$

Notice that because $P_{-1}(\lambda) = 0$, it follows from Eq. (28) that $c_0 = 0$, and as a consequence, an even branch c_{2N} always lies lower than an odd branch $c_{2N\pm1}$, so that $c_{2N} < c_{2N\pm1}$. Then, we may conclude from Eq. (141) that

$$c_N = \frac{\mathcal{D}_N^+ - (-1)^N \mathcal{D}_N^-}{2}, \tag{147}$$

and, as a consequence,

$$\Gamma_{2\sigma+1}^{(N)} = \frac{1}{\pi c_N}\int_{\mathcal{D}_N^-}^{\mathcal{D}_N^+} \frac{t^{2\sigma+1}\,dt}{\left[\left(\mathcal{D}_N^+\right)^2 - t^2\right]^{1/2}\left[t^2 - \left(\mathcal{D}_N^-\right)^2\right]^{1/2}}\left[t^2 - (-1)^N \mathcal{D}_N^-\mathcal{D}_N^+\right]. \tag{148}$$

C. Soft Edges in the Two–Band Phase

To find the equations determining the end points $\mathcal{D}_N^\pm$ where the Dyson spectral density goes to zero, we start with the following formula from the theory of orthogonal polynomials [63]

$$n = 2c_n \int d\alpha(t)\,\frac{dV}{dt} P_n(t)\,P_{n-1}(t), \tag{149}$$

also known as a "string equation". Let us use expansion Eq. (136) to evaluate the integral entering Eq. (149) in the limit $n = N \gg 1$. It is easy to see that

$$N = 2c_N \sum_{k=1}^{p} d_k \int d\alpha(t)\,P_N(t)\,P_{N-1}(t)\,t^{2k-1} = 2c_N \sum_{k=1}^{p} d_k \Gamma_{2k-1}^{(N)}, \tag{150}$$

where $\Gamma_{2k-1}^{(N)}$ is given by Eq. (146). Then, we immediately obtain the relationship

$$N = \frac{2}{\pi}\int_{\mathcal{D}_N^-}^{\mathcal{D}_N^+} \frac{dt}{\left[\left(\mathcal{D}_N^+\right)^2 - t^2\right]^{1/2}\left[t^2 - \left(\mathcal{D}_N^-\right)^2\right]^{1/2}}\frac{dV}{dt}\left[t^2 + c_N^2 - c_{N-1}^2\right]. \tag{151}$$

This result, rewritten for $n = N - 1$, yields in the large–N limit,

$$N = \frac{2}{\pi} \int_{\mathcal{D}_N^-}^{\mathcal{D}_N^+} \frac{dt}{\left[\left(\mathcal{D}_N^+\right)^2 - t^2\right]^{1/2} \left[t^2 - \left(\mathcal{D}_N^-\right)^2\right]^{1/2}} \frac{dV}{dt} \left[t^2 + c_{N-1}^2 - c_N^2\right]. \tag{152}$$

Equations (151) and (152) bring us two integral equations whose solutions determine the end points $\mathcal{D}_N^{\pm}$,

$$\int_{\mathcal{D}_N^-}^{\mathcal{D}_N^+} \frac{t^2 dt}{\left[\left(\mathcal{D}_N^+\right)^2 - t^2\right]^{1/2} \left[t^2 - \left(\mathcal{D}_N^-\right)^2\right]^{1/2}} \frac{dV}{dt} = \frac{\pi N}{2}, \tag{153}$$

and

$$\int_{\mathcal{D}_N^-}^{\mathcal{D}_N^+} \frac{dt}{\left[\left(\mathcal{D}_N^+\right)^2 - t^2\right]^{1/2} \left[t^2 - \left(\mathcal{D}_N^-\right)^2\right]^{1/2}} \frac{dV}{dt} = 0. \tag{154}$$

As $\mathcal{D}_N^- \to 0$, Eq. (153) coincides with the integral equation (66) for a single–band phase. In the same limit, Eq. (154) becomes equivalent to the assertion $\nu_D(0) = 0$, with ν_D being the spectral density in a single–band phase. This corresponds to the point of merging of two eigenvalue cuts.

References

1. Mehta, M. L. (1991) *Random Matrices*, Academic Press, Boston.
2. Guhr, T., Müller–Groeling, A., and Weidenmüller, H. A. (1998) Random matrix theories in quantum physics: Common concepts, *Phys. Rep.* **299**, 189 – 428.
3. Wigner, E. P. (1951) On the statistical distribution of the widths and spacings of nuclear resonance levels, *Proc. Cambridge Philos. Soc.* **47**, 790 – 798.
4. Dyson, F. J. (1962) Statistical theory of energy levels of complex systems I, II, III, *J. Math. Phys.* **3**, 140 – 156; **3**, 157 – 165; **3**, 166 – 175.
5. Dyson, F. J. (1962) The threefold way. Algebraic structure of symmetry groups and ensembles in quantum mechanics, *J. Math. Phys.* **3**, 1199 – 1215.
6. Brody, T. A., Flores, J., French, J. B., Mello, P. A., Pandey, A., and Wong, S. S. M. (1981) Random–matrix physics: Spectrum and strength fluctuations, *Rev. Mod. Phys.* **53**, 385 – 479.
7. Di Francesco, P., Ginsparg, P., and Zinn–Justin, J. (1995) $2D$ gravity and random matrices, *Phys. Rep.* **254**, 1 – 133.
8. Verbaarschot, J. J. M. and Zahed, I. (1993) Spectral density of the QCD Dirac operator near zero virtuality, *Phys. Rev. Lett.* **70**, 3852 – 3855; Verbaarschot, J. J. M. (1994) The spectrum of the QCD Dirac operator and chiral random matrix theory: the threefold way, *Phys. Rev. Lett.* **72**, 2531 – 2533; Verbaarschot, J. J. M. and Zahed, I. (1994) Random matrix theory and QCD, *Phys. Rev. Lett.* **73**, 2288 – 2291.
9. Gutzwiller, M. C. (1990) *Chaos in Classical and Quantum Mechanics*, Springer–Verlag, New York.
10. Altshuler, B. L., Lee, P. A., and Webb, R. A. (1991) *Mesoscopic Phenomena in Solids*, North–Holland, Amsterdam.

11. Beenakker, C. W. J. (1997) Random matrix theory of quantum transport, *Rev. Mod. Phys.* **69**, 731 - 808.
12. Gor'kov, L. P. and Eliashberg, G. M. (1965) Minute metallic particles in an electromagnetic field, *Zh. Eksp. Teor. Fiz.* **48**, 1407 - 1418 [*Sov. Phys. JETP* **21**, 940 - 947].
13. Porter, C. E. (1965) *Statistical theories of spectra: Fluctuations*, Academic Press, New York.
14. Efetov, K. B. (1983) Supersymmetry and theory of disordered metals, *Adv. Phys.* **32**, 53 - 127.
15. Berbenni–Bitsch, M. E., Meyer, S., Schäfer, A., Verbaarschot, J. J. M., and Wettig, T. (1998) Microscopic universality in the spectrum of the lattice Dirac operator, *Phys. Rev. Lett.* **80**, 1146 -1149.
16. Verbaarschot, J. J. M. (1998) Universal fluctuations in Dirac spectra, in *New Developments in Quantum Field Theory*, Plenum Press, New York.
17. Akemann, G., Damgaard, P. H., Magnea, U., and Nishigaki, S. (1997) Universality of random matrices in the microscopic limit and the Dirac operator spectrum, *Nucl. Phys. B* **487**, 721 - 738.
18. Dyson, F. J. (1972) A class of matrix ensembles, *J. Math. Phys.* **13**, 90 - 97.
19. Ambjørn, J., Jurkiewicz, J., and Makeenko, Yu. M. (1990) Multiloop correlators for two–dimensional quantum gravity, *Phys. Lett. B* **251**, 517 - 524.
20. Akemann, G. and Ambjørn, J. (1996) New universal spectral correlators, *J. Phys. A* **29**, L555 - L560.
21. Itoi, C. (1997) Universal wide correlators in non–Gaussian orthogonal, unitary and symplectic random matrix ensembles, *Nucl. Phys. B* **493**, 651 - 659.
22. Beenakker, C. W. J. (1993) Universality in the random–matrix theory of quantum transport, *Phys. Rev. Lett.* **70**, 1155 - 1158.
23. Beenakker, C. W. J. (1994) Universality of Brézin and Zee's spectral correlator, *Nucl. Phys. B* **422**, 515 - 520.
24. Brézin, E. and Zee, A. (1994) Correlation functions in disordered systems, *Phys. Rev. E* **49**, 2588 - 2596.
25. Hackenbroich, G. and Weidenmüller, H. A. (1995) Universality of random–matrix results for non–Gaussian ensembles, *Phys. Rev. Lett.* **74**, 4118 - 4121.
26. Verbaarschot, J. J. M., Weidenmüller, H. A., and Zirnbauer, M. R. (1985) Grassmann integration in stochastic quantum physics: The case of compound–nucleus scattering, *Phys. Rep.* **129**, 367 - 438.
27. Mehta, M. L. (1960) On the statistical properties of the level–spacings in nuclear spectra, *Nucl. Phys.* **18**, 395 - 419; Mehta, M. L. and Gaudin, M. (1960) On the density of eigenvalues of a random matrix, *Nucl. Phys.* **18**, 420 - 427.
28. Fox, D. and Kahn, P. B. (1964) Higher order spacing distributions for a class of unitary ensembles, *Phys. Rev.* **134**, B1151 - B1155.
29. Leff, H. S. (1964) Class of ensembles in the statistical theory of energy–level spectra, *J. Math. Phys.* **5**, 763 - 768.
30. Bronk, B. V. (1965) Exponential ensemble for random matrices, *J. Math. Phys.* **6**, 228 - 237.
31. Nagao, T. and Wadati, M. (1991) Correlation functions of random matrix ensembles related to classical orthogonal polynomials I, *J. Phys. Soc. Jpn.* **60**, 3298 - 3322; Nagao, T. and Wadati, M. (1992) Correlation functions of random matrix ensembles related to classical orthogonal polynomials II, III, *J. Phys. Soc. Jpn.* **61**, 78 - 88; **61**, 1910 - 1918.
32. Basor, E. L., Tracy, C. A., and Widom, H. (1992) Asymptotics of level spacing distribution functions for random matrices, *Phys. Rev. Lett.* **69**, 5 - 8.
33. Brézin, E. and Neuberger, H. (1991) Multicritical points of unoriented random surfaces, *Nucl. Phys. B* **350**, 513 - 553.
34. Tracy, C. A. and Widom, H. (1998) Correlations functions, cluster functions

and spacing distributions for random matrices, *Los Alamos preprint archive*, solv-int/9804004.
35. Mahoux, G. and Mehta, M. L. (1991) A method of integration over matrix variables: IV, *J. Phys. (France)* **1**, 1093 - 1108.
36. Brézin, E. and Zee, A. (1993) Universality of the correlations between eigenvalues of large random matrices, *Nucl. Phys. B* **402**, 613 - 627.
37. Freilikher, V., Kanzieper, E., and Yurkevich, I. (1996) Unitary random-matrix ensemble with governable level confinement, *Phys. Rev. E* **53**, 2200 - 2209.
38. Freilikher, V., Kanzieper, E., and Yurkevich, I. (1996) Theory of random matrices with strong level confinement: Orthogonal polynomial approach, *Phys. Rev. E* **54**, 210 - 219.
39. Shohat, J. A. and Tamarkin, J. D. (1963) *The problem of moments*, American Mathematical Society, Providence.
40. Szegö, G. (1921) Über die Randwerte analytischer Funktionen, *Math. Annalen* **84**, 232 - 244.
41. Slevin, K. and Nagao, T. (1993) New random matrix theory of scattering in mesoscopic systems, *Phys. Rev. Lett.* **70**, 635 - 638.
42. Altland, A. and Zirnbauer, M. R. (1997) Nonstandard symmetry classes in mesoscopic normal-superconducting hybrid structures, *Phys. Rev. B* **55**, 1142 - 1161.
43. Nagao, T. and Slevin, K. (1993) Nonuniversal correlations for random matrix ensembles, *J. Math. Phys.* **34**, 2075 - 2085; Laguerre ensembles of random matrices: Nonuniversal correlation functions, *J. Math. Phys.* **34**, 2317 - 2330.
44. Forrester, P. J. (1993) The spectrum of random matrix ensembles, *Nucl. Phys. B* **402**, 709 - 728.
45. Nagao, T. and Forrester, P. J. (1995) Asymptotic correlations at the spectrum edge of random matrices, *Nucl. Phys. B* **435**, 401 - 420.
46. Tracy, C. A. and Widom, H. (1994) Level-spacing distributions and the Bessel kernel, *Commun. Math. Phys.* **161**, 289 - 309.
47. Nishigaki, S. (1996) Proof of universality of the Bessel kernel for chiral matrix models in the microscopic limit, *Phys. Lett. B* **387**, 139 - 144.
48. Kanzieper, E. and Freilikher, V. (1998) Random-matrix models with the logarithmic-singular level confinement: Method of fictitious fermions, *Philos. Mag. B* **77**, 1161 - 1171.
49. Wigner, E. P. (1962) Distribution laws for the roots of a random Hermitean matrix, reprinted in: Porter, C. E. (1965) *Statistical theories of spectra: Fluctuations*, pp. 446 - 461 , Academic Press, New York.
50. Bowick, M. J. and Brézin, E. (1991) Universal scaling of the tail of the density of eigenvalues in random matrix models, *Phys. Lett. B* **268**, 21 - 28.
51. Kanzieper, E. and Freilikher, V. (1997) Universality in invariant random-matrix models: Existence near the soft edge, *Phys. Rev. E* **55**, 3712 - 3715.
52. Tracy, C. A. and Widom, H. (1994) Level-spacing distributions and the Airy kernel, *Commun. Math. Phys.* **159**, 151 - 174.
53. Kanzieper, E. and Freilikher, V. (1997) Novel universal correlations in invariant random-matrix models, *Phys. Rev. Lett.* **78**, 3806 - 3809.
54. Damgaard, P. H. and Nishigaki, S. M. (1998) Universal spectral correlators and massive Dirac operators, *Nucl. Phys. B* **518**, 495 - 512; Universal massive spectral correlators and QCD_3, *Phys. Rev. D* **57**, 5299 - 5302.
55. Akemann, G. (1996) Higher genus correlators for the Hermitian matrix model with multiple cuts, *Nucl. Phys. B* **482**, 403 - 430.
56. Kanzieper, E. and Freilikher, V. (1998) Two-band random matrices, *Phys. Rev. E* **57**, 6604 - 6611.
57. Shohat, J. (1930) *C. R. Hebd. Seances Acad. Sci.* **191**, 989; Shohat, J. (1939) A differential equation for orthogonal polynomials, *Duke Math. J.* **5**, 401 - 417.
58. Sener, M. K. and Verbaarschot, J. J. M. (1998) Universality in chiral random matrix theory at $\beta = 1$ and $\beta = 4$, *Los Alamos preprint archive*, hep-th/9801042.

59. Widom, H. (1998) On the relation between orthogonal, symplectic and unitary matrix ensembles, *Los Alamos preprint archive*, solv-int/9804005.
60. Szegö, G. (1967) *Orthogonal Polynomials*, American Mathematical Society, Providence.
61. Bonan, S. S. and Clark, D. S. (1986) Estimates of the orthogonal polynomials with weight $\exp(-x^m)$, m an even positive integer, *J. Appr. Theory* **46**, 408 - 410.
62. Sheen, R. C. (1987) Plancherel-Rotach type asymptotics for orthogonal polynomials associated with $\exp\left(-x^6/6\right)$, *J. Appr. Theory* **30**, 232 - 293; Bonan, S. S. and Clark, D. S. (1990) Estimates of the Hermite and the Freud polynomials, *J. Appr. Theory* **63**, 210 - 224; Bauldry, W. C. (1990) Estimates of asymmetric Freud polynomials, *J. Appr. Theory* **63**, 225 - 237; Mhaskar, H. N. (1990) Bounds for certain Freud type orthogonal polynomials, *J. Appr. Theory* **63**, 238 - 254.
63. Nevai, P. (1986) Géza Freud, orthogonal polynomials and Christoffel functions: A case study, *J. Appr. Theory* **48**, 3 - 167.
64. Lubinsky, D. S. (1993) An update on orthogonal polynomials and weighted approximation on the real line, *Acta Appl. Math.* **33**, 121 - 164.
65. Akemann, G. and Kanzieper, E. (1998) unpublished.
66. Akemann, G., Damgaard, P. H., Magnea, U., and Nishigaki, S. (1998) Multicritical microscopic spectral correlators of Hermitian and complex matrices, *Nucl. Phys. B* **519**, 682 - 714.
67. Gross, D. J. and Migdal, A. A. (1990) Nonperturbative two-dimensional quantum gravity, *Phys. Rev. Lett.* **64**, 127 - 130.
68. Dingle, R. B. (1973) *Asymptotic Expansions: Their Derivation and Interpretation*, Academic Press, New York.
69. Molinari, L. (1988) Phase structure of matrix models through orthogonal polynomials, *J. Phys. A* **21**, 1 - 6.
70. Douglas, M., Seiberg, N., and Shenker, S. (1990) Flow and instability in quantum gravity, *Phys. Lett. B* **244**, 381 - 386.
71. Sasaki, M. and Suzuki, H. (1991) Matrix realization of random surfaces, *Phys. Rev. D* **43**, 4015 - 4028.
72. Jurkiewicz, J., Nowak, M. A., and Zahed, I. (1996) Dirac spectrum in QCD and quark masses, *Nucl. Phys. B* **478**, 605 - 626.
73. Jackson, A. and Verbaarschot, J. (1996) Random matrix model for chiral symmetry breaking, *Phys. Rev. D* **53**, 7223 - 7230.
74. Cugliandolo, L. F., Kurchan, J., Parisi, G., and Ritort, F. (1995) Matrix models as solvable glass models, *Phys. Rev. Lett.* **74**, 1012 - 1015.
75. Morita, Y., Hatsugai, Y., and Kohmoto, M. (1995) Universal correlations in random matrices and one-dimensional particles with long-range interactions in a confinement potential, *Phys. Rev. B* **52**, 4716 - 4719. See also Ref. [80].
76. Shimamune, Y. (1982) On the phase structure of large N matrix models and gauge models, *Phys. Lett. B* **108**, 407 - 410.
77. Cicuta, G. M., Molinari, L., and Montaldi, E. (1990) Multicritical points in matrix models, *J. Phys. A* **23**, L421 - L425.
78. Jurkiewicz, J. (1991) Chaotic behaviour in one-matrix models, *Phys. Lett. B* **261**, 260 - 268.
79. Demeterfi, K., Deo, N., Jain, S., and Tan, C.-I. (1990) Multiband structure and critical behavior of matrix models, *Phys. Rev. D* **42**, 4105 - 4122.
80. Higuchi, S., Itoi, C., Nishigaki, S. M., and Sakai, N. (1997) Renormalization group approach to multiple-arc random matrix models, *Phys. Lett. B* **398**, 123 - 129.
81. Deo, N. (1997) Orthogonal polynomials and exact correlation functions for two cut random matrix models, *Nucl. Phys. B* **504**, 609 - 620.
82. Altshuler, B. L. and Shklovskii, B. I. (1986) Repulsion of energy levels and conductivity of small metal samples, *Zh. Eksp. Teor. Fiz.* **91**, 220 - 234 [*Sov. Phys. JETP* **64**, 127 - 135].

83. Jalabert, R. A., Pichard, J.-L., and Beenakker, C. W. J. (1993) Long-range energy level interaction in small metallic particles, *Europhys. Lett.* **24**, 1 - 6.
84. Brézin, E. and Deo, N. (1998) Smoothed correlators for symmetric double-well matrix models: Some puzzles and resolutions, *Los Alamos preprint archive*, cond-mat/9805096.
85. Brézin, E. and Zinn-Justin, J. (1992) Renormalization group approach to matrix models, *Phys. Lett. B* **288**, 54 - 58.
86. Muttalib, K. A., Chen, Y., Ismail, M. E. H., and Nicopoulos, V. N. (1993) New family of unitary random matrices, *Phys. Rev. Lett.* **71**, 471 - 475.
87. Bogomolny, E., Bohigas, O., and Pato, M. P. (1997) Distribution of eigenvalues of certain matrix ensembles, *Phys. Rev. E* **55**, 6707 - 6718.

OPTICS OF ORDERED AND DISORDERED ATOMIC MEDIA

RODNEY LOUDON
Department of Physics, University of Essex,
Colchester CO4 3SQ, England

Abstract. The optical properties of dispersive and lossy, ordered and disordered dielectric media are derived in the classical Lorentz model. Dielectric linear-response theory provides the spectrum of electric-field fluctuations, whose maxima define the polariton dispersion relation. The dependence of polariton radiative damping rates on the degree of disorder in the dielectric is calculated. Particular attention is given to the magnitudes of the scattered intensity for an incident light beam and the decay rates for material excitations. The quantum, semiclassical and classical theories of spontaneous emission are compared and the quantum theory is applied to the radiative decay of dilute impurity atoms in lossy homogeneous and inhomogeneous dielectrics. The phase, group and energy velocities for optical pulse propagation through a lossy dielectric are derived and their physical significances are discussed.

1. Introduction

The optical properties of a material that are accessible to measurement are mainly determined by the form of its dielectric function. The experiments of interest include measurements of refraction and attenuation that are directly related to the real and imaginary parts of the dielectric function, but they also include measurements of spontaneous emission rates, two-photon absorption and light-scattering cross sections whose relations to the dielectric function are less direct. The form of the dielectric function also governs the velocities associated with optical pulse propagation through media.

Much of the theory of these experiments is expressed in forms that apply generally to any dielectric material but it is useful to evaluate some of the expressions for a simple form of single-resonance dielectric function. The classical Lorentz theory and the quantum theory provide very similar expressions for the dielectric function that give good descriptions of a wide range of measurements In considerations of the effects of disorder, particularly on the radiative damping of dielectric excitations, it is necessary to make specific models of the spatial distribution of the atoms whose transitions provide the optical resonance of the dielectric. The atoms may have a random spatial distribution, as

J.-P. Fouque (ed.), Diffuse Waves in Complex Media, 213–247.

in a gas, or they may lie on a regular lattice, as in a perfect crystal, or they may have some intermediate distribution, as in a disordered or randomly diluted crystal.

It is assumed throughout the lectures that the mean separation between atoms is very small compared with the wavelength of the light and, for a gas, this implies a pressure greater than about 100 Pa. The atomic excitations in this long-wavelength regime are not confined to individual atoms but they take the forms of travelling waves, or excitons, whose excitation energies are shared beween all of the atoms.

2. Homogeneous media

Consider an isotropic dielectric medium that has a single optical resonance in the vicinity of the frequency of an incident light beam. For most purposes, it is permissible to ignore the optical resonances at more distant frequencies and to treat the atomic or ionic constituents of the medium as having a ground state and the single excited state that corresponds to the resonant transition. The excited state necessarily has threefold degeneracy because the transition dipole moment must have equal components in each of the three orthogonal coordinate directions. In the present section we summarize the theory of a homogeneous material in which the atoms are arranged on an ideal periodic lattice. The material excitation has the nature of an *exciton*, in which the excitation amplitude is equally shared by all of the dipole-active atoms or molecules and the phase of the excitation corresponds to a travelling wave of well-defined frequency and wavevector.

2.1. DIELECTRIC MODEL

The descriptions of the optical properties of dielectric media according to the classical and quantum theories are amost identical, and any distinctions between them are generally insignificant in the comparison of theory with experiment. It is convenient to begin with the classical Lorentz theory of dielectrics, where the excitonic transition dipole is represented as a harmonic oscillator of frequency ω_E. An excitation of the dielectric involves two coupled field variables, namely the dipole moment $e\mathbf{u}$ of the Lorentz oscillators and the local electric field $\mathbf{E}_{loc}$ at the atomic sites. The linear coupling between the dipole moment and the electric field ensures that these variables have the same frequency ω and wavevector $\mathbf{q}$. Thus with the common time and space dependence $\exp(-i\omega t + i\mathbf{q}.\mathbf{r})$ of the plane-wave fields suppressed, the Lorentz equation of motion is

$$m\left(\omega_E^2 - \omega^2\right)\mathbf{u} = e\mathbf{E}_{loc}, \tag{2.1}$$

where m and e are the effective mass and charge of the oscillator. The coupling of the oscillators to the electric field produces long-range forces that modify the excitation frequencies of the dielectric as a whole from the bare exciton transition frequency ω_E, as we show below.

The field $\mathbf{E}$ that enters Maxwell's equations is the electric field averaged over a macroscopic region of the dielectric rather than the local field $\mathbf{E}_{loc}$, and the two are related by local-field correction factors. These factors allow for the additional fields at the atomic sites produced by the electrical polarization of neighbouring atomic dipoles. The polarization associated with the oscillators is

$$\mathbf{P} = Ne\mathbf{u}/V, \tag{2.2}$$

where N is the number of atomic sites in the sample volume V. According to Lorentz himself [1] (see [2] for a more accessible discussion), the local and macroscopic electric fields are related by

$$\mathbf{E}_{\mathrm{loc}} = \mathbf{E} + (\mathbf{P}/3\varepsilon_0), \tag{2.3}$$

and this relation is indirectly confirmed by recent many-body theory derivations for a hard-sphere fluid [3].

Consider the dielectric excitations whose wavevectors q are much larger than ω / c. The electrostatic limit is valid in this case and the Maxwell equations for the electric field and the displacement give

$$\mathbf{q} \times \mathbf{E} = 0 \tag{2.4}$$

and

$$\mathbf{q}.\mathbf{D} = \mathbf{q}.(\varepsilon_0 \mathbf{E} + \mathbf{P}) = 0. \tag{2.5}$$

The above equations have three solutions that correspond to the three orientations of the electric field $\mathbf{E}$ relative to the excitation wavevector $\mathbf{q}$. The twofold degenerate *transverse* solutions have

$$\mathbf{q}.\mathbf{E} = 0 \tag{2.6}$$

so that Eq. (2.5) is automatically satisfied and it follows from Eq. (2.4) that the macroscopic field $\mathbf{E}$ must vanish. The frequency ω_{T} of the transverse excitations is thus obtained from Eqs. (2.1) to (2.3) as

$$\omega_{\mathrm{T}}^2 = \omega_{\mathrm{E}}^2 - (Ne^2/3\varepsilon_0 mV). \tag{2.7}$$

The nondegenerate *longitudinal* solutions have the macroscopic field parallel to the wavevector, so that Eq. (2.4) is automatically satisfied and the electric field is not required to vanish. The frequency ω_{L} of the longitudinal excitations is obtained from Eqs. (2.1) to (2.3) and (2.5) as

$$\omega_{\mathrm{L}}^2 = \omega_{\mathrm{E}}^2 + (2Ne^2/3\varepsilon_0 mV). \tag{2.8}$$

The long-range electrical forces in the dielectric thus lift the threefold exciton degeneracy to produce distinct transverse and longitudinal frequencies, whose separation is obtained by subtraction of Eqs. (2.7) and (2.8) as

$$\omega_L^2 - \omega_T^2 = Ne^2/\varepsilon_0 mV. \tag{2.9}$$

The transverse frequency is valid for wavevectors that satisfy

$$\omega / c << q << (N / V)^{1/3}, \tag{2.10}$$

corresponding to the right-hand side of Fig. 1.

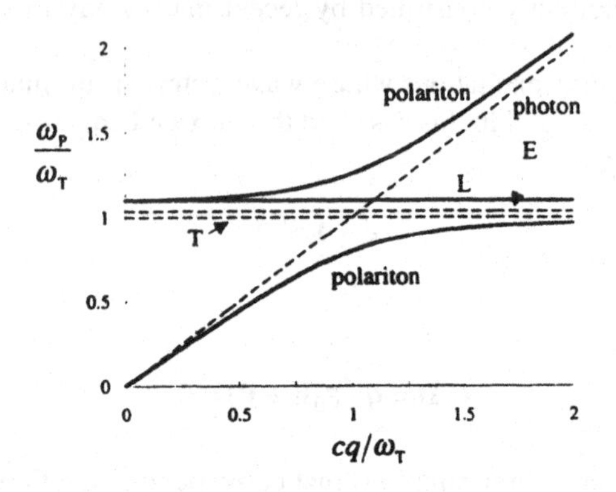

Figure 1. Dispersion relations of the longitudinal and two transverse polariton branches. The dashed lines show the exciton and transverse frequencies and the light line.

2.2. DIELECTRIC FUNCTION

The dielectric function $\varepsilon(\omega)$ is defined by

$$\mathbf{D} = \varepsilon_0 \mathbf{E} + \mathbf{P} = \varepsilon(\omega)\varepsilon_0 \mathbf{E}. \tag{2.11}$$

Its form is readily determined by elimination of $\mathbf{u}$, $\mathbf{E}_{loc}$ and $\mathbf{P}$ from Eqs. (2.1) to (2.3) and (2.11) as

$$\varepsilon(\omega) = \frac{\omega^2 - \omega_L^2}{\omega^2 - \omega_T^2}, \tag{2.12}$$

where Eq. (2.9) has also been used. It is seen that the dielectric function has a zero at the longitudinal frequency ω_L and an infinity, or pole, at the transverse frequency ω_T. These two frequencies are further related by the dielectric function at zero frequency

$$\varepsilon(0) = (\omega_L/\omega_T)^2, \tag{2.13}$$

which is a form of the *Lyddane–Sachs–Teller relation.*

The dielectric function derived from the quantum mechanics of a two-level atom (see for example [4]) has a form identical to that in Eq. (2.12), but the difference between the squares of the longitudinal and transverse frequencies given by the right-hand side of Eq. (2.9) is 'corrected' by multiplication by the *oscillator strength* of the transition,

$$\text{oscillator strength} = 2m\omega_T|d|^2/\hbar, \tag{2.14}$$

where ed is the quantum-mechanical transition dipole matrix element and a threefold-degenerate excited state is assumed. The same correction applies to the exciton transition and the modified form of Eq. (2.9) is thus

$$\omega_L^2 - \omega_T^2 = 2Ne^2\omega_T|d|^2/\varepsilon_0\hbar V. \tag{2.15}$$

Another relation obtainable from Eqs. (2.7) and (2.8),

$$\omega_L^2 + 2\omega_T^2 = 3\omega_E^2, \tag{2.16}$$

is unchanged by the correction.

The motions of the Lorentz dipole in the classical theory and the exciton amplitude in the quantum theory are subjected to various damping mechanisms that depend on the configurations of the atomic sites and on the nature of the dielectric host material. The dipole or exciton may be able to radiate electromagnetic waves in directions other than that of the incident light beam and there may also be significant nonradiative decay channels. The damping rate is represented by a function $\gamma \equiv \gamma(\omega)$ and it is useful to separate the radiative and nonradiative contributions as

$$\gamma(\omega) = \gamma_R(\omega) + \gamma_{NR}(\omega). \tag{2.17}$$

It is not difficult to insert damping or decay into the equations of motion and a convenient form of the modified quantum expression for the dielectric function is

$$\varepsilon(\omega) = \frac{\omega + \omega_L + i\gamma(-\omega)}{\omega + \omega_T + i\gamma(-\omega)} \frac{\omega - \omega_L + i\gamma(\omega)}{\omega - \omega_T + i\gamma(\omega)} = \varepsilon^*(-\omega), \tag{2.18}$$

where the damping function has the property

$$\gamma(-\omega) = 0 \quad \text{for} \quad \omega > 0. \tag{2.19}$$

The frequency ω is regarded as a real and positive quantity in the applications of dielectric theory to experiments but it is sometimes useful to include negative frequencies in formal calculations. However, the frequency is here taken to be posititve, so that

$$\varepsilon(\omega) = \frac{\omega + \omega_{\mathrm{L}}}{\omega + \omega_{\mathrm{T}}} \frac{\omega - \omega_{\mathrm{L}} + i\gamma(\omega)}{\omega - \omega_{\mathrm{T}} + i\gamma(\omega)}. \tag{2.20}$$

A broadly similar expression is obtained when damping is inserted into the classical theory.

The nature of the damping function is discussed in detail in section 3. An amplifying material, for example a laser gain medium, has a negative $\gamma(\omega)$ at the frequencies for which amplification occurs, but $\gamma(\omega)$ is positive at all frequencies in the lossy dielectric materials considered here. The radiative contribution to the damping vanishes in a homogeneous medium and $\gamma(\omega)$ then consists only of the nonradiative contribution. The zero-damping expression (2.12) is retrieved if γ is simply removed from Eq. (2.18) but if γ is allowed to tend to zero, the limiting form is

$$\lim_{\gamma \to 0} \varepsilon(\omega) = \frac{\omega^2 - \omega_{\mathrm{L}}^2}{\omega^2 - \omega_{\mathrm{T}}^2} + i\pi \frac{\omega_{\mathrm{L}}^2 - \omega_{\mathrm{T}}^2}{2\omega_{\mathrm{T}}} \delta(\omega - \omega_{\mathrm{T}}), \tag{2.21}$$

where a standard representation of the Dirac delta function has been used.

The frequency and wavevector of electromagnetic waves propagated through the dielectric are related to the dielectric function by

$$(qc / \omega)^2 = \varepsilon(\omega) = \varepsilon'(\omega) + i\varepsilon''(\omega) \tag{2.22}$$

or

$$qc / \omega = \sqrt{\varepsilon(\omega)} = \eta(\omega) + i\kappa(\omega). \tag{2.23}$$

Here $\eta(\omega)$ is the refractive index and $\kappa(\omega)$ is the extinction coefficient, and the complete function on the right-hand side is the complex refractive index. It follows for the form of the dielectric function in Eq. (2.20) that

$$\varepsilon'(\omega) = \eta^2(\omega) - \kappa^2(\omega) = \frac{\omega + \omega_{\mathrm{L}}}{\omega + \omega_{\mathrm{T}}} \frac{(\omega - \omega_{\mathrm{L}})(\omega - \omega_{\mathrm{T}}) + \gamma^2(\omega)}{(\omega - \omega_{\mathrm{T}})^2 + \gamma^2(\omega)} \tag{2.24}$$

and

$$\varepsilon''(\omega) = 2\eta(\omega)\kappa(\omega) = \frac{\omega + \omega_L}{\omega + \omega_T} \frac{(\omega_L - \omega_T)\gamma(\omega)}{(\omega - \omega_T)^2 + \gamma^2(\omega)}, \tag{2.25}$$

where the first factor on the right-hand side produces a small modification of the Lorentzian second factor.

The spatial dependences of the fields in the dielectric for propagation parallel to the z axis obtained with the use of Eq. (2.23) are governed by the common factor

$$\exp(iqz) = \exp\{[i\omega\eta(\omega) - \omega\kappa(\omega)](z/c)\}. \tag{2.26}$$

The presence of the extinction coefficient, which stems from the imaginary part of the dielectric function, thus causes an attenuation of, or loss of energy from, the waves. The *attenuation length* $L(\omega)$, the distance over which the intensity of the light falls to $1/e$ of its initial value, is given by

$$\frac{1}{L(\omega)} = \frac{2\omega\kappa(\omega)}{c} = \frac{\omega\varepsilon''(\omega)}{c\eta(\omega)}. \tag{2.27}$$

2.3. DISPERSION RELATIONS AND SUM RULES

The poles of the dielectric function $\varepsilon(\omega)$ of Eq. (2.18) clearly lie in the lower half of the complex ω plane. The absence of any poles in the upper half of the complex plane is one of the conditions that a function must satisfy in order that it may obey dispersion or Kramers–Kronig relations [5]. The other main condition is that the square modulus of the function should be integrable along the real ω axis, and this is clearly not satisfied by the dielectric functions in Eqs. (2.18) and (2.20), which tend to unity as ω tends to $\pm\infty$. However, it is readily verified that the function $\varepsilon(\omega) - 1$ is square integrable along the real axis. The resulting dispersion relations are [5,6]

$$\varepsilon''(\omega) = -\frac{2\omega}{\pi} \wp \int_0^\infty d\omega' \frac{\varepsilon'(\omega') - 1}{\omega'^2 - \omega^2} \tag{2.28}$$

and

$$\varepsilon'(\omega) - 1 = \frac{2}{\pi} \wp \int_0^\infty d\omega' \frac{\omega'\varepsilon''(\omega')}{\omega'^2 - \omega^2}, \tag{2.29}$$

where the principal-value integrals are defined in the usual way [7] and the dielectric function is separated into its real and imaginary parts according to Eq. (2.22). The Kramers–Kronig relations show that the real and imaginary parts of the dielectric function are very intimately connected. Indeed, a knowledge of one part at all positive frequencies provides a complete knowledge of the other part at all positive frequencies. The relation (2.28) shows that a frequency-dependent real part implies a nonzero imaginary part, that is, the dispersion and loss in a dielectric are closely related. The relation (2.29) shows that the real part of the dielectric function must equal unity at all frequencies if the imaginary part vanishes at all frequencies.

An important application of dispersion relations is the derivation of sum rules. For example, if ω is taken to be much larger than ω_T, then ω' in the denominator of the integrand of Eq. (2.29) can be neglected and use of the asymptotic limit of Eq. (2.24) gives the sum rule

$$\int_0^\infty d\omega\,\omega\varepsilon''(\omega) = \tfrac{1}{2}\pi\left(\omega_L^2 - \omega_T^2\right). \tag{2.30}$$

The analytic properties of the complex refractive index are more complicated than those of the dielectric function because of the square root in the definition of Eq. (2.22), but the sum rules

$$\int_0^\infty d\omega\{\eta(\omega) - 1\} = 0 \quad \text{and} \quad \int_0^\infty d\omega\,\omega\kappa(\omega) = \tfrac{1}{4}\pi\left(\omega_L^2 - \omega_T^2\right) \tag{2.31}$$

can be proved [8].

2.4. POLARITONS

The frequencies of the dielectric excitations with $\omega << cq$ are given in Eqs. (2.7) and (2.8). If this restriction on ω is now removed, the complete time-dependent Maxwell equations must be used to determine the excitation frequencies. For the single plane-wave fields assumed before, there is the usual relation

$$\mathbf{q} \times (\mathbf{q} \times \mathbf{E}) = -\mu_0\omega^2\mathbf{D}, \tag{2.32}$$

and, with the use of Eq. (2.11), this becomes

$$\mathbf{q}(\mathbf{q}.\mathbf{E}) - q^2\mathbf{E} = -(\omega / c)^2\varepsilon(\omega)\mathbf{E}. \tag{2.33}$$

Tranverse and longitudinal solutions can still be identified. The former solutions are specified by the condition (2.6) and it follows that the frequency and wavevector of the transverse waves, or polaritons, are related by Eq. (2.22). The damping is ignored in the present section and, with the form of dielectric function taken from Eq. (2.12), the polariton dispersion relation is $\omega = \omega_P$ with

$$\left(\frac{cq}{\omega_{\mathrm{P}}}\right)^2 = \frac{\omega_{\mathrm{P}}^2 - \omega_{\mathrm{L}}^2}{\omega_{\mathrm{P}}^2 - \omega_{\mathrm{T}}^2}. \tag{2.34}$$

The variation of ω_{P} with q is illustrated in Fig. 1. There are two polariton branches and for large q the upper branch tends to the linear relation $\omega_{\mathrm{P}} = cq$, known as the light line, while the lower branch tends to the transverse frequency ω_{T} derived in Eq. (2.7). There is a gap in the transverse solutions that extends from the transverse to the longitudinal frequency. The longitudinal solutions are specified by the condition (2.4) and it follows from Eq. (2.32) or (2.33) that

$$\varepsilon(\omega) = 0 \quad \text{or} \quad \omega = \omega_{\mathrm{L}}. \tag{2.35}$$

The longitudinal frequency derived in Eq. (2.8) thus occurs for all wavevectors, as is shown by the continuous horizontal line in Fig. 1.

The polariton dispersion relation (2.34) shows the typical behaviour of coupled-mode excitations in the region where there is a crossing of the bare dispersion relations, in this case of the exciton with $\omega = \omega_{\mathrm{E}}$ and the photon with $\omega = cq$. For each wavevector $\mathbf{q}$, the two polariton modes are combinations of the bare exciton and the bare photon of wavevector $\mathbf{q}$. The polariton has comparable amounts of exciton and photon contribution in the region of the crossover, but it mainly has the character of an exciton where its frequency is close to ω_{T} and it has the character of a photon where its frequency is close to cq (see [9] for a more detailed account of these and other polariton properties). It is emphasized that the theory presented here assumes the transition frequency ω_{E} to be independent of the exciton wavevector. This is usually a good approximation for wavevectors restricted by the inequality in Eq. (2.10) but dispersion in the bare exciton frequency becomes important at larger wavevectors.

Wavevector conservation in the homogeneous medium ensures that coupling occurs only between bare excitations with the same $\mathbf{q}$, and the polaritons are essentially the coupled modes that diagonalize the interaction between the two bare modes. The nature of the polariton removes any possibility of its radiative damping, as the spontaneous emission process requires the coupling of an electronic excitation to a continuous spatial density of emitted photons [10]. This topic is considered in detail in the following section.

3. Inhomogeneous media

The most striking effect of inhomogeneities or disorder in the arrangement of the atoms in a medium is the restoration of spontaneous emission as a mechanism for the damping of polaritons. The simplest example is that of an atomic gas, where density fluctuations introduce inhomogeneity, and we treat this system first. However, our main interest is in the effects of random distribution of the active atoms on a crystal lattice, which can also be evaluated straightforwardly, and most of the section is devoted to this problem. The prohibition of radiative damping of the polaritons in a homogeneous material strictly applies only to samples that fill all of space and it is broken for finite samples, for example thin crystal slabs [11], but these effects are not considered here.

3.1. ATOMIC GAS

The rate of spontaneous emission by a single atom in free space is governed by the Einstein *A* coefficient for the relevant transition, equal to twice the radiative contribution $\gamma_R(\omega)$ to the damping rate of Eq. (2.17). The standard expression for the rate of decay of a threefold-degenerate excited state to a nondegenerate ground state is (see for example [4])

$$2\gamma_R(\omega) = \frac{e^2\omega^3|d|^2}{3\pi\varepsilon_0\hbar c^3}, \tag{3.1}$$

where ed is again the transition dipole matrix element. This expression results from a summation over all of the propagation directions of the emitted photon in three dimensions. The spontaneous emission rate is usually presented with the frequency ω set equal to the atomic transition frequency. However, expressions for the dielectric function involve the damping rate at a general frequency ω, and this is left in place for the present.

Consider now a gas of N identical atoms in a volume V, with an incident parallel monochromatic light beam whose wavevector q continues to satisfy the right-hand part of the inequality (2.10). The theory of the dielectric function given in section 2.2 remains valid, except that we need to give further consideration to the radiative contribution to the damping rate. As the light beam propagates through the gas, photons are removed by excitation of the atoms. Some of the atoms decay again by stimulated emission and this process returns the photon energy to the light beam, without any net loss of optical intensity. However, the atoms that decay by the spontaneous process emit photons in random directions and their energy is lost from the beam, except for a negligible number of chance emissions parallel to the incident beam. This is the mechanism that generates the radiative contribution $\gamma_R(\omega)$ to the damping. Spontaneous emission thus causes an attenuation of the light beam and the lost energy reappears as photons of the same frequency distributed over all spatial directions. The process is essentially that of elastic or Rayleigh scattering, and the attenuation rate and the scattering cross-section are equivalent descriptions of the same physical process.

However, as discussed at the end of the previous section, the radiative decay, and hence the elastic scattering, by a homogeneous arrangement of atoms is totally inhibited, and this can be understood in simple terms as follows. We assume the incident light to be a classical electromagnetic wave of well-defined amplitude and phase. The components of the scattered field that arise from the atoms contained within a half-wavelength of the incident radiation are in phase with each other, but are out of phase with the fields arising from the atoms that lie within adjacent half-wavelengths. For a uniform distribution of atoms, the contributions to the total scattered field from the groups of atoms in the different half-wavelengths exactly cancel by destructive interference, except in the forward direction. The intensity of the light beam thus remains unchanged.

The same simple model also shows how density fluctuations in the gas restore the radiative damping. Suppose that a light beam of mean intensity $\bar{I}$ incident on a single atom in free space produces a mean intensity $\bar{I}_S$ in a certain direction of scattering. For an atomic gas, the volume V can be divided into smaller regions that contribute fields with the same sign of the phase to the scattered light. Suppose that a total of N_+ atoms lie in regions that contribute a scattered field of positive phase and N_- in regions that contribute a field of negative phase, with

$$N_+ + N_- = N. \tag{3.2}$$

The mean total scattered intensity in the given direction, proportional to the square of the field, is

$$\bar{I}_S\left\langle (N_+ - N_-)^2 \right\rangle = 4\bar{I}_S\left\langle \left(N_+ - \tfrac{1}{2}N\right)^2 \right\rangle = \bar{I}_S N. \tag{3.3}$$

Here the angle brackets represent averages over the random distribution of the N atoms between the regions of opposite sign of phase angle. It is assumed that each atom is equally likely to lie in one region or the other, so that a binomial probability distribution applies and the final expression uses the standard result for its variance.

It is clear from Eq. (3.3) that the total scattered intensity vanishes, as before, for a homogeneous atomic distribution with fixed and equal numbers of atoms in the regions of opposite phase, as is the case for a crystalline solid. The nonzero scattered intensity in a gas is a direct consequence of the fluctuating numbers of atoms in the two regions. The final outcome, of a mean scattered intensity equal to N times the single-atom intensity, reproduces the result found from a naïve theory in which the effects of phase coherence between the contributions of the different atoms are simply ignored. The total Rayleigh scattered intensity is N times that of a single atom, while the radiative contribution to the atomic damping rate is as given in Eq. (3.1). This result can be re-expressed with the help of Eq. (2.15) in the useful form

$$2\gamma_R(\omega) = \frac{\omega^3 V}{6\pi c^3 \omega_T N}\left(\omega_L^2 - \omega_T^2\right), \tag{3.4}$$

which involves only the atomic density and the characteristic frequencies of the material.

Results similar to those derived here also apply to glassy solids, where the static density fluctuations give rise to Rayleigh scattering, which can be an important source of loss in optical fibres. It should be mentioned that the spontaneous emission rate of Eq. (3.1) is only valid when the refractive index of the atomic surroundings is close to unity. The modification of the rate by immersion in an optically dense medium is discussed in section 4.2.

3.2. RANDOMLY-DILUTED CRYSTAL

Consider a regular lattice whose N sites are randomly populated by the active atoms with an occupation probability f, so that there are fN atoms in the volume V. The theory of the dielectric function given in section 2 remains valid to some extent for the diluted crystal, with suitable modifications, for wavevectors that satisfy the inequality

$$q << (fN/V)^{1/3}. \tag{3.5}$$

We first evaluate the modifications to the material frequencies and then the changes to the damping produced by the dilution.

The definitions of the transverse and longitudinal frequencies must be modified to take account of the dilution, so that Eqs. (2.7) and (2.8), after correction by the oscillator strength of Eq. (2.14), become

$$\omega_{Tf}^2 = \omega_E^2 - \left(2fNe^2\omega_T|d|^2/3\varepsilon_0\hbar V\right) \tag{3.6}$$

and

$$\omega_{Lf}^2 = \omega_E^2 + \left(4fNe^2\omega_T|d|^2/3\varepsilon_0\hbar V\right). \tag{3.7}$$

It follows that

$$\omega_{Lf}^2 - \omega_{Tf}^2 = 2fNe^2\omega_T|d|^2/\varepsilon_0\hbar V = f\left(\omega_L^2 - \omega_T^2\right), \tag{3.8}$$

where the longitudinal and transverse frequencies on the right-hand side are those of the undiluted crystal. The expression in Eq. (2.12) for the dielectric function in the absence of damping applies to the dilute crystal when the frequencies from Eqs. (3.6) and (3.7) are substituted.

Although the diluted crystal appears to be homogeneous to the extent that the undamped dielectric function is only modified by some simple scalings of the transverse and longitudinal frequencies, there is a more dramatic effect on the damping. We ignore the nonradiative damping for the remainder of the section and evaluate only the radiative contribution to the total damping rate of Eq. (2.17). It is convenient to consider first the scattering of light by the dilute crystal. The intensity of scattering of an incident parallel light beam is calculated by a development of the method used in section 3.1 for an atomic gas.

Suppose again that a single atom scatters light with a mean intensity $\bar{I}_S$ in a certain direction and consider the total volume of the scattering material to be divided into smaller regions that contribute scattered electric fields with the same sign of the phase. The mean total scattered intensity in the given direction is determined by the quantity on the left-hand side of Eq. (3.3), where N_+ is the number of atoms that contribute positive-phase fields and N_- the number that contribute negative-phase fields. Both of these numbers are governed by binomial distributions for the occupation of $N/2$ sites with probability f, and a simple calculation gives

$$\bar{I}_S\langle(N_+ - N_-)^2\rangle = \bar{I}_S f(1-f)N. \tag{3.9}$$

The scattering vanishes by destructive interference for $f = 1$, when the perfectly-ordered crystal lattice is restored, and also for $f = 0$, when there are no atoms to scatter the light.

The additional factor in Eq. (3.9) that modifies the light scattering cross-section is valid for frequencies well removed from the atomic resonance. In relation to the polariton dispersion curve shown in Fig. 1, these are the frequencies close to the light line $\omega = cq$, where the excitations are photon-like. Now consider the effect of the dilution on the radiative decay of the polaritons whose frequencies lie close to ω_{Tf}, where the excitations are exciton-like. Their emission of light is subject to the same destructive interference effects as the scattered intensity calculated above. However, in comparison to the exciton in the perfect lattice, the mean-square amplitude of the exciton at each occupied site is increased by a factor of $1/f$ to take account of the smaller number of sites that participate in the excitation. The modified radiative damping function is accordingly

$$\gamma_{Rf}(\omega) = (1-f)\gamma_R(\omega), \tag{3.10}$$

where $\gamma_R(\omega)$ is given by Eqs. (3.1) or (3.4).

The dielectric function for the dilute crystal is now obtained by substitution of the expressions from Eqs. (3.6), (3.7) and (3.10) for the corresponding quantities in Eq. (2.20). The results given in Eqs. (3.9) and (3.10) refer to the polaritons that are photon-like and exciton-like respectively. A more general derivation of the effects of disorder on the polariton damping requires a consideration of the linear response of the dielectric medium.

3.3. DIELECTRIC LINEAR RESPONSE

It is convenient to derive first the linear response theory for a perfect crystal. Consider the response of a dielectric to a fictitious applied polarization of the form

$$\mathbf{P}_{\text{app}} \exp(-i\omega t + i\mathbf{q}.\mathbf{r}). \tag{3.11}$$

The applied polarization induces an electric field $\mathbf{E}$ and an additional 'free' polarization $\mathbf{P}$ determined by the dielectric function $\varepsilon(\omega)$ in accordance with Eq. (2.11) as

$$\mathbf{P} = \varepsilon_0[\varepsilon(\omega) - 1]\mathbf{E}, \tag{3.12}$$

where all fields have the $\exp(-i\omega t + i\mathbf{q}.\mathbf{r})$ dependence. The relation (2.32) from Maxwell's equations is generalized with the use of Eq. (2.11) to

$$\mathbf{q}\times(\mathbf{q}\times\mathbf{E})+\left(\omega^2/c^2\right)\mathbf{E}=-\left(\omega^2/\varepsilon_0 c^2\right)\left(\mathbf{P}_{\mathrm{app}}+\mathbf{P}\right). \tag{3.13}$$

Expansion of the vector product and elimination of $\mathbf{P}$ with the use of Eq. (3.12) leads to

$$-\varepsilon_0\varepsilon(\omega)\left[c^2q^2-\omega^2\varepsilon(\omega)\right]\mathbf{E}=c^2\mathbf{q}\left(\mathbf{q}.\mathbf{P}_{\mathrm{app}}\right)-\omega^2\varepsilon(\omega)\mathbf{P}_{\mathrm{app}}. \tag{3.14}$$

The dielectric *Green function*, or *linear response function*, $G_{ij}(\mathbf{q},\omega)$ is defined by

$$E_i=\sum_j G_{ij}(\mathbf{q},\omega)P_{\mathrm{app}\,j}, \tag{3.15}$$

where i and j denote the three Cartesian components. Comparison of Eqs. (3.14) and (3.15) produces the solution [12,13]

$$G_{ij}(\mathbf{q},\omega)=-\frac{c^2q_iq_j-\omega^2\varepsilon(\omega)\delta_{ij}}{\varepsilon_0\varepsilon(\omega)\left[c^2q^2-\omega^2\varepsilon(\omega)\right]}, \tag{3.16}$$

and this is conveniently separated into transverse and longitudinal parts as

$$G_{ij}(\mathbf{q},\omega)=G_{ij}^{\mathrm{T}}(\mathbf{q},\omega)+G_{ij}^{\mathrm{L}}(\mathbf{q},\omega), \tag{3.17}$$

with

$$G_{ij}^{\mathrm{T}}(\mathbf{q},\omega)=\frac{1}{\varepsilon_0}\frac{\delta_{ij}-\bar{q}_i\bar{q}_j}{(cq/\omega)^2-\varepsilon(\omega)} \tag{3.18}$$

and

$$G_{ij}^{\mathrm{L}}(\mathbf{q},\omega)=-\frac{\bar{q}_i\bar{q}_j}{\varepsilon_0\varepsilon(\omega)}, \tag{3.19}$$

where $\bar{\mathbf{q}}$ is a unit vector parallel to $\mathbf{q}$.

The Green function is important because many experiments measure it fairly directly, for example inelastic light scattering and two-photon absorption, or are determined by its Fourier transform into coordinate space, $G_{ij}(\mathbf{r},\omega)$, for example spontaneous emission rates. In light-scattering or Raman scattering experiments, the polaritons of interest are those formed by coupling of light to the dipole-active phonons and the inelastic cross-section is determined in part by the electric-field fluctuations associated with the polariton [13]. The polariton electric-field fluctuations similarly determine the two-photon absorption coefficient [14], where the polaritons of interest are formed by coupling of dipole-active excitons to the light, as treated above. The strength of the field fluctuations is determined by the low-temperature form of the fluctuation-dissipation theorem [15]

$$\left\langle E_i(\mathbf{q},\omega)E_j^*(\mathbf{q},\omega')\right\rangle = 2\hbar\,\mathrm{Im}\,G_{ij}(\mathbf{q},\omega)\delta(\omega-\omega'), \tag{3.20}$$

where the restriction of the electric-field excitation to a given frequency and wavevector is now indicated explicitly. The imaginary part of the transverse Green function, Eq. (3.18), with the form of dielectric function from Eq. (2.20), is shown in Fig. 2. There are lines of ridges whose peaks lie very close to the two branches of the polariton dispersion relation shown in Fig. 1. The breadths of the ridges are evaluated below. The imaginary part of the longitudinal Green function, Eq. (3.19), with the same form of dielectric function, is proportional to

$$\mathrm{Im}\left(-\frac{1}{\varepsilon(\omega)}\right) = \frac{\omega_{\mathrm{L}}^2-\omega_{\mathrm{T}}^2}{2\omega_{\mathrm{L}}}\frac{\gamma(\omega_{\mathrm{L}})}{(\omega-\omega_{\mathrm{L}})^2+\gamma^2(\omega_{\mathrm{L}})} \quad (\omega \approx \omega_{\mathrm{L}}). \tag{3.21}$$

This is independent of the wavevector $\mathbf{q}$ and it has the form of a straight ridge peaked at the longitudinal frequency ω_{L} with a width of $2\gamma(\omega_{\mathrm{L}})$. It is assumed in Eq. (3.21) that the width of the ridge is much smaller than ω_{L}, so that ω can be set equal to ω_{L} except in the first factor of the Lorentzian denominator.

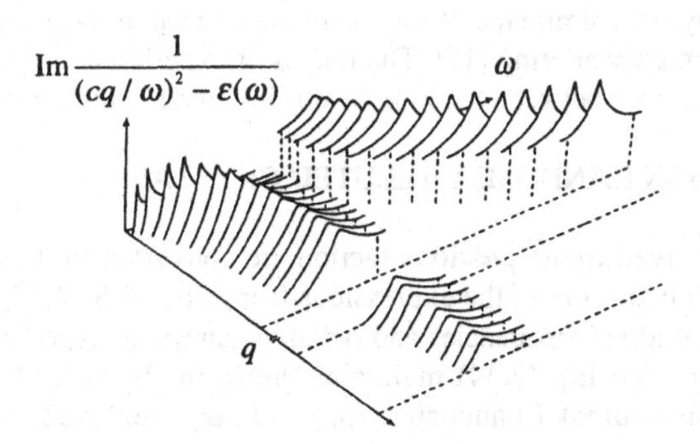

Figure 2. Imaginary part of the transverse Green function for $\omega_{\mathrm{L}} = 2\omega_{\mathrm{T}}$ and $2\gamma = 0.15\omega_{\mathrm{T}}$ plotted on a logarithmic vertical scale (after [13]).

The linewidth, or the damping, of the transverse polariton is determined by evaluation of the imaginary part of the Green function given by Eq. (3.18). Thus, with insertion of the form of dielectric function from Eq. (2.18) and after some algebra, the result is proportional to

$$\mathrm{Im}\frac{1}{(cq/\omega)^2-\varepsilon(\omega)}=\frac{\frac{1}{2}\omega_{\mathrm{P}}\left(\omega_{\mathrm{P}}^2-\omega_{\mathrm{T}}^2\right)^2}{\left(\omega_{\mathrm{P}}^2-\omega_{\mathrm{T}}^2\right)^2+\omega_{\mathrm{T}}^2\left(\omega_{\mathrm{L}}^2-\omega_{\mathrm{T}}^2\right)}\frac{\Gamma(\omega_{\mathrm{P}})}{(\omega-\omega_{\mathrm{P}})^2+\Gamma^2(\omega_{\mathrm{P}})}. \tag{3.22}$$

The spectrum has a Lorentzian profile centred on $\omega=\omega_{\mathrm{P}}$ with a damping function

$$\Gamma(\omega_{\mathrm{P}})=\frac{\frac{1}{2}\omega_{\mathrm{P}}(\omega_{\mathrm{P}}+\omega_{\mathrm{L}})(\omega_{\mathrm{P}}+\omega_{\mathrm{T}})(\omega_{\mathrm{L}}-\omega_{\mathrm{T}})}{\left(\omega_{\mathrm{P}}^2-\omega_{\mathrm{T}}^2\right)^2+\omega_{\mathrm{T}}^2\left(\omega_{\mathrm{L}}^2-\omega_{\mathrm{T}}^2\right)}\gamma(\omega_{\mathrm{P}}). \tag{3.23}$$

In deriving these results, the frequency ω has been set equal to the frequency ω_{P} of the undamped polariton defined in Eq. (2.34) except in the first factor of the denominator of the final Lorentzian function in Eq. (3.22). The polariton damping function $\Gamma(\omega_{\mathrm{P}})$ thus gives one half of the value of the width of the transverse ridges shown in Fig. 2 along a line of constant wavevector q. This approximation is, of course, valid only when the width is much smaller than the polariton frequency. In accordance with previous discussion, only the nonradiative damping contributes in the above expressions, which refer to polaritons in perfect crystals.

Both inelastic light-scattering, or Raman scattering, experiments and measurements of two-photon absorption probe the imaginary parts of the Green functions derived above along lines of almost constant q in the $\omega-q$ plane. Thus, for example, the longitudinal branch and both branches of the transverse polariton dispersion relations in CuCl have been determined by measurements of the positions of peak two-photon absorption [16] and two-photon Raman scattering [17]. The ranges of experimental points were produced by variation of the angles between the light beams involved in these processes.

3.4. POLARITONS IN RANDOMLY-DILUTED CRYSTALS

The expressions derived in the previous section are converted to those for a randomly-diluted crystal by substitution of the expressions from Eqs. (3.6), (3.7) and (3.10) for the transverse and longitudinal frequencies and radiative damping respectively. The polariton dispersion relation from Eq. (2.34) maintains the form shown in Fig. 1 but with the transverse and longitudinal frequencies ω_{T} and ω_{L} replaced by $\omega_{\mathrm{T}f}$ and $\omega_{\mathrm{L}f}$ respectively.

More drastic changes occur in the polariton damping function of Eq. (3.23). If the nonradiative contributions to the damping are again ignored, the modified damping is

$$\Gamma_f(\omega_{\mathrm{P}f})=\frac{\frac{1}{2}\omega_{\mathrm{P}f}(\omega_{\mathrm{P}f}+\omega_{\mathrm{L}f})(\omega_{\mathrm{P}f}+\omega_{\mathrm{T}f})(\omega_{\mathrm{L}f}-\omega_{\mathrm{T}f})}{\left(\omega_{\mathrm{P}f}^2-\omega_{\mathrm{T}f}^2\right)^2+\omega_{\mathrm{T}f}^2\left(\omega_{\mathrm{L}f}^2-\omega_{\mathrm{T}f}^2\right)}(1-f)\gamma_{\mathrm{R}}(\omega_{\mathrm{P}f}). \tag{3.24}$$

This complicated function has some simple special cases, with

$$\Gamma_f(\omega_{Tf}) = (1-f)\gamma_R(\omega_{Tf}) \quad \text{and} \quad \Gamma_f(\omega_{Lf}) = (1-f)\gamma_R(\omega_{Lf}), \tag{3.25}$$

and the same factor of $1-f$, as derived in Eq. (3.10), modifies the radiative damping of the exciton-like polaritons at both the transverse and longitudinal frequencies.

Again, for very low frequencies, where the polariton is photon-like,

$$\Gamma_f(\omega_{Pf}) = \frac{\omega_{Pf}(\omega_{Lf} - \omega_{Tf})}{2\omega_{Tf}\omega_{Lf}}(1-f)\gamma_R(\omega_{Pf}) \quad (\omega_{Pf} << \omega_{Tf}). \tag{3.26}$$

Thus with the use of Eqs. (3.6) to (3.8), the *energy decay time* $T(\omega_P)$ of the photon-like excitation, the time after which the energy decays to 1 / e of its initial value, is given by

$$\frac{1}{T_f(\omega_{Pf})} = 2\Gamma_f(\omega_{Pf}) = \frac{\omega_{Pf}(\omega_L^2 - \omega_T^2)}{2\omega_E^3} f(1-f)\gamma_R(\omega_{Pf}) \quad (\omega_{Pf} << \omega_{Tf}), \tag{3.27}$$

to a good approximation. The damping rate in the diluted crystal thus acquires the same factor of $f(1-f)$ as derived in Eq. (3.9) for the scattering of incident light. Note that insertion of the radiative damping rate from Eq. (3.1) or (3.4) produces a damping proportional to the fourth power of the frequency, which is a familiar feature of the Rayleigh scattering.

The same expression as in Eq. (3.27) is also derived straightforwardly from Eq. (2.27) by conversion of the attenuation length to a decay time, according to

$$\frac{1}{T(\omega)} = \frac{c}{\eta(\omega)L(\omega)} = \frac{\omega\varepsilon''(\omega)}{\eta^2(\omega)}. \tag{3.28}$$

The imaginary part of the dielectric function is obtained from the low-frequency limit of Eq. (2.25) and the square of the low-frequency refractive index is given by Eq. (2.13). It is easily verifed that the expression for the decay time in Eq. (3.27) is reproduced when the same substitutions and approximations are made.

4. Spontaneous emission

Spontaneous emission is the radiation of light by an excited atom into the surrounding space, in which no electromagnetic radiation is present initially. It is one of the three radiative processes postulated by Einstein [18] and in some respects it is the most fundamental as, unlike absorption and stimulated emission, it does not depend on any arbitrary excitation of applied radiation fields. The rate of spontaneous emission does however depend on the environment of the excited atom and its free-space value can be seriously modified by immersion in a dielectric material or by the presence of nearby objects. Several practical applications rely on such modifications of the spontaneous emission rate, for example in optical sensors of the environment or in the achievement of laser threshold for reduced values of the pumping.

The previous sections considered spontaneous emission by the constituent atoms of the medium, whose excited states of frequency ω_E formed travelling-wave collective excitations, or excitons, of the dielectric as a whole. The present section considers spontaneous emission by individual atoms inserted into the medium with sufficient dilution that they can be treated as isolated emitters.

4.1. QUANTUM VERSUS CLASSICAL CALCULATIONS

Spontaneous emission, including the modifications caused by the atomic environment, can be considered within the frameworks of three varieties of theory. Different calculations have used different combinations of the classical and quantum theories and it is important to understand the restrictions on their validities.

In the *quantum* theory, where both the atom and the electromagnetic field are treated quantum-mechanically, spontaneous emission is a consequence of the coupling of the atomic transition to the vacuum electromagnetic field. The decay rate, or Einstein A coefficient, for an atom in free space is obtained from time-dependent perturbation theory as the expression given in Eq. (3.1) with ω set equal to the atomic transition frequency ω_A,

$$2\gamma_{RQ}(\omega_A) = e^2\omega_A^3|d_A|^2/3\pi\varepsilon_0\hbar c^3, \qquad (4.1)$$

where an additional subscript Q is here introduced to identify the quantum result. The decay rate depends on the transition dipole matrix element ed_A. The expression (4.1) agrees very well with measured decay rates and it appears to be universally valid for atoms in free space.

In the *semiclassical* theory, where the atom is treated quantum-mechanically but the electromagnetic field is treated classically, spontaneous emission does not occur and the atom remains in its excited state. The absence of any electromagnetic field in the classical vacuum surrounding the atom nullifies the strength of the coupling between the atomic transition and the field, so that the emission rate vanishes. The Einstein B coefficient for the rate of stimulated emission can of course be calculated by semiclassical theory, and the Einstein A coefficient can then be deduced from the relation between the two obtained by comparison of the radiation equilibrium state with that given by Planck's law [4]. However, this procedure essentially introduces the full quantum theory by a back door, and the result cannot be regarded as a derivation from semiclassical theory.

In the *classical* theory, where both the atom and the electromagnetic field are treated classically, spontaneous emission is equivalent to the radiation by an oscillating dipole moment. The fields of the emitted radiation, proportional to the acceleration of the dipole charges, are routinely calculated in textbooks on electromagnetic theory [19]. For a dipole of mass m and oscillation frequency ω_A in free space, the rate of decay of its energy caused by the radiation is

$$2\gamma_{RC}(\omega_A) = e^2\omega_A^2/6\pi\varepsilon_0 mc^3, \tag{4.2}$$

where the additional C subscript identifies the classical result. The phenomenon of spontaneous emission thus occurs in both classical and quantum theories, but the atomic decay rates are different. In particular, the classical rate does not depend on the dipole moment and the theory provides no valid means of calculating the rates for specific atomic transitions.

The results (4.1) and (4.2) of the quantum and classical theories for the energy decay rates are quite different. However, the corresponding expressions for the rates of change of the energy of the radiating atom or dipole are more similar. The initial energy $W(\omega_A)$ of the radiator and the radiative decay rate are related by

$$\mathrm{d}W(\omega_A)/\mathrm{d}t = -2\gamma_R(\omega_A)W(\omega_A), \tag{4.3}$$

with atomic energies

$$W_Q(\omega_A) = \hbar\omega_A \tag{4.4}$$

in the quantum theory and

$$W_C(\omega_A) = m\omega_A^2|d_A|^2 \tag{4.5}$$

in the classical theory, where ed_A is the effective dipole moment of the classical oscillator. It is seen that substitutions of Eqs. (4.1) and (4.4), or (4.2) and (4.5), into Eq. (4.3) produce essentially the same expressions, differing only by a factor of 2 (see [20] for a discussion of the origin of this factor). Similarly, the quantum rates of change of the energy for highly-excited Rydberg states of atoms approach those calculated classically for equivalent circularly-orbiting charges. Nevertheless, the differences between the corresponding decay rates $2\gamma_R(\omega_A)$ remain, because of their scaling by the quantum or classical energies (4.4) or (4.5) respectively.

A further aspect of the comparison between quantum and classical methods emerges when the modifications of the emission rate by the atomic environment are considered, including the effects of surrounding dielectric media of arbitrary spatial configurations. The quantum theory differs from the classical theory in its representation of the fields by operators instead of algebraic variables and in the treatment of their time dependences. In contrast to these differences, the spatial variations of the fields are the same in both theories. Thus the field boundary conditions are the same and these produce the same spatial modes in the two theories. The boundary conditions also determine the linear response properties of the fields, and the response functions, or Green functions, are again the same in quantum and classical theories. The spontaneous emission rate is proportional to the imaginary part of a suitable Green function and the difference between the two theories is limited to the constants of proportionality, which essentially scale as the free-space decay rates $2\gamma_{RQ}(\omega_A)$ and $2\gamma_{RC}(\omega_A)$. Thus the result of a classical calculation of the spontaneous emission rate in a particular atomic environment can be converted to the corresponding quantum-mechanical result upon multiplication by the ratio

$$\frac{\gamma_{RQ}(\omega_A)}{\gamma_{RC}(\omega_A)} = \frac{2|d_A|^2 m\omega_A}{\hbar} = \left(\frac{|d_A|}{a_B}\right)^2 \frac{\omega_A}{\omega_{Ryd}}. \tag{4.6}$$

The form of the ratio on the right shows that the quantum and classical rates are the same only when the atomic transition occurs at the Rydberg frequency ω_{Ryd} and the transition dipole moment equals ea_B, where a_B is the Bohr radius.

4.2. SPONTANEOUS EMISSION IN HOMOGENEOUS DIELECTRICS

The spontaneous emission rate according to Fermi's golden rule is

$$2\gamma_R(\mathbf{r},\omega_A) = \frac{2\pi}{\hbar^2}\sum_k |\langle k|e\mathbf{d}.\mathbf{E}(\mathbf{r},t)|0\rangle|^2 \delta(\omega_k - \omega_A), \tag{4.7}$$

where $|0\rangle$ is the vacuum state of the electromagnetic field and $|k\rangle$ is its final state after the radiative decay has taken place. Here $\mathbf{r}$ is the position of the decaying atom and the relation (4.7) allows for inhomogenous atomic environments where the spontaneous emission rate varies with position. The emission rate is determined essentially by the square of the Fourier component of the radiative electric field at the atomic transition frequency, with both fields evaluated at the atomic position. A suitable development of Eq. (4.7) with use of the standard closure relation puts it into the form [21]

$$2\gamma_R(\mathbf{r},\omega_A) = \frac{2e^2}{\hbar}\sum_{i,j} \mathrm{Im}\{d_i G_{ij}(\mathbf{r},\mathbf{r},\omega_A) d_j\} \quad i,j = x,y,z, \tag{4.8}$$

where the Green function is introduced via the coordinate-space version of the fluctuation–dissipation theorem of Eq. (3.20).

We shall see, however, that it is difficult to evaluate this Green function directly and it is better to begin with the field correlation at two different positions that are then made to tend to the single position of the atom. The relation between the corresponding cordinate-space and wavevector-space Green functions is

$$G_{ij}(\mathbf{r},\mathbf{r}+\mathbf{s},\omega_A) = (2\pi)^{-3}\int d\mathbf{q}\, G_{ij}(\mathbf{q},\omega_A)\exp(-i\mathbf{q}.\mathbf{s}), \tag{4.9}$$

where the vector $\mathbf{s}$ tends to zero for substitution of the Green function into the spontaneous emission rate of Eq. (4.8). Although this calculation relies on quantum-mechanical transition-rate theory, any environmental modifications of the free-space rate are determined by the Green function in Eq. (4.8) which can be calculated classically, as discussed in the previous section.

The spontaneous emission rate is now evaluated by insertion of the Green function for the atomic environment of interest into Eq. (4.8). The procedure is straightforward in principle but somewhat complicated in practice, even for the simplest material surroundings of the atom. We consider here an atom in the homogeneous isotropic dielectric treated in previous sections. The transverse and longitudinal components of the Green function are given by Eqs. (3.18) and (3.19) respectively. A somewhat tedious calculation [21,22] produces a power-series expansion of the required transverse Green function as

$$G_{ij}^{\mathrm{T}}(\mathbf{r},\mathbf{r}+\mathbf{s},\omega_{\mathrm{A}})=\frac{\omega_{\mathrm{A}}^{2}}{4\pi\varepsilon_{0}c^{2}}\left\{\frac{\delta_{ij}+\bar{s}_{i}\bar{s}_{j}}{2s}+\frac{2i\omega_{\mathrm{A}}}{3c}\left[\eta(\omega_{\mathrm{A}})+i\kappa(\omega_{\mathrm{A}})\right]\delta_{ij}+O(s)\right\}, \quad (4.10)$$

where $\bar{\mathbf{s}}$ is a unit vector parallel to $\mathbf{s}$. It is seen that, fortunately, the imaginary part of the transverse Green function is well-behaved in the limit $\mathbf{s}\rightarrow 0$, with

$$\operatorname{Im} G_{ij}^{\mathrm{T}}(\mathbf{r},\mathbf{r},\omega_{\mathrm{A}})=\frac{\omega_{\mathrm{A}}^{3}}{6\pi\varepsilon_{0}c^{3}}\eta(\omega_{\mathrm{A}})\delta_{ij}. \quad (4.11)$$

The transverse contribution to the spontaneous emission rate is accordingly [23]

$$2\gamma_{\mathrm{R}}^{\mathrm{T}}(\mathbf{r},\omega_{\mathrm{A}})=\frac{e^{2}\omega_{\mathrm{A}}^{3}|d_{\mathrm{A}}|^{2}}{3\pi\varepsilon_{0}\hbar c^{3}}\eta(\omega_{\mathrm{A}})=2\gamma_{\mathrm{R}}(\omega_{\mathrm{A}})\eta(\omega_{\mathrm{A}}), \quad (4.12)$$

where $2\gamma_{\mathrm{R}}(\omega_{\mathrm{A}})$ is the quantum expression for the free-space emission rate, given by Eq. (4.1). The effect of immersion in the dielectric is thus a modification of the free-space rate by the material refractive index $\eta(\omega_{\mathrm{A}})$ at the transition frequency. The transverse contribution corresponds to atomic decay processes in which the excitation energy is transferred to the transverse radiation field. The energy radiated in this way may subsequently be absorbed in a lossy dielectric but its initial form is that of radiative electromagnetic energy propagating away from the atom.

The longitudinal component of the Green function is obtained by a similar calculation as

$$G_{ij}^{\mathrm{L}}(\mathbf{r},\mathbf{r}+\mathbf{s},\omega_{\mathrm{A}})=-\frac{1}{4\pi\varepsilon_{0}\varepsilon(\omega_{\mathrm{A}})}\left\{\frac{\delta_{ij}-3\bar{s}_{i}\bar{s}_{j}}{s^{3}}+\frac{4\pi}{3}\delta_{ij}(\mathbf{s})\right\}, \quad (4.13)$$

where the delta function is now of the Dirac variety. The $\mathbf{s}\rightarrow 0$ limit is not at all well-behaved and its divergent nature results essentially from the absence of any fall-off with wavevector of the Green function in Eq. (3.19), in contrast to the transverse Green function in Eq. (3.18). It is necessary to apply a regularization procedure [24] to the longitudinal function, which is here averaged over a sphere of radius R, representing the volume of the atom or molecule, with the result

$$\overline{G_{ij}^{L}(\mathbf{r},\mathbf{r},\omega_A)} = -\frac{\delta_{ij}}{4\pi\varepsilon_0\varepsilon(\omega_A)R^3}. \tag{4.14}$$

The longitudinal contribution to the spontaneous emission rate is accordingly [21]

$$2\gamma_{NR}^{L}(r,\omega_A) = 2\gamma_R(\omega_A)\frac{3\varepsilon''(\omega_A)}{2|\varepsilon(\omega_A)|^2}\left(\frac{c}{\omega_A R}\right)^3. \tag{4.15}$$

This contribution clearly vanishes for a material with negligible loss at the atomic transition frequency and it is also very small for a medium with very high loss. However, in general it can make a significant contribution to the atomic decay rate. It corresponds to nonradiative atomic decay processes in which the excitation energy is transferred to electrical currents in the dielectric, and thence to Joule heating of the material. The longitudinal expression is proportional to the cube of the emission wavelength divided by the ill-defined radius of the atomic volume, which gives a large factor of uncertain magnitude.

The decay rates derived above are determined by the fluctuation spectrum of the macroscopic electric field $\mathbf{E}$. In practice, an atom in a dielectric host interacts with its own local field and the expressions should be multipled by local-field correction factors. The relation between local and macroscopic fields obtained from Eqs. (2.3) and (2.11) is

$$\mathbf{E}_{loc} = \frac{\varepsilon(\omega)+2}{3}\mathbf{E}. \tag{4.16}$$

Conversion of the fields from macroscopic to local thus results in additional multiplicative factors of

$$\left|[\varepsilon(\omega_A)+2]/3\right|^2 \quad \text{(LL)} \tag{4.17}$$

in Eqs. (4.12) and (4.15). This correction factor is based on the Lorentz–Lorenz model in which the radiating atom is assumed to occupy an otherwise empty spherical cavity, whose dimension is much smaller than the transition wavelength. The cavity is considered as virtual in the sense that the macroscopic field in the surrounding dielectric is assumed to be undisturbed by the presence of the cavity. Alternative correction factors are obtained by other assumptions about the nature of the cavity. In particular, the Glauber-Lewenstein model treats an empty cavity that is real in the sense that the field in the surrounding dielectric is modified by its presence [25]. The resulting correction factor is

$$\left|3\varepsilon(\omega_A)/[2\varepsilon(\omega_A)+1]\right|^2 \quad \text{(GL)}. \tag{4.18}$$

Figs. 3 shows the transverse decay rate normalized by the free-space rate of Eq. (4.1) as a function of the location of the atomic transition frequency ω_A relative to the transverse frequency ω_T of the dielectric host. The full curve shows the rate from Eq. (4.12) with no local-field correction while the broken and dotted curves show the rates with inclusion of the Lorentz–Lorenz and Glauber-Lewenstein factors from Eqs. (4.17) and (4.18) respectively. Fig. 4 shows the longitudinal decay rate for the same dielectric function but the curves are now normalized by the rate from Eq. (4.15) at frequency $\omega_A = \omega_T$. The three rates in each figure have similar outline shapes but the variations are

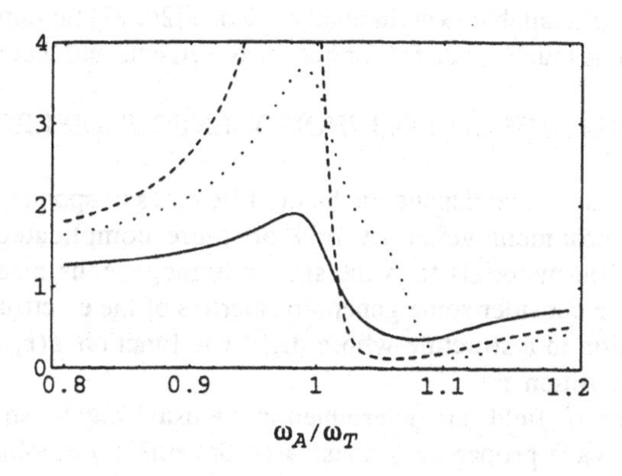

Figure 3. Transverse spontaneous emission rate as a function of the atomic transition frequency. The dielectric function has $\omega_L = 1.1\omega_T$ and $2\gamma(\omega_T) = 0.05\omega_T$ (after [21]).

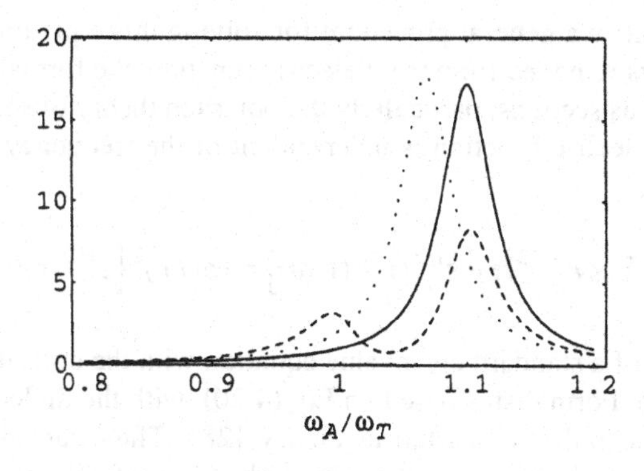

Figure 4. Same as Fig. 3 but for the longitudinal emission rate (after [21]).

different in detail. The variations of the decay rates reflect the densities of states of the polariton modes that are degenerate with the atomic transition frequency. Thus, in the transverse rate, the relatively small values for atomic frequencies between the transverse and longitudinal frequencies of the host material, with no local-field correction, are consequences of the gap shown in Figs. 1 and 2. Again, in the longitudinal rate, the divergence that occurs without regularization is a consequence of the infinite density of states associated with the wavevector-independent Green function in Eq. (3.21).

The appropriate modelling of the dielectric and the local fields for various materials and dopants could in principle be resolved by measurements. However, it is not usually possible to vary the transition frequency ω_A of the same atom through the dielectric resonance and the use of a series of different atoms brings not only a range of transition frequencies but also a variation in their free-space emission rates that obscures the desired information. The little available experimental evidence [26, 27] favours the GL local-field correction, but there is much scope for further measurements and theory.

4.3. SPONTANEOUS EMISSION IN INHOMOGENEOUS DIELECTRICS

The practical applications that depend on the modifications of spontaneous emission rates by the atomic environment generally rely on more complicated arrangements of surrounding dielectric materials than the simple homogeneous medium treated in the previous section. We consider some general properties of the electromagnetic fields and spontaneous emission in a structure whose dielectric function $\varepsilon(\mathbf{r},\omega)$ has an arbitrary dependence on the position $\mathbf{r}$.

The electromagnetic fields are determined in the usual way by solution of Maxwell's equations and, for wave propagation, these are conveniently combined to produce the Helmholtz equation

$$\nabla\times(\nabla\times\mathbf{E}(\mathbf{r},\omega))=\nabla(\nabla.\mathbf{E}(\mathbf{r},\omega))-\nabla^2\mathbf{E}(\mathbf{r},\omega)=(\omega/c)^2\varepsilon(\mathbf{r},\omega)\mathbf{E}(\mathbf{r},\omega). \quad (4.19)$$

It is difficult to identify a general procedure for solving this equation. Of course, if the spatial dependence is removed from the dielectric function, the formalism reduces to that considered in previous sections, particularly the polariton theory of sections 2.4 and 3.3.

Again, if the dielectric function is independent of the frequency, Eq. (4.19) can be rearranged as

$$\varepsilon^{-1/2}(\mathbf{r})\nabla\times\nabla\times\varepsilon^{-1/2}(\mathbf{r})\left\{\varepsilon^{1/2}(\mathbf{r})\mathbf{E}(\mathbf{r},\omega)\right\}=(\omega/c)^2\left\{\varepsilon^{1/2}(\mathbf{r})\mathbf{E}(\mathbf{r},\omega)\right\}, \quad (4.20)$$

which has the form of a standard eigenvalue equation with the curly-bracketted quantity as the eigenfunction. Formalisms based on Eq. (4.20), with the dielectric function taken as real, are used in multiple scattering theory [28]. They can also be used in the calculation of photonic band structures, although it is preferable to use an analogous eigenvalue equation based on the magnetic field vector [29]. Such calculations provide good approximations for electromagnetic waves whose frequency components lie in regions of fairly constant dielectric function. However, as is discussed in section 2.3, the dielectric function is necessarily dispersive and complex, such that it can have real and constant values only over limited ranges of the frequency.

Another technique for the solution of Eq. (4.19) with a properly dispersive and complex dielectric function applies when the system of interest can be divided into spatial regions labelled by an index i, each of which has a position-independent dielectric function $\varepsilon_i(\omega)$. The standard theory applies within each region and the usual boundary conditions connect the solutions in the different regions. The method has been applied to the electromagnetic field quantization in a range of dielectric structures constructed from stacks of parallel slabs of homogeneous and isotropic materials, so that $\varepsilon(\mathbf{r},\omega)$ is a function of only one spatial coordinate. Examples include two media in contact at a flat interface and dielectric slabs, cavities with parallel and transversely-infinite mirrors, and multilayer stacks of alternating materials [30-34].

For calculations of spontaneous emission rates, it is pointed out in section 4.1 that a completely classical calculation can be converted to quantum mechanics by application of the connection formula (4.6). A noteworthy application of this procedure is the spontaneous emission rate of a atom placed adjacent to the flat surface of a dispersive and lossy dielectric. The corresponding classical calculation of the emission rate by a dipole antenna supported above the surface of the partially-conducting earth has been treated in several papers dating back to 1909 [35-37]. The relations between spontaneous emission rates in free space and in the presence of the dielectric surface derived in this way [38] agree with entirely quantum-mechanical calculations [39]. Spontaneous emission rates have been measured for atoms in transparent liquids close to dielectric surfaces [40] and the close correspondence between classical and quantum theories is again demonstrated.

Irrespective of the detailed spatial variation of the dielectric function $\varepsilon(\mathbf{r},\omega)$, it is possible to derive a sum rule that limits the changes in transverse spontaneous emission rate caused by immersion of the atom in an arbitrary environment [41]. The Green function that enters the emission rate formula of Eq. (4.8) is itself the solution of an inhomogeneous Helmholtz equation, given by

$$-\left\{\nabla^2+(\omega/c)^2\varepsilon(\mathbf{r},\omega)\right\}G_{ij}^{\mathrm{T}}(\mathbf{r},\mathbf{r}',\omega)=\left(1/\varepsilon_0 c^2\right)\delta_{ij}^{\mathrm{T}}(\mathbf{r}-\mathbf{r}'), \qquad (4.21)$$

where the transverse part of the Dirac delta-function [20] ensures that solution of this equation produces the transverse part of the Green function. The Green function has a general property that, similar to the dielectric function, its poles all lie in the lower half of the complex frequency plane. It is therefore a candidate for the derivation of frequency sum rules but in order to qualify, as discussed in section 2.3, it must also satisfy the requirement of square integrability along the real axis. This, however, it fails to do. It is evident from the expression in Eq. (4.10) for the special case of a homogeneous dielectric that the Green function diverges as ω^3 for large ω, given that $\eta(\omega)$ tends to unity.

These difficulties can, however, be overcome by working with the electromagnetic vector potential instead of the electric field and by subtracting the free-space Green function from that in the arbitrary dielectric environment. The procedure is straightforward and the detailed derivation given in [41] produces the sum rule

$$\int_0^\infty d\omega_A \frac{\gamma_R^T(\mathbf{r},\omega_A)-\gamma_R(\omega_A)}{\gamma_R(\omega_A)}=0. \tag{4.22}$$

The rule applies to the rate of spontaneous emission from a fixed atomic excited state in a thought experiment where the frequency of the transition is swept from zero to infinity, while keeping the dipole matrix element constant. It expresses a kind of conservation for the spontaneous emission rate in the arbitrary environment relative to that in free space. Thus any environmental increase in the spontaneous emission rate over a range of frequencies must be compensated by a decrease in the rate over other frequency ranges, relative to free-space values.

It is again impossible in practice to vary the transition frequency ω_A of the same atom over the complete range of frequencies but the sum rule in Eq. (4.22) provides a check on model calculations. The arbitrary environment is completely general and it can include, for example, metallic mirrors, Bragg reflectors, photonic band-gap materials, microcavities, and absorptive dielectrics or semiconductors. Complete results for these structures are either not available or, if they are available, it is difficult to extract the transverse contribution to the atom decay rate. However, an immediate test can be made of the result from Eq. (4.12) for a homogeneous dielectric, where substitution in the sum rule of Eq. (4.22) gives

$$\int_0^\infty d\omega_A\{\eta(\omega_A)-1\}=0. \tag{4.23}$$

This is identical to the well-known refractive-index sum rule in Eq. (2.31).

5. Optical propagation

The propagation of optical pulses through absorptive and dispersive dielectrics displays a wealth of phenomena associated with a variety of velocities that can be identified. There is much interest in apparent superluminal, even infinite, pulse velocities and, at the opposite extreme, propagation that ceases altogether in the phenomenon of optical localization. The calculations given in the present section assume that the frequency components of the optical pulse lie close to the transverse resonance frequency ω_T of the homogeneous dielectric treated in section 2.

5.1. PHASE AND GROUP VELOCITIES

Consider the propagation along the z axis of an optical pulse described by an electric field with spectrum $E(\omega)$. The spectrum is taken to be a Gaussian,

$$E(\omega) = E_0 \exp\left\{-(\omega - \omega_C)^2/4\Delta^2\right\}, \tag{5.1}$$

where the bandwidth Δ is very much smaller than the carrier frequency ω_C. The field at a general time and position is given by

$$E(z,t) = \int_0^\infty d\omega E(\omega) \exp(-i\omega t + iqz), \tag{5.2}$$

where the complex wavevector q is a function of the frequency in accordance with Eq. (2.23),

$$q = q(\omega) = q'(\omega) + iq''(\omega) = (\omega / c)\{\eta(\omega) + i\kappa(\omega)\}. \tag{5.3}$$

The pulse field at $z = 0$ is easily determined as

$$E(0,t) = 2\pi^{1/2} \Delta E_0 \exp\left(-i\omega_C t - \Delta^2 t^2\right). \tag{5.4}$$

The pulse field at $z > 0$ is difficult, or indeed impossible, to determine in general as it depends on the forms of the functions $\eta(\omega)$ and $\kappa(\omega)$. The integration in Eq. (5.2) could, of course, be carried out numerically for given functions but it is more instructive to consider limiting values of the pulse width and dielectric function that correspond to the main experiments of interest.

Many of the experiments are performed on materials that have a resonance well described by the dielectric function of Eq. (2.20) but added to a relatively large background contribution from more distant resonances that is almost real and constant, described here by a real refractive index η_B. The pulse carrier frequency ω_C is assumed to be close to the transverse frequency ω_T of the dielectric and the pulse width Δ is assumed to be much smaller than the resonance damping rate γ, whose frequency dependence is here ignored. The refractive index and extinction coefficient in this case are given by

$$\eta(\omega) + i\kappa(\omega) = \eta_B - \frac{\omega_L^2 - \omega_T^2}{4\eta_B \omega_T} \frac{1}{\omega - \omega_T + i\gamma} \tag{5.5}$$

to a good approximation. Furthermore, the above assumptions ensure that a Taylor expansion of the wavevector around the pulse carrier frequency converges rapidly and we truncate it after the first two terms. Thus, with insertion of the pulse shape from Eq. (5.1), the field from Eq. (5.2) is

$$E(z,t) = E_0 \int_0^\infty d\omega \exp\left\{ -i\omega t + iq'_C z - q''_C z + i(\omega - \omega_C) z \left.\frac{\partial q'(\omega)}{\partial \omega}\right|_C \right.$$
$$\left. -(\omega - \omega_C) z \left.\frac{\partial q''(\omega)}{\partial \omega}\right|_C - \frac{(\omega - \omega_C)^2}{4\Delta^2} \right\}, \tag{5.6}$$

where the C subscripts indicate quantities evaluated at the carrier frequency.

The approximate evaluation of this integral is considered in detail in [42] and we here indicate only the main steps. The final factor in the integrand of Eq. (5.6) ensures that the range of integration can be extended to negative frequencies with negligible change in the value of the integral, provided that the propagation distance is sufficiently short for the penultimate term to remain relatively small. This latter term can be neglected altogether in the conditions of many experiments and the integration is then performed straightforwardly, with the result

$$E(z,t) = 2\pi^{1/2} \Delta E_0 \exp\left\{ -i\omega_C \left(t - \frac{z}{v_P(\omega_C)} \right) - \frac{\omega_C \kappa(\omega_C) z}{c} - \Delta^2 \left(t - \frac{z}{v_G(\omega_C)} \right)^2 \right\}. \tag{5.7}$$

Here Eq. (5.3) has been used to introduce the refractive index and extinction coefficient. The points of constant phase in the electric-field wavefronts travel with the phase velocity, defined by

$$\frac{c}{v_P(\omega)} = \eta(\omega) = \eta_B - \frac{\omega_L^2 - \omega_T^2}{4\eta_B \omega_T} \frac{\omega - \omega_T}{(\omega - \omega_T)^2 + \gamma^2}, \tag{5.8}$$

where the form of the refractive index is taken from Eq. (5.5). Bearing in mind that the second term is assumed to be much smaller than the first, the phase velocity is always positive. Indeed, the phase velocity is a positive quantity in general, although it often takes values greater than c, in no conflict with relativity principles as the planes of constant phase do not carry information.

The measured quantity in pulse propagation is usually more closely related to the Poynting vector, proportional to the square modulus of the electric field and obtained from Eq. (5.7) as

$$S(z,t) = S(0,0) \exp\left\{ -\frac{2\omega_C \kappa(\omega_C) z}{c} - 2\Delta^2 \left(t - \frac{z}{v_G(\omega_C)} \right)^2 \right\}. \tag{5.9}$$

The peak of the Gaussian term in this expression travels with the group velocity, defined by

$$\frac{c}{v_G(\omega)} = c\frac{\partial q'(\omega)}{\partial\omega} = \frac{\partial[\omega\eta(\omega)]}{\partial\omega} = \frac{c}{v_P(\omega)} + \omega\frac{\omega_L^2 - \omega_T^2}{4\eta_B\omega_T}\frac{(\omega-\omega_T)^2 - \gamma^2}{\left[(\omega-\omega_T)^2 + \gamma^2\right]^2}. \tag{5.10}$$

Insertion of the phase velocity from Eq. (5.8) gives

$$\frac{c}{v_G(\omega)} = \eta_B + \frac{\omega_L^2 - \omega_T^2}{4\eta_B}\frac{(\omega-\omega_T)^2 - \gamma^2}{\left[(\omega-\omega_T)^2 + \gamma^2\right]^2}, \tag{5.11}$$

again with the assumption that ω is close to ω_T. For a narrow resonance with $\gamma << \omega_T$, the second term in this expression can be much larger than the first for $\omega \approx \omega_T$, leading to a negative group velocity. With increasing separation between ω and ω_T, the expression in Eq. (5.11) passes through zero and becomes positive. The zeros on either side of ω_T correspond to infinities in the group velocity. Fig. 5 shows the frequency variations of the inverse phase and group velocities in the vicinity of a narrow resonance.

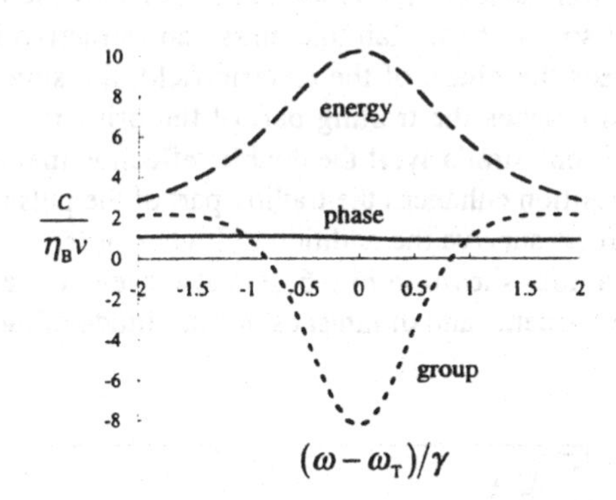

Figure 5. Frequency dependences of the inverse phase, group and energy (or transport) velocities in the vicinity of a dielectric resonance for $\eta_B = 1$, $\omega_L = 1.1\omega_T$ and $2\gamma = 0.05\omega_T$.

Although seemingly in conflict with relativity principles, pulse propagation with velocities larger than the free-space velocity of light has been observed experimentally [43]. The measurements determine the time delay between the entry of the peak of an optical pulse into a dielectric slab and the exit of the peak of the pulse from the far side of the slab, as a function of the detuning between the carrier frequency ω_C and the dielectric resonance ω_T. The observed delays are consistent with the variation of the group velocity in Eq. (5.11), including the occurrence of the zero delays that correspond to infinite group velocity. These observations do not in fact disagree with relativity principles. The apparent superluminal propagation occurs because, for the conditions assumed in the above derivations, the leading part of the pulse is transmitted with little modification but the trailing part of the pulse suffers a destructive interference that diminishes its intensity. The different fates of the pulse front and rear result in a forward shift of the pulse peak in the direction of propagation. The complete theory of the effect shows that the transmitted pulse intensity alway lies under that of the incident pulse propagated with the free-space velocity of light, so that Einstein causality is satisfied.

The measurements in [43] confirm the characteristic features of resonant optical pulse propagation through a bulk dielectric. Further contributions to the shifts in peak position arise from the transmission of optical pulses through the surfaces of a dielectric slab [44] or a multilayer structure [45]. The surface effects are most readily understood for a parallel slab of nonresonant dielectric when the length $L = c / \Delta$ of the incident optical pulse is greater than the slab thickness $2l$. Fig. 6 shows the spatial shift δz in the peak of the transmitted Gaussian pulse as a function of the slab refractive index at the carrier frequency. Dispersion is here ignored and the shift in peak position associated solely with the optical thickness of the slab is $2l[1 - \eta(\omega_C)]$. The oscillations of the shifts in peak position around this value result entirely from the multiple reflections within the slab. Thus the transmitted pulse is a superposition of a primary component that passes through the slab with no reflections and components that suffer two internal reflections, four internal reflections, and so on. If the slab thickness and refractive index are such that a double reflection reverses the phase of the electric field, the superposition of the first reflected component diminishes the trailing part of the primary component to give an advance in the peak position. Similarly, if the double reflection maintains the phase of the electric field, the superposition enhances the trailing part of the pulse to give a delay in the peak position. These effects survive the addition of higher-order reflected components to give the oscillatory behaviour shown in Fig. 6. It is also seen how attenuation in the slab reduces the reflected components and diminishes the amplitude of the oscillations.

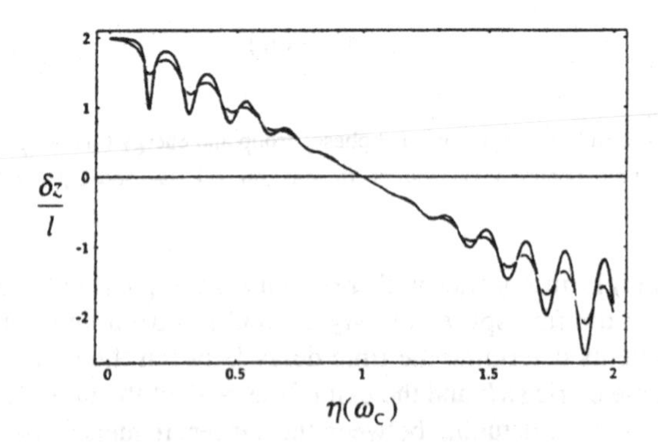

Figure 6. Shift in the pulse peak after transmission through a slab relative to free-space propagation as a function of the refractive index for $\kappa(\omega_C) = 0$ (black curve) and 0.02 (grey curve) (after [46]).

5.2. ENERGY VELOCITY

The group velocity, as defined by Eq. (5.10), is commonly considered to be the velocity with which the energy of a pulse travels through a dispersive system. There is, however, an alternative definition of the velocity of energy flow that does not explicity assume a pulse format. In the presence of attenuation, this energy velocity differs from the group velocity except for carrier frequencies well removed from a dielectric resonance. The form of the energy velocity is derived here for the same model of an atomic medium as used in previous sections. The medium is assumed to be one in which radiative decay occurs and nonradiative contributions to the decay are assumed to be small.

Consider a light beam with radiative energy density $W_R(\omega)$ in a medium whose constituent atomic transitions are described by Einstein coefficients A and B, including the corrections from their free-space values A_0 and B_0 caused by immersion in the medium. The transition rates are

$$\text{absorption rate } = BW_R(\omega), \text{ emission rate } = A + BW_R(\omega), \tag{5.12}$$

where both spontaneous and stimulated contributions are included in the latter. The fraction of the time that an atom spends in its excited state is easily shown to be [4]

$$\frac{BW_R(\omega)}{A + 2BW_R(\omega)} \approx \frac{BW_R(\omega)}{A} \quad \text{for} \quad BW_R(\omega) << A, \tag{5.13}$$

when the atomic transition is far from saturation as is assumed henceforth. The lineshape of the atomic transition is described by a normalized spectral profile function

$$F(\omega) = \frac{\gamma_R/\pi}{(\omega - \omega_T)^2 + \gamma_R^2}, \tag{5.14}$$

where the shift of the atomic, or exciton, transition frequency from ω_E to ω_T in the medium is included but the frequency dependence of γ_R is assumed to be negligible over the narrow range of frequencies considered.

The atomic excitation provides an additional material contribution to the radiative energy density, so that the total energy density associated with the light beam is

$$W(\omega) = W_R(\omega) + \frac{NBW_R(\omega)\hbar\omega F(\omega)}{VA}. \tag{5.15}$$

The part of the energy density associated with the atomic excitation does not propagate. The magnitude of the energy flow associated with the radiative energy density is determined by Poynting's vector and for monochromatic fields

$$\text{energy flow } = v_P(\omega)W_R(\omega). \tag{5.16}$$

The energy velocity $v_E(\omega)$ is defined by

$$v_E(\omega) = \frac{\text{energy flow}}{\text{energy density}} = \frac{v_P(\omega)W_R(\omega)}{W(\omega)}. \tag{5.17}$$

Thus, with the use of Eq. (5.15), the energy velocity is given by

$$\frac{1}{v_E(\omega)} = \frac{1}{v_P(\omega)} + \frac{NB\hbar\omega F(\omega)}{VAv_P(\omega)}. \tag{5.18}$$

This expression can be put into a variety of more useful forms. Thus, with the relation between the Einstein coefficients B in the medium and B_0 in free space [47] and the standard expression for the latter [4]

$$B = \frac{B_0}{\eta^2(\omega)} = \frac{\pi e^2 |d|^2}{\varepsilon_0 \hbar^2 \eta^2(\omega)} = \frac{\pi V}{\hbar N \eta^2(\omega)} \frac{\omega_L^2 - \omega_T^2}{2\omega_T}, \tag{5.19}$$

where Eq. (2.15) is used in the final step. Insertion of this expression into Eq. (5.17) together with the spectrum from Eq. (5.14) and the relation $A = 2\gamma_R$ gives

$$\frac{c}{v_E(\omega)} = \frac{c}{v_P(\omega)} + \frac{\omega_L^2 - \omega_T^2}{4\eta(\omega)\omega_T} \frac{\omega}{(\omega - \omega_T)^2 + \gamma_R^2}. \tag{5.20}$$

Insertion of the phase velocity from Eq. (5.8) produces the final form

$$\frac{c}{v_E(\omega)} = \eta_B + \frac{\omega_L^2 - \omega_T^2}{4\eta_B} \frac{1}{(\omega - \omega_T)^2 + \gamma_R^2}. \tag{5.21}$$

It is seen by comparison with Eq. (5.11) that the group velocity is only the same as the velocity of energy propagation for frequencies well away from the centre of the resonance, with $|\omega - \omega_T| >> \gamma$. In particular, the energy velocity does not have the infinities that occur in the group velocity for detunings of order γ, and it is generally smaller than the free-space velocity of light, as in the example shown in Fig. 5.

The inverse energy velocity can also be written in a more general form. Thus from Eqs. (2.25) and (2.27), for frequencies close to resonance, the attenuation length is given by

$$\frac{1}{L(\omega)} = \frac{\omega}{c} \frac{\omega_L^2 - \omega_T^2}{2\eta(\omega)\omega_T} \frac{\gamma_R}{(\omega - \omega_T)^2 + \gamma_R^2} \tag{5.22}$$

and Eq. (5.20) can be written [48,49]

$$\frac{1}{v_E(\omega)} = \frac{1}{v_P(\omega)} + \frac{1}{2\gamma_R L(\omega)}. \tag{5.23}$$

Although this result is derived here for a specific model of a homogeneous dielectric, it is found to apply to a wide range of systems, with insertion of the temporal damping rate $2\gamma_R$ and the attenuation length $L(\omega)$ appropriate to the problem of interest. It has been shown [50] that the energy velocity takes over from the group velocity in the propagation of optical pulses through a resonant dielectric after distances much greater than the attenuation length. Relations equivalent to Eq. (5.23) also occur in signal propagation through a resonant amplifier [51] and in self-induced transparency in two-level atoms [52]. The same form of relation is derived [28,53] in detailed calculations of the velocity of energy transport through media containing randomly-distributed scatterers. Measurements of the propagation of ultrasonic signals [54] have been shown to be well described by a relation similar to Eq. (5.23), which is ascribed to previous calculations valid in the regime of propagation over distances much larger than the attenuation length [55].

6. Summary

The propagation of light through dielectric media displays a wealth of phenomena that can largely be described in terms of classical electromagnetic theory. The dielectric function is the main tool in the description and interpretation of measurements. The theory presented here is valid for materials in which the dielectric function has discrete resonances and many of the results assume that only a single resonance close to the frequency of the incident light need be included in the calculations. Very similar dielectric functions are obtained from classical and quantum theories and, for most purposes, the classical Lorentz theory of a dielectric provides a simple but adequate model.

The linear response of a dielectric is explored in greater detail by measurements that can vary the frequency and wavevector of the electromagnetic field independently. This is achieved in experiments that use two or more light beams at variable relative angles, such as inelastic light scattering and two-photon absorption. The results of such measurements determine the electric-field Green function of the dielectric, which is easily calculated and is again almost the same in classical and quantum theories. The peaks of the Green function determine the polariton dispersion relation, which characterizes the excited states of the dielectric.

The radiative damping of long-wavelength excitations in dielectric media requires special consideration. Any damping process depends on the coupling to the excitation of interest to a continuum of modes that form a reservoir for the irreversible decay. No such reservoir is available for the radiative decay of the excitons in a homogeneous material, as each exciton is coupled only to the electromagnetic field component of the same frequency and wavevector. The coupled exciton and photon that form the polariton diagonalize the system Hamiltonian and there is no coupling to other modes. Radiative damping is restored in inhomogeneous or disordered systems, where the decay rate depends on the nature and degree of the disorder.

Other aspects of the dielectric function govern the velocities of propagation of optical pulses. The flow of energy proceeds with the energy, or transport, velocity which has a simple form that is valid for a wide variety of physical systems over a broad range of parameters. However, the peak of an optical pulse travels with the group velocity over distances up to several attenuation lengths in an lossy dielectric, before transferring towards the energy velocity for greater path lengths. The group velocity can be negative or larger than the free-space velocity of light close to a resonant frequency of the medium but any superluminal behaviour is apparent rather than real, as the transmitted pulse is merely reshaped such that it always lies inside the envelope of a twin causal pulse propagated through free space.

Acknowledgements

This work owes a great deal to the contributions over many years of all of my colleagues named in the joint references below. I should also like to thank Ad Lagendijk for asking many pertinent questions, answers to some of which are attempted in these lectures.

References

1. Lorentz, H.A. (1880) *Wiedem. Ann.* **9**, 641.
2. Born, M. and Wolf, E. (1980) *Principles of Optics*, 6th edition, Cambridge University Press, Cambridge, pp. 84-104.
3. Lagendijk, A., Nienhuis, B., van Tiggelen, B.A., and de Vries, P. (1997) *Phys. Rev. Lett.* **79**, 657.
4. Loudon, R. (1983) *The Quantum Theory of Light*, 2nd edition Oxford University Press, Oxford.
5. Nussenzveig, H.M. (1972) *Causality and Dispersion Relations*, Academic Press, New York.
6. Loudon, R. (1973) *The Quantum Theory of Light*, 1st edition Oxford University Press, Oxford, pp. 64-8.
7. Boas, M.L. (1983) *Mathematical Methods in the Physical Sciences*, 2nd edition Wiley, New York, p. 606.
8. Altarelli, M., Dexter, D.L., Nussenzweig, H.M., and Smith, D.Y. (1972) *Phys. Rev. B* **6**, 4502.
9. Mills, D.L. and Burstein, E. (1974) *Rep. Prog. Phys.* **37**, 817.
10. Hopfield, J.J. (1958) *Phys. Rev.* **112**, 1555.
11. Knoester, J. (1992) *Phys. Rev. Lett.* **68**, 654.
12. Abrikosov, A.A., Gor'kov, L.P., and Ye Dzyaloshinskii, I. (1965) *Quantum Field Theoretical Methods in Statistical Physics*, Pergamon, Oxford, Chap. 6.
13. Barker, A.S. and Loudon, R. (1972) *Rev. Mod. Phys.* **44**, 18.
14. Boggett, D. and Loudon, R. (1972) *Phys. Rev. Lett.* **28**, 1051; (1973) *J. Phys. C* **6**,1763.
15. Landau, L.D. and Lifshitz, E.M. (1969) *Statistical Physics*, (Pergamon, Oxford, 1969).
16. Fröhlich, D., Mohler, E., and Wiesner, P. (1971) *Phys. Rev. Lett.* **26**, 554.
17. Hönerlage, B., Bivas, A., and Vu Duy Phach (1978) *Phys. Rev. Lett.* **41**, 49.

18. Einstein, A. (1917) *Phys. Z.* **18**, 121.
19. Jackson, J.D. (1975) *Classical Electrodynamics*, Wiley, New York.
20. Milonni, P.W. (1994) *The Quantum Vacuum*, Academic Press, New York.
21. Barnett, S.M., Huttner, B., Loudon, R., and Matloob, R. (1996) *J. Phys. B* **29**, 3763.
22. Cohen-Tannoudji, C., Dupont-Roc, J., and Grynberg, G. (1989) *Photons and Atoms: Introduction to Quantum Electrodynamics*, Wiley, New York.
23. Barnett, S.M., Huttner, B., and Loudon, R. (1992) *Phys. Rev. Lett.* **68**, 3698.
24. de Vries, P., van Coevorden, D., and Lagendijk, A. (1998) *Rev. Mod. Phys.* **70**, 447.
25. Glauber, R.J. and Lewenstein, M. (1991) *Phys. Rev. A* **43**, 467.
26. Rikken, G.L.J.A. and Kessener, Y.A.R.R. (1995) *Phys. Rev. Lett.* **74**, 880.
27. Lavallard, P., Rosenbauer, M., and Gacoin, T. (1996) *Phys. Rev.* **54**, 5450.
28. Lagendijk, A. and van Tiggelen, B.A. (1996) *Phys. Reps.* **270**, 143.
29. Joannopoulos, J.D., Meade, R.D., and Winn, J.N. (1995) *Photonic Crystals*, Princeton University Press, Princeton.
30. Matloob, R., Loudon, R., Barnett, S.M., and Jeffers, J. (1995) *Phys. Rev. A* **52**, 4823.
31. Gruner, T. and Welsch, D.G. (1996) *Phys. Rev. A* **53**, 1818.
32. Matloob, R. and Loudon, R. (1996) *Phys. Rev. A* **53**,4567.
33. Gruner, T. and Welsch, D.G. (1996) *Phys. Rev. A* **54**, 1661.
34. Dung, H.T., Knöll, L., and Welsch, D.G. (1998) *Phys. Rev. A* **57**, 3931.
35. Sommerfeld, A. (1909) *Ann. Phys.* **28**, 665; (1949) *Partial Differential Equations*, Academic Press, New York, chap. 6.
36. Stratton, J.A. (1941) *Electromagnetic Theory*, McGraw-Hill, New York, chap. 9.
37. Banos, A. (1966) *Dipole Radiation in the Presence of a Conducting Half Space*, Pergamon, New York.
38. Chance, R.R., Prock, A., and Silbey, R. (1974) *J. Chem. Phys.* **60**, 2184; (1974) *ibid.* **60**, 2744; (1975) *ibid.* **62**, 2245.
39. Yeung, M.S. and Gustafson, T.K. (1996) *Phys. Rev. A* **54**, 5227.
40. Snoeks, E., Lagendijk, A., and Polman, A. (1995) *Phys. Rev. Lett.* **74**, 2459.
41. Barnett, S.M. and Loudon, R. (1996) *Phys. Rev. Lett.* **77**, 2444.
42. Garrett, C.G.B. and McCumber, D.E. (1970) *Phys. Rev. A* **1**, 305.
43. Chu, S. and Wong, S. (1982) *Phys. Rev. Lett.* **48**, 738; (1982) *ibid.* **49**, 1293.
44. Gaspar, J.A. and Halevi, P. (1994) *Phys. Rev. B* **49**, 10 742.
45. Steinberg, A.M., Kwiat, P.G., and Chiao, R.Y. (1993) *Phys. Rev. Lett.* **71**, 708.
46. Artoni, M. and Loudon, R. (1997) *Phys. Rev. A* **55**, 1347.
47. Milonni, P.W. (1995) *J. Mod. Opt.* **42**, 1991.
48. Loudon, R. (1970) *J. Phys. A* **3**, 233.
49. Loudon, R., Allen, L., and Nelson, D.F. (1997) *Phys. Rev. E* **55**, 1071.
50. Oughstun, K.E. and Balictsis, C.M. (1996) *Phys. Rev. Lett.* **77**, 2210.
51. Bolda, E.L., Garrison, J.C., and Chiao, R.Y. (1994) *Phys. Rev. A* **49**, 2938.
52. Allen, L. and Eberly, J.H. (1987) *Optical Resonance and Two-Level Atoms*, Dover, New York.
53. van Tiggelen, B.A., Lagendijk, A., van Albada, M.P., and Tip, A. (1992) *Phys. Rev. B* **45**, 12 233.
54. Shiren, N.S. (1962) *Phys. Rev.* **128**, 2103.
55. Baerwald, H.G. (1930) *Ann. Physik* **7**, 731.

COLD ATOMS AND MULTIPLE SCATTERING

ROBIN KAISER
Institut Non Linéaire de Nice
1361, route des Lucioles
F-06560 Valbonne
kaiser@inln.cnrs.fr

1. Introduction

Laser cooling of atoms has developed rapidly since 1985. Milestones of this development have been the first realization of molasses of atoms[1], the first magneto-optical trap [2], the sub-Doppler [3] and sub-recoil [4] cooling techniques and more recently the observation of Bose-Einstein condensation of dilute atomic gases [5][6]. A review on milestones in this field can be found in the 1997 Nobel lectures by S.Chu [7], C.Cohen-Tannoudji [8] and W.Phillips [9]. The interaction between laser light and atoms have been used in a very elegant way to modify the momentum of the atoms in such a way to reduce the spread of the velocity distribution of the atoms. This narrowing of the velocity distribution is generally called laser cooling of atoms, although strictly speaking there is no thermal equilibrium between the atoms. Each atoms interacts with the laser field and interactions between atoms are usually neglected.

One main reason to use situations where such interactions are indeed negligible is the fact that the theoretical description of atom-light interaction is much easier if one can consider the light field as known and the atoms to interact with this light field with their intrinsic response function (energy level, cross sections) also known a priori. Theoretical complication, even in this approximation, arise from the complex internal structure of the atoms and the need to quantify the external motion for the colder atoms. There are a few exceptions to this generally used approximation : collisions of cold atoms [10] have been studied partly taking into account many body aspects in presence of light. Evaporation of atoms[11] leading to the Bose-Einstein condensation also takes into account many body effects, but usually with no laser field applied.

J.-P. Fouque (ed.), Diffuse Waves in Complex Media, 249–288.

We are particularly interested in the regime where light scattered by one atom is not escaping the system directly, but can be scattered again by another atom. Laser cooled atoms in this regime of optical thick systems have been studied in order to increase the density of atoms on the way to Bose-Einstein condensation. However, even though the density in real space has been large, it is difficult to combine the lowest temperatures with highest spatial densities, limiting thus the density in phase space to values below the condensation threshold. For higher densities of laser cooled atoms, the temperature of the atoms increases [12] and spatial structures such as rings have also been observed [13]. The reason for this new features the multiple scattering of light in the sample. Each atoms thus does not only see the external applied laser field, with a controlled wave vector and polarization, but also the light scattered by neighboring atoms, with "random" wavevectors and polarization. A rigorous theory of atom-laser interaction in the multiple scattering regime would need to take into account the interferences of the multiple scattered light. Laser cooled atoms are indeed sensitive to spatially varying light shifts which depend on the value of local electric field. It is thus of interest for the laser cooling community to learn from other systems where multiple scattering of light is dominant.

On the other hand it is also possible to study the properties of the light scattered by laser cooled atoms. Such a sample has very particular scattering properties and in this sense these new samples can provide an interesting tool to study precisely the multiple scattering of light. The parameters for the sample are so different from what is usually available in light scattering media, that it is worth analyzing atoms as new scatterers.

In this article we want to give a classical way of understanding the main cooling and trapping features of atoms, without using the optical Bloch equations or other extensive quantum treatment (section 2). This will allow us to describe the main features of the atoms as very special scatterers (section 3). In a last part (section 4) we will then present the situation where the scattered wave will no longer be the electric field, but the atoms themselves considered as matter waves. We do not aim here at giving a detailed or rigorous description of laser cooling of atoms. We rather try to give a simple description of basic features involved in laser manipulation of atoms in order to open this new tool to research outside of the atomic physics community.

2. Laser manipulation of atoms

The interaction between atoms and quasiresonant light is usually treated by considering the coupling between a classical electric field and an atom described either as a two level system, or for more sophisticated effects as a multi level system with hyperfine and magnetic (Zeeman) structure. In the section we want to use a classical model, both for the light (no quantization of the electric field in terms of photons will be required) and for the atom. It is indeed possible to describe most basic effects (such as Doppler cooling of atoms and the magneto-optical trap) in such classical terms. Obviously there will be some approximations done in order to achieve this goal and the main purpose is not to give a very precise description of these effects, but rather a simple method to understand some of the main features of this physics without having to learn about optical Bloch equations and the like. Scientists from other communities (especially from the multiple scattering community addressed in this collection of articles) are thus invited to use this models to evaluate new effects on such rather particular scatterers.

2.1. INTERNAL MOTION : ELASTICALLY BOUND ELECTRON

We want to study the center of mass motion of atoms interacting with quasiresonant light. The radiative forces experienced by the atoms will depend on the detuning δ between the laser frequency ω_L and the atomic resonant frequency ω_{at}[14]. If for example one wants to compute the radiation pressure one needs to know the scattering cross section, which, in the case of particles with an internal resonance, can be much larger than the geometrical size of the particle (Mie resonances e.g.). In order to take into account these internal resonance effects, we will model the atom as a kernel surrounded by an elastically bound electron, with a resonance frequency ω_{at}. The laser light drives the electron and thus induces a dipole $\vec{d} = q\vec{r} = q\left(\vec{r}_e - \vec{R}\right)$ which, in the driven regime, oscillates at the driving frequency ω_L. It will be the interaction between this driven dipole and the electro-magnetic field of the laser which acts on the center of mass of the atom.

We will hence proceed in two steps: first compute the dipole induced by the laser light and second study the motion of this oscillating dipole with the electromagnetic field.

In this model we will suppose that the distance $\vec{r} = \vec{r}_e - \vec{R}$ between the electron and the kernel of the atom follows the equation [14]:

$$\frac{d^2}{dt^2}\vec{r} + \Gamma\frac{d}{dt}\vec{r} + {\omega_{at}}^2\vec{r} = \frac{\vec{f}_{ext}}{m_e} \tag{1}$$

The total force acting on the electron is composed by the force $\vec{f}_E$ due to the electric field at the position $\vec{r}_e = \vec{R} + \vec{r}$ of the electron :

$$\vec{f}_E = q\,\vec{E}(\vec{r}_e, t) \tag{2}$$

and by a component due to the magnetic field :

$$\vec{f}_B = q\,\frac{d\vec{r}_e}{dt} \wedge \vec{B}(\vec{r}_e, t) \tag{3}$$

The ratio between the amplitude of these two forces is of the order of

$$\frac{f_B}{f_E} \simeq \frac{dr_e}{dt}\frac{1}{c} = \frac{v_e}{c} \ll 1 \tag{4}$$

and we hence can neglect the effect of the magnetic field for computing the relative motion of the electron. Furthermore, the mass of the kernel being much larger than that of the electron, the distance $\vec{r} = \vec{r}_e - \vec{R}$ between the electron and the kernel of the atom is determined by the motion of the electron. We will use the complex notation for the electric field for a monochromatic linear polarized light :

$$\vec{E}(\vec{r}, t) = E_0(\vec{r})\vec{\varepsilon_x} \exp{(-i\omega_L t)} \tag{5}$$

Using eqs. (1) and (5) and only taking into account the electric field force on gets a driven solution $\vec{r}(t) = \vec{r}_0 \exp{(-i\omega_L t)}$ with :

$$-\omega_L^2\vec{r}_0 - i\omega_L\Gamma\vec{r}_0 + {\omega_{at}}^2\vec{r}_0 = \frac{qE_0(\vec{r}_e)}{m_e}\vec{\varepsilon_x} \tag{6}$$

Defining the polarizability $\alpha(\omega_L)$ of the atomic dipole by :

$$\vec{d} = q\vec{r} = \varepsilon_o\,\alpha(\omega_L)\vec{E} \tag{7}$$

one thus obtains :

$$\alpha(\omega_L) = \frac{1}{({\omega_{at}}^2 - \omega_L^2 - i\omega_L\Gamma)}\frac{q^2}{\varepsilon_o m_e} \tag{8}$$

We will use the real and the complex part of $\alpha(\omega_L)$: $\alpha = \alpha' + i\alpha''$:

$$\alpha' = \frac{{\omega_{at}}^2 - \omega_L^2}{({\omega_{at}}^2 - \omega_L^2)^2 + (\omega_L\Gamma)^2}\frac{q^2}{\varepsilon_o m_e} \tag{9}$$

and :

$$\alpha'' = \frac{\omega_L\Gamma}{({\omega_{at}}^2 - \omega_L^2)^2 + (\omega_L\Gamma)^2}\frac{q^2}{\varepsilon_o m_e} \tag{10}$$

With real notations for the electric field and for the dipole $\vec{\mathcal{D}}$ one thus has:

$$\vec{\mathcal{D}} = \mathrm{Re}\,\vec{d} = \mathrm{Re}\left[\varepsilon_o\, \alpha(\omega_L)\vec{E}\right] \tag{11}$$

For a wave propagating along Oz such as:

$$\vec{E}(\vec{r},t) = E_0(\vec{r})\vec{\varepsilon_x}\exp\left[-i\left(\omega_L t - kz\right)\right] \tag{12}$$

one gets :

$$\vec{\mathcal{D}} = \varepsilon_o\left|\alpha\right| E_0(\vec{r})\vec{\varepsilon_x}\cos\left[-\left(\omega_L t - kz\right) + \varphi_\alpha\right] \tag{13}$$

where $\alpha = |\alpha|\exp\left(i\varphi_\alpha\right)$. The induced dipole follows the driving electric field with some delay. This delay depend on the detuning between the laser frequency and the resonance frequency of the dipole. If the electric field oscillates very slowly compared to the dipole resonance frequency ($\omega_L << \omega_{at}$), we have $\alpha' \gg \alpha$" and the induced dipole almost immediately follows the electric field excitation with a static polarizability :

$$\alpha_{stat} = \frac{q^2}{\varepsilon_o m_e {\omega_{at}}^2} = \alpha_0 \tag{14}$$

For a resonant excitation ($\omega_L = \omega_{at}$) we have $\alpha' = 0$, i.e. a purely imaginary polarizability, and the dipole is in quadrature phase with the driving field. Defining the detuning $\delta = \omega_L - \omega_{at}$ as the difference between the laser frequency ω_L and the atomic resonance frequency ω_{at} one gets for a quasi-resonant excitation ($\delta << \omega_L$, $\omega_L \simeq \omega_{at}$) (figure 1) :

$$\alpha' = \frac{-\delta\left(\delta + 2\omega_{at}\right)}{\delta^2\left(\delta + 2\omega_{at}\right)^2 + \left(\omega_L\Gamma\right)^2}\frac{q^2}{\varepsilon_o m_e} \simeq \frac{-\delta}{\delta^2 + \frac{\Gamma^2}{4}}\frac{q^2}{2\varepsilon_o m_e \omega_{at}} \tag{15}$$

and

$$\alpha" = \frac{\omega_L\Gamma}{\delta^2\left(\delta + 2\omega_{at}\right)^2 + \left(\omega_L\Gamma\right)^2}\frac{q^2}{\varepsilon_o m_e} \simeq \frac{\frac{\Gamma}{2}}{\delta^2 + \frac{\Gamma^2}{4}}\frac{q^2}{2\varepsilon_o m_e \omega_{at}} \tag{16}$$

or :

$$\alpha' = \frac{-\delta\omega_L}{\delta^2 + \frac{\Gamma^2}{4}}\frac{\alpha_o}{2} \tag{17}$$

$$\alpha" = \frac{\frac{\Gamma}{2}\omega_L}{\delta^2 + \frac{\Gamma^2}{4}}\frac{\alpha_o}{2} \tag{18}$$

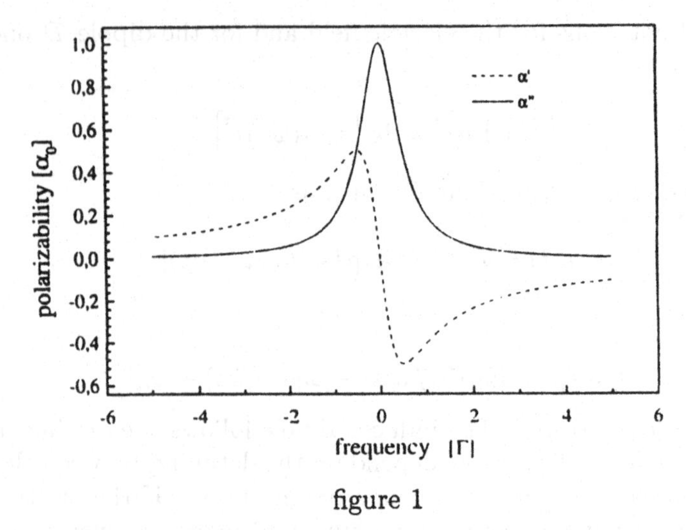

figure 1

For the motion of the electron $\vec{r}_e(t) = \vec{r}_0 \exp(-i\omega_L t)$ one finally gets :

$$\vec{r}_0 = \alpha \frac{\varepsilon_0}{q} E_0(\vec{r}_e) \vec{\varepsilon}_x = (\alpha' + i\alpha'') \frac{\varepsilon_0}{q} E_0(\vec{r}_e) \vec{\varepsilon}_x \tag{19}$$

Remarks :

i) In order to compute this polarizability we have approached the electric field $E_0(\vec{r}_e)$ at the position of the electron with the field at $\vec{R}$ of the center of mass M of the atom. This approximation is valid if the distance between the electron and the center of mass of the atom is small compared to the scale on which the electric field varies, i.e. small compared to the wavelength of the laser : $|\vec{r}| = \left|\vec{r}_e - \vec{R}\right| \ll \lambda$. This approximation is called electric dipole approximation as one can consider the atom as a point dipole on the scale of the wavelength λ. Note that even in this dipole approximation, real atoms have a more complex internal structure (Zeeman sublevels e.g.) and exhibit some features which cannot be described by a classical dipole oscillation.

ii) The damping of the dipole can be explained by the radiation of the oscillating dipole. This radiation depends on the frequency of the oscillation and strictly speaking one should replace Γ by $\Gamma\frac{\omega_L^2}{\omega_{at}^2}$. But we will only be interested in frequencies close to the resonant frequency and we will thus neglect the change of the damping on the scale of δ.

2.2. RADIATION FORCE ACTING ON THE ATOM : "CLASSICAL APPROACH"

The force acting on the center of mass of atom (considered as a dipole) has two components : one due to the electric field and one due to the magnetic field of the incident laser field, propagating along the Oz axes. The force $\vec{f}_E$ due to the electric field, which we take polarized along the Ox axes :

$$\vec{E}(\vec{r},t) = E_0(\vec{r})\vec{\varepsilon_x}\exp(-i\omega_L t) \tag{20}$$

is directed parallel to this electric field :

$$\vec{f}_E = \sum q\vec{E}(\vec{r},t) \propto \vec{\varepsilon}_x \tag{21}$$

The magnetic force $\vec{f}_B$ on the contrary will be directed along the axes of propagation of the laser (for a linear polarized plane wave)! The electron driven by the electric field has a velocity along the axes Ox : $\frac{d\vec{r}_e}{dt} = \left|\frac{d\vec{r}_e}{dt}\right|\vec{e_x}$ and for a magnetic field along Oy : $\vec{B}(\vec{r}_e,t) = B_0(\vec{r}_e)\vec{e}_y\exp(-i\omega_L t)$ one gets :

$$\vec{f}_B = q\frac{d\vec{r}_e}{dt} \wedge \vec{B}(\vec{r}_e,t) = q\left|\frac{d\vec{r}_e}{dt}\right| B_0(\vec{r}_e)\exp(-i\omega_L t)\vec{\varepsilon}_z \tag{22}$$

It is thus clear that one cannot neglect the effect of the magnetic field for computing the force acting on the center of mass of the atom! One can keep in mind the model of the electron driven by the electric field and it is the magnetic force which acts on this moving charged particle (figure 2). Although this is not a complete description, it allows one e.g. to understand why the radiation pressure force is along the axes of propagation of the laser light, without using a quantum treatment for the electric field of the laser.

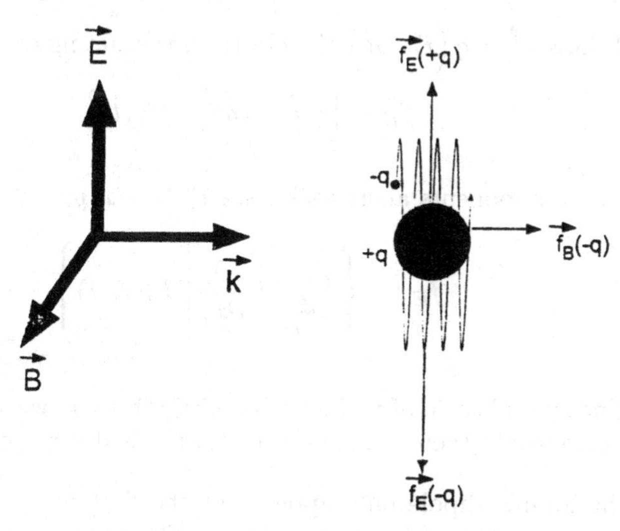

figure 2

Despite $\frac{f_B}{f_E} \simeq \frac{v}{c} \ll 1$ one cannot neglect the magnetic force when evaluating the force acting on the center of mass of the atom. The magnetic and the electric field indeed yield forces which are not directed along the same axes. For a plane wave polarized along the Ox axes propagating along Oz the magnetic force seems to give the radiation pressure. It is the sum of the electric and magnetic forces that give rise to the radiation pressure force acting on atoms [15].

The electric field will yield a force on both charged parts of the atom: the electron and the nucleus :

$$\overrightarrow{f_E} = |q|\,\vec{E}(\overrightarrow{R},t) - |q|\,\vec{E}(\vec{r}_e,t) \tag{23}$$

The two components of the electric force have opposite signs. However the net electric force is not zero as the electric field is not the same at the location $\vec{r}_e$ of the electron and $\overrightarrow{R}$ of the nucleus. The electric force is hence a differential effect.

The magnetic force on the other hand depends on the velocities of the charged particles. As the electron does move much faster than the nucleus, one only needs to consider the magnetic force acting on the electron. But one has to consider this force together with the electric force acting on both charges.

Electric force :

The electric force can be evaluated by making a first order expansion of $\vec{E}(\vec{r}_e,t)$:

$$\vec{E}(\vec{r}_e,t) \simeq \vec{E}(\overrightarrow{R},t) + \left\{\left[\left(\vec{r}_e - \overrightarrow{R}\right)\cdot\overrightarrow{grad}\right]\vec{E}(\vec{r}_e,t)\right\}_{\vec{r}_e=\overrightarrow{R}} \tag{24}$$

Taking $\vec{d} = q\left(\vec{r}_e - \overrightarrow{R}\right)$ the electric force acting on the atom is :

$$\overrightarrow{f_E} = \left\{\left[\vec{d}\cdot\overrightarrow{grad}\right]\vec{E}(\vec{r}_e,t)\right\}_{\vec{r}_e=\overrightarrow{R}} \tag{25}$$

with components along each axes $\vec{e_i}$ $(i = x, y, z)$:

$$f_{E,i} = \left\{\left[\sum_j d_j \frac{\partial}{\partial x_j}\right] E_i(\vec{r}_e,t)\right\}_{\vec{r}_e=\overrightarrow{R}} \tag{26}$$

The spatial scale of variation for $\vec{E}(\vec{r}_e,t)$ is the wavelength λ. The electric force would then be zero if the electric field were uniform and its effect on the atomic dipole only appears at the first order in $\frac{\left|\vec{r}_e - \overrightarrow{R}\right|}{\lambda}$. We now have computed the instantaneous force $\overrightarrow{f_E}$ which needs to be averaged over the fast optical frequency in order to describe the slow motion of the center of mass of the atom.

Magnetic force :

For the magnetic force we restrict ourselves to :

$$\overrightarrow{f_B} = q\frac{d\overrightarrow{r}_e}{dt} \wedge \overrightarrow{B}(\overrightarrow{r}_e, t) \tag{27}$$

and at the lowest order one has :

$$\overrightarrow{f_B} = q\frac{d\overrightarrow{r}}{dt} \wedge \overrightarrow{B}(\overrightarrow{R}, t) \tag{28}$$

As $\vec{d} = q\overrightarrow{r} = \varepsilon_o\, \alpha(\omega_L)\vec{E}$ one can write :

$$\overrightarrow{f_B} = \frac{d}{dt}\left(\overrightarrow{d} \wedge \overrightarrow{B}(\overrightarrow{R}, t)\right) - \overrightarrow{d} \wedge \frac{d}{dt}\overrightarrow{B}(\overrightarrow{R}, t) \tag{29}$$

Taking $\frac{d}{dt}\overrightarrow{B}(\overrightarrow{R}, t) \simeq \frac{\partial}{\partial t}\overrightarrow{B}(\overrightarrow{R}, t)$ (the velocities of the charges are small compared to c) and using Maxwell's equation $\frac{\partial \overrightarrow{B}}{\partial t} = -\overrightarrow{rot}\overrightarrow{E}$ one can express the magnetic force as a function of the electric field :

$$\overrightarrow{f_B} = \frac{d}{dt}\left(\overrightarrow{d} \wedge \overrightarrow{B}(\overrightarrow{R}, t)\right) + \overrightarrow{d} \wedge \overrightarrow{rot}\overrightarrow{E} \tag{30}$$

or, for the $\overrightarrow{e_i}$ component :

$$f_{B,i} = \frac{d}{dt}\left(\overrightarrow{d} \wedge \overrightarrow{B}(\overrightarrow{R}, t)\right)_i + \sum_j \left(d_j\frac{\partial}{\partial x_i}E_j - d_j\frac{\partial}{\partial x_j}E_i\right) \tag{31}$$

The time average of the first term is zero and we hence neglect this part in the following.

Total force :

The total average force $\left\langle \overrightarrow{f} \right\rangle$ on a atom by a light wave is:

$$\left\langle \overrightarrow{f} \right\rangle = \left\langle \overrightarrow{f_E} \right\rangle + \left\langle \overrightarrow{f_B} \right\rangle = \left\{\left[\overrightarrow{d} \cdot \overrightarrow{grad}\right] \vec{E}(\vec{r}_e, t)\right\}_{\vec{r}_e = \overrightarrow{R}} + \overrightarrow{d} \wedge \overrightarrow{rot}\overrightarrow{E} \tag{32}$$

with components along $\overrightarrow{e_i}$:

$$\langle f_i \rangle = \left\{\left[\sum_j d_j\frac{\partial}{\partial x_j}\right] E_i(\vec{r}_e, t)\right\}_{\vec{r}_e = \overrightarrow{R}} + \left\{\sum_j \left(d_j\frac{\partial}{\partial x_i}E_j - d_j\frac{\partial}{\partial x_j}E_i\right)\right\}_{\vec{r}_e = \overrightarrow{R}} \tag{33}$$

$$\langle f_i \rangle = \left\{ \sum_j d_j \frac{\partial}{\partial x_i} E_j \right\}_{\vec{r}_e = \vec{R}} \tag{34}$$

This force seems to derive from a time average potential :

$$W = -\left\langle \vec{d}\,\vec{E} \right\rangle = -\left\langle \sum_j d_j E_j \right\rangle \tag{35}$$

where the gradient is taken only on the electric field $\vec{E}$ and not on the dipole $\vec{d}$:

$$\left\langle \vec{f} \right\rangle = -\overrightarrow{grad} W = \left\langle \sum_j d_j \overrightarrow{grad} E_j \right\rangle \tag{36}$$

Average radiation force :
Let us consider the following electric field $\vec{E}_0(\vec{r}) = \vec{e}_x E_0(\vec{r}) \exp{(ikz - i\omega_L t)}$, with $E_0(\vec{r})$ real.
Coming back to real notations for the fields and the dipoles one has :

$$\vec{\mathcal{D}} = \mathrm{Re}\,\vec{d} = \mathrm{Re}\left(\varepsilon_o\, \alpha \vec{E}_0(\vec{r}) \exp{(-i\omega_L t)}\right) \tag{37}$$

For a wave propagating along Oz one gets : $\vec{E}_0(\vec{r}) = \vec{e}_x \left|\vec{E}_0(\vec{r})\right| \exp{(ikz)}$ and :

$$\vec{\mathcal{D}} = \varepsilon_o \left|\vec{E}_0(\vec{r})\right| \vec{e}_x \left(\alpha' \cos{(\omega_L t - kz)} + \alpha'' \sin{(\omega_L t - kz)} \right) \tag{38}$$

To calculate the average force let's first take

$$\overrightarrow{grad}\,\mathrm{Re}\,[E_0(\vec{r})(-i\omega_L t)] = \mathrm{Re}\left\{ \overrightarrow{grad}\,[|E_0(\vec{r})| \exp{(ikz)}]\,(-i\omega_L t) \right\} \tag{39}$$

$$\overrightarrow{grad}\,\mathrm{Re}\,[E_0(\vec{r})(-i\omega_L t)] = \overrightarrow{grad}\,[|E_0(\vec{r})|] \cos{(\omega_L t - kz)} + k\vec{e_z}\,|E_0(\vec{r})| \sin{(\omega_L t - kz)} \tag{40}$$

We then obtain the instantaneous force :

$$\vec{f} = \left[\varepsilon_o |E_0(\vec{r})| \left(\alpha' \cos{(\omega_L t - kz)} + \alpha'' \sin{(\omega_L t - kz)} \right) \right] * \tag{41}$$

$$\left[\overrightarrow{grad}\,[|E_0(\vec{r})|] \cos{(\omega_L t - kz)} + k\vec{e_z}\,|E_0(\vec{r})| \sin{(\omega_L t - kz)} \right] \tag{42}$$

The time average force is thus :

$$\left\langle \vec{f} \right\rangle = \varepsilon_o \left(\frac{1}{4}\alpha' \overrightarrow{grad} \left[|E_0(\vec{r})|^2 \right] + \frac{1}{2}\alpha'' k\vec{e_z} \, |E_0(\vec{r})|^2 \right) \tag{43}$$

Resonant radiation pressure :

The second term of (43) is called the resonant radiation pressure $\vec{f}_{rad}$. It is aligned along the direction of propagation of the laser ($\vec{e_z}$) and it is proportional to the laser intensity : $I_{inc} = \frac{1}{2}\varepsilon_o c \left| \vec{E} \right|^2$:

$$\vec{f}_{rad} = \alpha'' k\vec{e_z} \frac{I_{inc}}{c} \tag{44}$$

One can thus define a scattering cross section σ_{at} for the atoms. By taking : $\vec{f}_{rad} = \alpha'' k\vec{e_z} \frac{I_{inc}}{c} = \sigma_{at} \frac{I_{inc}}{c} \vec{e_z}$ one gets : $\sigma_{at} = \alpha'' k$, which depends on the detuning δ (figure 3) :

$$\sigma_{at}(\delta) = \frac{\frac{\Gamma}{2}}{\delta^2 + \frac{\Gamma^2}{4}} \frac{q^2}{2\varepsilon_o c m_e} \tag{45}$$

At resonance one has $\sigma_{at}^{res} = \alpha_0 k \frac{\omega_{at}}{\Gamma}$. Taking for the damping constant the radiation losses due to the oscillating electron ($\Gamma = \frac{4}{6mc^3} \frac{q^2}{4\pi\varepsilon_0} \omega_{at}^2$), one finds at resonance :

$$\sigma_{at}^{res} = \frac{3\lambda_{at}^2}{2\pi} \tag{46}$$

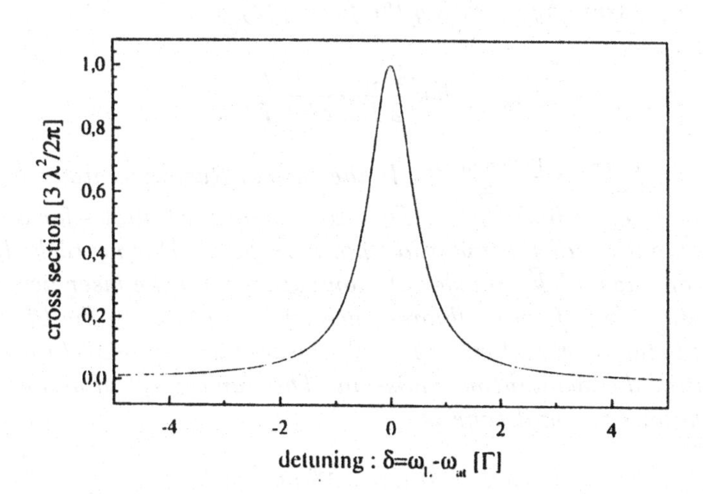

figure 3

Remarks :

"Quantum approach"

It is possible to evaluate the radiation pressure force of a plane monochromatic wave acting on an atom by a linear momentum conservation argument. As a classical electric field is not a eigenstate of the momentum operator, we use a quantum description of the light field in terms of photons. For a wave propagating along the Oz axes, each absorption process give rise to a momentum transfer of $\Delta\vec{p} = \hbar k\vec{e}_z$. *The emission of photons will occur in a random direction such that, on average, the momentum transfer after several fluorescence cycles will be zero for the emission processes. One thus gets an average momentum transfer of* $\langle\Delta\vec{p}\rangle = \hbar k\vec{e}_z$ *per fluorescence cycle. The average force* $\vec{f}_{av}$, *depending on the number of fluorescence cycles per second* $\frac{dN}{dt}$ *(hence also on the laser intensity and detuning) is thus given by :*

$$\vec{f}_{av} = \frac{dN}{dt}\hbar k\vec{e}_z \tag{47}$$

directed along the axes of propagation of the incident laser light. This argument is based on a quantum treatment of the laser light (the electric field being quantized in terms of photons) and is not along the main line of the calculation followed in this article. However it clearly shows that the direction of the force acting on the atom can be along the direction of propagation of the laser light and hence transverse to the electric field! A similar argument could give the force acting on a mirror when reflecting light.

It is possible to reconcile the "classical" and "quantum" description of the radiation pressure by rewriting the force (44) as :

$$\vec{f}_{rad} = \hbar\vec{k}\,\frac{\Gamma}{2}\frac{I_{inc}}{I_{sat}}\frac{\Gamma^2/4}{\delta^2 + \Gamma^2/4} \tag{48}$$

with $I_{sat} = \frac{\Gamma^2\hbar c}{2\alpha_o\omega_L} = \frac{\Gamma^2\hbar c\varepsilon_0 m_e\omega_L}{2q^2}$. *In the case of Rubidium atoms one gets for example :* $I_{sat} = 1.6mW/cm^2$. *This expression (48) allows for a simple physical explanation of the radiation pressure force. During each fluorescence cycle one has a* $\hbar\vec{k}$ *transfer of momentum from the laser field to the atom. The time scale for one fluorescence cycle depends on the lifetime* $\frac{1}{\Gamma}$ *of the excited state of the atom and on the laser intensity needed to reexcite the atoms after the spontaneous emission. The number* $\frac{dN}{dt}$ *of absorbed and reemitted photons per unit time is :*

$$\frac{dN}{dt} = \frac{\Gamma}{2}\frac{I_{inc}}{I_{sat}}\frac{\Gamma^2/4}{\delta^2 + \Gamma^2/4} \tag{49}$$

which has the resonant frequency dependance.

Dipole force :

The first term of (43) is called dipole force $\overrightarrow{f}_{dip}$. A particle with an real polarizability α' is attracted to a spatial region where its potential energy $W = -\overrightarrow{d}\,\overrightarrow{E}$ is minimum. With $d = \varepsilon_o \alpha' \overrightarrow{E}$ one has :

$$W = -\varepsilon_o \alpha' \left|\overrightarrow{E}\right|^2 \tag{50}$$

One thus obtains a force oriented towards regions of high electric field in the case of $\alpha' > 0$ (high field seekers, such as dielectric spheres in air) and towards low electric field for $\alpha' < 0$ (low field seekers, such as air bubbles in champagne).

Remark :

Whereas the radiation pressure force can be explained by fluorescence cycles of absorption followed by spontaneous emission, the dipole force can be expressed in a quantum approach in terms of absorption followed by stimulated emission processes.

Let us consider the case of a standing wave obtained by two contra-propagating plane waves $\left|\overrightarrow{E}\right|^2 = 4\left|\overrightarrow{E_0}\right|^2 \cos^2(kz)$.The dipole force for an atom located at z is in this case (figure 4):

$$\overrightarrow{f}_{dip} = -8\hbar\overrightarrow{k}\,\delta \frac{I_{inc}}{I_{sat}} \frac{\Gamma^2/4}{\delta^2 + \Gamma^2/4} \sin(2kz) \tag{51}$$

where I_{inc} is the incident intensity of each of two planes waves producing the standing wave. Note that this force changes sign when moving along Oz.

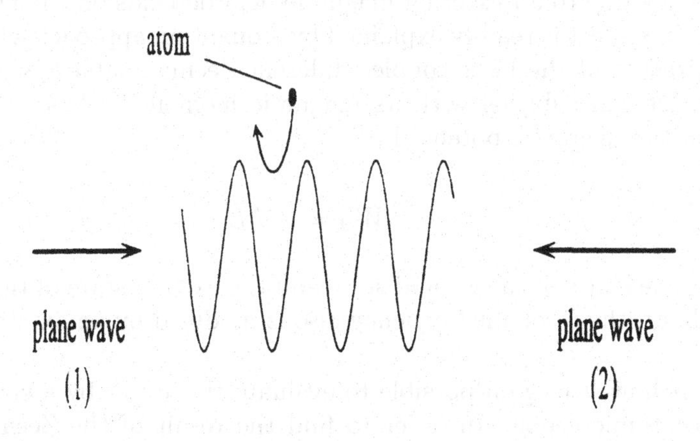

figure 4

Order of magnitude :

For an on-resonant laser ($\delta = 0$) with an intensity of $I = 1mW/cm^2$ the radiation pressure force is $f_{rad} = \alpha'' k \frac{I}{c} \simeq 10^{-20} N$. This force is 10^4 time larger than the gravitational force : $f_g = Mg = 10^{-24} N$! Even though each momentum transfer is quite small the radiation pressure forces are huge because after the emission of photons the atoms can quickly be reexcited to the upper state and up to 10^7 cycles of fluorescence per second can be achieved. One condition for this to happen is that, after a spontaneous emission, the atoms falls back into the initial ground state from where it quickly can be reexcited (so-called closed transitions). In the elastically bound electron model, this condition does not appear explicitly, except for the fact that we suppose one single atomic frequency to be present. In some atoms this condition can be fulfilled (sometimes at the expense of an additional "repumping" laser), but in the case of molecules, it is much more difficult to find closed transitions useful for laser cooling.

2.3. ZEEMAN EFFECT

In the preceding section we have considered a model for the atomic polarizability in absence of an external magnetic field. As we will see later, applying external magnetic field (in particular inhomogeneous fields) are commonly used in laser trapping of cold atoms. It is thus important to include the effect of such magnetic fields on the atoms response to an laser excitation.

The effect of an external magnetic field on the atomic polarizability is known as the Zeeman effect and as a result the resonant frequencies of the atoms are modified. When discussing Zeeman effects the polarization of the incident laser beam is important and the frequency shift of the atomic frequency due to an external magnetic field depends on the laser polarization. This effect is readily explained by a quantum approach where different polarization of the laser couples different Zeeman sublevels of the atoms. Each Zeeman sublevel with its magnetic moment $\vec{\mu}$ has an energy shift due to the magnetic potential :

$$W_B = -\vec{\mu}\vec{B} \tag{52}$$

As the shifts of the ground state and the excited state of the atoms usually differ, the resonant frequency also is modified by an external magnetic field.

It is however again possible to evaluate this effect by a classical model of the atomic dipole. In order to find the result of the Zeeman effect in our model, we calculate the modified dipole resonance due to the presence of a static external magnetic field. Neglecting the damping effect for this purpose, the equation of motion of the electron is then given by :

$$m\frac{d^2\overrightarrow{r}}{dt^2} = q\frac{d\overrightarrow{r}}{dt} \wedge \overrightarrow{B_0} - m\omega_{at}^2\overrightarrow{r} \tag{53}$$

Consider first an electron moving in a plane orthogonal to the external magnetic field. We will later assume that this motion is induced by the electric field of the applied laser light. Changing to a frame rotating at a frequency $\overrightarrow{\Omega}$ around the direction of the static external magnetic field one gets :

$$m\left(\frac{d^2\overrightarrow{r}}{dt^2}\right)_{rotating} = -\left(q\overrightarrow{B_0} + 2m\overrightarrow{\Omega}\right) \wedge \left(\frac{d\overrightarrow{r}}{dt}\right)_{rotating} \tag{54}$$

$$+\left(m\left|\overrightarrow{\Omega}\right|^2 + qB_0\Omega - m\omega_{at}^2\right)\overrightarrow{r} \tag{55}$$

as $q\overrightarrow{B_0} \wedge \left(\overrightarrow{\Omega} \wedge \overrightarrow{r}\right) = -qB_0\Omega\overrightarrow{r}$.

By choosing a frame rotating at $\overrightarrow{\Omega} = -\frac{q\overrightarrow{B_0}}{2m}$, equation (54) can be written as :

$$\left(\frac{d^2\overrightarrow{r}}{dt^2}\right)_{rotating} = -\left(\omega_{at}^2 + \left(\frac{\omega_c}{2}\right)^2\right)\overrightarrow{r} \tag{56}$$

with $\omega_c = -\frac{qB_0}{2m}$. The atomic resonance frequency is thus modified to ω_{res} :

$$\omega_{res}^2 = \omega_{at}^2 + \left(\frac{\omega_c}{2}\right)^2 \tag{57}$$

Coming back to the laboratory frame one has 2 frequencies :

$$\omega_1 = |\Omega + \omega_L| = \left|\frac{\omega_c}{2} + \sqrt{\omega_{at}^2 + \left(\frac{\omega_c}{2}\right)^2}\right| \tag{58}$$

$$\omega_2 = |\Omega - \omega_L| = \left|\frac{\omega_c}{2} - \sqrt{\omega_{at}^2 + \left(\frac{\omega_c}{2}\right)^2}\right| \tag{59}$$

In usual atomic cooling experiments the cyclotron frequency ω_c is small compared to the atomic resonance frequency ω_{at} (even for a magnetic field of $B = 1T$ the cyclotron frequency ω_c for an electron is $\frac{\omega_c}{2\pi} = 28GHz$ compared to atomic resonance frequencies of the order of $\frac{\omega_{at}}{2\pi} = 10^{+14}Hz$). To the first order in $\frac{\omega_c}{\omega_{at}}$ one thus obtains :

$$\omega_1 = \omega_{at} + \frac{\omega_c}{2} \tag{60}$$

$$\omega_2 = \omega_{at} - \frac{\omega_c}{2} \tag{61}$$

One thus finds that applying a external magnetic field leads to a splitting of the atomic resonance frequency. This Zeeman effect can thus be described by a classical model of the atomic dipole. The two frequencies ω_1 and ω_2 correspond to different electron motion. For one frequency the motion will have to be induced by a $\sigma^{(+)}$ circular polarized light and by a $\sigma^{(-)}$ circular polarized light for the other frequency.

Remark :

1) We only found a splitting into two frequencies. It is however known that a orbital angular momentum of the electron, as considered here, can only give rise to an odd number of Zeeman levels. In order to find an even number of Zeeman level one needs to take into account the spin of the electron which we obviously have not done here. In the description of the Zeeman effect done above we have considered the atom to move in a plane perpendicular to the external magnetic field. allowing for a motion along this direction would give rise to an unchanged atomic frequency, as $\frac{d\vec{r}}{dt} \wedge \vec{B_0} = \vec{0}$.

2) We can link the classical result of the Zeeman effect to the quantum result by using as magnetic moment μ *the Bohr magneton's value :* $\mu_B = \frac{q\hbar}{2m_e}$. *The energy shift* $W_B = -\vec{\mu}\vec{B}$ *from eq. (52) has to compared to a frequency shifts* $\pm\frac{\omega_c}{2}$ *¿from eq. (60),(61)* ($\omega_c = -\frac{qB_0}{2m}$) *multiplied by Planck's constant. The quantum result yields thus* $\mu_B B = \hbar\frac{\omega_c}{2}$, *the same as the classical one.*

2.4. DOPPLER COOLING

We now apply the radiation pressure forces to study laser cooling of atoms, i.e. in order to reduce the width of the velocity distribution of a sample of atoms. The simplest idea for such a cooling has been proposed in 1975 by T. Hänsch and A. Shawlow[16] for neutral atoms and by D. Wineland and H. Dehmelt[17] for ions. Consider the case of two laser counterpropagating along Oz with the same frequency ω_L (figure 5). This arguments can be generalized to three dimensions, but for simplicity we will restrict ourselves to one dimension. An atom interacting with such a laser configuration will be submitted to the radiation pressure forces calculated in section 2.2. A detailed analysis of this situation has to include both effects of the resonant radiation pressure and of the dipole force. But a basic explanation of the so-called Doppler cooling can be given by simply adding independently the resonant radiation pressure forces of the two propagating laser fields.

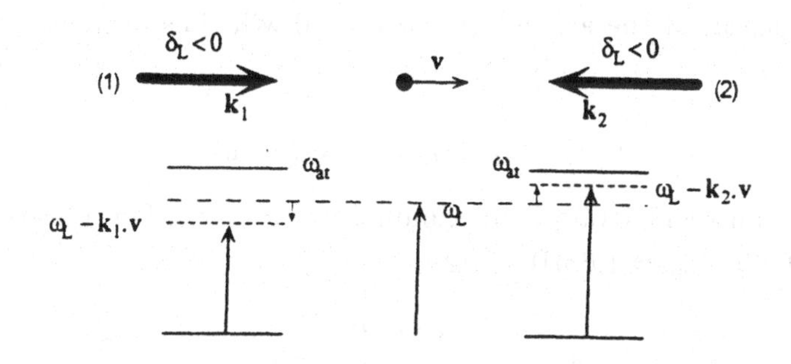

figure 5

Let us consider the case of the laser frequency being detuning below the atomic resonant frequency : $\delta = \omega_L - \omega_{at} < 0$ ("red detuning"). For an atom at rest ($v_z = 0$), the excitation by the laser light will not be efficient as the resonance condition for neither laser will be fulfilled. If now an atom is moving ($v_z \neq 0$) it will experience different frequencies from the two laser fields. In the atom's rest frame, one of the laser frequencies will be shifted towards higher frequencies, the other one towards lower frequencies. For a "red detuning" ($\delta < 0$) the atom will shift into resonance ($\delta - k_z v_z = 0$) with the laser propagating opposite to the atoms velocity : $\delta = k_z v_z < 0$. The other laser on the contrary be shifted further out of resonance.

The net force for an atom moving towards $+Oz$ will be directed along $-Oz$, i.e. opposite to its velocity. In the same way, an atoms moving towards $-Oz$ will experience a force along $+Oz$. By summing the two velocity dependent forces one gets with $\vec{k_1} = -\vec{k_2} = k\vec{e_z}$:

$$\vec{f}_{tot}(v_z) = \left(\hbar\vec{k_1}\frac{\Gamma}{2}\frac{I_{inc}}{I_{sat}}\frac{\Gamma^2/4}{(\delta - k_{1,z}v_z)^2 + \Gamma^2/4} + \hbar\vec{k_2}\frac{\Gamma}{2}\frac{I_{inc}}{I_{sat}}\frac{\Gamma^2/4}{(\delta - k_{2,z}v_z)^2 + \Gamma^2/4} \right) \tag{62}$$

$$\vec{f}_{tot}(v_z) = \hbar\vec{k_1}\frac{\Gamma}{2}\frac{I_{inc}}{I_{sat}}\left(\frac{\Gamma^2/4}{(\delta - kv_z)^2 + \Gamma^2/4} + \frac{\Gamma^2/4}{(\delta + kv_z)^2 + \Gamma^2/4}\right) \tag{63}$$

For small velocities one can get a linearized expression of this force :

$$f_z(v_z) = -\gamma m v_z \tag{64}$$

with a friction coefficient γ :

$$\gamma = -4\delta \frac{\hbar k^2}{M} \frac{\Gamma}{2} \frac{I_{inc}}{I_{sat}} \frac{\Gamma^2/4}{(\delta^2 + \Gamma^2/4)^2} \tag{65}$$

The velocities of the atoms around $v_z = 0$ will thus decrease exponentially

$$v_z(t) = v_z(t_0) \exp\left[-\gamma(t - t_0)\right] \tag{66}$$

The friction coefficient γ is maximum for $\delta = -\frac{\Gamma}{2\sqrt{3}}$. For a laser with an intensity of $I = I_{sat} = 1.6mW/cm2$:

$$\gamma_{\max} = \frac{9}{4\sqrt{3}} \frac{\hbar k^2}{M} = 25\mu s \tag{67}$$

This is a very fast processes. Because of this friction this type of cooling has been called "optical molasses". If one would change the frequency of the laser to positive ("blue") detuning, one would get a heating process for the atoms (increasing their velocities).

The limit of such a cooling process is given by the fluctuations of the forces. These fluctuations are of quantum nature and depend on the recoil velocity $\hbar k$. At each cycle of fluorescence a photon is emitted in a random direction. This yields a random walk in momentum space with a step of size $\delta p = \hbar k$. One thus gets an increase in the cinetic energy of the atomic distribution :

$$\frac{d(\Delta p)^2}{dt} = 2D \tag{68}$$

The diffusion coefficient D is given by the ratio of the step size δp and the time scale for a fluorescence cycle τ :

$$D = \frac{(\hbar k)^2}{\tau} \tag{69}$$

The average time τ between two spontaneous emissions is given by :

$$\frac{1}{\tau} = \frac{\Gamma}{2} \frac{I_{inc}}{I_{sat}} \frac{\Gamma^2/4}{\delta^2 + \Gamma^2/4} \tag{70}$$

This diffusion gives rise to an increase in the cinetic energy, i.e. a heating of the atoms.

At equilibrium the heating due to the fluctuations and the cooling due to the friction effect compensate :

$$\left[\frac{d(\Delta p)^2}{dt}\right]_{éq} = -2\gamma (\Delta p)^2_{éq} + 2D = 0 \tag{71}$$

and one obtains the temperature (figure 6) :

$$\frac{1}{2}k_B T = \frac{(\Delta p)^2_{éq}}{2M} = -\hbar\Gamma \frac{(\delta^2 + \Gamma^2/4)}{8\delta\Gamma} \tag{72}$$

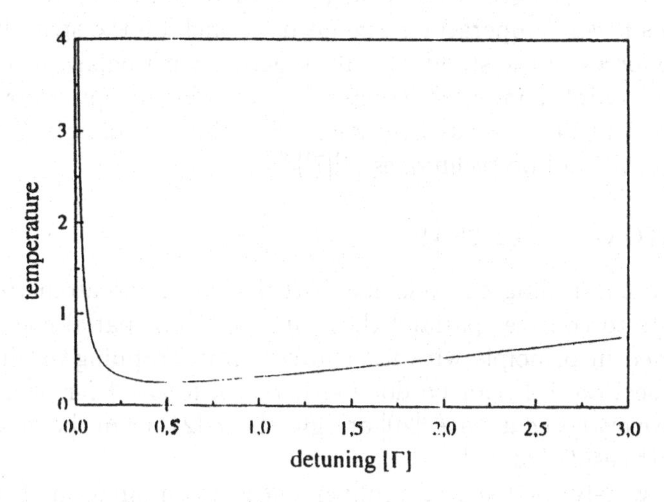

figure 6

The minimum temperature is hence :

$$k_B T_{\min} = \frac{\hbar\Gamma}{4} \tag{73}$$

which is obtained for $\delta = -\frac{\Gamma}{2}$.

Remark :

A precise calculation needs to take into account the fluctuations of the number of fluorescence cycles and not only the random direction of the spontaneous emitted photons and the three dimensional aspect of the random walk, slightly changed the numerical value of the above result, yielding in a one dimension configuartion a so-called Doppler limit of

$$k_B T^{(1D)}_{Dopp} = \frac{\hbar\Gamma}{3} \tag{74}$$

and in a standard three dimensional situation a Doppler tempreature of:

$$k_B T_{Dopp} = \frac{\hbar\Gamma}{2} \tag{75}$$

For Rubidium atoms this Doppler temperature is :

$$T_{Dopp}^{Rb} = \frac{1}{k_B}\frac{\hbar\Gamma}{2} = 120\mu K \tag{76}$$

These are extremely low temperatures, below what has been obtained by other techniques before. In addition, experiments by W.D.Phillips and coworkers in 1988 [9] have resulted in even lower temperatures. In order to explain this lower temperatures one needs to include the spatially modulated dipole forces in the standing wave together with polarization effects and the more complex internal structure of the atoms (optical pumping) [7][8]. But we will not discuss here such "Sisyphus" cooling[18] or other elegant sub-recoil cooling techniques [4][7][8].

2.5. MAGNETO-OPTICAL TRAP

In addition to controlling the velocity distribution of the atoms (cooling) one also wants to confine spatially these atoms. Such spatial confinement can be achieved in principle with laser forces only. Trapping in dipole potentials (see section 2.2) can be done either in a focused far off resonant trap[19], above evanescent waves[20] or light sheets[21] or in the much studied optical lattices[22][23][24].

The most widely used spatial confinement is a combination of the light pressure forces and the Zeeman effect. This so-called magneto-optical traps have first been observed in 1985[2]. As we have seen in section 2.3 the resonance frequency of the atoms are shifted by $0, \pm\frac{\omega_c}{2}$, depending on the polarization of the incident light. Consider now two counterpropagating laser fields with opposite circular polarizations placed in a gradient of a magnetic field along Oz : $B(z) = \frac{\partial B}{\partial z} z$ (figure 7).

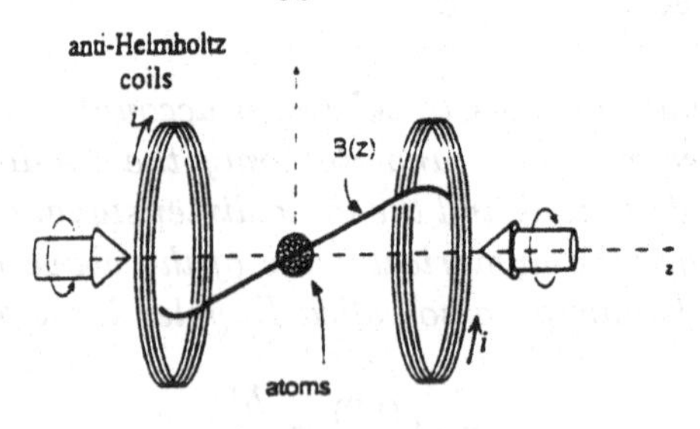

figure 7

Adding the two light pressure forces with spatial dependant shifts of the resonances, one gets a similar result to the case of the Doppler cooling :

$$\vec{f}_{tot}(z) = \hbar \vec{k_1} \frac{\Gamma}{2} \frac{I_{inc}}{I_{sat}} \left(\frac{\Gamma^2/4}{(\delta - \kappa z)^2 + \Gamma^2/4} + \frac{\Gamma^2/4}{(\delta + \kappa z)^2 + \Gamma^2/4} \right) \tag{77}$$

with $\kappa = \left|\frac{\partial B}{\partial z}\frac{\omega_c}{2}\right|$. Instead of the Doppler shift of eq. 63 we now have a spatial shift of the resonances. If one wants to get a confining force ($\vec{f}_{tot}(z) \cdot \vec{r} < 0$) in the case of negative detuning around $z = 0$ one needs to assure that the beam propagating along $+Oz$ shifts into resonance for $z < 0$ whereas the beam propagating along $-Oz$ shifts into resonance at $z < 0$. The Zeeman effect in some way now selects the laser with which the atoms interacts: if the atoms is placed on the right side of the trap ($z > 0$) then the Zeeman effect selects the $\sigma^{(-)}$ polarized laser coming from the right pushing the atom back to $z = 0$. On the contrary an atoms on the placed left of the trap will be pushed to the trap center by the $\sigma^{(+)}$ polarized laser coming from the left (figure 8).

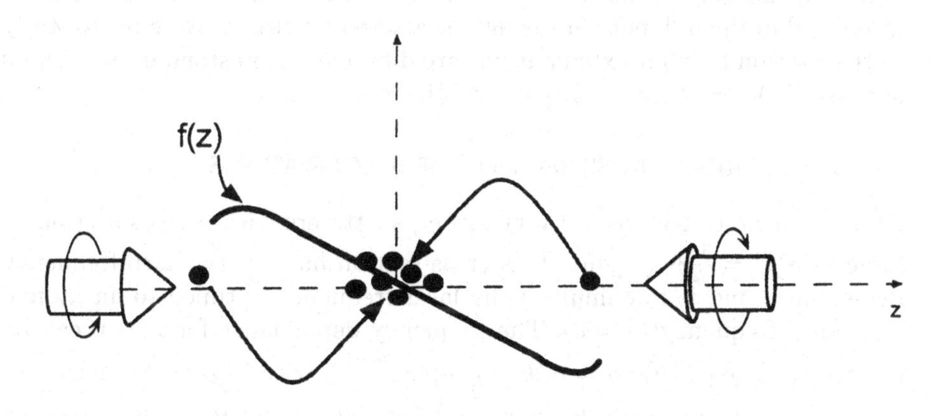

figure 8

This mechanism is very robust and can be extended to a three dimensional situation. The adequate magnetic field gradients are usually obtained by two coils in an anti-Helmholtz configuration (opposite currents in two coils separated by twice their radius). Another very important factor for the success of this traps is that one can confine and cool the atoms with the same laser light by choosing a negative detuning. The total force can then be written as :

$$\vec{f}_{tot}(z) = \hbar\vec{k}_1 \frac{\Gamma}{2} \frac{I_{inc}}{I_{sat}} \left(\frac{\Gamma^2/4}{(\delta - kv_z - \kappa z)^2 + \Gamma^2/4} + \frac{\Gamma^2/4}{(\delta + kv_z + \kappa z)^2 + \Gamma^2/4} \right) \tag{78}$$

An expansion around the trap center for small velocities gives an damped oscillator type equation :

$$f_z(v_z) = -\gamma m v_z - \beta z \tag{79}$$

With such magneto-optical traps it is now possible to trap more than 10^8 atoms in a volume of a few mm^3 and to cool them down to a the order of $100\mu K$. As the realization of such magneto-optical traps is very robust in respect to laser intensity fluctuations and other experimental imperfections, it is now a starting point of many new experiments in atomic physics.

3. Atoms as scatterer of light

In the preceding section we have studied the effect of atom-laser interaction in respect of its effect on the atomic momentum and position distribution. This is the main purpose of the community of laser manipulation of atoms. When considering atoms as scatterers, one will on the other side be more interested in the influence of the light scattered by atoms. We want to study in this section to what extend atoms are different ¿from standard scatterers such as TiO_2 or other Rayleigh and Mie scatterers.

3.1. SCATTERING CROSS SECTION OF SINGLE ATOMS

Atoms can be considered as very strong scatterers with a cross section as large as $\sigma_{at}^{res} = \frac{3\lambda_{at}^2}{2\pi}$ (eq.46). This cross section has a very sharp frequency dependance and is maximum if the laser frequency is tuned to an atomic resonance frequency ($\delta = 0$). The frequency dependance for $\omega \to 0$ can be deduced from eq. (10) : $\sigma = \alpha'' k$. By replacing Γ with $\Gamma\frac{\omega^2}{\omega_{at}^2}$.(see remark 2 of section 2.2) one recovers the known ω^4 dependance for Rayleigh scatterers at low frequencies. As the frequency increases one usually gets into the Mie regime with more or less sharp resonances and a cross section with tends to twice the geometrical cross section :

$$\sigma \underset{\omega \to \infty}{\longrightarrow} 2\pi r^2 \tag{80}$$

From the scattering cross section one defines a mean free path l_{MFP} as the average distance between two successive scattering events :

$$l_{MFP} = \frac{1}{n\sigma} \tag{81}$$

where n is the density of scatterers in the medium.

In this respect one can consider the atomic resonance frequency as a first sharp resonance of the scattering cross section. In real atoms there are many transitions corresponding to different resonances in the cross section. An atomic spectrum is thus equivalent to a frequency dependant cross section. In these terms the finesse of a resonance in the case of atoms is extremely large. For Rubidium atoms e.g. one has :

$$\frac{\omega}{\delta\omega} = \frac{\omega}{\Gamma} \tag{82}$$

With $\omega = \frac{2\pi c}{\lambda}$ one gets for $\lambda = 780nm$ and with a natural linewidth of $\Gamma = 6MHz$ a finesse of

$$\frac{\omega}{\delta\omega} \simeq 6\ 10^7 \tag{83}$$

Such large values have been obtained in other types of scatterers[25][26] but in order to make use of this large values of the finesse in a multiple scattering experiment one needs to be able to produce many scatterers with the same resonance frequency (with a precision of the width of the resonance). This seems to be unrealistic with scatterers such as microspheres. Atoms of the same element however have all exactly the same resonant frequency and are in this respect an unique way of studying high finesse cross section in multiple scattering[27].

If one does not only want to study an extinction cross section (attenuation of the incident beam) it is also of interest to analyze the differential scattering cross section and the absorption cross section. Let us only consider the far field radiation by a dipole excited by an incident electric field of a linear polarized laser $E_0 \overrightarrow{\varepsilon_x} \exp(-i\omega_L t)$. Using the results of section 2.1, one obtains a dipole $\vec{d}$:

$$\vec{d} = \varepsilon_o \left(\alpha' + i\alpha"\right) E_0 \overrightarrow{\varepsilon_x} \exp\left(-i\omega_L t\right) \tag{84}$$

The scattered field by a dipole is given by :

$$\vec{E}_{scat} = \frac{1}{4\pi\varepsilon_0}\frac{\exp(ikr)}{kr}k^3\left\{\left(1+\frac{i}{kr}-\frac{1}{(kr)^2}\right)\vec{d}+\left(-1-\frac{3i}{kr}+\frac{3}{(kr)^2}\right)\left(\vec{e}_r\cdot\vec{d}\right)\vec{e}_r\right\} \tag{85}$$

which in the far field limit gives :

$$\vec{E}_{scat} = \frac{1}{4\pi\varepsilon_0}\frac{\exp(ikr)}{kr}k^3\left\{\vec{d} - \left(\vec{e}_r \cdot \vec{d}\right)\vec{e}_r\right\} \tag{86}$$

or :

$$\vec{E}_{scat} = \frac{1}{4\pi\varepsilon_0}\frac{\exp(ikr)}{kr}k^3\left\{\left(\vec{e}_r \times \vec{d}\right) \times \vec{e}_r\right\} \tag{87}$$

The scattered field will not always be in phase with the incident field, because of the complex polarizability α of the dipole. The forward scattered field interferes with the incident field. In this direction an imaginary polarizability (on resonance $\alpha = \alpha''$) will correspond to a destructive interference as the dipole is then in quadrature phase compared to the incident field. For a collection of scatterers this can be compared to a complex part of the index of refraction and to an attenuation of the incident beam. In any other direction however there will be no interference with the incident beam and only the scattered field has to be considered, with a frequency dependant phase shift.

3.2. MULTIPLE SCATTERING SAMPLES IN ATOMIC PHYSICS

Atom vapors can be used for multiple scattering experiments and both for room temperature and laser cooled samples an optical thickness larger than one can be obtained. One has to pay care to the various situations one can produce in atomic vapors. As examples let us consider 3 accessible situations.

First, an oven of sodium atoms[28] (actually many other elements can be used) at a temperature of $\simeq 200°$ Celsius can yield a density of $10^{13} atoms/cc$ and an with a sample tickness of $L = 1cm$ on-resonant optical thickness of several 10^3. In this case however the Doppler shifts of the atoms leads to an inhomogeneous broadening of the cross section. For a detuning of 1 GHz however, the Doppler broadening can be neglected. The optical thickness is reduced but remains still larger than unity. In this situation one might have to consider dependant scattering events[27][42][43] which will modify the multiple scattering properties.

A magneto-optical trap of 10^{11} Rubidium atoms in a volume at a few $100\mu K$ can be obtained in a volume of a few mm^3 An optical thickness of several 10 has been obtained in several experiments, in particular those who aim at reaching Bose-Einstein condensation of cold atoms. This samples have the advantage of negligible Doppler effect. If the cooling and trapping light is however present, then the multiple scattering leads to a repulsion between the atoms inducing a correlation between the position of scatterers.

A Bose-Einstein condensate of cold atoms[5][6] has considerable higher densities of atoms, of the order of a few $10^{14} atoms/cc$. In this case however the sample does not have the same local fluctuations in the dielectric constant as all atoms are in the same quantum state. This is an extreme case of dependant scattering and transparency. Optical excitation of such condensates is subject to many recent theoretical investigation[6].

3.3. DWELL TIME

Another aspect of scattering by a atom with an internal resonance is that there will be a frequency dependant phase shift of the scattered light. This effect can also be put in the time domain as a retardation effect. Different formulations of such retardation times are being used in the community of multiple scattering, such as Wigner time, delay time[27][29][30][31] [32]. We will use a simple delay time interpretation by defining the dwell time as :

$$\tau_D = \frac{\partial \varphi(\omega)}{\partial \omega} \tag{88}$$

For a slab of glass with an index of refraction of $n = 1.5$ e.g. the transmitted light through a thickness L is phase shifted by :

$$\exp(i\varphi(\omega)) = \exp\left(i\frac{\omega}{c}nL\right) \tag{89}$$

In this simple case the dwell time is :

$$(\tau_D)_{glass} = \frac{\partial \varphi(\omega)}{\partial \omega} = \frac{nL}{c} \tag{90}$$

and is n times longer than through free space. In a simple approach this would be the travel time for a light pulse through the sample. Applying this idea to a delay time for light scattering by a damped dipole, we will get with $\alpha = |\alpha| \exp(i\varphi_\alpha(\omega))$:

$$\tan(\varphi_\alpha) = \frac{\alpha''}{\alpha'} = -\frac{\Gamma}{2\delta} \tag{91}$$

a dwell time of :

$$(\tau_D)_{dip} = \frac{2}{\Gamma}\frac{(\Gamma^2/4)}{\delta^2 + (\Gamma^2/4)} \tag{92}$$

For Rubidium atoms e.g. with $\Gamma^{-1} = 25ns$ one has an on resonance ($\delta = 0$) delay time of

$$(\tau_D)_{res} = \frac{2}{\Gamma} = 50ns \tag{93}$$

which would correspond to $15m$ travel distance in free space! When using the radiative transfer equation in multiple scattering, various velocities have to be defined, such as group velocity v_g and transport velocity v_E which can strongly depend on the dwell time[27][33]. Atomic samples with large dwell times thus seem to be an good testing ground for studying the influence of dwell time in multiple scattering experiments.

3.4. COHERENT BACKSCATTERING OF LIGHT

When discussing sample parameters for multiple scattering experiments additional aspects have to be studied. One can for example look for a difference between the mean free path l_{MFP} (mean distance between two scattering events) and the transport mean free path l^* (mean distance to loose the initial direction of propagation) :

$$l^* = \frac{l_{MFP}}{1 - \langle \cos\theta \rangle} \tag{94}$$

where θ is the angle between the incident and scattered light. Even though the dipole radiation pattern is not isotropic, one has $\langle \cos\theta \rangle = 0$ and hence $l^* = l_{MFP}$.

One particular effect in multiple scattering has been the subject of detailed studies in recent years : coherent backscattering of waves by a random medium. Scattering of wave by a static random medium of scatterers gives rise to a so-called speckle pattern[34]. Such speckle pattern are observed whether the medium is optically thin, with single scattering being dominant, or optically thick in the multiple scattering regime. When averaging over different realizations of the scatterer distribution, the scattered intensity will have a smooth angular dependance. The main argument in this explanation is that the detected field is the coherent sum of electric fields with random phases :

$$\vec{E} = \sum_j \vec{E}_j \exp(i\varphi_j) \tag{95}$$

The average detected intensity will then be :

$$\langle I \rangle = \left\langle \left| \sum_j \vec{E}_j \exp(i\varphi_j) \right|^2 \right\rangle \tag{96}$$

A first approach will be to suppose the interference terms to average to zero and thus obtain :

$$\langle I \rangle = \left\langle \sum_j I_j \right\rangle \tag{97}$$

This argument however is neglecting the particular case of backscattering. Let us group two by two all scattering paths giving a contribution to the detected field, by taking for each path its reverse path (figure 9) :

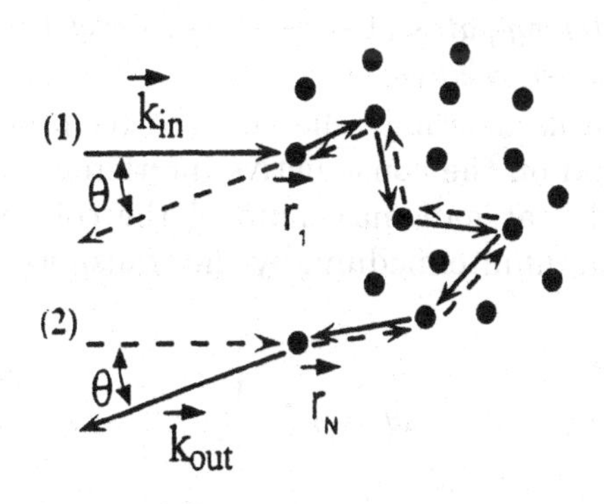

figure 9

Assuming that the dephasing for the forward and the reverse path are identical[35] the phase difference of the two paths will be :

$$\Delta\varphi = \left(\vec{k}_{inc} + \vec{k}_{out}\right)\left(\vec{r}_1 - \vec{r}_N\right) \tag{98}$$

One can thus see that if the relative position of the scatterers is randomly changing the phase difference is generally also a random parameter and interference terms in eq. (96) will be cancelled. However for the particular case of backscattering :

$$\vec{k}_{inc} + \vec{k}_{out} = 0 \tag{99}$$

the two reverse paths have exactly the same phase shift regardless of the position of the scatterers. Always having such a constructive interference will give rise to an enhanced backscattering peak when averaging over the sample distribution (figure 10).

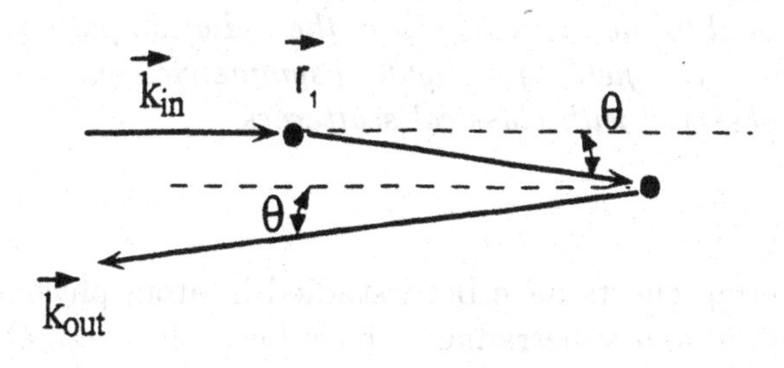

figure 10

Remark :

Note that for a static sample one does not always have a maximum intensity in backward direction. Indeed, even though paths interfere constructively two by two in this direction, the relative phase shift between the various multiple scattering paths ($1-2-....-N$ and $1'-2'-....-N'$ e.g.) do not have a fixed phase relation.

This enhanced backscattering relies on the reciprocity of the reverse paths of scattering and on the constructive interference between these two paths. The total width at half maximum of the coherent backscattering cone is given for a half infinite medium by the transport mean free path[36] [37][38][39] :

$$\Delta\theta = 0.7\,\frac{1}{kl^*} \tag{100}$$

where k is the wavevector in the scattering medium. This results usually holds for kl^*. When kl^* become of the order of unity, the so-called Joffe-Regel criterium for strong localization will be obtained[27][40]. The coherent backscattering cone is a signature of interference effects in multiple scattering. It has been observed with many classical scatterers, but up to date not in atomic samples.

Remark :

One has to be careful when using this result in e.g. the width of a coherent backscattering cone. If all scattering orders ($N = 1, .., \infty$) contribute to the coherent backscattering cone, such results might apply. Consider however a case where only double scattering contributes to the detected signal. In such a situation, a symmetric radiation pattern but with a very narrow peak in the forward and backward direction will give $\langle\cos\theta\rangle \ll 1$ double scattering contributing to a cone will arise from one forward and one backward scattering (see figure 10).

The width $\Delta\theta$ of the coherent backscattering cone in the plane of incident for a linear polarized light will then be larger than $0.77kl^*$. Using $\langle|\cos\theta|\rangle$ for a dipole radiation pattern $I(\theta) \propto \cos^2(\theta)$ one would get $\widetilde{l} \simeq \frac{2}{3} l_{MFP}$. In a plane orthogonal to the incident plane, the radiation pattern of a dipole is uniform ($I(\varphi) = cte$) yielding a slightly asymmetric cone. Such anisotropy effects can be observed with classical scatterers.

3.5. ALBEDO

Multiple scattering effects have been studied in atom physics[41] [42] and radiation trapping and superradiance have been observed. One important aspect in multiple scattering of light by atoms is the frequency spectrum of the scattered light. The radiation spectrum has a complex shape[50] [51]

[52] and does not only show an elastic component at the drive frequency. In the case of a two level atom an inelastic component appears for larger intensities as the upper state population becomes more and more important and features such as the Mollow triplet have been observed[44]. These inelastic components have a spectral width of the order of the natural linewidth of the atomic transition and it seems interesting to investigate what will be the influence of these components on multiple scattering properties such as the coherent backscattering.

For a two level atom the total scattering rate is given by :

$$\Gamma' = \frac{\Gamma}{2}\frac{s}{1+s} \tag{101}$$

where Γ is the width of the excited state and s the saturation parameter

$$s = \frac{I_{inc}}{I_{sat}}\frac{\Gamma^2/4}{\delta^2+\Gamma^2/4} \tag{102}$$

This total rate can be separated in an elastic component Γ'_{elas}, having the same frequency spectrum as the incident laser, and an inelastic component Γ'_{inelas} with a broadened spectrum and a triplet structure. One can show that [45] (figure 11) :

$$\Gamma'_{elas} = \frac{\Gamma}{2}\frac{s}{(1+s)^2} \tag{103}$$

$$\Gamma'_{inelas} = \frac{\Gamma}{2}\frac{s^2}{(1+s)^2} \tag{104}$$

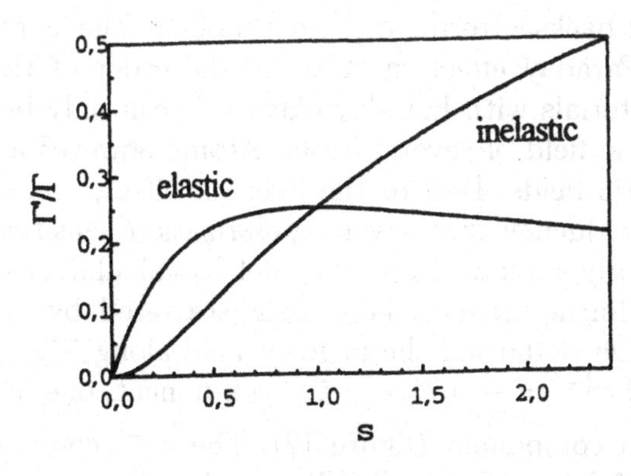

figure 11

If we suppose that only the elastically scattered light contributes to a coherent backscattering cone (a hypothesis which is under present investigation) one could define an equivalent of an albedo for standard scatterers by :

$$a = \frac{\Gamma'_{elas}}{\Gamma'_{elas} + \Gamma'_{inelas}} = \frac{1}{1+s} \tag{105}$$

which decreases as the transition rate of the atom becomes saturated. Several parameters will be of interest in such a study. Take for example an incident light with a broad spectrum (of $\Delta v = 6MHz$ e.g.). The coherence length for such a laser is of the order of $\Delta x \simeq \frac{c}{\Delta v} \simeq 8m$ which will have to be compared to the length scales of the problem. Another aspect will be the time dependance of the scattered light. In time correlation experiments[24] for example this spectral width has to be compared to the detection bandwidth, which varies by orders of magnitude for standard CCD cameras or photomultipliers.

3.6. FARADAY EFFECT

One condition to observe the coherent backscattering of light is reciprocity ("I can see you, so you can see me"). Several factors can break reciprocity. One of them is the Faraday effect. Experiments in strong magnetic fields have shown that the coherent backscattering cone does indeed vanish in increasing magnetic field[46]. The relevant parameter is the Verdet constant V and the magnetic field B. A linear polarized light will see its axes of polarization tilted by θ after propagating for a distance L in a Faraday medium by an angle :

$$\theta = VBL \tag{106}$$

For the coherent backscattering cone to disappear, the length scale $l_F = \frac{1}{VB}$ given by the Faraday effect must be of the order of the mean free path. For usual materials with Faraday effect this can only be achieved in very strong magnetic field of several Tesla. Atoms however are extremely sensitive to magnetic fields. Due to the Zeeman effect, one can split the atomic resonance frequency into several polarization sensitive lines. One can use the elastically bound electron model to calculate the tilt of the polarization of an linear incident laser light scattered by an atom in a magnetic field[14]. Decomposing the incident field along $\vec{e}_x$ into a $\sigma^{(+)} = \frac{1}{\sqrt{2}}(\vec{e}_x + i\vec{e}_y)$ and $\sigma^{(-)} = \frac{1}{\sqrt{2}}(\vec{e}_x - i\vec{e}_y)$ component, one has a different phase shift for each component (figure 12). The $\sigma^{(+)}$ component has an effective detuning of $\delta^{(+)} = \delta - \mu_B B$, whereas the $\sigma^{(-)}$ component has an effective detuning of $\delta^{(-)} = \delta + \mu_B B$. Neglecting the modification of the radiation pattern for the two components[47] this leads to a phase shift by the scattering event for the two components :

$$\tan\left(\varphi^{(+)}\right) = \frac{\alpha^{(+)''}}{\alpha^{(+)'}} = -\frac{\Gamma}{2\delta^{(+)}}$$

$$\tan\left(\varphi^{(-)}\right) = \frac{\alpha^{(-)''}}{\alpha^{(-)'}} = -\frac{\Gamma}{2\delta^{(-)}}$$

In the limit $\Gamma \ll \mu_B B \ll \delta$ one has small phase shifts $\varphi^{(+)}$ and $\varphi^{(-)}$ and one gets :

$$\varphi^{(+)} - \varphi^{(-)} \simeq -\frac{\Gamma}{2\delta^{(+)}} + \frac{\Gamma}{2\delta^{(-)}} \simeq -\frac{\Gamma}{2\delta}\left(1 - \frac{\delta - \mu_B B}{\delta + \mu_B B}\right) \tag{107}$$

$$\varphi^{(+)} - \varphi^{(-)} \simeq -\frac{\Gamma}{\delta}\frac{\mu_B B}{\delta} \propto \frac{B}{\delta^2} \tag{108}$$

But one can also try to get a maximum phase difference by choosing the detuning such as to have a resonant scattering for one component ($\varphi^{(+)} = \frac{\pi}{2}$) whereas the other component is out of resonance ($\varphi^{(-)} \simeq 0$).

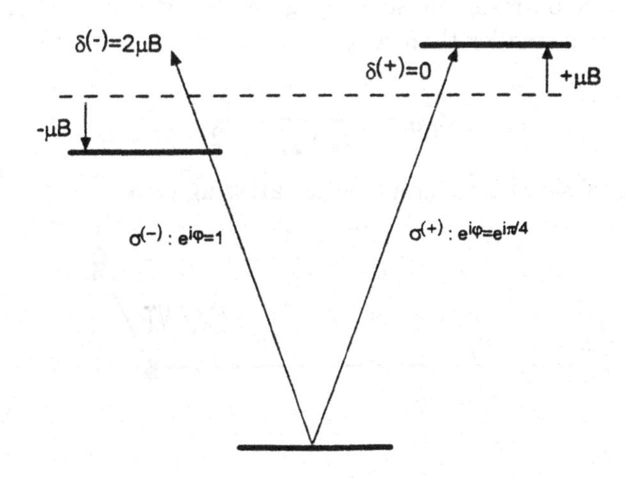

figure 12

One can thus get large phase shift for each scattering event realizing effectively :

$$l_F = l \tag{109}$$

having a considerable Faraday effect for each scattering event. For cold atoms it is possible with a magnetic field of a few gauss to create a Zeeman shift of several MHz. It is thus possible to resolve the two Zeeman components and induce such large phase shifts. For atoms at room temperature, the Zeeman shift would need to compete against the Doppler broadening, of the order of several 100MHz. In this case magnetic fields of a 1000gauss (0.1Tesla) would be needed to obtain similar results.

3.7. MOTION OF THE ATOMS

Another effect can destroy reciprocity in multiple scattering experiments. If the motion of the scatterers is fast compared to the time for the scattered wave to pass through the medium, one path and the reverse paths will experience different configurations, resulting in a "dynamic" break down of reciprocity. For non resonant scatterers, this requires extremely fast motion, almost relativistic particles[48]. In the case of quasi-resonant scattering by atoms however, the time scale is considerably increased by the large dwell time. Taking as a criterium for the break down of the coherent backscattering cone that each scatterer has moved by one wavelength during that typical time scale (figure 13):

$$\Delta x = v\tau \sim \lambda \tag{110}$$

and taking a on-resonant scattering dwell time of $(\tau_D)_{res} = \frac{2}{\Gamma}$, one requires velocities smaller than v_{crit} :

$$v_{crit} \sim \frac{\lambda}{(\tau_D)_{res}} = \frac{\lambda\Gamma}{2} \tag{111}$$

in order to observe a coherent backscattering cone.

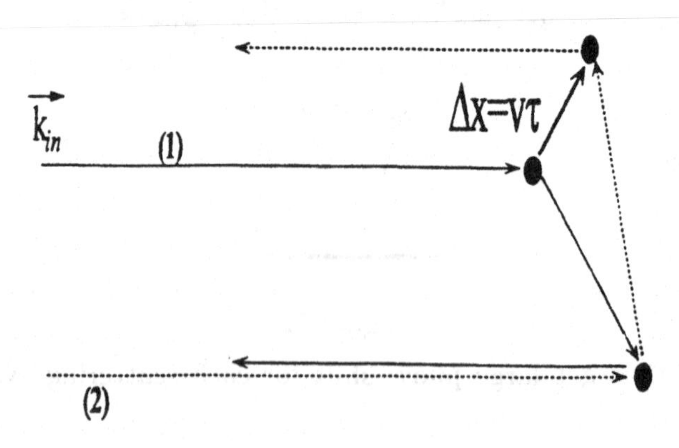

figure 13

In terms of Doppler broadening this corresponds to :

$$kv_{crit} \sim \Gamma \tag{112}$$

This shows that using a resonant laser light (to increase the optical thickness of the sample and to reduce the mean free path) on a dilute atomic gas, laser cooled atoms should to be used. For Rubidium atoms e.g. the cinetic energy required by this condition (112) needs to be smaller than $T_{crit} = 0.25K$.

One could however consider the possibility of using atomic gases at room temperature. But in order to fulfill (111) one would have to detune the laser frequency from resonance in order to lower the dwell time. This would yield the condition :

$$\tau_D = \frac{2}{\Gamma}\frac{(\Gamma^2/4)}{\delta^2 + (\Gamma^2/4)} \lesssim \frac{\lambda}{v} \tag{113}$$

or

$$\left(\frac{\delta}{\Gamma}\right)^2 \gtrsim \frac{kv}{\Gamma} \tag{114}$$

On the other hand, increasing the detuning will decrease the scattering cross section and hence the optical thickness D of the sample. If one requires the optical thickness $D(\delta) = n\sigma(\delta)L$ to be larger than unity, one has to fulfill :

$$D(\delta) = D(0)\frac{(\Gamma^2/4)}{\delta^2 + (\Gamma^2/4)} \gtrsim 1 \tag{115}$$

or

$$\left(\frac{\delta}{\Gamma}\right)^2 \lesssim D(0) \tag{116}$$

These two conditions (114) and (116) can be fulfilled simultaneously if

$$D(0) \gtrsim \frac{kv}{\Gamma} \tag{117}$$

i.e. if the optical thickness on resonance of the medium is larger than the Doppler broadening in units of Γ. It seems to be possible to be realize such situations in hot atomic vapors[28], but up to now no coherent backscattering of light by either cold or hot atoms has been reported.

Another aspect to be considered when using cold atoms is the possible feedback between the position and velocity of the atoms and the scattered light. Such atoms are extremely sensitive to spatial light field variations and if the scattered light field yields dipole forces of the order of the cinetic energy of the atoms, one has to take into account such feedback. Depending on the detuning of the laser it could then be possible that the atoms tend to stabilize around maxima of electric field intensity (red detuning) or minima (blue detuning). This would lead to spatial structures with an optical feedback and could be of interest in spatio-temporal structure formation[49].

3.8. INTERNAL STRUCTURE

The internal structure of the atoms makes it a complex and rich system to study. At the beginning of the laser cooling of atoms, a simple two level model gave first ideas what could be achieved. The complex internal structure of the atoms gave additional parameters to adapt (polarization of the laser in that case) and yielded in better results than previously thought to be achievable. In the case of laser cooling of atoms, it was experimental results by W.Phillips[9] which first showed that the internal structure could have such a dominant effect.

In the case of multiple scattering experiments it also seems obvious that the internal structure will play an important role. If one prepares the atomic sample by optical pumping into one Zeeman sublevel, the scattering cross section will have a strong polarization dependance and the radiation pattern (the phase function) will also be affected. Aligning all atomic magnetic moments in one direction should give rise to an anisotropic scattering matrix, as is expected for liquid crystals for example.

Another important consequence of the internal structure is the possibility of Raman scattering : in one single scattering event, the atoms can for example be excited by a linear polarized light and scatter a circular polarized light (figure 14).

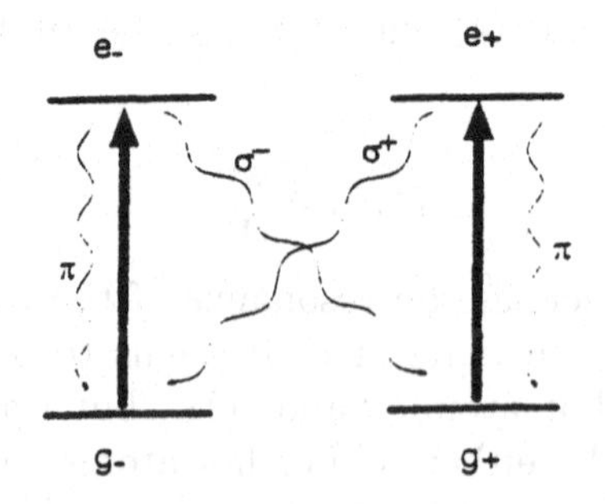

figure 14

In addition to the two level case, where one has to distinguish the elastic and the inelastic components, Raman scattering (which also a has a broad and narrow spectral component[50][51][52]) might behave differently in respect to interference properties with a reference beam. It thus seems to be interesting to investigate what the influence of Raman scattering will be on multiple scattering experiments.

3.9. GAIN MEDIA

Another interesting feature for atoms used in multiple scattering experiments is the possibility of producing a situation with a population inversion leading to gain. It is not necessary for this purpose to excite most atoms into the excited state. As shown by pump/probe experiments by detuning the probe frequency in respect to a pump (trapping) frequency gain or absorption can arise from stimulated Raman and Rayleigh scattering[53]. Random media with gain have been in the focus of theoretical and recent experimental work[54][55][56][57]. Again it seems that if it is possible to observe multiple scattering effects, such as the coherent backscattering cone, in cold atomic vapors, new experimental investigations will be accessible.

4. Atoms as matter waves

In the preceding part of this article, we have discussed the scattering properties of atoms. The atoms were taken as the scatterers and the electromagnetic field as the wave which is scattered. As already mentioned, in case of atomic scatterers, it is not always possible to consider the atomic sample to be independent of the scattered light, as spatial correlations and large dipole forces can appear.

But there is another possibility of using cold atoms in scattering experiments. One can consider the atoms to scatter off a regular or random potential which can be formed by the light shift induced by the electromagnetic field. We then have to consider the atoms as matter waves scattered by a light potential.

4.1. HEISENBERG INCERTITUDE PRINCIPLE

A result from quantum mechanics is that one can associate to a particle with mass M and a velocity v a so-called De Broglie wavelength $\lambda_{DB} = \frac{h}{Mv}$. The larger the velocity if the atom, the smaller its De Broglie wavelength. For room temperature atoms, typical De Broglie wavelengths are of the order of 0.1Å smaller than the distance between the electron and the nucleus. But even when the De Broglie wavelength is larger than 1Å, i.e. if the center of mass of the atom is delocalized over a distance larger than the atomic scale, the waveproperties of the atoms do usually not need to be considered and one can used the limit of "geometrical optics". Only if effects on the scale of the De Broglie wavelength are considered does this approximation fall down. This is in particular the case for ultracold (subrecoil) atoms[4][7][8], where the De Broglie wavelength is larger than the optical wavelength. When considering scattering of atoms by light potentials, hot atoms can be considered as classical particles moving around in a random potential. Note that even on this situation interference effects, such as achromatic fringes, can take place. But the wave properties of ultracold atoms are much more striking. Diffraction of matter waves and interferometers with cold atoms have been developed in the last years. In this context it is important to distinguish between the spread of the wavepacket and coherence length. Starting from a minimal wavepacket with $\Delta x \Delta p_x \simeq \frac{\hbar}{2}$ free space evolution will lead to a spread of the wavepacket. The momentum distribution and hence the coherence length however will not evolve in free space. Even though coherence is maintained over the whole wavepacket, the coherence length is of the order of $\frac{\hbar}{\Delta p}$. In optics this effect can be compared to the wavelength of a laser beam and the size of the beam waist.

4.2. REFLECTION OF ATOMS OFF EVANESCENT WAVE

As an example of scattering by atoms off a light potential we can look at the reflection of atoms off an evanescent wave[7][8][20] (figure 15). The spatially decreasing electric field of a blue detuned laser creating an evanescent wave above a prism can repel atoms released from a magneto-optical trap located above this potential barrier. The dipole potential has to be added up to the gravitational potential and if the laser intensity is large enough atoms can be bounced off the evanescent wave without hitting the prism surface. Several groups have studied evanescent wave mirrors and diffraction in time [60][61] and in space[58][59] have been observed in recent years.

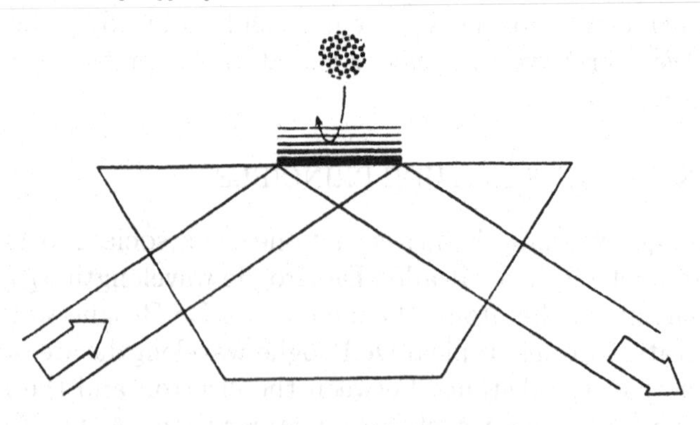

figure 15

4.3. DIFFUSE REFLECTION, SINGLE AND MULTIPLE SCATTERING

Most experiments studying reflection of atoms off an evanescent wave try to obtain specular reflection. In a detailed study of the spatial spread of the reflected atomic cloud, we have found that due to surface imperfections, atoms are not reflected in the specular direction. Interference between the main evanescent light wave and light scattered by surface roughness, leads to a spatially dependant light field with random fluctuations[63]. As the repulsive potential no longer is invariant along the surface of the prism, atoms acquire an additional transverse momentum after the reflection. This effect can be analyzed in a "thin film grating" approach[62], which can be compared to a single scattering situation. Using prism with better surface quality has in the following allowed to observe diffraction of atoms by a spatially modulated evanescent wave[58]. If one want to study multiple scattering of matter waves, one can use a similar setup as used for the atom mirror experiments (figure 16).

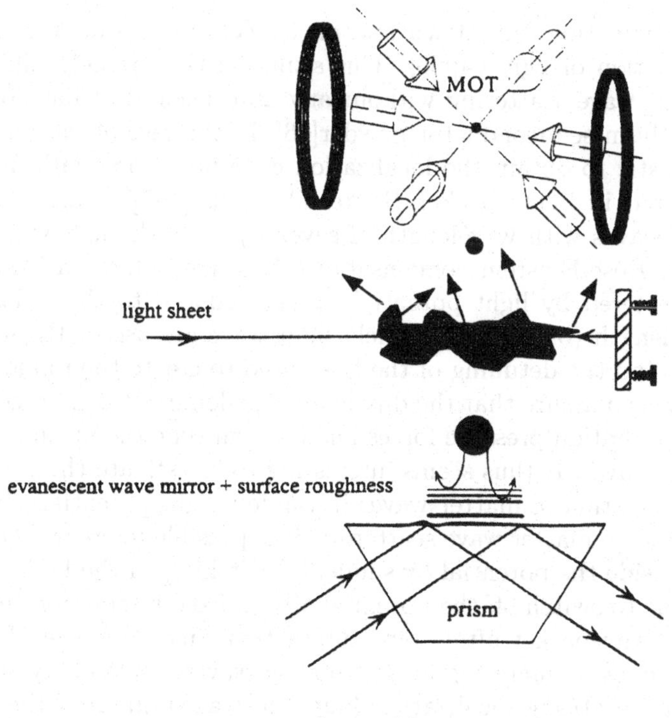

figure 16

In order to be in the multiple scattering regime one either could use a very rough prism surface, or a speckle pattern imaged below the magneto-optical trap. A very precise analysis of the reflected atomic momentum distribution can be obtained by a sequence of Raman pulses. In this configuration looking for coherent backscattering of matter waves does not imply the use of a beamsplitter. The analog for usual coherent backscattering schemes would be a pulsed experiment and putting the detection apparatus in the backward direction after releasing the initial wave pulse.

4.4. JOFFE-REGEL CRITERIUM

In most multiple scattering experiments, the wave scattered by one scatterer can expand into the far field before being scattered by the next particle. In this limit one can consider the multiple scattering as a sequence of single scattering events, which might interfere at some point. This approximation seems to be reasonable if the mean free path is large compared to the wavelength (and if dependant scattering can be excluded). When approaching the case :

$$l_{MFP} \simeq \lambda \tag{118}$$

this approximation falls down and on expects to be one the onset of strong localization of light[64][65]. This situation is extremely difficult to obtain in light wave scattering without any absorption, but have been reported recently in semiconductor powder[66]. In the case of matter waves, it seems realistic to obtain the localization criterium (118) with ultracold atoms scattered in light speckle. Subrecoil cooling[4][7][8] allows to produce matter waves with wavelength of several μm (without having to use techniques of Bose-Einstein condensation). The mean free path for matter waves scattered by light potential can be reduced to the order of an optical wavelength (or less if near field light wave are used). By adapting the intensity and the detuning of the laser used to create the random light potential, one can assure that the dipole force is dominant compared to the spontaneous radiation pressure forces and one can tune the mean free path of the matter waves. It thus seems interesting to investigate the possibility of strong localization of matter waves in random light potentials. In addition, in the case of matter wave scattering, it is possible to create the initial wavepacket inside the potential by suddenly switching on the laser light. It is also possible to switch off the potential after a fixed interaction time and to image the wavepacket after some interaction time. However the most precise techniques of analyzing atomic matter waves is a velocity analysis. The signature of strong localization (and the transition from the diffuse regime to the localization regime) on the momentum distribution remains to be calculated. The dynamics of strong localization, which seems to be difficult to study with light waves, and the impact of non linear effects or of noise are yet other interesting topics to be studied.

5. Acknowledgments

The GDR POAN has allowed for very useful meetings. I would in particular like to thank A.Aspect, D.Delande, G.Labeyrie, R.Maynard, Ch.Miniatura, S.Reynaud, F.de Tomasi, B.v.Tiggelen for many discussion during the preparation of this notes.

References

1. S.Chu, L.Hollberg, J.Bjorkholm, A.Cable, A.Ashkin, Phys.Rev.Lett. **55**, 48 (1985).
2. A.Midgall, J.Prodan, W.Phillips, T.Bergeman, H.Metcalf, Phys.Rev.Lett. **54**, 2596 (1985).
3. P.D.Lett, R.N.Watts, C.I.Westbrook, W.D.Phillips, P.L.Gould, H.J.Metcalf, Phys.Rev.Lett. **61**, 169 (1988).
4. A.Aspect, E.Arimondo, R.Kaiser, N.Vansteenkiste, C.Cohen-Tannoudji, Phys.Rev.Lett. **61**, 826-829 (1988).
5. M.H. Anderson, J.R. Ensher, M.R. Matthews, C.E. Wieman and E.A. Cornell, Science **269**, 198 (1995).
6. http://amo.phy.gasou.edu:80/bec.html/
7. S.Chu, Rev.Mod.Phys. **70**, 685 (1998).
8. C.Cohen-Tannoudji, Rev.Mod.Phys. **70**, 707 (1998).
9. W.D.Phillips, Rev.Mod.Phys. **70**, 721 (1998).
10. J.Weiner, V.Bagnato, S.Zilio, P.Julienne, Rev. Mod. Phys. (in press).
11. C. J. Joachain, Quantum collision theory, (North-Holland, Amsterdam, 1983).
12. D.Boiron, A.Michaud, P.Lemonde, Y.Castin, C.Salomon, S.Weyers, K.Szymaniec, L.Cognet, A.Clairon, Phys.A **53**, R3734 (1996).
13. T.Walker, D.Sesko, C.Wieman, Phys.Rev.Lett. **64**, 408 (1990).
14. G.Grynberg, A.Aspect, C.Fabre, "Introduction aux laser et à l'optique quantique" (Ellipses 1997).
15. C.Cohen-Tannoudji, College de France lectures 1982.
16. T.W.Hänsch, A.Schawlow, Opt. Comm. **13**, 68 (1975).
17. D.Wineland, H.Dehmelt, Bull. Am. Phys. Soc. **20**, 637 (1975).
18. J. Dalibard, C. Cohen-Tannoudji, J.O.S.A. **B 6**, 2023 (1989).
19. J.D.Miller, R.Cline, D.Heinzen, Phys.Rev. **A47**, R4567 (1993).
20. V.I.Balykin, V.S.Lethokov, Y.B.Ovchinnikov, A.I.Sidorov, Phys. Rev. Lett., **60**, 2137 (1988).
21. N.Davidson, H.J.Lee, C.S.Adams, M.Kasevich, S.Chu, Phys.Rev.Lett. **74**, 1311 (1995).
22. G.Birkl, M.Gatzke, I.H.Deutsch, S.L.Rolston, W.D.Phillips, Phys.Rev.Lett. **75**, 2823 (1995).
23. M.Weidemüller, A.Hemmerich, A.Görlitz, T.Esslinger, T.W.Hänsch, Phys.Rev.Lett. **75**, 4583 (1995).
24. C. Jurczak, K. Sengstock, R. Kaiser, N. Vansteenkiste, C.I. Westbrook, A. Aspect, Opt. Commun. **115**, 480 (1995).
25. H.Mabuchi, H.J.Kimble, Opt. Lett. **19**, 749 (1994).
26. J.C.Knight, N.Dubreuil, V.Sandoghdar, J.Hare, V.Lefèvre-Seguin, J.M.Raimond, S.Haroche, Opt. Lett. **20**, 1515 (1995).
27. A.Lagendijk, B.v.Tiggelen, Phys.Rep. **270**,143 (1996).
28. G.L.Lippi, G.P.Barozzi, S.Barbay, J.R.Tredicce, Phys. Rev. Lett. **76**, 2452 (1996).
29. A.Steinberg, P.G.Kwiat, R.Ciao, Phys. Rev. Lett. **71**, 708 (1993).
30. H.M.Brodwosky, W.Heitman, G.Nimtz, Phys. Rev. **A 222**, 125 (1996).
31. E.H.Hauge, J.A.Stövneng, Rev. Mod. Phys. **61**, 917 (1984).
32. V.S.Olkhovsky, E.Recami, Phys. Rep. **214**, 339 (1992).

33. "New Aspects of Electromagnetic and Acoustic Wave Diffusion", Springer Tracts in Modern Physics, vol.144, POAN Research group (ed.), 1998.
34. "Laser Speckle and Related Phenomena", Topcs in Applied Physics, vol 9, J.C.Dainty (ed.)Springer-Verlag (1975).
35. This assumption will break down in the case of a Faraday effect or in the case of moving scatterers, leading to a decrease of the coherent backscattering cone.
36. Y. Kuga and A. Ishimaru, J. Opt. Soc. Am. **A1**, 831 (1984).
37. P.E. Wolf and G. Maret, Phys. Rev. Lett. **55**, 2696 (1985).
38. M.P. van Albada, A.Lagendijk, Phys. Rev. Lett., **55**, 2692 (1985).
39. D. Wiersma, M. van Albada, B. van Tiggelen and A. Lagendijk, Phys. Rev. Lett. **74**, 4193 (1995).
40. A.F.Joffe, A.R.Regel, Progress in semiconductors **4**, 237 (1960).
41. R.H.Dicke, Phys. Rev. **93**, 99 (1954).
42. L.Mandel, E.Wolf "Optical Coherence and Quantum Optics", Cambridge University Press (1995).
43. A.E.Siegman, "Lasers", University Science Books (1986).
44. B.R.Mollow, Phys. Rev. **188**, 1969 (1969). B.R.Mollow, Progress in Optics, vol XIX, 1, (E.Wolf ed.)North Holland (1981).
45. C.Cohen-Tannoudji, J.Dupont-Roc, G.Grynberg, "Processus d'interaction entre photons et atomes" Interéditions 1988.
46. F.A.Erbacher, R.Lenke, G.Maret, Europhysics Letters **21**, 551 (1993).
47. D.Lacoste, B.v.Tiggelen, G.Rikken, A.Sparenberg, JOSA **A15**, 1636, 1998.
48. A.A. Golubentsev, Sov. Phys. JETP **59** , 26 (1984).
49. R.Neubecker, G.L.Oppo, B.Thuerung, T.Tschudi, Phys. Rev. **A52**, 791 (1995).
50. D.Polder, M.F.H.Schuurmans, Phys. Rev. **A14**, 1468 (1976).
51. C.Cohen-Tannoudji, S.Reynaud, J.Phys.**38**, L173 (1977).
52. C.Cohen-Tannoudji, S.Reynaud, J.Phys.B **10**, 365 (1977).
53. P.Verkerk, B.Lounis, C.Salomon, C.Cohen-Tannoudji, J.Y.Courtois, G.Grynberg, Phys. Rev. Lett. **68**, 3861 (1992).
54. V. S. Letokhov, Sov. Phys. JETP **26**, 835 (1968).
55. C. Gouedard, D. Husson, C. Sauteret, F. Auzel and A. Migus, J. Opt. Soc. Am. **B10**, 2358 (1993).
56. N.M.Lawandy, R.M.Balachandran, A.S.L.Gomes, E.Sauvain, Nature **368**, 436 (1994).
57. D.S.Wiersma, M.P.v.Albada, A.Lagendijk, Nature **373**, 203 (1995).
58. A. Landragin, L. Cognet, G.Z.K. Horvath, C. I. Westbrook, N. Westbrook, A. Aspect, Europhys. Lett., **39**, 485 (1997).
59. C. Henkel, K. Mölmer, R. Kaiser, N. Vansteenkiste, C. I. Westbrook, A. Aspect., Phys. Rev. **A55**, 1160 (1996).
60. C. Henkel, A. Steane, R. Kaiser, J. Dalibard, J. Phys. II France **4**, 1877 (1994),
61. A.Steane, P.Szriftgiser, P.Desbiolles, J.Dalibard, Phys. Rev. Lett. **74**, 4972 (1995).
62. C. Henkel, K. Mölmer, R. Kaiser, C. I. Westbrook, Phys. Rev. **A 56**, 1 (1997)
63. A. Landragin, G. Labeyrie, R. Kaiser, N. Vansteenkiste, C. I. Westbrook, A. Aspect, Opt. Lett. **21**, 1591 (1996).
64. Anderson, P. W. Phil. Mag. **B 52**, 505-509 (1985).
65. Ping Sheng, "Introduction to Wave Scattering, Localization, and Mesoscopic Phenomena" (Academic, San Diego, 1995).
66. D.Wiersma, P.Bartolini, A.Lagendijk, R.Righini, Nature **390**, 671 (1997).

WAVE SCATTERING FROM ROUGH SURFACES

M. NIETO-VESPERINAS AND A. MADRAZO
Instituto de Ciencia de Materiales de Madrid,
Consejo Superior de Investigaciones Científicas
Cantoblanco, Madrid 28049, Spain.

Abstract. We discuss the scattering of classical waves from randomly rough surfaces separating two media of different dielectric constant. The boundary conditions leading to a solution to this problem are established. Then, both analytical and numerical procedures are addressed for the field and the angular distribution of mean scattered intensity, these leading to the observation of single and multiple scattering phenomena of current interest. These include the retrieval of evanescent waves, and their underlying subwavelength structure resolution in the near fields scattered from surfaces, as well as the detection of objects hidden under rough surfaces by means of the increase of the enhanced backcattering peak that their presence produce.

1. Introduction

The subject of wave scattering from rough surfaces is of interest in many areas of science such as optics, acoustics, condensed matter, biophysics, nuclear physics or electromagnetism. There exist several reviews on this area [1]-[6]. The analytical approaches to tackle this problem lay in two main cathegories: the Kirchhoff approximation [7] and the small perturbation methods, (both amplitude and phase perturbations [8] - [10]). The known Rayleigh method is related to the latter [11]-[12]. In the last years, however, great progress has been made in the knowledge of the range of validity of the above approximations [13]-[16] as well as in the understanding of multiple scattering phenomena such as *enhanced backscattering* [17], [18], transmission and total internal reflection effects in corrugated dielectric interfaces [19], and angular correlations [20], [21]. This has been achieved by means of numerical simulations based on integral methods, of which

J.-P. Fouque (ed.), Diffuse Waves in Complex Media, 289–317.

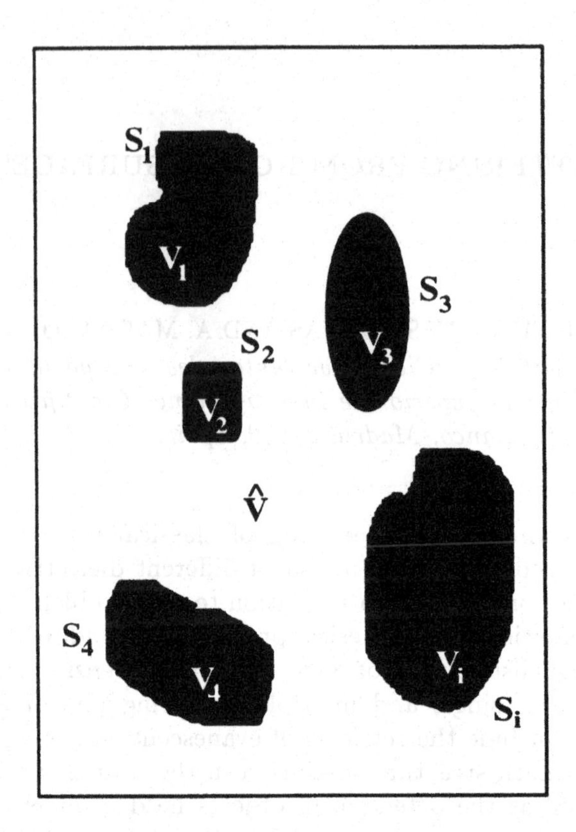

Figure 1. Illustration for a multiply connected domain.

the best known is that based on the second Green's identity or *extinction theorem* [5]. This procedure is of use in dealing with scattering from surfaces of both single and multiple bodies.

Since all the methods and effects quoted above have been extensively reported and reviewed in the current literature. we shall concentrate in this review with two derivations and applications of use in very recent times; these pertain to the theory of near field detection and excitation of surface waves at corrugated surfaces, and to the detection of objects hidden beneath rough interfaces. The formulation of these problems and solutions are based on extensions of the extinction theorem to multiply connected domains.

2. Integral Equations in Multiply Connected Scattering Volumes. The Extinction Theorem

Let us consider an electromagnetic field, with electric and magnetic vectors $\mathbf{E}^{(i)}(\mathbf{r})$ and $\mathbf{H}^{(i)}(\mathbf{r})$, respectively, incident on a medium of permittivity ϵ occupying a volume V, constituted by a multiply connected domain, namely,

V is composed of parts V_1, V_2,..., V_N, each of which is limited by a surface S_1, S_2,..., S_N, respectively, (see Fig. 1). We shall denote by $\mathbf{r}^<$ the position vector of a generic point inside the volume V_i, and by $\mathbf{r}^>$ that of a generic point in the volume $\hat{V}$, which is outside all volumes V_i.

The electric and magnetic vectors of a monochromatic field, satisfy, respectively, the wave equations [5]:

$$\nabla \times \nabla \times \mathbf{E}(\mathbf{r}) - k^2\mathbf{E}(\mathbf{r}) = \mathbf{F}_e(\mathbf{r}), \tag{1}$$

$$\nabla \times \nabla \times \mathbf{H}(\mathbf{r}) - k^2\mathbf{H}(\mathbf{r}) = \mathbf{F}_m(\mathbf{r}), \tag{2}$$

where k is the wavenumber, and $\mathbf{F}_e$ and $\mathbf{F}_m$ are the source terms that characterize the generation of electromagnetic waves. These terms are zero in the volume $\hat{V}$ outside the medium; and inside are:

$$\mathbf{F}_e(\mathbf{r}) = 4\pi[\frac{ik}{c}\mathbf{j}(\mathbf{r}) + k^2\mathbf{P}(\mathbf{r}) + ik\nabla \times \mathbf{M}(\mathbf{r})], \tag{3}$$

$$\mathbf{F}_m(\mathbf{r}) = 4\pi[\frac{1}{c}\nabla \times \mathbf{j}(\mathbf{r}) - ik\nabla \times \mathbf{P}(\mathbf{r}) + k^2\mathbf{M}(\mathbf{r})], \tag{4}$$

In Eqs.(3) and (4) $\mathbf{P}$ and $\mathbf{M}$ are the polarization and magnetization vectors, respectively; and $\mathbf{j}$ represents the electric current density.

Let $\mathcal{G}(\mathbf{r}, \mathbf{r}')$ represent the dyadic Green function:

$$\mathcal{G}(\mathbf{r}, \mathbf{r}') = (\mathcal{I} + \frac{1}{k^2}\nabla\nabla)G(\mathbf{r}, \mathbf{r}'). \tag{5}$$

$G(\mathbf{r}, \mathbf{r}')$ being the outgoing spherical wave: $\exp(ik|\mathbf{r} - \mathbf{r}'|)/|\mathbf{r} - \mathbf{r}'|$.

$\mathcal{G}(\mathbf{r}, \mathbf{r}')$ satisfies the equation:

$$\nabla \times \nabla \times \mathcal{G}(\mathbf{r}, \mathbf{r}') - k^2\mathcal{G}(\mathbf{r}, \mathbf{r}') = 4\pi\delta(\mathbf{r} - \mathbf{r}')\mathcal{I}. \tag{6}$$

For $\mathbf{P}$ and $\mathbf{Q}$ well behaved in a volume V surrounded by a surface S the vector form of Green's theorem reads [22]:

$$\int_V (\mathbf{Q} \cdot \nabla \times \nabla \times \mathbf{P} - \mathbf{P} \cdot \nabla \times \nabla \times \mathbf{Q})dv = \int_S (\mathbf{P} \times \nabla \times \mathbf{Q} - \mathbf{Q} \times \nabla \times \mathbf{P}) \cdot \mathbf{n}ds, \tag{7}$$

$\mathbf{n}$ being the unit outward normal.

Let us now apply Eq. (7) to the vectors: $\mathbf{P} = \mathcal{G}(\mathbf{r}, \mathbf{r}')\mathbf{C}$, ($\mathbf{C}$ being a constant vector), and $\mathbf{Q} = \mathbf{E}(\mathbf{r})$. (The singularity of $\mathcal{G}(\mathbf{r}, \mathbf{r}')$ at the origin is integrable for s-waves, whereas it is not for p-waves, (cf. e.g. Refs. [5] and [23] , this will be addressed in Section 3). Taking Eqs.(1) and (6) into account, we obtain:

$$\int_V \mathbf{E}(\mathbf{r}')\delta(\mathbf{r} - \mathbf{r}')dv = \frac{1}{4\pi}\int_V \mathbf{F}_e(\mathbf{r}') \cdot \mathcal{G}(\mathbf{r}, \mathbf{r}')dv - \frac{1}{4\pi}\mathbf{S}_e(\mathbf{r}). \tag{8}$$

Where $\mathbf{S}_e$ is:

$$\mathbf{S}_e(\mathbf{r}) = \nabla \times \nabla \times \int_S (\mathbf{E}(\mathbf{r}')\frac{\partial G(\mathbf{r},\mathbf{r}')}{\partial \mathbf{n}} - G(\mathbf{r},\mathbf{r}')\frac{\partial \mathbf{E}(\mathbf{r}')}{\partial \mathbf{n}})ds. \tag{9}$$

Eq.(8) adopts different forms depending on where the points $\mathbf{r}$ and $\mathbf{r}'$ are considered. The wavenumber k entering in Eq.(8) is: $k = k_0\sqrt{\epsilon}$ or $k = k_0$ according to whether $\mathbf{r}'$ is in V or in $\hat{\mathrm{V}}$, respectively, $k_0 = 2\pi/\lambda$. By means of straightforward calculations one obtains:

a) If $\mathbf{r}$ and $\mathbf{r}'$ belong to any of the volumes V_i, $(i = 1, 2, ..., N)$, namely, V becomes either of the volumes V_i:

$$\mathbf{E}(\mathbf{r}^<) = \frac{1}{4\pi}\int_{V_i} \mathbf{F}_e(\mathbf{r}') \cdot \mathcal{G}(\mathbf{r}^<,\mathbf{r}')dv - \frac{1}{4\pi}\mathbf{S}_i^{(in)}(\mathbf{r}^<). \tag{10}$$

Where:

$$\mathbf{S}_i^{(in)}(\mathbf{r}^<) = \nabla \times \nabla \times \int_{S_i} (\mathbf{E}_{in}(\mathbf{r}')\frac{\partial G(\mathbf{r}^<,\mathbf{r}')}{\partial \mathbf{n}} - G(\mathbf{r}^<,\mathbf{r}')\frac{\partial \mathbf{E}_{in}(\mathbf{r}')}{\partial \mathbf{n}})ds. \tag{11}$$

In Eq.(11) $\mathbf{E}_{in}$ represents the limiting value of the electric vector on the surface S_i taken from inside the volume V_i. Eq.(10) shows that the field inside each of the scattering volumes V_i does not depend on the sources generated in the other volumes.

b) If $\mathbf{r}$ belongs to any of the volumes V_i, namely, V becomes V_i, and $\mathbf{r}'$ belong to $\hat{\mathrm{V}}$:

$$0 = \mathbf{S}_{ext}(\mathbf{r}^<). \tag{12}$$

In Eq.(12) $\mathbf{S}_{ext}$ is:

$$\mathbf{S}_{ext}(\mathbf{r}^<) = \sum_i \mathbf{S}_i^{(out)}(\mathbf{r}^<) - \mathbf{S}_\infty(\mathbf{r}^<), \tag{13}$$

where:

$$\mathbf{S}_i^{(out)}(\mathbf{r}^<) = \nabla \times \nabla \times \int_{S_i} (\mathbf{E}(\mathbf{r}')\frac{\partial G(\mathbf{r}^<,\mathbf{r}')}{\partial \mathbf{n}} - G(\mathbf{r}^<,\mathbf{r}')\frac{\partial \mathbf{E}(\mathbf{r}')}{\partial \mathbf{n}})ds. \tag{14}$$

In Eq.(14) the surface values of the electric vector are taken from the volume $\hat{\mathrm{V}}$ outside the volumes V_i. The normal $\mathbf{n}$ now points towards inside each of the volumes V_i.

Also, $\mathbf{S}_\infty$ has the same meaning as Eq.(14), but with the surface of integration now being a large sphere whose radius will eventually tend to infinity. It is not difficult to see that $-\mathbf{S}_\infty$ in Eq.(13) is equal to 4π times the incident field $\mathbf{E}^{(i)}(\mathbf{r}^<)$ (cf. Refs. [5] and [24]). Therefore Eq.(13) becomes finally:

$$0 = \mathbf{E}^{(i)}(\mathbf{r}^<) + \frac{1}{4\pi}\sum_i \mathbf{S}_i^{(out)}(\mathbf{r}^<). \tag{15}$$

Eq.(15) is the *extinction theorem* for a multiply connected domain consisting of several bodies each of which is limited by the surface S_i. Note that when this equation is used as a non-local boundary condition, the unknown *sources* to be determined, given by the limiting values of $\mathbf{E}(\mathbf{r}')$ and $\partial\mathbf{E}(\mathbf{r}')/\partial\mathbf{n}$ on each of the surfaces S_i, (cf. Eq.(14)), appear coupled to those corresponding sources on all the other surfaces S_j, $j \neq i$.

Following similar arguments, one obtains:

c) For $\mathbf{r}$ belonging to $\hat{\mathrm{V}}$ and $\mathbf{r}'$ belonging to any of the volumes V_i, namely, V becoming V_i:

$$0 = \frac{1}{4\pi}\int_{V_i} \mathbf{F}_e(\mathbf{r}') \cdot \mathcal{G}(\mathbf{r}^>, \mathbf{r}')dv - \frac{1}{4\pi}\mathbf{S}_i^{(in)}(\mathbf{r}^>), \tag{16}$$

with $\mathbf{S}_i^{(in)}$ given by Eq.(11), this time evaluated in $\mathbf{r}^>$.

d) For both $\mathbf{r}$ and $\mathbf{r}'$ belonging to $\hat{\mathrm{V}}$:

$$\mathbf{E}(\mathbf{r}^>) = \mathbf{E}^{(i)}(\mathbf{r}^>) + \frac{1}{4\pi}\sum_i \mathbf{S}_i^{(out)}(\mathbf{r}^>), \tag{17}$$

Hence, the exterior field is the sum of the fields emitted from each scattering surface S_i with sources resulting from the coupling involved in the ET Eq.(15).

The other important case corresponds to a penetrable, optically homogeneous, isotropic, non-magnetic and spatially nondispersive medium, (this applies to a real metal or a pure dielectric). In this case, Eqs. (10) and (15) become, respectively:

$$\begin{aligned}\mathbf{E}(\mathbf{r}^<) &= -\frac{1}{4\pi k_0^2\epsilon}\nabla \times \nabla \times \\ &\int_{S_i}(\mathbf{E}_{in}(\mathbf{r}')\frac{\partial G^{(in)}(\mathbf{r}^<, \mathbf{r}')}{\partial \mathbf{n}} - G^{(in)}(\mathbf{r}^<, \mathbf{r}')\frac{\partial \mathbf{E}_{in}(\mathbf{r}')}{\partial \mathbf{n}})ds, \end{aligned} \tag{18}$$

$$\begin{aligned}0 &= \mathbf{E}^{(i)}(\mathbf{r}^<) + \frac{1}{4\pi k_0^2}\nabla \times \nabla \times \\ &\sum_i \int_{S_i}(\mathbf{E}(\mathbf{r}')\frac{\partial G(\mathbf{r}^<, \mathbf{r}')}{\partial \mathbf{n}} - G(\mathbf{r}^<, \mathbf{r}')\frac{\partial \mathbf{E}(\mathbf{r}')}{\partial \mathbf{n}})ds. \end{aligned} \tag{19}$$

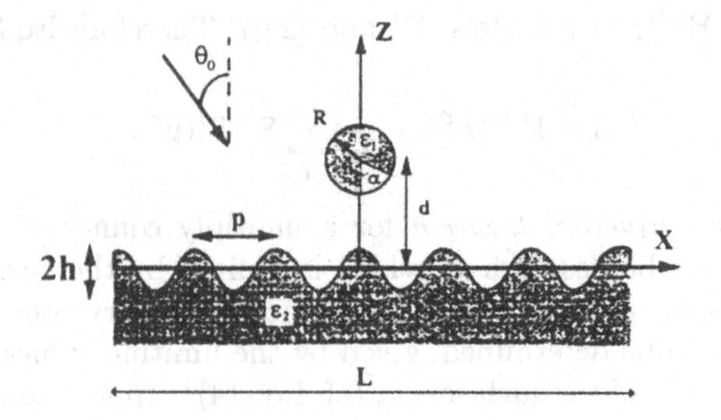

Figure 2. Scattering geometry.

Whereas Eqs.(16) and (17) yield:

$$0 = \frac{1}{4\pi k_0^2}\nabla\times\nabla\times \int_{S_i}(\mathbf{E}_{in}(\mathbf{r}')\frac{\partial G^{(in)}(\mathbf{r}^>,\mathbf{r}')}{\partial \mathbf{n}} - G^{(in)}(\mathbf{r}^>,\mathbf{r}')\frac{\partial \mathbf{E}_{in}(\mathbf{r}')}{\partial \mathbf{n}})ds, \quad (20)$$

$$\mathbf{E}(\mathbf{r}^>) = \mathbf{E}^{(i)}(\mathbf{r}^>) + \frac{1}{4\pi k_0^2 \epsilon}\nabla\times\nabla\times \sum_i \int_{S_i}(\mathbf{E}(\mathbf{r}')\frac{\partial G(\mathbf{r}^>,\mathbf{r}')}{\partial \mathbf{n}} - G(\mathbf{r}^>,\mathbf{r}')\frac{\partial \mathbf{E}(\mathbf{r}')}{\partial \mathbf{n}})ds. \quad (21)$$

In Eqs.(18) and (20) "in" means that the limiting values are taken on the surface from inside the volume V_i; note that this implies for both $G^{(in)}$ and $\mathbf{E}_{in}$ that $k = k_0\sqrt{\epsilon}$.

The saltus conditions:

$$\mathbf{n}\times[\mathbf{E}_{in}(\mathbf{r}^<) - \mathbf{E}(\mathbf{r}^>)] = 0, \quad \mathbf{n}\times[\mathbf{H}_{in}(\mathbf{r}^<) - \mathbf{H}(\mathbf{r}^>)] = 0 \quad (22)$$

across the surface S_i permit to find both $\mathbf{E}$ and $\partial\mathbf{E}/\partial\mathbf{n}$ from either the pair Eqs. (20) and (21), or, equivalently, from the pair Eqs.(18) and (19), as both $\mathbf{r}^>$ and $\mathbf{r}^<$ tend to a point in S_i. Then the scattered field outside the medium is given by the second term of Eq.(21).

3. Formulation of the Scattering Problem.

Let us consider the 2-D physical system shown in Fig. 2. This consists of a cylinder of radius a in vacuum with its center at a distance d from the plane

$z = 0$. The axis of the cylinder is along OY. A corrugated interface $z = D(x)$ of mean plane $z = 0$ separates the vacuum $z > D(x)$ from an arbitrary medium $z < D(x)$. Both the cylinder and the medium at $z < D(x)$ have linear, spatially uniform and isotropic, either metal or dielectric, frecuency-dependent dielectric constants, namely, $\epsilon_i(\omega) = (Re\,[\epsilon_i(\omega)]\,, Im\,[\epsilon_i(\omega)])$, where the index $i = 1, 2$ characterizes either of these media, and $Re\,[...]$, $Im\,[...]$ denotes the real and imaginary parts of $\epsilon_i(\omega)$, respectively.

Let the plane of incidence be the xz-plane. Then, there is no depolarization in the scattering of either S (electric field along the Y axis) or P waves (magnetic field along the Y axis). Therefore, the incident electric and magnetic vectors $\mathbf{E}^{(i)}$ and $\mathbf{H}^{(i)}$ are:

$$\mathbf{E}^{(i)}(\mathbf{r}) = E^{(i)} \exp\left[i\left(K_0 x - q_0 z\right)\right] \widehat{\jmath} \text{ for } S \text{ polarization,} \tag{23}$$

$$\mathbf{H}^{(i)}(\mathbf{r}) = H^{(i)} \exp\left[i\left(K_0 x - q_0 z\right)\right] \widehat{\jmath} \text{ for} P \text{ polarization,} \tag{24}$$

where $\widehat{\jmath}$ is the unit vector along the OY axis, $E^{(i)}$ and $H^{(i)}$ are complex constant amplitudes. K_0 and q_0 are the x and z components of the incident wavevector $\mathbf{k}_0$, respectively. $K_0 = k_0 \sin\theta_i$, $q_0 = k_0 \cos\theta_i$, θ_i being the angle of incidence, formed by $\mathbf{k}_0$ and the OZ axis. $\mathbf{r}$ is a generic position vector with Cartesian components x and z. In what follows, we shall address each polarization separately.

3.1. S POLARIZATION.

Tha scattered electric fields in each media can be expressed as, (see [25]):

$$\begin{aligned} E^{(0)}(\mathbf{r}) &= E^{(i)}(\mathbf{r}) \\ &+ \frac{1}{4\pi} \int_C ds' \left[\frac{\partial G_0(\mathbf{r},\mathbf{r}')}{\partial n'} E^{(1)}(\mathbf{r}') - G_0(\mathbf{r},\mathbf{r}') \frac{\partial E^{(1)}(\mathbf{r}')}{\partial n'} \right] \\ &+ \frac{1}{4\pi} \int_D ds' \left[\frac{\partial G_0(\mathbf{r},\mathbf{r}')}{\partial n'} E^{(2)}(\mathbf{r}') - G_0(\mathbf{r},\mathbf{r}') \frac{\partial E^{(2)}(\mathbf{r}')}{\partial n'} \right] \end{aligned} \tag{25}$$

$$E^{(1)}(\mathbf{r}) = -\frac{1}{4\pi} \int_C ds' \left[\frac{\partial G_1(\mathbf{r},\mathbf{r}')}{\partial n'} E^{(1)}(\mathbf{r}') - G_1(\mathbf{r},\mathbf{r}') \frac{\partial E^{(1)}(\mathbf{r}')}{\partial n'} \right], \tag{26}$$

$$E^{(2)}(\mathbf{r}) = -\frac{1}{4\pi} \int_D ds' \left[\frac{\partial G_2(\mathbf{r},\mathbf{r}')}{\partial n'} E^{(2)}(\mathbf{r}') - G_2(\mathbf{r},\mathbf{r}') \frac{\partial E^{(2)}(\mathbf{r}')}{\partial n'} \right]. \tag{27}$$

The quantities $E^{(j)}(\mathbf{r})$ in Eqs. (25-27) stand for either the total electric field in the vacuum ($j = 0$), inside the cylinder ($j = 1$) or in the half-space $z < D(x)$ ($j = 2$), respectively. They are expressed as surface integrals of the limiting values of both the electric field and its normal derivate, taken either on the cylinder surface C or on the corrugated surface D from the vacuum.

The functions $G_0(\mathbf{r}, \mathbf{r}'), G_1(\mathbf{r}, \mathbf{r}'), G_2(\mathbf{r}, \mathbf{r}')$ are the Green's functions:

$$\begin{aligned} G_0(\mathbf{r}, \mathbf{r}') &= \pi i H_0^{(1)}(k_0 |\mathbf{r} - \mathbf{r}'|), \\ G_1(\mathbf{r}, \mathbf{r}') &= \pi i H_0^{(1)}(\sqrt{\epsilon_1} k_0 |\mathbf{r} - \mathbf{r}'|), \\ G_2(\mathbf{r}, \mathbf{r}') &= \pi i H_0^{(1)}(\sqrt{\epsilon_2} k_0 |\mathbf{r} - \mathbf{r}'|), \end{aligned} \tag{28}$$

where $H_0^{(1)}(...)$ is the zeroth order Hankel's function of the first kind.

Once the values of the electric field and its normal derivate on both the cylinder and the D surface are known, one calculates the scattered electric field in either medium from the set of Eqs.(25)-(27) (see, for instance, Refs. [23], [26]). Taking into account the continuity of both E and $\partial E / \partial n'$ across C and D, one obtains from Eqs. (25) and (26) (cf. [23]):

For points $\mathbf{r}$ on the cylinder surface C:

$$\begin{aligned} E^{(1)}(\mathbf{r}) \;=\; & E^{(i)}(\mathbf{r}) + \frac{i}{4} \int_C ds' \left[\frac{\partial H_0^{(1)}(k_0 |\mathbf{r} - \mathbf{r}'|)}{\partial n'} E^{(1)}(\mathbf{r}') - \right. \\ & \left. H_0^{(1)}(k_0 |\mathbf{r} - \mathbf{r}'|) \frac{\partial E^{(1)}(\mathbf{r}')}{\partial n'} \right] \\ & + \frac{i}{4} \int_D ds' \left[\frac{\partial H_0^{(1)}(k_0 |\mathbf{r} - \mathbf{r}'|)}{\partial n'} E^{(2)}(\mathbf{r}') - \right. \\ & \left. H_0^{(1)}(k_0 |\mathbf{r} - \mathbf{r}'|) \frac{\partial E^{(2)}(\mathbf{r}')}{\partial n'} \right], \end{aligned} \tag{29}$$

$$\begin{aligned} 0 = -\frac{i}{4} \int_C ds' & \left[\frac{\partial H_0^{(1)}(\sqrt{\epsilon_1} k_0 |\mathbf{r} - \mathbf{r}'|)}{\partial n'} E^{(1)}(\mathbf{r}') - \right. \\ & \left. H_0^{(1)}(\sqrt{\epsilon_1} k_0 |\mathbf{r} - \mathbf{r}'|) \frac{\partial E^{(1)}(\mathbf{r}')}{\partial n'} \right], \end{aligned} \tag{30}$$

where $\mathbf{r} = (a \sin \alpha, d - a \cos \alpha)$, α being the polar angle of the generic point $\mathbf{r}$ on the cylinder. The first integral in Eq.(29) is taken on the cylinder surface C so that a point $\mathbf{r}'$ on this surface has the Cartesian components:

$$\mathbf{r}' = (a \sin\alpha', d - a\cos\alpha') , \tag{31}$$

whereas the second integral in Eq.(29) is taken over the surface D, having $\mathbf{r}'$ the components:

$$\mathbf{r}' = (x', D(x')) . \tag{32}$$

When $\mathbf{r}$ is taken on the surface D, one obtains:

$$\begin{aligned} E^{(2)}(\mathbf{r}) \;=\;& E^{(i)}(\mathbf{r}) + \frac{i}{4}\int_C ds' \left[\frac{\partial H_0^{(1)}\left(k_0 \left|\mathbf{r}-\mathbf{r}'\right|\right)}{\partial n'} E^{(1)}(\mathbf{r}') - \right. \\ & \left. H_0^{(1)}\left(k_0 \left|\mathbf{r}-\mathbf{r}'\right|\right) \frac{\partial E^{(1)}(\mathbf{r}')}{\partial n'} \right] \\ & + \frac{i}{4}\int_D ds' \left[\frac{\partial H_0^{(1)}\left(k_0 \left|\mathbf{r}-\mathbf{r}'\right|\right)}{\partial n'} E^{(2)}(\mathbf{r}') - \right. \\ & \left. H_0^{(1)}\left(k_0 \left|\mathbf{r}-\mathbf{r}'\right|\right) \frac{\partial E^{(2)}(\mathbf{r}')}{\partial n'} \right] , \end{aligned} \tag{33}$$

$$\begin{aligned} 0 = -\frac{i}{4}\int_D ds' & \left[\frac{\partial H_0^{(1)}\left(\sqrt{\epsilon_2} k_0 \left|\mathbf{r}-\mathbf{r}'\right|\right)}{\partial n'} E^{(2)}(\mathbf{r}') - \right. \\ & \left. H_0^{(1)}\left(\sqrt{\epsilon_2} k_0 \left|\mathbf{r}-\mathbf{r}'\right|\right) \frac{\partial E^{(2)}(\mathbf{r}')}{\partial n'} \right] . \end{aligned} \tag{34}$$

It is worth noting that the unknown sources on both C and D appear coupled in Eqs. (29)-(30) and (33)-(34). This fact shows the multiple scattering process between the cylinder and the sample. The system of inhomogeneus integral equations constituted by the set of Eqs. (29) (30), (33) and (34) are numerically solved [23].

The reflected and transmitted fields in the far zone are:

$$\begin{aligned} E^{(r)}(r_>, \theta_r) \;=\;& \frac{\exp\left[i\left(k_0 r_> - \pi/4\right)\right]}{\left(8\pi k_0 r_>\right)^{1/2}} \left[\int_C ds' \left[(\mathbf{n}'.\mathbf{k}_r)\, E^{(1)}(\mathbf{r}') \right.\right. \\ & \left. - i \frac{\partial E^{(1)}(\mathbf{r}')}{\partial n'} \right] \exp\left(-i\mathbf{k}_r.\mathbf{r}'\right) \\ & + \int_D ds' \left[(\mathbf{n}'.\mathbf{k}_r)\, E^{(2)}(\mathbf{r}') \right. \\ & \left.\left. - i \frac{\partial E^{(2)}(\mathbf{r}')}{\partial n'} \right] \exp\left(-i\mathbf{k}_r.\mathbf{r}'\right) \right] , \end{aligned} \tag{35}$$

for the reflected field. And:

$$E^{(t)}(r_<,\theta_t) = \frac{\exp\left[i\left(\sqrt{\epsilon_2}k_0 r_< - \pi/4\right)\right]}{\left(8\pi\sqrt{\epsilon_2}k_0 r_<\right)^{1/2}} \int_D ds' \left[(\mathbf{n}'.\mathbf{k}_t)\, E^{(2)}(\mathbf{r}') -i\frac{\partial E^{(2)}(\mathbf{r}')}{\partial n'}\right] \exp\left(-i\mathbf{k}_t.\mathbf{r}'\right) \quad (36)$$

for the transmitted field.

In Eqs. (35) and (36) $\mathbf{k}_r$ and $\mathbf{k}_t$ are the reflected and transmitted wavevectors,

$$\begin{aligned} \mathbf{k}_r &= k_0\left(\sin\theta_r, \cos\theta_r\right), \\ \mathbf{k}_t &= \sqrt{\epsilon_2}k_0\left(\sin\theta_t, -\cos\theta_t\right) \end{aligned} \quad (37)$$

θ_r and θ_t being the observation angles above and bellow the surface $z = D(x)$, respectively (see Fig. 2).

Depending on whether the integration is made over the C or D surface, the unit normal $\mathbf{n}'$ is:

$$\mathbf{n}' = \begin{cases} (\sin\alpha', -\cos\alpha') & \text{if } \mathbf{r}' \text{ belongs to } C, \\ \frac{(-D'(x'),1)}{\sqrt{(1+D'(x')^2)}} & \text{if } \mathbf{r}' \text{ belongs to } D. \end{cases} \quad (38)$$

3.2. *P* POLARIZATION.

The scattered magnetic fields have the expressions:

$$\begin{aligned} H^{(0)}(\mathbf{r}) &= H^{(i)}(\mathbf{r}) \\ &+\frac{1}{4\pi}\int_C ds' \left[\frac{\partial G_0(\mathbf{r},\mathbf{r}')}{\partial n'} H^{(1)}(\mathbf{r}') - G_0(\mathbf{r},\mathbf{r}')\frac{\partial H^{(1)}(\mathbf{r}')}{\partial n'}\right] \\ &+\frac{1}{4\pi}\int_D ds' \left[\frac{\partial G_0(\mathbf{r},\mathbf{r}')}{\partial n'} H^{(2)}(\mathbf{r}') - G_0(\mathbf{r},\mathbf{r}')\frac{\partial H^{(2)}(\mathbf{r}')}{\partial n'}\right] \end{aligned} \quad (39)$$

$$H^{(1)}(\mathbf{r}) = -\frac{1}{4\pi}\int_C ds' \left[\frac{\partial G_1(\mathbf{r},\mathbf{r}')}{\partial n'} H^{(1)}(\mathbf{r}') - G_1(\mathbf{r},\mathbf{r}')\frac{\partial H^{(1)}(\mathbf{r}')}{\partial n'}\right], \quad (40)$$

$$H^{(2)}(\mathbf{r}) = -\frac{1}{4\pi}\int_D ds' \left[\frac{\partial G_2(\mathbf{r},\mathbf{r}')}{\partial n'} H^{(2)}(\mathbf{r}') - G_2(\mathbf{r},\mathbf{r}')\frac{\partial h^{(2)}(\mathbf{r}')}{\partial n'}\right], \quad (41)$$

where $H^{(0)}(\mathbf{r}), H^{(1)}(\mathbf{r})$ and $H^{(2)}(\mathbf{r})$ are the magnetic fields in the vacuum, inside the cylinder and below the surface D, respectively. The source functions $H^{(1)}(\mathbf{r}')$, $H^{(2)}(\mathbf{r}')$ and their normal derivates are the limiting values, taken from the vacuum, of the magnetic field on the surfaces C and D, respectively.

In a similar fashion to the case of S-polarization, we obtain:

For $\mathbf{r}$ on C :

$$
\begin{aligned}
H^{(1)}(\mathbf{r}) &= H^{(i)}(\mathbf{r}) + \frac{i}{4}\int_C ds' \left[\frac{\partial H_0^{(1)}\left(k_0\left|\mathbf{r}-\mathbf{r}'\right|\right)}{\partial n'} H^{(1)}(\mathbf{r}') - \right. \\
&\left. H_0^{(1)}\left(k_0\left|\mathbf{r}-\mathbf{r}'\right|\right)\frac{\partial H^{(1)}(\mathbf{r}')}{\partial n'}\right] \\
&+ \frac{i}{4}\int_D ds' \left[\frac{\partial H_0^{(1)}\left(k_0\left|\mathbf{r}-\mathbf{r}'\right|\right)}{\partial n'} H^{(2)}(\mathbf{r}') - \right. \\
&\left. H_0^{(1)}\left(k_0\left|\mathbf{r}-\mathbf{r}'\right|\right)\frac{\partial H^{(2)}(\mathbf{r}')}{\partial n'}\right],
\end{aligned}
\tag{42}
$$

$$
\begin{aligned}
0 = -\frac{i}{4}\int_C ds' &\left[\frac{\partial H_0^{(1)}\left(\sqrt{\epsilon_1}k_0\left|\mathbf{r}-\mathbf{r}'\right|\right)}{\partial n'} H^{(1)}(\mathbf{r}') - \right. \\
&\left. \epsilon_1 H_0^{(1)}\left(\sqrt{\epsilon_1}k_0\left|\mathbf{r}-\mathbf{r}'\right|\right)\frac{\partial H^{(1)}(\mathbf{r}')}{\partial n'}\right],
\end{aligned}
\tag{43}
$$

For $\mathbf{r}$ on $z = D(x)$:

$$
\begin{aligned}
H^{(2)}(\mathbf{r}) &= H^{(i)}(\mathbf{r}) + \frac{i}{4}\int_C ds' \left[\frac{\partial H_0^{(1)}\left(k_0\left|\mathbf{r}-\mathbf{r}'\right|\right)}{\partial n'} H^{(1)}(\mathbf{r}') - \right. \\
&\left. H_0^{(1)}\left(k_0\left|\mathbf{r}-\mathbf{r}'\right|\right)\frac{\partial H^{(1)}(\mathbf{r}')}{\partial n'}\right] \\
&+ \frac{i}{4}\int_D ds' \left[\frac{\partial H_0^{(1)}\left(k_0\left|\mathbf{r}-\mathbf{r}'\right|\right)}{\partial n'} H^{(2)}(\mathbf{r}') - \right. \\
&\left. H_0^{(1)}\left(k_0\left|\mathbf{r}-\mathbf{r}'\right|\right)\frac{\partial H^{(2)}(\mathbf{r}')}{\partial n'}\right],
\end{aligned}
\tag{44}
$$

$$
0 = -\frac{i}{4}\int_C ds' \left[\frac{\partial H_0^{(1)}\left(\sqrt{\epsilon_2}k_0\left|\mathbf{r}-\mathbf{r}'\right|\right)}{\partial n'} H^{(2)}(\mathbf{r}') - \right.
$$

$$\epsilon_2 H_0^{(1)}\left(\sqrt{\epsilon_2}k_0\left|\mathbf{r}-\mathbf{r}'\right|\right)\frac{\partial H^{(2)}(\mathbf{r}')}{\partial n'}\Bigg], \tag{45}$$

From Eqs. (39) - (41) one obtains the far fields,

$$\begin{aligned}H^{(r)}(r_>,\theta_r) &= \frac{\exp\left[i\left(k_0 r_> - \pi/4\right)\right]}{\left(8\pi k_0 r_>\right)^{1/2}}\Bigg[\int_C ds'\left[\left(\mathbf{n}'.\mathbf{k}_r\right)H^{(1)}(\mathbf{r}')-\right.\\ &\left. i\frac{\partial H^{(1)}(\mathbf{r}')}{\partial n'}\right]\exp\left(-i\mathbf{k}_r.\mathbf{r}'\right)\\ &+\int_D ds'\left[\left(\mathbf{n}'.\mathbf{k}_r\right)H^{(2)}(\mathbf{r}')-\right.\\ &\left. i\frac{\partial H^{(2)}(\mathbf{r}')}{\partial n'}\right]\exp\left(-i\mathbf{k}_r.\mathbf{r}'\right)\Bigg],\end{aligned} \tag{46}$$

$$\begin{aligned}H^{(t)}(r_<,\theta_t) &= \frac{\exp\left[i\left(\sqrt{\epsilon_2}k_0 r_< - \pi/4\right)\right]}{\left(8\pi\sqrt{\epsilon_2}k_0 r_<\right)^{1/2}}\int_D ds'\left[\left(\mathbf{n}'.\mathbf{k}_t\right)H^{(2)}(\mathbf{r}')-\right.\\ &\left. i\epsilon_2\frac{\partial H^{(2)}(\mathbf{r}')}{\partial n'}\right]\exp\left(-i\mathbf{k}_t.\mathbf{r}'\right).\end{aligned} \tag{47}$$

4. Evanescent wave detection. Numerical examples

We carry out numerical simulations with the x-integrals on the grating surface restricted to a finite interval of length L. Note that, due to the presence of the cylinder, the system has no translation invariance like the grating alone, and thus one cannot use periodic boundary conditions. In the case of absence of cylinder, the results with finite lenght L are similar to those with periodic boundary conditions, this indicating the validity of the calculation done here. For far field calculations we have used $L = 50\lambda$ with a sampling interval Δx that depends on the corrugation of the grating. Specifically, we have used a sampling interval $\Delta x = 0.1\lambda$ and 0.05λ for the two grating periods addressed, namely, 1.25λ and 0.6λ, respectively. In the near field calculations, several surface lengths have been considered depending on each specific case. For the grating of period 1.25λ the surface length is $L = 50\lambda$ and the sampling interval is $\Delta x = 0.05\lambda$. In the case of the grating with period 0.6λ, two larger surface lengths have been used, i.e., $L = 90\lambda$ and $L = 110\lambda$. The choice of larger surface records L together with the use of focused incident beams, prevents spureous effects in the scattered fields arising from edge effects, specially when resonance

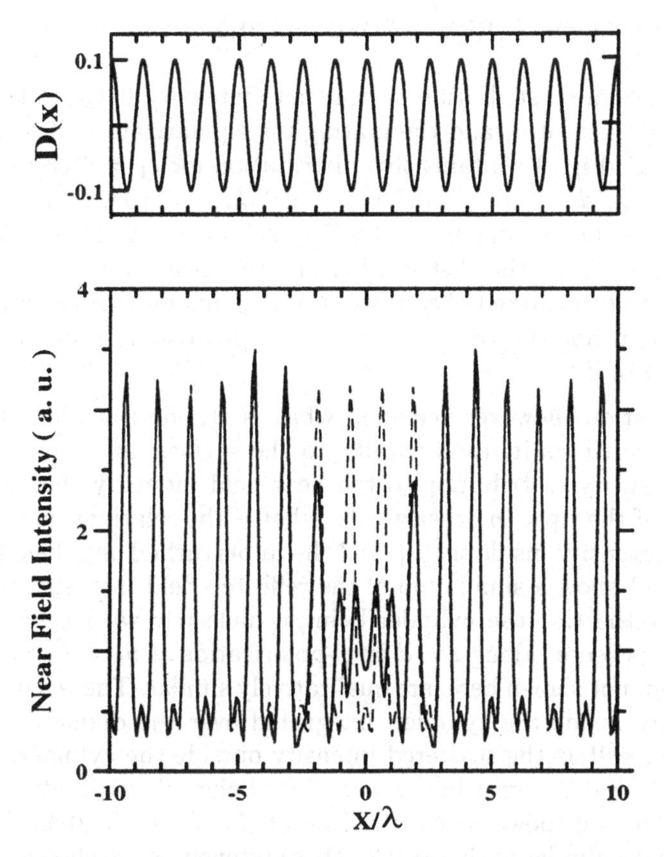

Figure 3. Normalized near field scattered intensity in arbitrary units (a.u.) at $z = 0.25\lambda$ for the grating with parameters: $b = 1.25\lambda$, $h = 0.1\lambda$, $\epsilon_2 = (-17.2, 0.498)$. S polarization. Thin solid line: $a = 0.05\lambda, \epsilon_1 = 2.12, d = 10\lambda$. Thick solid line: $a = 0.25\lambda$, $\epsilon_1 = 2.12$, $d = 0.6\lambda$. The grating profile is shown on the top.

plasmons are excited. In addition, we have taken a constant number of $M = E[ka/\Delta s] = 64$ sampling points in the cylinder, $E[\cdot]$ being "the integer part of". This means different samplig intervals Δs on the cylinder surface depending on its radius. In the calculations presented here we have used sampling intervals $\Delta s = 0.05\lambda$, 0.02λ, 0.01λ and 0.005λ for each of the radii considered: $a = 0.5\lambda$, 0.2λ, 0.1λ and 0.05λ, respectively. The convergence of the results with the number of sampling points has been checked.

We consider a grating of corrugation $z = D(x) = h\cos(2\pi x/b)$, separating vacuum from silver with dielectric permittivity $\epsilon_2 = -17.2 + i0.498$ at $\lambda = 652.6$ nm.

4.1. RETRIEVAL OF SURFACE TOPOGRAPHY

Next we show the distribution of near field intensity between the cylinder of dielectric constant ϵ_1 and the grating for s-polarization at angle of incidence $\theta_i = 0^0$. Fig. 3 contains this distribution for $\epsilon_1 = 2.12$ at $z = 0.25\lambda$ for: $a = 0.25\lambda$, $d = 0.6\lambda$, and for $a = 0.05\lambda$, $d = 10\lambda$, respectively. The parameters of the grating are $b = 1.25\lambda$ and $h = 0.1\lambda$. These figures show the perturbation on the distribution of near field by the tip just below it. It should be remarked that for $a \leq 0.1\lambda$, this perturbation is so small that, although not shown explicitely here, this distribution is practically independent of d.

One question, however remains: what is the distribution of intensity "seen" by the tip as it scans parallel to the surface, and what connection has this intensity distribution to the near field intensity that would exist in absence of the tip?. In other words, what is the relationship between the perturbed intensity inside the tip and the unperturbed near field intensity?. Fig. 4(a)-4(c) show a simulation of the reflected field intensity distribution detected by the tip (the cylinder being a dielectric with $\epsilon_1 = 2.12$) as it scans with its center along $z = d$ for p-polarization at $\theta_i = 0^0$. Results for s-polarization, not shown here, are qualitatively similar. These figures contain the intensity inside the cylinder, integrated over the diameter $2a$ parallel to $z = 0$, as well as the scattered intensity outside the cylinder, integrated over the interval $2a$ on a line $z = z_0$ just below this cylinder . For a tip with $a = 0.5\lambda$ we show these intensities at $d = 0.8\lambda$ $z = 0.2\lambda$, (Fig. 4(a)), whereas for a cylinder with $a = 0.2\lambda$ these intensities are shown at $d = 0.5\lambda$ and $z = 0.2\lambda$, (Fig. 4(b)), and at $d = \lambda$ and $z = 0.7\lambda$, (Fig. 4(c)). These figures show that, in general, the intensity resulting from the scan either as detected inside the tip, or the scattered intensity below it, does not follow the surface profile. The intensity distribution oscillates with both d and z, and its contrast can be inverted, depending on the distance of scanning to the grating. This is due to the standing wave along the OZ-axis that exists between the cylinder and the grating. Thus, if the cylinder represents a tip of a photon scanning tunnelling microscope (PSTM), when multiple scattering takes place like in the range of parameters addressed here, the images of the surface do not exhibit its actual topography. This fact, that affects the interpretation of images of the surface in near field microscopy, had been pointed out before [27]-[29], although in those works the presence of the tip had not been taken into account. Also, Ref. 30 points out this problem for the case of a model of scanning near field optical microscope (SNOM), namely, when the tip illuminates the surface and one collects the far field.

Simulations done for the illuminating wave incident on the grating at

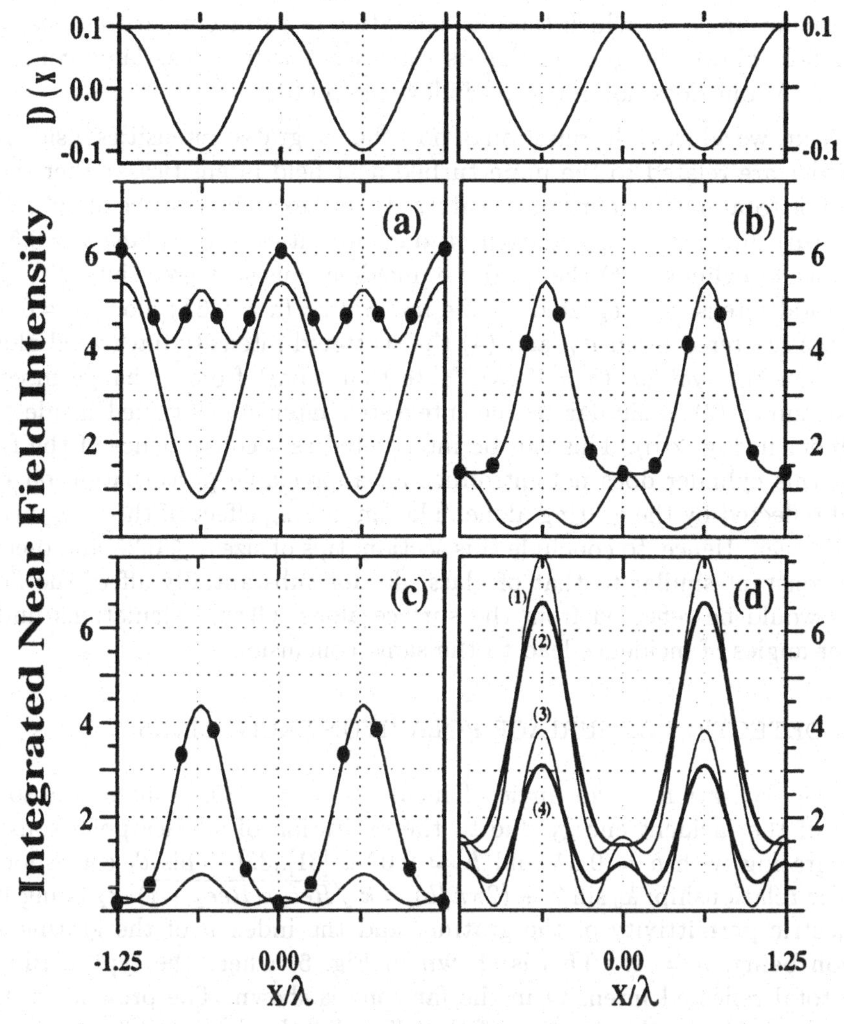

Figure 4. Variation of the integrated intensity with the cylinder position moved along $z = d$ for several radii a of the cylinder and distances d cylinder-plane. The cylinder has a dielectric permitivity $\epsilon_1 = 2.12$. Parameters of the grating: $b = 1.25\lambda, h = 0.1\lambda, \epsilon_2 = (-17.2, 0.498)$. $\theta_0 = 0^0$. p-polarization 4(a): $a = 0.5\lambda, d = 0.8\lambda, z_0 = 0.2\lambda$. 4(b): $a = 0.2\lambda, d = 0.5\lambda, z_0 = 0.2\lambda$. 4(c): $a = 0.2\lambda, d = 1\lambda, z_0 = 0.7\lambda$. Solid line with circles: Intensity inside the cylinder and integrated over its diameter. Solid line: Scattered intensity outside the cylinder and integrated on a segment of length $2a$ on a line below it at distance $z = z_0$. 4(d): Intensity distribution on the plane $z = 0.5\lambda$ due to the grating alone. (1) Total, i.e. incident plus scattered, intensity. (2) Total intensity integrated on the interval $2a$. (3) Scattered intensity. (4) Scattered intensity integrated on the interval $2a$. The grating profile is shown on the top.

a resonant angle, at which there is excitation of surface polaritons, show a near field above the grating that is evanescent with z, has large contrast for $z < \lambda$, but does not generally follow the profile.

Next, we address the question of how the integrated intensities inside the cylinder are related to the unperturbed near field intensities, (either scattered or total, i.e. incident plus scattered) that exist in absence of cylinder. Fig. 4(d) illustrates a calculation of four intensities in the plane $z = 0.5$ in absence of cylinder: (1) the total, i.e. incident plus scattered, intensity, (2) the total intensity integrated on the interval $2a$ ($a = 0.2\lambda$) along $z = 0.5\lambda$, (3) the scattered intensity, and (4) the scattered intensity integrated along the same interval $2a$, ($a = 0.2\lambda$). Note that, apart from a change in contrast, curve (2) is similar to the integrated intensity obtained inside the cylinder in Fig. 2(b). This can be interpreted as a consequence of the fact that this cylinder does not introduce an appreciable perturbation on the field reflected by the grating alone. The smoothing effect of the integration is also seen. Hence. to conclude this section, tips of size $a \lesssim 0.2\lambda$ and dielectric constant similar to that of glass, do not substantially affect the field that would be reflected from the surface alone. Other calculations under other angles of incidence lead to the same conclusion.

4.2. DETECTION OF SURFACE POLARITON EXCITATIONS

For p-polarization, at an angle of incidence $\theta_i = -39.1^0$ there is absorption of the incident energy due to the excitation of surface polaritons at the grating with $b = 0.6\lambda$ and $h = 0.02\lambda$, [31],[32]. This θ_i corresponds to the relationship: $k_0 \sin\theta_i + (2\pi n/b) = k\sqrt{Re\epsilon_2/(Re\epsilon_2 + 1)}$, ϵ_2 being the dielectric permittivity of the grating, and the index n of the grating relation being: $n = 1$. This is shown in Fig. 5, where the typical dip in the total reflected intensity in the far zone is shown. The presence of the cylinder perturbs the position of this dip, slightly shifting it to larger angles of incidence as shown in Fig. 5 for dielectric tips of $\epsilon_1 = 2.12$ and 3.55, respectively, at distance $d = 0.8\lambda$ from its center. In addition, due to the perturbation of the tip on the field scattered by the surface, the minimum of this dip fluctuates with the distance from the cylinder to the surface, this is illustrated by Figs. 6(a) and 6(b) which show the variation of this minimum with the distance d. Fig. 6(a) corresponds to a cylinder with $\epsilon_1 = 2.12$. These variations are shown for four different radius of the cylinder: $a = 0.5\lambda$, 0.2λ, 0.1λ, and 0.05λ. Also, it is observed that when the cylinder is close to the surface, the reflectance at the dip sharply increases, this is due to the coupling of the surface polaritons to propagating waves via the cylinder. This constitutes in fact a signal manifesting the existence of the surface waves, (see Refs.[33] and [34] for NFO detection of plasmons

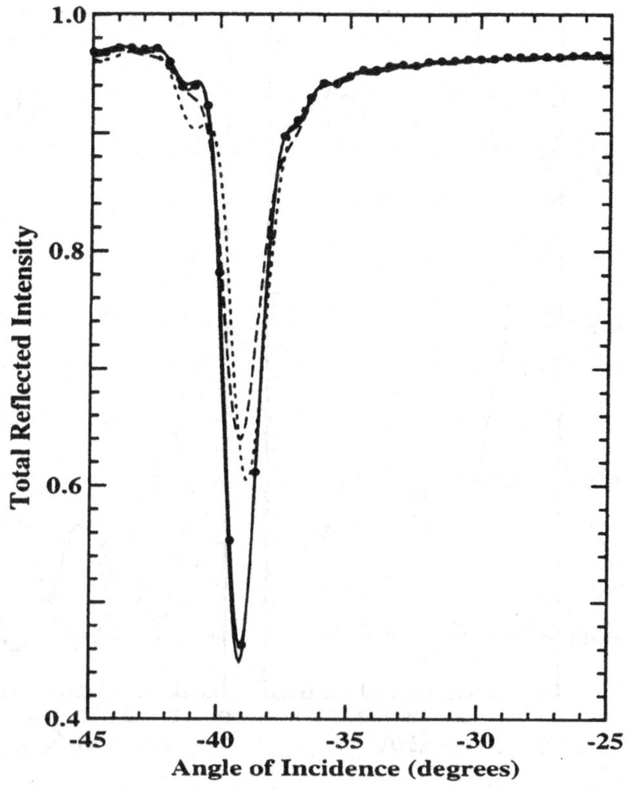

Figure 5. Total reflected intensity near resonance for the grating alone and the grating in the presence of a cylinder of radius a at distance $d = 0.8\lambda$. p-polarization. $b = 0.6\lambda, h = 0.02\lambda$, $\epsilon_2 = (-17.2, 0.498)$. Solid line: without cylinder. Solid line with circles: $a = 0.2\lambda, \epsilon_1 = 2.12$. Broken line: $a = 0.5\lambda, \epsilon_1 = 3.55$. Dotted line: $a = 0.5\lambda, \epsilon_1 = 2.12$.

in thin films). Of course, the smaller the tip is, the weaker this signal is, in other words, the smaller is the cylinder interface available to produce this coupling. Also, the coupling is more effective the larger the permittivity is. As the permittivity increases, both the amplitude of the fluctuations and their average become larger, as well as the signal at close distances to the surface. This signal can be further increased by using tiny metallic tips ($a \leq 0.1\lambda$), as shown in Fig.6(b) which corresponds to a cylinder with $\epsilon_1 = -17.2 + i0.498$. However, with larger metallic tips the absorption effects by the metal obliterate this coupling and the signal at low d is weaker than for tips with a large positive dielectric constant. Metallic tips also give rise to larger background fluctuations as d increase.

The presence of the cylinder distorts this field distribution in the region just below it. This distortion being almost negligible when $a \leq 0.2\lambda$. The oscillations of this field distribution have a period about 0.45λ. The expected

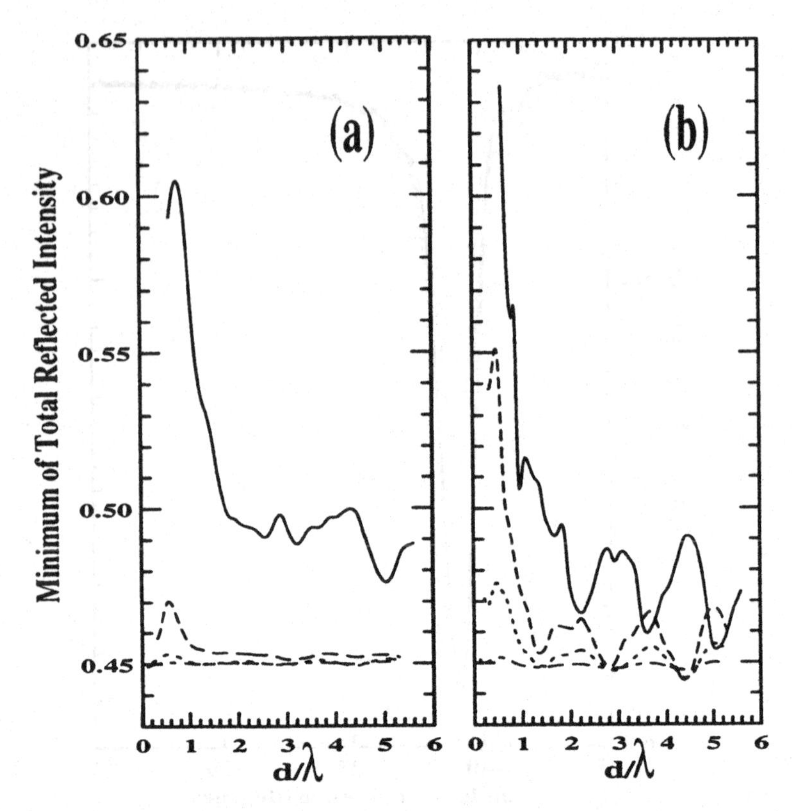

Figure 6. Variation of the minimun of the total reflected intensity with the distance d from the cylinder to the grating for several radii of the cylinder and dielectric permitivities. p-polarization. $b = 0.6\lambda, h = 0.02\lambda$, $\epsilon_2 = (-17.2, 0.498)$. (a) $\epsilon_1 = 2.12$. (b) $\epsilon_1 = (-17.2, 0.498)$. Solid line: $a = 0.5\lambda$. Dashed line: $a = 0.2\lambda$. Dotted line: $a = 0.1\lambda$. Dotted-dashed line: $a = 0.05\lambda$

value from diffraction grating theory is close to [35]: $(Re\ \alpha_{sp} + (\lambda/b))^{-1}$, where for $Im\ \epsilon_2 << |Re\ \epsilon_2|$ one has for the real part of the surface polariton propagation constant $Re\ \alpha_{sp}$: $Re\ \alpha_{sp} = \sqrt{Re\epsilon_2/(Re\epsilon_2 + 1)}$, [36]. This yields a period about 0.4λ.

The effect when a p-polarized incident Gaussian beam is used is shown in Fig. 7. This figure shows the intensity of the total (i.e., incident plus scattered) field for p-polarization at $z = 0.12\lambda$, corresponding to $a = 0.01\lambda$, $d = 18\lambda$, $\epsilon_1 = 1.1$, $\theta_i = -39.1^0$ and $\theta_i = 0^0$, the half-width of the intercept of the beam with the surface $g = w/\cos(-39.1^0)$ is $g = 10.3\lambda$, w being the beam waist. It should be remarked that, given these parameters of the cylinder, this case is practically equivalent to absence of it. At the resonant angle $\theta_i = -39.1^0$, the near field distribution is enhanced, it has the beam shape, and it is elongated towards $x > 0$ due to the propagation and decay of the polariton in this direction. As before, the oscillation period is about

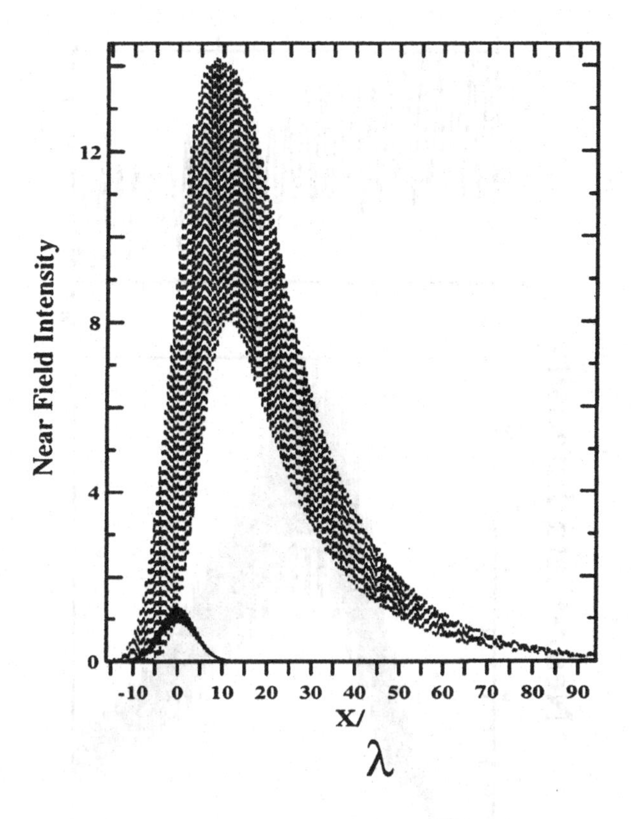

Figure 7. Scattered near field intensity, (arbitrary units), at $z = 0.12\lambda$ for a grating with $b = 0.6\lambda, h = 0.02\lambda, \epsilon_2 = (-17.2, 0.498)$. The incident field is a Gaussian beam with $w = 8\lambda$. Dashed line: $\theta_i = -39.1^0$. Solid line: $\theta_i = 0^0$.

0.45λ. This peak also decreases exponentially with z due to the binding of the surface wave to the interface. On the other hand, at $\theta_i = 0^0$ there is no such enhancement, and the intensity distribution closely follows the beam shape, with oscillations varying about the grating period (0.6λ). When a cylinder with $\epsilon_1 = 2.12$ and radius $a = 0.1\lambda$ scans parallel to the mean plane of the grating, with its center on the line $z = d$, then we show in Fig. 8 the corresponding enhanced total field intensity inside the cylinder, integrated along the diameter $2a$ in the line $z = d$. The detail of this distribution in the interval $(10\lambda, 25\lambda)$ is illustrated on the top of this figure, showing that the detection with this tip through integration conserves the subwavelength resolution of the oscillations. This intensity constitutes the detected NFO *image* of the surface polariton propagation on the grating, and it has been observed in thin films [37], although not yet in gratings.

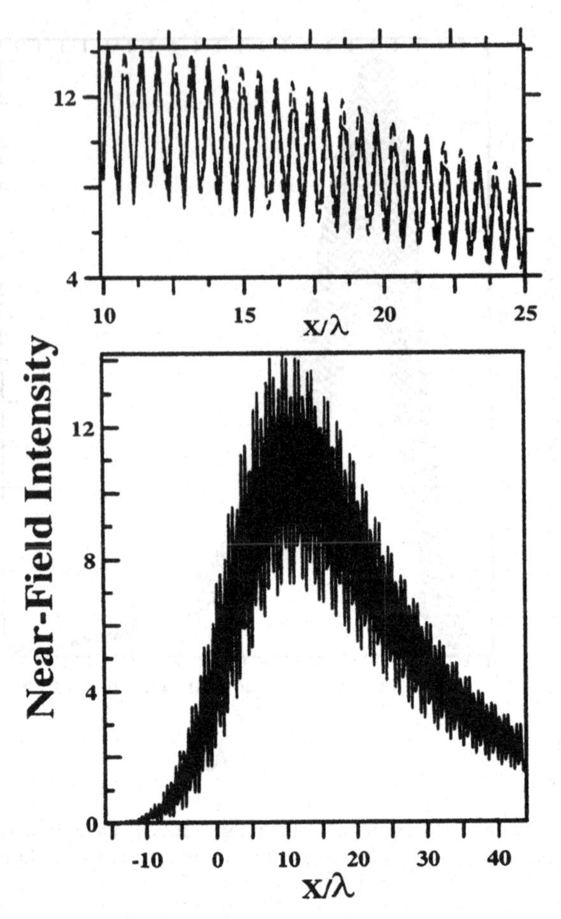

Figure 8. Integrated intensity over the interval $2a$ inside a cylinder with $a = 0.1\lambda$ and $\epsilon_1 = 2.12$, as this cylinder moves with its center on $z = d = 0.17\lambda$. The incident wave and the grating are both as for the dashed line of Fig. 7. The scan has been limited to the interval $(-15\lambda, 45\lambda)$. Details of this distribution for x in the interval $(10\lambda, 25\lambda)$ are shown on the top.

5. Objects hidden under rough surfaces. Illustration results

The physical system adressed in this section is shown in Fig. 9 We simulate random rough surface samples of length $L = 50\lambda$. Each sample being restricted to 500 sampling points taken out of a sequence of random numbers with Gaussian statistics, zero mean and a Gaussian correlation function: $c(\tau) =< D(x)D(x+\tau) > /\sigma^2 = \exp(-\tau^2/T^2)$, σ being the rms deviation of the surface height and T standing for its correlation length. Averages are taken over $N = 800$ samples. Unitarity is satisfied to within 99%. The Gaussian beam has a half-width, $W = 8\lambda_0$. At this point, it should be remarked that the performance of ensamble averaging may be difficult to accomplish

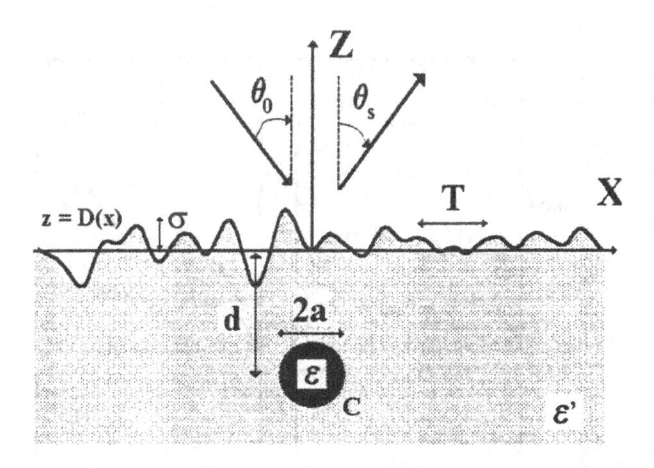

Figure 9. Geometry of a hidden cylinder below a random rough surface.

in many practical instances as this implies to illuminate different portions of the surface, several of which may not contain the object behind. However, there is yet a large number of situations in which the speckle is fine enough so that this ensemble average is close to an average of one intensity realization over a narrow range of angles of scattering without a serious lost of resolution, or even to an average over a range of frequencies. Nevertheless, we shall present at the end of this section more realistic statistical averages (i.e. average over frequencies, instead of different realizations of the random rough surface) supporting the same conclusions in a qualitative way.

Figs. 10(a) and 10(c) show the angular distribution of mean scattered intensity, normalized to the total intensity of the incident beam, on reflection at the dielectric surface under s-polarization and p-polarization, respectively, for normal incidence, $\theta_0 = 0^0$. The parameters are: $T = 3.16\lambda_0$, $\sigma = 1.9\lambda_0$, $\epsilon = 1$, $\epsilon' = 2.04$. Three different cases are considered, i.e., either without cylinder, or with this body at $d = 10\lambda_0$, having radii: $a = \lambda_0$, $2\lambda_0$ and $5\lambda_0$, respectively. The results corresponding to $\theta_0 = 30^0$ are shown in Figs. 10(b) and 10(d), respectively. As seen, the presence of the body dramatically enhances the far zone mean scattered intensity. In fact, in absence of body, no noticeable backscattering peak is detected for this dielectric surface[38, 39]. Although it should be remarked that, with these statistical parameters σ and T, a metallic surface would yield enhanced backscattering [40]. Also, these figures show that this peak decreases as the angle of incidence increases.

The distribution of transmitted intensities are shown in Figs. 11(a)-11(d). It is worth noting that it does not change so dramatically due to

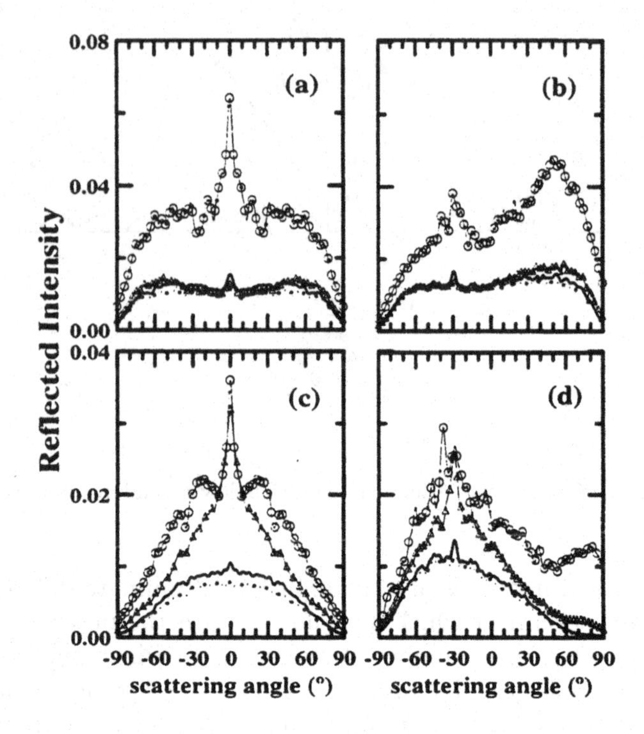

Figure 10. Angular distribution of the mean reflected intensity averaged over $N = 800$ realizations. $T = 3.16\lambda_0$, $\sigma = 1.9\lambda_0$, $\epsilon' = 2.04$. $d = 10\lambda_0$, $\epsilon = 1$, and $W = 8\lambda_0$. Dotted line: no cylinder. Solid line: $a = 1\lambda_0$. Solid line with triangles: $a = 2\lambda_0$. Solid line with circles: $a = 5\lambda_0$. 2(a): s-polarization, $\theta_0 = 0^0$. 2(b): s-polarization, $\theta_0 = 30^0$. 2(c): p-polarization, $\theta_0 = 0^0$. 2(d): p-polarization, $\theta_0 = 30^0$.

the presence of the cylinder. It has the well known narrow shape around the forward direction[39], with no appreciable difference between s and p-polarization, as expected. In fact, it is well known [39] that the Kirchhoff approximation accounts well for the transmitted wave by dielectric rough surfaces with these low permittivities. The presence of the body slightly increases this intensity at large angles of observation at the expense of diminishing the central part, i.e. it produces a sort of "hat" shape distribution. All these effects are more dramatic as the contrast ϵ/ϵ' increases.

As the distance d increases, the effect of the buried body on the backscattering enhancement also diminishes. For instance, Figs. 12(a) and 12(b) show the mean scattered intensity of s-waves from a surface as before, with

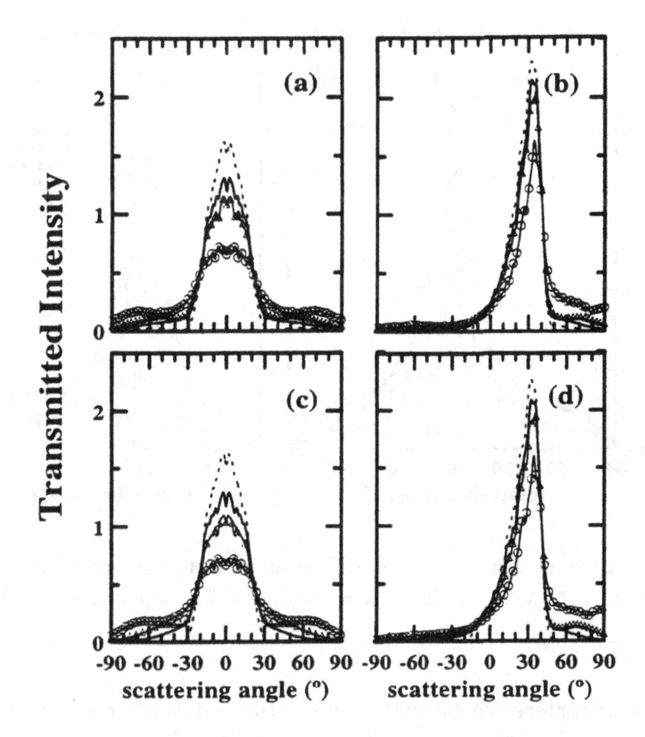

Figure 11. Angular distribution of the mean transmitted intensity averaged over $N = 800$ realizations. $T = 3.16\lambda_0$, $\sigma = 1.9\lambda_0$, $\epsilon' = 2.04$. $d = 10\lambda_0$, $\epsilon = 1$, and $W = 8\lambda_0$. Dotted line: no cylinder. Solid line: $a = 1\lambda_0$. Solid line with triangles: $a = 2\lambda_0$. Solid line with circles: $a = 5\lambda_0$. 3(a): *s*-polarization, $\theta_0 = 0^0$. 3(b): *s*-polarization, $\theta_0 = 30^0$. 3(c): *p*-polarization, $\theta_0 = 0^0$. 3(d): *p*-polarization, $\theta_0 = 30^0$.

a buried cylinder of radius $a = 5\lambda_0$ and $\epsilon = 7.5$ at incidence $\theta_0 = 0^0$ and $\theta_0 = 30^0$, respectively. The incident beam has a $W = 8\lambda_0$ half width and several distances d cylinder-surface are shown, i.e., $d = 6\lambda_0$, $8\lambda_0$, $10\lambda_0$, $20\lambda_0$, $30\lambda_0$, and $40\lambda_0$, labeled in the plots with the numbers (1), (2), (3), (4), (5), and (6), respectively. The peak of enhanced backscattering now narrows and weakens as the distance d of the cylinder to the surface increases. Also, this peak decreases as the width of the incident beam gets smaller.

This enhancement effect has a certain analogy with the case of reflection from a phase screen with a highly reflecting surface behind it [41]. The double pass through the surface due to reflection at the cylinder produces

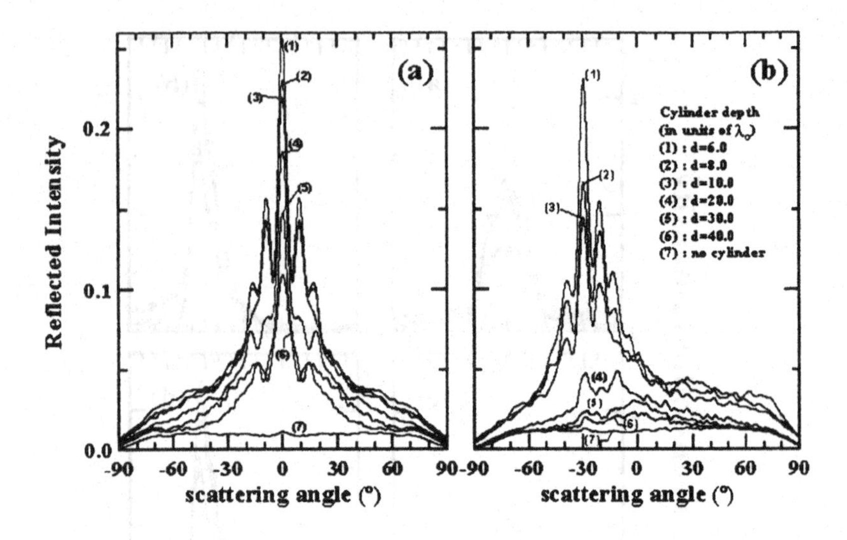

Figure 12. Angular distribution of the mean reflected intensity versus distance d. $T = 3.16\lambda_0$, $\sigma = 1.9\lambda_0$, $\epsilon' = 2.04$. $a = 5\lambda_0$, $\epsilon = 7.5$. s-polarization. 9(a): $\theta_0 = 0^0$. 9(b): $\theta_0 = 30^0$.

constructive interference between reversible paths of the wave that traverse different points of the surface. Then, on averaging, a maximum enhancement is produced in the backscattering direction. As either W/a or the distance d between the cylinder and the surface increase, the proportion of reversible paths decreases, and , in addition, due to spreading on transmission, the fraction of wave energy that interacts with the cylinder and is reflected back to the surface decreases, therefore, the peak of enhancement becomes weaker. This effect is even more noticeable at oblique incidence.

Figs. 13(a) and 13(b) show the variation, for s and p-waves respectively, of the backscattering peak versus d for a cylinder with radius $a = \lambda_0$ and $\epsilon = 1$. They exhibit a monotonous decrease at larger values of d. The non uniform variation at small d indicates that the oscillation of the scattered intensity dominates in this range of values of d over the aforementioned decrease of reversed paths as d increases, namely, this decrease is not much noticeable at small values of d.

If one attempted to obtain an image of the cylinder through the rough surface, one would get a replica of the wavefront distribution of the scattered field at planes in the air side, close to the interface. To see the aspect of such image, we shall evaluate the intensity distribution at one of such planes. Figs. 14(a)-14(d) show three different portions of intensity distributions, between $x = -10\lambda_0$ and $x = 10\lambda_0$, computed at the plane $z = 8\lambda_0$, for s and p-waves, at $\theta_0 = 0^0$ and 30^0, respectively. All of them correspond

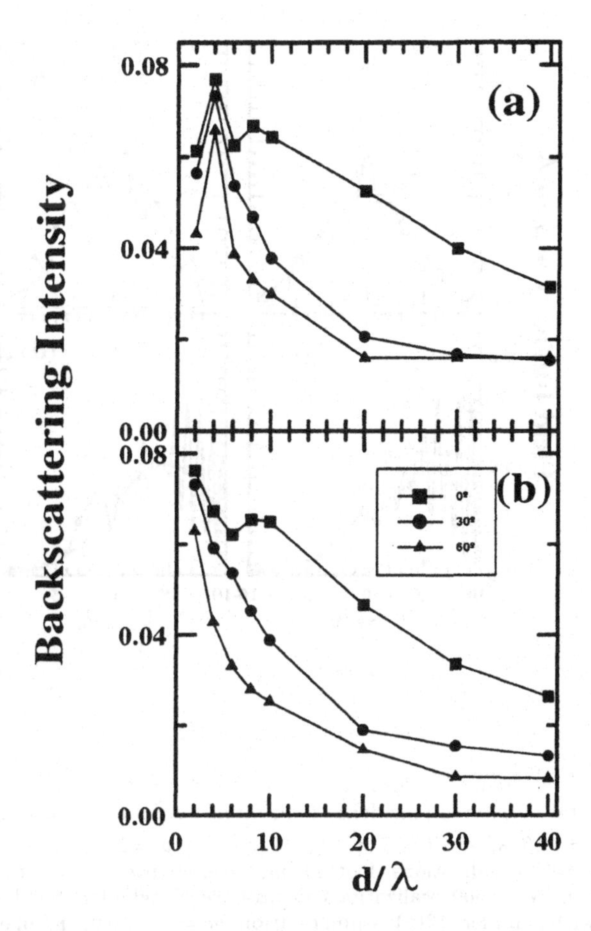

Figure 13. Backscattering peak versus d. $\sigma = 1.9\lambda_0$, $T = 3.16\lambda_0$, $\epsilon' = 2.04$. $a = 1\lambda_0$, $\epsilon = 1$. Squares: $\theta_0 = 0^0$. Circles: $\theta_0 = 30^0$. Triangles: $\theta_0 = 60^0$. 10(a): s-polarization. 10(b): p-polarization.

to a cylinder buried in either a random or a flat interface. The parameters of the random surface are: $\sigma = 1.9\lambda_0$, $T = 3.16\lambda_0$, $d = 10\lambda_0$, $a = 5\lambda_0$, $\epsilon = 7.5$ and $\epsilon' = 2.04$. Four different situations have been addressed: the mean total (scattered+incident) field intensity, averaged over 500 realizations, the intensity averaged over just two realizations, the intensity when the surface is flat, and the average intensity in absence of cylinder. The averaged intensity distribution is very smoothed and has no resemblance to the near zone diffraction pattern of the object. It is in fact very similar to the one obtained in the absence of cylinder. Thus, indicating that the object cannot be discerned from these images. The one averaged over just two realizations shows a clear speckle pattern, whereas the one corresponding to a flat interface has information on the diffraction by the object, but

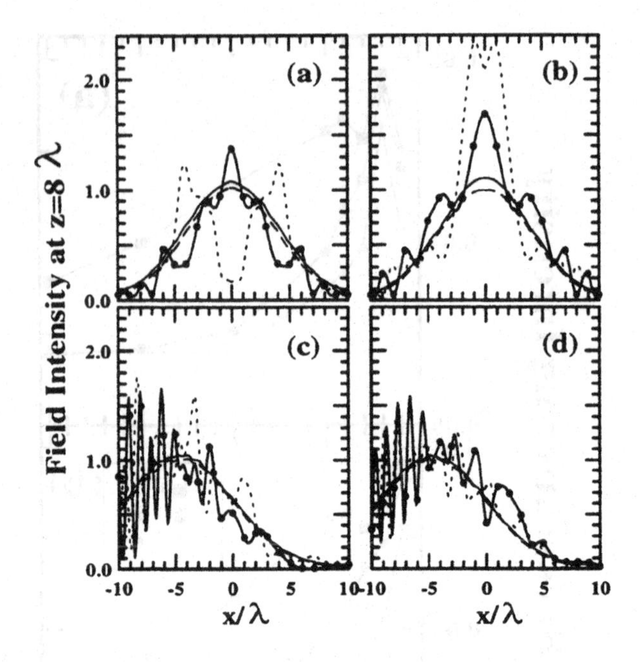

Figure 14. Total (scattered+incident) field intensity distribution at $z = 8\lambda_0$. Surface parameters: $T = 3.16\lambda_0$, $\sigma = 1.9\lambda_0$. $\epsilon' = 2.04$, $\epsilon = 7.5$. $a = 5\lambda_0$ and $d = 10\lambda_0$. Solid line: average over $N = 500$ realizations. Dotted line: average over $N = 2$ realizations. Broken line: average over $N = 500$ realizations in absence of cylinder. Solid line with circles: Flat interface with cylinder. 13(a): *s*-polarization; $\theta_0 = 0^0$. 13(b): *p*-polarization; $\theta_0 = 0^0$. 13(c): *s*-polarization; $\theta_0 = 30^0$. 13(d): *p*-polarization; $\theta_0 = 30^0$.

the presence of the interface greatly alters this pattern with respect to the one that would be given at this plane by the cylinder alone.

6. The question of averaging

As mentioned at the beginning of Section 5, the ensamble average, which has an interpretative value on the fundamentals of the phenomenon under study, may be difficult to implement in practice in several situations. In this section we present results in an attempt to seek for practical alternatives to that kind of average. To this end, we have studied the possibility

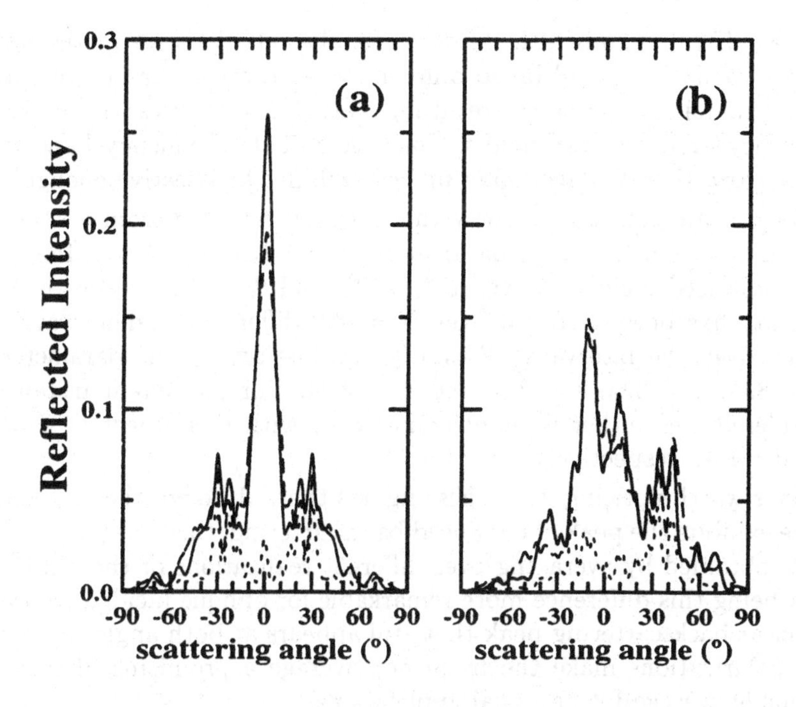

Figure 15. Angular distribution of the mean reflected intensity averaged over $N = 400$ incident wavelengths taken from the interval $[0.95\lambda_0, 1.05\lambda_0]$. s-polarization. $T = 3.16\lambda_0$, $\sigma = 1.9\lambda_0$, $\epsilon' = 2.04$. $a = 2\lambda_0$, $\epsilon = 7.5$, and $W = 8\lambda_0$. Dotted line: no cylinder. Solid line: $d = 10\lambda_0$. Broken line: $d = 8\lambda$. 14(a): $\theta_0 = 0^0$. 14(b): $\theta_0 = 10^0$.

of replacing the ensemble average by a frequency average, which is much easier to carry out in practical instances. This frequency average has been considered for memory effects [42]. In this connection it should be remarked that when a frequency average is performed, a trade-off has to be taken into account between the frequency bandwidth and the meaningfulness of the average. If this bandwidth is too large, the relative sizes of the different surface realizations will be so different to each other that there will be a large dispersion between the resulting scattered intensities, this making the frequency average meaningless. Conversely, if the frequency bandwidth is too narrow, then the averaging process will not be effective enough.

Figs. 15(a) and 15(b), show the intensity reflected from a single realization of the surface with parameters: $\sigma = 1.9\lambda_0$, and $T = 3.16\lambda_0$, under an s polarized incident wave at $\theta_0 = 0^0$, and $\theta_0 = 10^0$, respectively. The intensity has been averaged over $N = 400$ different incident wavelengths taking from the interval $[0.95\lambda_0, 1.05\lambda_0]$. The rest of the parameters are: $W = 8\lambda_0$, $a = 2\lambda_0$, $\epsilon = 7.5$, and $\epsilon' = 2.04$. Three different instances are considered, i.e., either without cylinder, or with this object at $d = 10\lambda_0$, and $d = 8\lambda_0$, respectively.

It is worth noticing from these figures that, although the frequency average exhibits the peak of enhanced backscattering, these results differ from those obtained by averaging over different realizations of the random surface, being this difference more remarkable for oblique incidence. However, the clear backscattering peak that still appears at both angles of incidence and polarizations make the frequency average a promising alternative to ensemble average for practical applications.

Acknowledgments

This research has been supported by Comision Interministerial de Ciencia y Tecnologia of Spain under grant PB 95-0061 and by the Fundación Ramón Areces.

References

1. P. Beckmann and A. Spizzichino, *The Scattering of Electromagnetic Waves from Rough Surfaces*, Pergamon Press, New York, 1963.
2. F.B. Bass and I.M. Fuks, *Wave Scattering from Statistically Rough Surfaces*, Pergamon press, Oxford, 1980.
3. R.F. Wallis and G.I. Stegeman, eds., *Electromagnetic Surface Excitations*, Springer-Verlag, Berlin, 1986.
4. J.A. Ogilvy, *Theory of Wave Scattering from Random Rough Surfaces*, Adam Hilger, Bristol, 1991.
5. M. Nieto-Vesperinas, *Scattering and Diffraction in Physical Optics*, J. Wiley, New York, 1991, Chapter 7.
6. J.M. Bennett, *Surface Finish and its Measurement*, Optical Society of America, Washington D.C., 1992.
7. A. Ishimaru, *Wave Propagation and Scattering in Random Media* Vol II, Academic press, New York, 1978.
8. S.O. Rice, *Commun. Pure Appl. Math.* **4**, 351 (1951).
9. M. Nieto-Vesperinas and N. Garcia, it Opt. Acta **28**, 1651 (1981).
10. J. Shen and A.A. Maradudin, *Phys. Rev. B* **22**, 4234 (1980).

11. Lord Rayleigh, *Theory of Sound*. Vol.2, Dover, New York, 1945.
12. R.F. Millar, *Proc. Camb. Phil. Soc.* **65**, 773 (1969; *Ibid.* **69**, 217 (1971).
13. E.I. Thorsos, *J. Acoust. Soc. Am.* **83**, 78 (1988).
14. J.M. Soto-Crespo and M. Nieto-Vesperinas, *J. Opt. Soc. Am. A* **6**, 367 (1989).
15. E.I. Thorsos and D.R. Jackson, *J. Acoust. Soc. Am.* **86**, 261 (1989).
16. J.M. Soto-Crespo, M. Nieto-Vesperinas and A.T. Friberg, *J. Opt. Soc. Am. A* **7**, 1185 (1990).
17. K.A. OĎonnell and E.R. Mendez, *J. Opt. Soc. Am. A* **4**, 1194 (1987).
18. M. Nieto-Vesperinas and J.M. Soto-Crespo, *Opt. Lett.* **12**, 979 (1987).
19. A.J. Sant, J.C. Dainty and M.J. Kim, *Opt. Lett.* **14**, 1183 (1989).
20. T.R. Michel and K.A. O'Donnell, *J. Opt. Soc. Am. A* **9**, 1374 (1992).
21. M. Nieto-Vesperinas and J.A. Sanchez-Gil, *Phys. Rev. B* **48**, 4132 (1993).
22. P.M. Morse and H. Feshbach, *Methods of Theoretical Physics*, Mc Graw Hill, New York, 1953.
23. M. Nieto-Vesperinas and J.M. Soto-Crespo, *Opt. Lett.* **12**, 979 (1987); J.A. Sanchez-Gil and M. Nieto-Vesperinas, *J. Opt. Soc. Am. A* **8**, 1270 (1991).
24. D.N. Pattanayak and E. Wolf, *Opt. Comm.* **6**, 217 (1972).
25. A. Madrazo and M. Nieto-Vesperinas, *J. Opt. Soc. Am. A* **12**, 1298 (1995).
26. A.A. Maradudin, E.R. Mendez and T. Michel, *Opt. Lett.* **14**, 151 (1989); A.A. Maradudin, T. Michel, A.R. Mc Gurn and E.R. Mendez, *Ann. Phys.* (New York) **203**, 255 (1990).
27. N. Garcia and M. Nieto-Vesperinas, *Opt. Lett.* **24**, 2090-2092, (1993).
28. R. Carminati, A. Madrazo and M. Nieto-Vesperinas, *Opt. Comm.* **111**, 26-33 (1994).
29. N. Garcia and M. Nieto-Vesperinas, *Opt. Lett.* **20**, 949-951 (1995); N. Garcia and M. Nieto-Vesperinas, *Appl. Phys. Lett.* **66**, 3399-3341 (1995).
30. L. Novotny, D.W. Pohl and P. Regli, *J. Opt. Soc. Am. A* **11**, 1768-1779 (1994).
31. M.C. Hutley and D. Maystre, *Opt. Comm.* **19**, 431-436 (1976).
32. N. Garcia, *Opt. Comm.* **45**, 307-310 (1983).
33. U. Ch. Fischer and D.W. Pohl, *Phys. Rev. Lett.* **62**, 458-461 (1989).
34. P.M. Adam, L. Salomon, F. de Fornel and J.P. Goudonnet, *Phys. Rev. B* **48**, 2680-2683 (1993).
35. D. Maystre and M. Neviere, *J. Optics (Paris)* **8**, 165-174 (1977).
36. H. Raether, *Surface Plasmons on Smooth and Rough Surfaces and Gratings*, Springer Tracts on Modern Physics, Vol. III, (Springer-Verlag, Berlin, 1988), Ch.1.
37. P. Dawson, F. de Fornel and J.P. Goudonnet, *Phys. Rev. Lett.* **72**, 2927-2930 (1994).
38. M.J. Kim, J.C. Dainty, A.T. Friberg and A.J. Sant, *J. Opt. Soc. Am. A* **7**, 569-577 (1990).
39. J.A. Sanchez-Gil and M. Nieto-Vesperinas, *J. Opt. Soc. Am. A* **8**, 1270-1286 (1991).
40. J.A. Sanchez-Gil and M. Nieto-Vesperinas, *Phys. Rev. B* **45**, 8623-8633 (1992).
41. E. Jakeman, *J. Opt. Soc. Am. A* **5**, 1638-1648 (1988); Ref. 1, pp. 111-123.
42. L. Tsang, G. Zhang, and K. Pak, *Micr. and Opt. Tech. Lett.* **11**, 300-304 (1996).

ACOUSTIC PULSES PROPAGATING IN RANDOMLY LAYERED MEDIA

J.P. FOUQUE
Department of Mathematics
North Carolina State University
Raleigh, NC 27695-8205

Abstract. We present mathematical results which have been obtained in the last fifteen years for waves propagating in randomly layered media in the framework of separation of scales introduced by G. Papanicolaou and his coauthors. Basic facts on diffusion-approximation and on the use of stochastic calculus will be recalled. We show how a multi-frequency asymptotic analysis gives a precise description of the probability distribution of the transmitted front generated by an incoming pulse. An inverse problem consists in reconstructing the macroscopic variations of the medium from the noisy reflected signal. We present the solution proposed by G. Papanicolaou and his coauthors and show how it depends on statistical estimates for the local power spectral densities of the non stationary reflected signal. We conclude by another approach to this estimation problem which uses a time reversal method introduced in the context of ultrasounds by M. Fink and his group.

1. Introduction

Waves propagating in layered media can be studied by using the propagator technique which consists in decomposing the wave into right and left going waves traveling in the direction where the coefficients of the medium are varying. By using a Fourier transform in the other directions it is then possible to reduce the problem to an initial value problem in the frequency domain. Assuming that the medium is random with a correlation length much smaller than the typical distances of propagation, pulses containing intermediate wavelengths can be used to probe the medium. This is done by an asymptotic analysis as these scales separate and by using diffusion-

J.-P. Fouque (ed.), Diffuse Waves in Complex Media, 319–345.

approximation results. This approach has been initiated by G. Papanicolaou and his coauthors. We refer to the review paper [1] for a detailed presentation of the method. The goal of this contribution is to present in a condensed way the results obtained in these recent years using this technique.

In Section 2 we present the simplest case of a one-dimensional random medium. We recall the basic facts needed to describe Markovian coefficients, we introduce the different scales and we write carefully the boundary conditions. Simulations borrowed from [1] show the transmitted pulse and the refected signal. Integral representations of these quantities of interest are given.

Section 3 is devoted to the study of the transmitted front. The basic diffusion-approximation result needed in this study is presented and applied to deduce the shape of the front. Generalizations to the three-dimensional layered case are indicated at the end of the section.

The reflected signal is studied in Section 4. We review the asymptotic analysis developed in [1], the limiting spectral densities and the system of transport equations used in the inverse problem.

In Section 5 we present recent results obtained by using time reversal mirror techniques. We show that sending back into the same medium a piece of the reflected signal after time reversal will produce a refocalization at the initial source in time and in space.

2. A One-Dimensional Model

2.1. ACOUSTIC WAVE EQUATIONS

We consider a simple one-dimensional model where the space variable is denoted by x and the density and the bulk modulus of the medium are respectively $\rho(x)$ and $K(x)$. We assume that these coefficients vary only in the slab $\{0 \leq x \leq L\}$ and remain constant outside. The velocity and the pressure will be denoted respectively by $u(x,t)$ and $p(x,t)$. The linearized acoustic equations are:

$$\left\{ \begin{array}{ll} \rho(x)\frac{\partial u}{\partial t}(x,t) + \frac{\partial p}{\partial x}(x,t) & = 0 \\ \frac{1}{K(x)}\frac{\partial p}{\partial t}(x,t) + \frac{\partial u}{\partial x}(x,t) & = 0, \end{array} \right. \tag{1}$$

with appropriate boundary conditions which will be added later on.

2.2. RANDOM FLUCTUATIONS OF THE COEFFICIENTS

We assume that the coefficients ρ and $1/K$ are made of macroscopic and microscopic variations. The macroscopic variations will be described by the smooth positive functions $\rho_0(x)$ and $1/K_0(x)$ and the microscopic random fluctuations will be described by the stationary centered stochastic processes $\eta(x/\varepsilon^2)$ and $\nu(x/\varepsilon^2)$. The "microscopic" correlation length of the medium is represented by the small parameter ε^2 (since we will need an intermediate scale it is convenient to denote this small scale by ε^2 instead of ε). More comments on the role of the small parameter ε will be given in Section 2.4.2. Our coefficients can be written as follows:

$$\rho(x) = \begin{cases} \rho_0(x)(1+\eta(x/\varepsilon^2)) & \text{for } 0 \le x \le L \\ 1 & \text{for } x < 0 \text{ or } x > L, \end{cases} \tag{2}$$

and

$$\frac{1}{K(x)} = \begin{cases} \frac{1}{K_0(x)}(1+\nu(x/\varepsilon^2)) & \text{for } 0 \le x \le L \\ 1 & \text{for } x < 0 \text{ or } x > L. \end{cases} \tag{3}$$

We assume that $|\eta|$ and $|\nu|$ are bounded by a constant C strictly smaller than 1 so that the density ρ and the bulk modulus K are bounded and positive. Denoting by $\mathbb{E}\{\cdots\}$ the ensemble average with respect to the invariant distribution of the random fluctuations, the correlation length of the medium is:

$$\int_0^\infty \mathbb{E}\{\eta(0)\eta(x/\varepsilon^2)\}dx = \varepsilon^2\alpha_\eta, \tag{4}$$

where α_η is the integrated covariance of the random fluctuations of ρ given by:

$$\alpha_\eta = \int_0^\infty \mathbb{E}\{\eta(0)\eta(x)\}dx, \tag{5}$$

and a similar expression for α_ν.

The fluctuations η and ν are not supposed to be small which means that we are in the "high-contrast" situation.

The macroscopic impedance $I_0(x)$ and sound velocity $C_0(x)$ are defined as follows:

$$I_0(x) = \sqrt{\rho_0(x)K_0(x)}, \tag{6}$$

and

$$C_0(x) = \sqrt{\frac{K_0(x)}{\rho_0(x)}}. \tag{7}$$

Finally we shall assume that the random processes η and ν have nice decorrelation (or mixing) properties which can be described as follows: let

x and y be two points such that $0 < x < y$ and denote by A an event which depends on $(\eta(z), \nu(z))$ only for $0 \leq z \leq x$ and by B an event which depends on $(\eta(z), \nu(z))$ only for $z \geq y$.

We assume that $|\mathbb{P}(A \cap B) - \mathbb{P}(A)\mathbb{P}(B)| \to 0$ exponentially fast as $|y - x| \nearrow +\infty$ and uniformly in the events A and B.

2.3. MARKOVIAN MODELS

Markovian models are models which depend on an auxiliary Markov process taking its values in a state space S. These models are fairly general and are well-suited for explicit computations. To avoid technical difficulties we shall restrict ourself to compact state spaces S and even more than that to finite spaces. In the usual context of Markov processes the variable x represents time while in our situation it is space. We may still call it time when dealing with the theory of Markov processes which we summarize now.

We denote by $(q_x)_{x \geq 0}$ a Markov process with values in S and we assume that q is homogeneous which means that the probability to be in the state s' at time y given that q is in the state s at the earlier time x depends only on $y - x$. This probability will be denoted as follows:

$$\mathbb{P}(q_y = s' / q_x = s) = P_{y-x}(s, s'). \tag{8}$$

For functions $f : S \to \mathbb{R}$, $P_x f(s) = \mathbb{E}\{f(q_x)/q_0 = s\}$ defines a *semi-group* $(P_x)_{x \geq 0}$. Its *infinitesimal generator* Q is obtained as the limit of $x^{-1}(P_x - Id)$ as $x \downarrow 0$. For an infinitesimal time Δx one can write:

$$\Delta x Q(s, s') = \mathbb{P}\{\text{jump from } s \text{ to } s' \text{ in time } \Delta x\},$$

and

$$1 + \Delta x Q(s, s) = \mathbb{P}\{\text{no jump from } s \text{ in time } \Delta x\},$$

since $Q(s, s) = -\sum_{s'=s}$. In the simplest case of two states, $S = \{s_1, s_2\}$, the infinitesimal generator Q is the 2×2 matrix

$$Q = \begin{pmatrix} -\lambda & +\lambda \\ +\lambda & -\lambda \end{pmatrix},$$

where $\lambda > 0$ is the jump rate. In that case we have for an infinitesimal time Δx:

$$\begin{aligned} \mathbb{P}\{\text{jump from } s_1 \text{ to } s_2 \text{ in time } \Delta x\} &= \\ \mathbb{P}\{\text{jump from } s_2 \text{ to } s_1 \text{ in time } \Delta x\} &= \lambda \Delta x, \end{aligned}$$

and also

$$1 - \lambda\Delta x = \mathbb{P}\{\text{no jump from } s_1 \text{ (resp. } s_2) \text{ in time } \Delta x\}.$$

The *invariant*, or *equilibrium*, distribution is the unique probability distribution $\bar{\mathbb{P}}$ on S such that if the initial distribution of q_0 is $\bar{\mathbb{P}}$ then the distribution of q_x at later times $x > 0$ remains equal to $\bar{\mathbb{P}}$. $\bar{\mathbb{P}}$ satisfies the equilibrium equation $\bar{\mathbb{P}}Q = 0$ and it can be shown that $P_x f(s)$ converges exponentially fast to $\bar{\mathbb{P}}f = \int f d\bar{\mathbb{P}}$ as $x \nearrow \infty$ for every initial state s.

In the two states case $\bar{\mathbb{P}}$ is simply $(1/2, 1/2)$. The jump times are Poisson distributed: that is the differences between consecutive jumps are independent and exponentially distributed with the probability density function $\lambda \exp(-\lambda x)$, $x > 0$, λ being the jump rate. Taking $s_1 = s$ and $s_2 = -s$ for some s such that $|s| < 1$, one can construct a fluctuation process by setting $\eta(x) = q_x$, initially distributed according to the invariant probability $\bar{\mathbb{P}} = (1/2, 1/2)$. This process can be represented as follows: let $(X_1, X_2, \ldots)$ be a sequence of independent identically distributed exponential(λ) random variables. $\eta(0)$ being uniformly distributed over $\{-s, s\}$, $\eta(x)$ remains constant until X_1, the first jump time, it then jumps to the other value and remains constant until $X_1 + X_2$, the second jump time; it jumps like that between s and $-s$ at the successive jump times $X_1 + \cdots + X_k$ as in the following Figure:

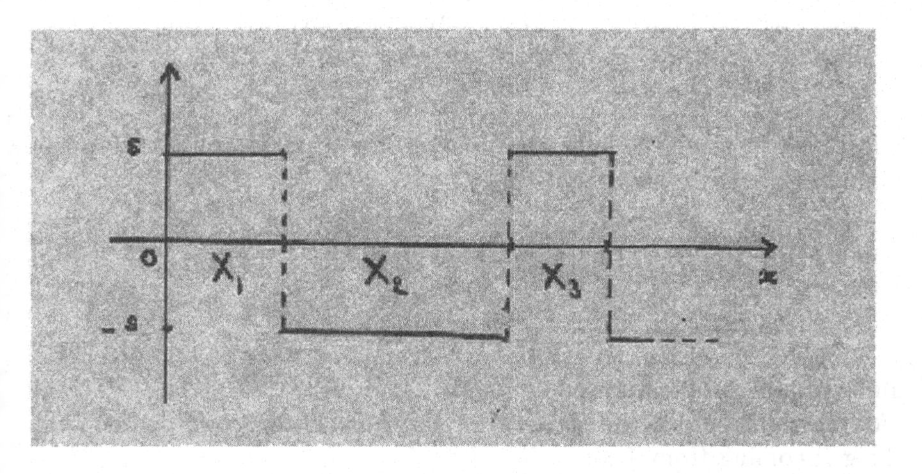

Figure 1. One realization of a two-state Markov process.

The fluctuation process $\eta(x/\varepsilon^2)$ is simply obtained by replacing the X_i's by $\varepsilon^2 X_i$. A simple exercise shows that α_η given by (5) is equal to $s^2/2\lambda$.

More generally the fluctuation processes η and ν will be obtained by setting $\eta(x/\varepsilon^2) = \eta(q_{x/\varepsilon^2})$ and $\nu(x/\varepsilon^2) = \nu(q_{x/\varepsilon^2})$ where q is a Markov process on S under its invariant distribution $\bar{\mathbb{P}}$ and η and ν are two real functions defined on S such that $|\eta|, |\nu| \leq C < 1$. We assume that η and ν are centered: $\mathbb{E}\{\eta(x/\varepsilon^2)\} = \int \eta d\bar{\mathbb{P}} = 0$ and $\mathbb{E}\{\nu(x/\varepsilon^2)\} = \int \nu d\bar{\mathbb{P}} = 0$.

2.4. BOUNDARY CONDITIONS

2.4.1. *Right and left going waves*

In order to provide (1) with appropriate boundary conditions we introduce the right and left going waves A and B as follows:

$$\left\{ \begin{array}{ll} A & = I_0^{-\frac{1}{2}} p + I_0^{\frac{1}{2}} u \\ B & = -I_0^{-\frac{1}{2}} p + I_0^{\frac{1}{2}} u. \end{array} \right. \tag{9}$$

They satisfy the following equations:

$$\frac{\partial}{\partial x}\begin{pmatrix} A \\ B \end{pmatrix} = -\frac{1}{C_0(x)}\left[\begin{pmatrix} 1 & 0 \\ 0 & -1 \end{pmatrix} + \begin{pmatrix} m^\varepsilon(x) & n^\varepsilon(x) \\ -n^\varepsilon(x) & -m^\varepsilon(x) \end{pmatrix}\right]\frac{\partial}{\partial t}\begin{pmatrix} A \\ B \end{pmatrix} + \frac{d}{dx}\left(\log\sqrt{I_0(x)}\right)\begin{pmatrix} 0 & 1 \\ 1 & 0 \end{pmatrix}\begin{pmatrix} A \\ B \end{pmatrix}, \tag{10}$$

where we have defined:

$$\begin{array}{ll} m^\varepsilon(x) & = \frac{1}{2}(\eta(x/\varepsilon^2) + \nu(x/\varepsilon^2)), \\ n^\varepsilon(x) & = \frac{1}{2}(\eta(x/\varepsilon^2) - \nu(x/\varepsilon^2)). \end{array} \tag{11}$$

For instance in the particular case where only $1/K(x)$ contains random fluctuations we have $\eta = 0$ and:

$$\begin{pmatrix} m^\varepsilon(x) & n^\varepsilon(x) \\ -n^\varepsilon(x) & -m^\varepsilon(x) \end{pmatrix} = \frac{1}{2}\nu(x/\varepsilon^2)\begin{pmatrix} 1 & -1 \\ 1 & -1 \end{pmatrix}. \tag{12}$$

The boundary conditions we impose on A and B correspond to a two-point boundary value problem for equation (10).

At the right end of the slab we impose a radiation condition so that nothing is coming from $+\infty$:

$$B(L,t) = 0 \quad \text{for every } t \quad \text{(radiation condition).} \tag{13}$$

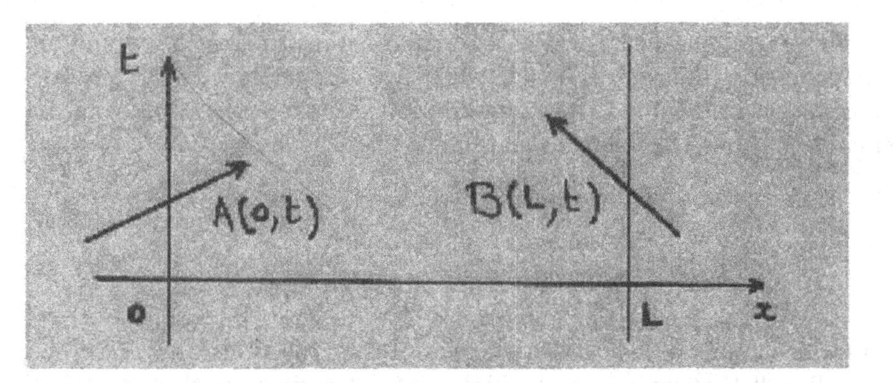

Figure 2. Waves entering the slab.

At the left end of the slab we have an incoming right going wave corresponding to a pulse entering the slab at $x = 0$. The shape of this pulse will be given by a smooth compactly supported function f. It is rescaled such that the typical wavelength is small of order ε and a factor $\varepsilon^{-\gamma}$ is introduced in order to rescale also the amplitudes:

$$A(0,t) = \varepsilon^{-\gamma} f(t/\varepsilon) \quad \text{(scaled pulse)}. \tag{14}$$

The equations being linear, the multiplicative factor $\varepsilon^{-\gamma}$ plays no particular role for a constant ε. The case $\gamma = 0$ corresponds to constant amplitudes while the case $\gamma = \frac{1}{2}$ corresponds to a constant energy entering the slab since in that case $\int |A(0,t)|^2 dt = \int |f(t)|^2 dt$ which is independent of ε.

2.4.2. *Separation of scales*

We shall perform an asymptotic analysis of the problem as $\varepsilon \downarrow 0$ in the **regime of separation of scales** which can be summarized as follows:

$$\begin{pmatrix} \text{correlation length} \\ \approx \ \varepsilon^2 \end{pmatrix} \ll \begin{pmatrix} \text{wavelength} \\ \approx \ \varepsilon \end{pmatrix} \ll \begin{pmatrix} \text{distance of propagation} \\ \approx \ 1 \end{pmatrix}. \tag{15}$$

We would like to insist here on the role of the small parameter ε which is only a convenient mathematical tool to embed our problem in a family of similar problems (indexed by ε) for which we can perform an asymptotic analysis and use the limit as an approximation for the original problem. In practical situations the following parameters will be given:

$$\begin{cases} l: & \text{correlation length,} \\ L: & \text{distance of propagation} \approx 1, \\ \lambda: & \text{wavelength.} \end{cases}$$

where we use a unit of length of the order of the typical distances of propagation. Recall that we are in the high-contrast regime which means that the fluctuations are not supposed to be weak.

If the following condition is satisfied:

$$\frac{\lambda}{L} \approx \frac{l}{\lambda} \ll 1,$$

one can define a small parameter ε by

$$\frac{\lambda}{L} \approx \frac{l}{\lambda} \approx \varepsilon,$$

which implies $\lambda \approx \varepsilon$ and $l \approx \varepsilon^2$ since $L \approx 1$. In our model this is achieved by taking a pulse of the form $f(t/\varepsilon)$ and fluctuations of the form $\eta(x/\varepsilon^2)$. For instance, in the context of Geophysics, typical frequencies are 20Hz, the sound speed being about 2km/s, we get $\lambda = 100\text{m} = 10^{-1}\text{km}$. The correlation length being like 2-3m, our condition will be satisfied for L's of order a few kilometers and ε will be like 10^{-2} which is small.

2.4.3. *Simulations*

Figure (3) shows a simulation which we borrowed from [1], Figure 1.1a (we like to thank the authors of this review paper here). The bottom part shows one realization of the one-dimensional velocity profile $C_0(x)$ with its macroscopic variation (on the scale of kilometers in the context of Geophysics) and the random fluctuations with a small correlation length (of order a few meters) as described in (3), η, the random fluctuations in the density $\rho(x)$, being 0 in this model. The top part shows the propagation of a pulse into this medium. The width of the pulse is chosen such that we are in the regime of separation of scales (15) which is of the order of a hundred meters. One can observe the spreading of the front of the pulse which moves with the mean speed and the formation of a *coda* due to multiple scattering. There is no attenuation in this model.

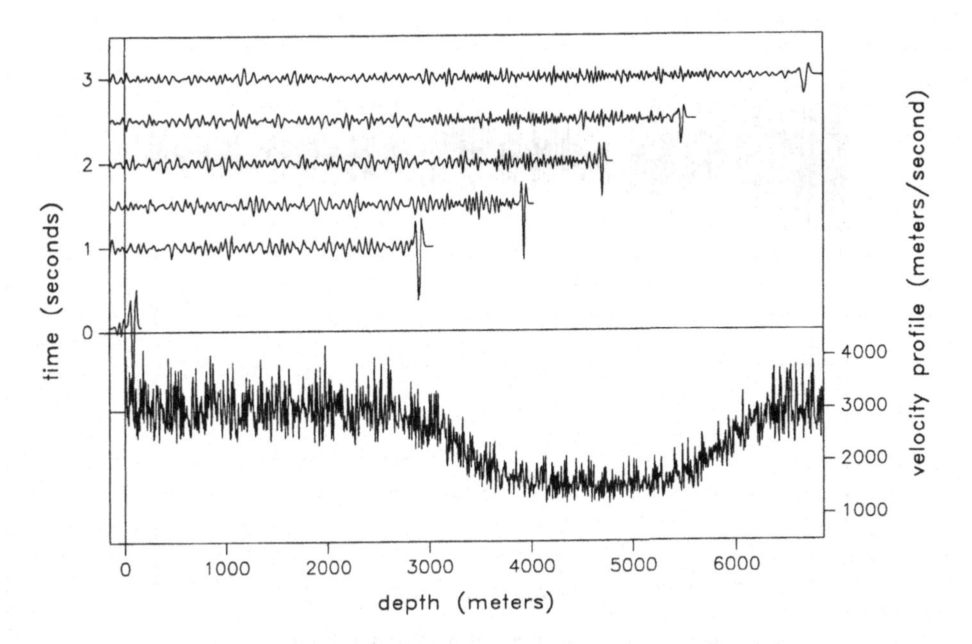

Figure 3. Velocity profile (bottom) - transmitted pulse (top)

Figure (4) (also borrowed from [1], Figure 1.1b) shows the reflected signal at the surface $x = 0$ as a function of the travel time which will be defined in the following paragraph. The bottom part shows the reflected signal when the fluctuations in the sound speed are removed (the scale is on the right). The top part shows the very fluctuating reflected signal when random fluctuations are present (the scale is on the left). One can observe the small signal to noise ratio (of order 10^{-3}) and guess that extracting information from the noisy reflected signal is not easy.

The goal of this paper is to present recent complements to the theory elaborated in [4], [3], [2], [16] and presented in the review paper [1].

2.5. QUANTITIES OF INTEREST: TRANSMISSION AND REFLECTION

2.5.1. *Transmission*

The transmitted signal is the right going wave at $x = L$, namely $A(L, t)$. Defining the *travel time* in the deterministic macroscopic medium by

$$\tau(x) = \int_0^x \frac{dy}{C_0(y)}, \tag{16}$$

our quantity of interest is the pulse front observed around the arrival time

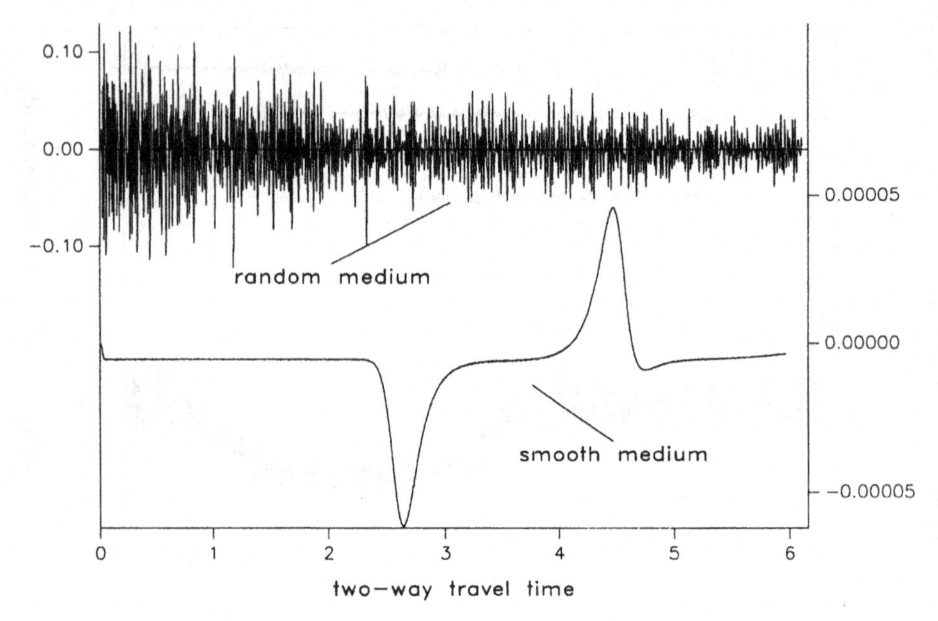

Figure 4. Reflected signal with and without noise

$\tau(L)$ in the original scale ε of the initial pulse (14). This consists in introducing a new scaled time variable σ and setting $t = \tau(L)+\varepsilon\sigma$. The quantity to be studied is

$$a^{\varepsilon}(x,\sigma) = A(x,\tau(x)+\varepsilon\sigma), \tag{17}$$

and the pulse front is $a^{\varepsilon}(L,\sigma)$.

2.5.2. *Reflection*

The reflected signal is the left going wave at $x = 0$, namely $B(0,t)$. This signal can be observed locally by opening a time window centered at time t_0 and in the scale ε. The quantity of interest is then $B(0,t_0+\varepsilon\sigma)$. The quantity to be studied is

$$b^{\varepsilon}(x,\sigma) = B(x,-\tau(x)+\varepsilon\sigma), \tag{18}$$

and the reflected signal becomes $b^{\varepsilon}(0,\frac{t_0}{\varepsilon}+\sigma)$.

The wave functions a^{ε} and b^{ε} are A and B along the characteristics $t = \tau(x)$ and $t = -\tau(x)$ in the scale ε in time.

2.5.3. *Integral representations*

The medium being time-independent it is natural to perform a Fourier transform in time. We define $\hat{a^{\varepsilon}}(x,\omega) = \int e^{i\omega\sigma}a^{\varepsilon}(x,\sigma)d\sigma$ and similarly for

$\hat{b^\varepsilon}(x,\omega)$. Taking a Fourier transform of equations (9) and using the definitions (17) and (18) we get

$$\frac{d}{dx}\begin{pmatrix} \hat{a^\varepsilon} \\ \hat{b^\varepsilon} \end{pmatrix} = \left(\frac{1}{\varepsilon}P^\varepsilon(x,\omega) + Q^\varepsilon(x,\omega)\right)\begin{pmatrix} \hat{a^\varepsilon} \\ \hat{b^\varepsilon} \end{pmatrix}, \tag{19}$$

where

$$P^\varepsilon(x,\omega) = \frac{i\omega}{C_0(x)}\begin{pmatrix} m^\varepsilon(x) & n^\varepsilon(x)e^{-2i\omega\frac{\tau(x)}{\varepsilon}} \\ -n^\varepsilon(x)e^{2i\omega\frac{\tau(x)}{\varepsilon}} & -m^\varepsilon(x) \end{pmatrix}, \tag{20}$$

and

$$Q^\varepsilon(x,\omega) = \frac{d}{dx}\left(\log\sqrt{I_0(x)}\right)\begin{pmatrix} 0 & e^{-2i\omega\frac{\tau(x)}{\varepsilon}} \\ e^{2i\omega\frac{\tau(x)}{\varepsilon}} & 0 \end{pmatrix}. \tag{21}$$

The boundary conditions for (19) are simply

$$\hat{a^\varepsilon}(0,\omega) = \varepsilon^{-\gamma}\hat{f}(\omega)\ ,\ \hat{b^\varepsilon}(L,\omega) = 0. \tag{22}$$

Denoting by $(\hat{a_1^\varepsilon}, \hat{b_1^\varepsilon})$ the solution of (19) where the first boundary condition is replaced by $\hat{a^\varepsilon}(0,\omega) = 1$ in (22), $\varepsilon^{-\gamma}\hat{f}(\omega)$ being a simple multiplicative factor we obtain the following integral representations for our quantities of interest:

$$A(L,\tau(L)+\varepsilon\sigma) = a^\varepsilon(L,\sigma) = \frac{\varepsilon^{-\gamma}}{2\pi}\int e^{-i\omega\sigma}\hat{f}(\omega)\hat{a_1^\varepsilon}(L,\omega)d\omega, \tag{23}$$

and

$$B(0,t_0+\varepsilon\sigma) = b^\varepsilon(0,\frac{t_0}{\varepsilon}+\sigma) = \frac{\varepsilon^{-\gamma}}{2\pi}\int e^{-i\omega(\frac{t_0}{\varepsilon}+\sigma)}\hat{f}(\omega)\hat{b_1^\varepsilon}(0,\omega)d\omega. \tag{24}$$

2.5.4. *Hyperbolicity and the role of L*

By the asumptions we have made on the coefficients of equation (1) we know that there is a finite maximum speed of propagation C_{max}. Therefore, for any given time $T > 0$, the reflected signal before T, $(B(0,t), t \leq T)$, does not depend on L as soon as $L > C_{max}T$. It is interesting to notice that nevertheless $\hat{b_1^\varepsilon}(0,\omega)$ in (24) does depend on L. Of course the transmitted front depends on L.

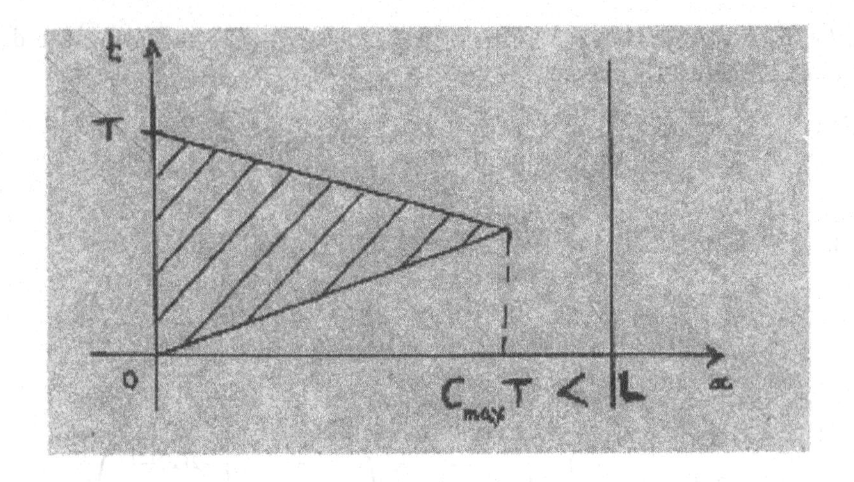

Figure 5. Hyperbolicity

3. The Transmitted Front

3.1. ONE-DIMENSIONAL MODEL

The transmitted front $A(L, \tau(L) + \varepsilon\sigma)$ has the integral representation (23) where $(\hat{a}_1^\varepsilon, \hat{b}_1^\varepsilon)$ is the solution of (19) with the bondary conditions $\hat{a}_1^\varepsilon(0,\omega) = 1$ and $\hat{b}_1^\varepsilon(L,\omega) = 0$. This is a linear two-point boundary value problem well centered and scaled in the sense that the random coefficients appearing in P^ε, such as $\eta(x/\varepsilon^2)$, are centered ($\mathbb{E}\{P^\varepsilon\} = 0$) and $\frac{1}{\varepsilon}\eta(x/\varepsilon^2)$ is approximately a white noise.

It is convenient to introduce the *propagator* for this problem. It is a complex 2×2 matrix $Y^\varepsilon(x,\omega)$ solution of (19) with the initial condition $Y^\varepsilon(0,\omega) = Id$. Noting that (α, β) is a solution of (19) implies that $(\bar{\beta}, \bar{\alpha})$ is also a solution one can show that the propagator can be written

$$Y^\varepsilon = \begin{pmatrix} \alpha & \bar{\beta} \\ \beta & \bar{\alpha} \end{pmatrix}. \tag{25}$$

The trace of $\frac{1}{\varepsilon}P^\varepsilon + Q^\varepsilon$ being 0, the determinant of Y^ε is constant equal to 1. In other words $|\alpha|^2 - |\beta|^2 = 1$ for every x and ω. By the definition of the propagator and the fact that $\hat{a}_1^\varepsilon(0,\omega) = 1$ we have

$$\begin{pmatrix} \hat{a}_1^\varepsilon(x,\omega) \\ \hat{b}_1^\varepsilon(x,\omega) \end{pmatrix} = Y^\varepsilon(x,\omega) \begin{pmatrix} 1 \\ \hat{b}_1^\varepsilon(0,\omega) \end{pmatrix}, \tag{26}$$

which applied at $x = L$ gives

$$\hat{a_1^\varepsilon}(L,\omega) = \frac{1}{\bar{\alpha}(L,\omega)}, \quad \hat{b_1^\varepsilon}(0,\omega) = -\frac{\beta(L,\omega)}{\bar{\alpha}(L,\omega)}. \tag{27}$$

The property $\det Y^\varepsilon = 1$ is the *conservation of energy* relation

$$|\hat{a_1^\varepsilon}(L,\omega)|^2 + |\hat{b_1^\varepsilon}(0,\omega)|^2 = 1 \tag{28}$$

which implies in particular that $\hat{a_1^\varepsilon}(L,\omega)$ appearing in (23) is uniformly bounded by 1.

3.2. CHARACTERIZATION OF THE FRONT

In the regime of constant amplitudes, $\gamma = 0$ in (23), the transmitted front admits the integral representation

$$a^\varepsilon(L,\sigma) = \frac{1}{2\pi}\int e^{-i\omega\sigma}\hat{f}(\omega)\hat{a_1^\varepsilon}(L,\omega)d\omega. \tag{29}$$

In order to characterize its limiting probability distribution as $\varepsilon \downarrow 0$ one needs to study the joint distribution of $\hat{a_1^\varepsilon}(L,\omega)$ simultaneously for all the ω's. This is a complicated infinite dimensional problem. One way to treat it is to study all the moments of the front and reduce it to an infinite number of finite dimensional problems by the following relation

$$\begin{aligned}\mathbb{E}\left\{a^\varepsilon(L,\sigma_1)^{p_1}\cdots a^\varepsilon(L,\sigma_k)^{p_k}\right\} = \left(\frac{1}{2\pi}\right)^p \int e^{-i\Sigma\sigma_j\omega_j^l}\prod\left(\hat{f}(\omega_j^l)\right) \\ \times\mathbb{E}\left\{\prod\left(\hat{a_1^\varepsilon}(L,\omega_j^l)\right)\right\}\prod(d\omega_j^l)\end{aligned} \tag{30}$$

where $p_1,\ldots,p_k$ are integers, $p = p_1 + \cdots + p_k$ and

$$\left(\omega_j^l;\ j = 1,\ldots,k;\ l := l_j = 1,\ldots,p_j\right)$$

forms a finite set of frequencies. The asymptotic limit of this moment requires to study the joint distribution of $\left\{\hat{a_1^\varepsilon}(L,\omega_1),\ldots,\hat{a_1^\varepsilon}(L,\omega_p)\right\}$ for p distinct frequencies $(\omega_1,\ldots,\omega_p)$. This amounts to the study of a system of p equations like (19) for which we define a $2p \times 2p$ propagator matrix $Y^\varepsilon(x,\omega_1,\ldots,\omega_p)$ obtained by puting the $Y^\varepsilon(x,\omega_j)$'s on the diagonal. This propagator is still the solution of an equation like (19) with

$P^\varepsilon(x,\omega_1,\ldots,\omega_p)$ (resp. $Q^\varepsilon(x,\omega_1,\ldots,\omega_p)$) obtained by puting the $P^\varepsilon(x,\omega_j)$'s (resp. $Q^\varepsilon(x,\omega_j)$'s) on the diagonal. The initial condition remains

$$Y^\varepsilon(0,\omega_1,\ldots,\omega_p) = Id$$

for the $2p \times 2p$ Identity matrix.

The limit as $\varepsilon \downarrow 0$ of the solution of equations of this type is described in the theory of "diffusion-approximation" that we summarize in the next paragraph.

3.3. A LIMIT THEOREM

3.3.1. *The main result*

Let (q_x) be a Markov process as in Section 2.3 and (Y^ε_x) defined as the unique $\mathbb{R}^d$-valued solution of the differential equation

$$\frac{dY^\varepsilon_x}{dx} = \frac{1}{\varepsilon}F\left(x, \frac{x}{\varepsilon}, q_{x/\varepsilon^2}, Y^\varepsilon_x\right) + G\left(x, \frac{x}{\varepsilon}, q_{x/\varepsilon^2}, Y^\varepsilon_x\right), \tag{31}$$

starting with the initial condition $Y^\varepsilon_0 = y_0 \in \mathbb{R}^d$. F and G are smooth functions, periodic in the second variable (the fast phase $\frac{x}{\varepsilon}$), and F is centered in the sense that

$$\mathbb{E}\left\{F(x', \tau, q_x, y)\right\} = 0 \tag{32}$$

when the first, second and fourth variables are fixed and the expectation is with respect to the invariant distribution of the Markov process (q_x). Note that (19) is a linear equation of this form, the first variable x appearing in $C_0(x)$ and $I_0(x)$ and the second variable, the fast phase, appearing in the exponentials. q is the Markov process driving the fluctuations η and ν on the scale ε^2.

The **main result** of this Section asserts that the process $(Y^\varepsilon_x)_{x\geq 0}$ (which is not Markovian by itself) converges in probability distribution, as $\varepsilon \downarrow 0$, to the d-dimensional Markov diffusion process $(Y_x)_{x\geq 0}$ which admits the following infinitesimal generator:

$$L_x = \int_0^\infty \langle \mathbb{E}\left\{F(x,\tau,q_0,y)\cdot\nabla_y F(x,\tau,q_t,y)\cdot\nabla_y\right\}\rangle_\tau dt \\ +\langle \mathbb{E}\left\{G(x,\tau,q_0,y)\right\}\rangle_\tau, \tag{33}$$

where $\langle\cdots\rangle_\tau$ denotes an average over a period with respect to the variable τ.

3.3.2. *A simple case*

In order to give an idea of how a formula like (33) is derived, let us consider the simple case where F only depends on q and y and $G = 0$. Equation (31) reduces to

$$\frac{dY_x^\varepsilon}{dx} = \frac{1}{\varepsilon} F\left(q_{x/\varepsilon^2}, Y_x^\varepsilon\right), \tag{34}$$

where $Y_0^\varepsilon = y_0 \in \mathbb{R}^d$ and $\mathbb{E}\{F(q_x, y)\} = 0$.

Denoting q_{x/ε^2} by q_x^ε, the pair $(q_x^\varepsilon, Y_x^\varepsilon)$ is a Markov process on $S \times \mathbb{R}^d$ which admits the infinitesimal generator

$$L^\varepsilon = \frac{1}{\varepsilon^2} Q + \frac{1}{\varepsilon} F(q, y) \cdot \nabla_y, \tag{35}$$

easily obtained by taking a limit as $x \downarrow 0$ in

$$\frac{1}{x}\left(\mathbb{E}\{f(q_x^\varepsilon, Y_x^\varepsilon)/q_0 = q, Y_0^\varepsilon = y_0\} - f(q, y)\right),$$

for any smooth function f and Q being the infinitesimal generator of the Markov process (q_x) as described in Section 2.3.

For a given real smooth function f we define a perturbed test function f^ε by

$$f^\varepsilon(q, y) = f(y) + \varepsilon f_1(q, y) + \varepsilon^2 f_2(q, y), \tag{36}$$

for functions f_1 and f_2 to be determined as follows: find an infinitesimal generator L such that $|L^\varepsilon f^\varepsilon - Lf|$ goes to 0 as $\varepsilon \downarrow 0$. When computing $L^\varepsilon f^\varepsilon$ there is no $\mathcal{O}(1/\varepsilon^2)$ term since Q in (35) acts on the variable q and f is independent of q. The condition for the $\mathcal{O}(1/\varepsilon)$ term to vanish is

$$Q f_1 + F \cdot \nabla_y f = 0, \tag{37}$$

which is a Poisson equation which admits the solution $f_1 = (-Q)^{-1}(F \cdot \nabla_y f)$ since F is centered with respect to the invariant distribution of the Markov process (q_x). This solution can also be written

$$f_1(q, y) = \int_0^\infty \mathbb{E}\{F(q_t, y) \cdot \nabla_y f / q_0 = q\} dt. \tag{38}$$

The $\mathcal{O}(1)$ term is equal to $Q f_2 + F \cdot \nabla_y f_1$. By substracting and adding $\mathbb{E}\{F \cdot \nabla_y f_1\}$, f_2 can be chosen such that

$$Q f_2 + F \cdot \nabla_y f_1 - \mathbb{E}\{F \cdot \nabla_y f_1\} = 0,$$

since this is again a Poisson equation in Q with a centered second term. It follows that the $\mathcal{O}(1)$ term is reduced to $\mathbb{E}\{F \cdot \nabla_y f_1\}$ which is the desired infinitesimal generator L which, acting on f, can also be written

$$Lf = \int_0^\infty \mathbb{E}\{F(q_0, y) \cdot \nabla_y (F(q_t, y) \cdot \nabla_y f)\} dt \tag{39}$$

which is formula (33) in this simple case. The averaging with respect to the variable τ in the general case (33) is done after the averaging with respect to q since $\tau = x/\varepsilon$ is a fast variable but slower than x/ε^2.

In the simple linear case

$$\frac{dY_x^\varepsilon}{dx} = \frac{1}{\varepsilon}\eta(x/\varepsilon^2)F(Y_x^\varepsilon),$$

the infinitesimal generator L reduces to the simple second order operator

$$L = \alpha_\eta F(y) \cdot \nabla_y (F(y) \cdot \nabla_y)$$

where α_η is defined in (5).

In general the limiting process (Y_x) is a Markov diffusion with infinitesimal generator given by (33). This process can be represented as solution to a *stochastic differential equation* involving *Brownian motions.* Ito's stochastic calculus can be used in the computation of expressions such as $\mathbb{E}\{\phi(Y_x)\}$ which will approximate $\mathbb{E}\{\phi(Y_x^\varepsilon\}$ for ε small.

3.4. APPLICATION TO THE FRONT SHAPE

The expectation appearing in (30) can be written as the expectation of a functional ϕ of the propagator:

$$\mathbb{E}\left\{\prod \left(\hat{a_1^\varepsilon}(L, \omega_j^l)\right)\right\} = \mathbb{E}\left\{\phi\left(Y_L^\varepsilon(\omega_1, \ldots, \omega_p)\right)\right\},$$

which will be approximated, for ε small, by $\mathbb{E}\left\{\phi\left(Y_L(\omega_1, \ldots, \omega_p)\right)\right\}$ as explained in the previous Section. In this case the multi-frequency limiting propagator $Y_x(\omega_1, \ldots, \omega_p)$ is obtained as $Y_x^\varepsilon(\omega_1, \ldots, \omega_p)$ in Section 3.2 by putting the $Y_x(\omega_j)$'s on the diagonal where each $Y_x(\omega_j)$ can be written as in (25). Using (27) we see that the functional ϕ is given by

$$\phi(Y(\omega_1, \ldots, \omega_p)) = \prod_{j=1}^{p} \left(\frac{1}{\bar{\alpha}(\omega_j)}\right).$$

The use of Ito's stochastic calculus enables us to derive explicit formulas for expectations of such functionals. We refer to [6] for details where it is shown that $a^\varepsilon(L,\sigma)$ converges in probability distribution as $\varepsilon \downarrow 0$ to $a(L,\sigma)$ which admits the following integral representation similar to (29)

$$a(L,\sigma) = \frac{1}{2\pi} \int e^{-i\omega\sigma} \hat{f}(\omega) \hat{a_1}(L,\omega) d\omega, \tag{40}$$

where

$$\hat{a_1}(L,\omega) = \exp\left(i\omega Z_L - \omega^2 \alpha_n \int_0^L \frac{dy}{C_0(y)^2} \right), \tag{41}$$

and Z_L is a centered Gaussian random variable with variance $2\alpha_m \int_0^L \frac{dy}{C_0(y)^2}$. The coefficients α_n and α_m are associated by (5) to m and n given by (11). From (40) and (41) one can easily deduce

$$a(L,\omega) = (f \star G_L)(\sigma + Z_L), \tag{42}$$

where G_L is the centered Gaussian density with variance $2\alpha_n \int_0^L \frac{dy}{C_0(y)^2}$.

The result (42) tells us that the transmitted front, in the same scale as the orinal pulse, has approximately a deterministic shape obtained by a convolution of the original pulse f with a Gaussian kernel G_L but shifted by a Gaussian random variable Z_L. This *stabilization effect* , the deterministic shape, has also been obtained independently in [14]. As observed in [6], this is the only coherent energy exiting the slab in the sense that $A(L, t + \varepsilon\sigma)$ for $t \neq \tau(L)$ or $B(0, t + \varepsilon\sigma)$ for any t, go to 0 as $\varepsilon \downarrow 0$ in this regime of constant amplitudes $\gamma = 0$ in (14).

3.5. THE 3-D LAYERED CASE

We briefly summarize here the result obtained in [5]. The space is now three-dimensional and we assume that the coefficients of the acoustic equations depend only on the "depth", the vertical direction z, the transverse directions being denoted by x and y. We assume that a point source is placed at the surface $z = 0$ and that the pressure is observed at the bottom of the random slab $z = -L$ as shown in Figure (6). We suppose that only the bulk modulus contains random fluctuations so that we have the same description of the coefficients than in (2) and (3) with $0 \leq x \leq L$ replaced by $-L \leq z \leq 0$ and η being set to 0.

Using the specific Fourier transform

$$\hat{\phi}(\omega, h, k) = \int\int\int e^{i w(t - hx - ky)} \phi(t, x, y) dt dx dy \tag{43}$$

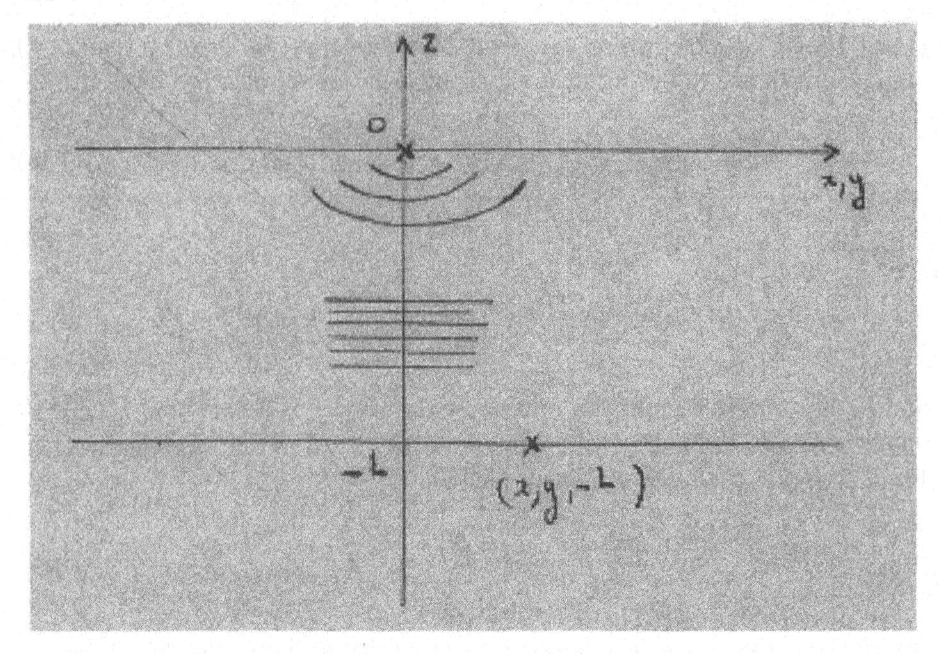

Figure 6. A point source over a 3-D randomly stratified medium.

and defining the slowness $\kappa = \sqrt{h^2 + k^2}$, the travel time

$$\tau(z, \kappa) = \int_0^z \frac{\sqrt{(1 - C_0(s)^2\kappa^2)}}{C_0(s)} ds,$$

and the impedance

$$I_0(z, \kappa) = \frac{\rho_0(z)C_0(z)}{\sqrt{1 - C_0(s)^2\kappa^2)}},$$

it is possible to obtained an integral representation similar to (23) for the transmitted pressure $p(t_0 + \varepsilon\sigma, x, y, -L)$ in the ε-scale of the initial pulse, around a time t_0, at the bottom $z = -L$ and at the offset (x, y). This integral representation in very similar to (23) but involves also a fast phase $e^{i\phi/\varepsilon}$.

In [5] it is shown how to apply simultaneously a diffusion-approximation result as in Section 3.3 and a *stationary phase* argument to derive the asymptotic transmitted pressure field.

The position $(x, y, -L)$ being given, we solve $\nabla\phi = 0$ to get the values (h_0, k_0) where the integral concentrate. Solving $\phi = 0$ for these values one can deduce the right time t_0 where the front arrives at this given position. For this value t_0 and jointly in (x, y) , the transmitted pressure $p(t_0 +$

$\varepsilon\sigma, x, y, -L)$ converges in probability distribution to

$$\beta(x, y, -L)\,(f \star G'_L(x, y))\,(\sigma + \alpha(x, y)Z_L). \tag{44}$$

The coefficients β and α are described in details in [5], β being the coefficient obtained in the geometric acoustic theory. We simply remark that, like in the one-dimensional case (42), the stabilization of the front is still there but now the convolution is with the derivative of a Gaussian kernel. It is also noticable that only one Gaussian random variable Z_L is needed to described the pressure field for all the offsets (x, y).

4. Reflected Signals

4.1. THE REFLECTION COEFFICIENT

Reformulating in the slab $[x, L]$, with the left extremity varying, the one-dimensional problem described in Section 3.1 , we define a reflection coefficient $R^\varepsilon(x, \omega)$ and a transmission coefficient $T^\varepsilon(x, \omega)$ corresponding respectively to $\hat{b_1^\varepsilon}$ and $\hat{a_1^\varepsilon}$ defined in Section 2.5.3.

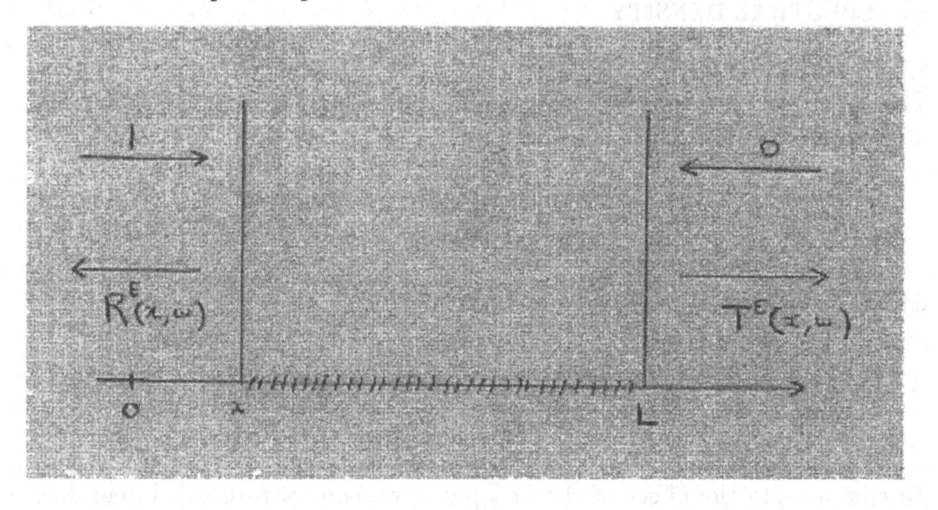

Figure 7. Relection coefficient; invariant embedding.

This technique is known as *invariant embedding.* The conservation of energy relation (28) becomes

$$|R^\varepsilon|^2 + |T^\varepsilon|^2 = 1. \tag{45}$$

It turns out that the reflection coefficient R^ε satisfies a *Riccati equation* which we write in the case $C_0(x) = 1$ and $\nu = 0$ for simplicity

$$\frac{dR^\varepsilon}{dx} = \frac{i\omega}{2\varepsilon}\eta(x/\varepsilon^2)\left(e^{2i\omega\frac{x}{\varepsilon}} + 2R^\varepsilon + (R^\varepsilon)^2 e^{-2i\omega\frac{x}{\varepsilon}}\right), \tag{46}$$

with the final condition $R^\varepsilon(L,\omega) = 0$. At $x = 0$ we have $R^\varepsilon(0,\omega) = \hat{b}_1^\varepsilon(0,\omega)$ and the integral representation for the reflected signal

$$B(0, t_0 + \varepsilon\sigma) = \varepsilon^{-\gamma}\frac{1}{2\pi}\int e^{-i\omega\sigma}e^{-i\omega\frac{t_0}{\varepsilon}}\hat{f}(\omega)R^\varepsilon(0,\omega)d\omega. \tag{47}$$

For $\gamma = \frac{1}{2}$, in the regime of constant energy, it has been shown that the stochastic signal $\{B(0, t_0 + \varepsilon\sigma)\}_{\sigma\in\mathbf{R}}$ converges in probability distribution to a stationary centered Gaussian process whose covariance is described in the following Section (we refer to [1] for this result). That is in this sense that the reflected signal $B(0,t)$ is locally stationary.

4.2. SPECTRAL DENSITY

4.2.1. *Covariance*

The covariance of the reflected signal in the ε scale, around the time t and for the pulse f is given by

$$C_{t,f}^\varepsilon(\sigma) = \mathbb{E}\left\{\overline{B^\varepsilon(0,t)}B^\varepsilon(0, t+\varepsilon\sigma)\right\}, \tag{48}$$

which, by using the integral representation (47), can be written

$$\varepsilon^{-2\gamma}\frac{1}{4\pi^2}\int\int e^{-i\omega_2\sigma}e^{i(\omega_1-\omega_2)\frac{t}{\varepsilon}}\overline{\hat{f}}(\omega_1)\hat{f}(\omega_2)\mathbb{E}\left\{\overline{R^\varepsilon(0,\omega_1)}R^\varepsilon(0,\omega_2)\right\}d\omega_1 d\omega_2. \tag{49}$$

To compensate the effect of the fast phase we concentrate this integral on nearby frequencies by the change of variables $\omega_1 - \omega_2 = \varepsilon h$ and $\omega_2 = \omega$. The covariance (48) becomes

$$\varepsilon^{-2\gamma+1}\frac{1}{4\pi^2}\int\int e^{-i\omega_2\sigma}e^{ith}\overline{\hat{f}}(\omega+\varepsilon h)\hat{f}(\omega)\mathbb{E}\left\{\overline{R^\varepsilon(0,\omega+\varepsilon h)}R^\varepsilon(0,\omega)\right\}d\omega dh. \tag{50}$$

Denoting by $\Lambda^\varepsilon(\omega,t)$ the inverse Fourier transform with respect to h of

$\mathbb{E}\left\{\overline{R^\varepsilon(0,\omega+\varepsilon h)}R^\varepsilon(0,\omega)\right\}$ and using $\overline{\hat{f}}(\omega+\varepsilon h)\hat{f}(\omega) \approx |\hat{f}(\omega)|^2$, the covariance (48) can be written

$$\varepsilon^{-2\gamma+1}\frac{1}{2\pi}\int e^{-i\omega\sigma}|\hat{f}(\omega)|^2\Lambda^\varepsilon(\omega,t)d\omega. \tag{51}$$

Choosing $\gamma = \frac{1}{2}$, it can be proved that $\Lambda^\varepsilon(\omega,t)$ converges to a limit $\Lambda(\omega,t)$ as $\varepsilon \downarrow 0$ (see [1] for details). The covariance of the limiting centered stationary Gaussian process is then equal to

$$\frac{1}{2\pi}\int e^{-i\omega\sigma}|\hat{f}(\omega)|^2\Lambda(\omega,t)d\omega. \tag{52}$$

where $|\hat{f}(\omega)|^2\Lambda(\omega,t)$ is its spectral density.

4.2.2. *The uniform background case*
In the uniform background case, $C_0(x) = 1$, the limiting spectral density can be explicitely computed and is given by

$$\Lambda(\omega,t) = \frac{\alpha\omega^2}{4(1+\omega^2\alpha t)^2} \tag{53}$$

where $\alpha = \alpha_\eta$, defined in (5), in the case $\nu = 0$ for instance.

This explicit case is important for the validation of numerical studies.

4.2.3. *The general case*
In the general case, C_0 varying, the limiting spectral density is no longer explicitely known.

From the Riccati equation (46) for R^ε we can deduce an infinite system of equations for the quantities

$$\overline{R^\varepsilon(x,\omega+\varepsilon h)}^M R^\varepsilon(x,\omega)^N, \quad M,N \geq 0 \quad \text{integers}.$$

It is then possible to study the limit, as $\varepsilon \searrow 0$, of the expectations of these quantities. It turns out that the diagonal terms obtained for $M = N$ form, in the limit, a closed system. Denoting these limiting expectations by

$$\widetilde{W}^N(x,h) = \lim_{\varepsilon\downarrow 0}\mathbb{E}\left\{\overline{R^\varepsilon(x,\omega+\varepsilon h)}^N R^\varepsilon(x,\omega)^N\right\},$$

their Fourier transforms in h, denoted by $\{W^N(x,t), N \geq 0\}$, satisfy a system of *transport equations* described in the following Section.

4.3. TRANSPORT EQUATIONS

It is shown in [1] that the $\{W^N(x,t), N \geq 0\}$ satisfy the following system of transport equations referred as the *W-equations*:

$$\frac{\partial W^N}{\partial x} - \frac{2N}{C_0(x)}\frac{\partial W^N}{\partial t} - \frac{2\omega^2\alpha N^2}{C_0(x)^2}\left(W^{N+1} - 2W^N + W^{N-1}\right) = 0 \qquad (54)$$

with the condition $W^N|_{x=L} = \delta(t)\delta_{N,0}$ at $x = L$.

Estimating $\Lambda(\omega,t) = W^1(0,t)$ from the observed reflected signal, an inverse problem consists in reconstructing the sound speed $C_0(x)$ which appears in the coefficients of (54). We refer to [1] and [16] for a complete treatment of the inverse problem. We simply remark that it involves a statistical estimation of the spectral density $\Lambda(\omega,t)$. A windowed Fourier transform is used in [1] and a wavelet transform is proposed in [7]. In th next Section we summarize a method using a time reversal mirror technique and developped in [8].

5. Time Reversal Mirror Techniques

A time reversal mirror is, roughly speaking, a device which is capable of receiving an acoustic signal in time, keeping it in memory, and sending it back into the medium in the reversed direction of time. In the context of ultrasounds, time-reversal mirrors have been developed and their effects have been studied experimentally by Mathias Fink and his team at the Laboratoire Ondes et Acoustique (ESPCI-Paris). We refer to [9]. We shall exploit here the possibility of sending back only a part of the reflected signal after time reversal. The idea being that the content of the reflected random signal around the arrival time t is related in some way to the macroscopic properties of the medium at a depth corresponding to a travel time $t/2$. Sending back, in the time-independent medium, this portion of the signal, the new reflected signal should contain more information about the properties of the medium at this precise depth. It will turn out that this method enables us to have direct access to quantites such as local correlations of the reflected signal without any computation, these correlations being made by the medium itself and therefore very robust.

In the next Section, we present this idea in the simple one-dimensional model containing the main property of separation of scales described in Section 2.4.2.

We recall that, from (47), the reflected signal has the integral representation

$$B(0,t) = \varepsilon^{-\gamma}\frac{1}{2\pi}\int e^{-i\omega\frac{t}{\varepsilon}}\hat{f}(\omega)R^{\varepsilon}(0,\omega)d\omega. \tag{55}$$

We use a scaled time-window $G_{\varepsilon}(t) = G(t/\varepsilon^{\rho})$ where G is anything like an indicator function $1_{(-M,M)}(t)$ or a Gaussian density. The reflected signal $B(0,t)$ is "cut" by $G_{\varepsilon}(t)$ around the time t_0 and time-reversed to obtain the new signal $B(0,t_0-t)G_{\varepsilon}(t)$. In the ε-scale this new signal is

$$g(\sigma) = B(0,t_0-\varepsilon\sigma)G(-\frac{\sigma}{\varepsilon^{\rho}}), \tag{56}$$

to be sent in the same medium.

In order to obtain an integral representation for the reflection from this new pulse, we need its Fourier transform which is given by the convolution

$$\hat{g}(\omega) = \varepsilon^{\rho-\gamma-1}\int e^{i\omega'\frac{t_0}{\varepsilon}}\overline{\hat{f}}(\omega')\hat{G}((\omega-\omega')\varepsilon^{\rho-1})d\omega'. \tag{57}$$

Using (47) with the pulse f replaced by the new pulse g we have the integral representation for the new reflected signal denoted by $S^{\varepsilon}_{t_0}$:

$$S^{\varepsilon}_{t_0}(\sigma) = \frac{1}{2\pi}\int e^{-i\omega\sigma}e^{-i\omega\frac{t_0}{\varepsilon}}\hat{g}(\omega)R^{\varepsilon}(0,\omega)d\omega. \tag{58}$$

Using (57) in (58) we get

$$\begin{aligned} S^{\varepsilon}_{t_0}(\sigma) = \varepsilon^{\rho-\gamma-1}\tfrac{1}{2\pi}\int\int\ & e^{-i\omega\sigma}e^{-i(\omega'-\omega)\frac{t_0}{\varepsilon}} \\ & \overline{\hat{f}}(\omega')\overline{R^{\varepsilon}(0,\omega')}R^{\varepsilon}(0,\omega)\hat{G}((\omega-\omega')\varepsilon^{\rho-1})d\omega'd\omega. \end{aligned} \tag{59}$$

As in Section 4.2.1, in order to compensate the fast phase, we perform the change of variables $\omega'-\omega = \varepsilon h$, keeping ω, to obtain

$$S^{\varepsilon}_{t_0}(\sigma) = \frac{1}{2\pi}\int e^{-i\omega\sigma}\widehat{S^{\varepsilon}_{t_0}}(\omega)d\omega, \tag{60}$$

with

$$\widehat{S^{\varepsilon}_{t_0}}(\omega) = \varepsilon^{\rho-\gamma}\int e^{iht_0}\overline{\hat{f}}(\omega+\varepsilon h)\overline{R^{\varepsilon}(0,\omega+\varepsilon h)}R^{\varepsilon}(0,\omega)\hat{G}(\varepsilon^{\rho}h)dh. \tag{61}$$

We are now ready to perform an asymptotic analysis as $\varepsilon \downarrow 0$.

5.2. ASYMPTOTICS

In the regime of constant amplitudes, $\gamma = 0$, we set $\rho = 0$ in (61). This means that the window $G_\varepsilon(t)$ is independent of ε but can be chosen small compared to 1.

The first moment $\mathbb{E}\{\widehat{S^\varepsilon_{t_0}}(\omega)\}$ involves quantities such as

$$\mathbb{E}\{\overline{R^\varepsilon(0,\omega+\varepsilon h)}R^\varepsilon(0,\omega)\}.$$

Applying the asymptotic analysis described in Section 4.2.1 one can show that

$$\lim_{\varepsilon\downarrow 0} \mathbb{E}\{\widehat{S^\varepsilon_{t_0}}(\omega)\} = 2\pi\overline{\hat{f}}(\omega)\left(\Lambda(\omega,t)\star G(t)\right)(t_0)$$

Using second moments one can actually prove that $\widehat{S^\varepsilon_{t_0}}(\omega)$ itself converges to this same deterministic limit. We refer to [8] for more details.

To summarize, the Fourier transform of the new reflected signal $S^\varepsilon_{t_0}(\sigma)$ leads to a consistent estimator of the spectral density $\Lambda(\omega,t)$, up to a convolution by the window $G(t)$.

We conclude this Section by two remarks.

This convergence has been obtained when the time-reversed reflected signal is observed at time t_0, the same time around which we have cut the original reflected signal. Let us suppose that it is observed at time $t_0+\Delta t_0$. Then a fast phase $e^{i\omega\Delta t_0/\varepsilon}$ appears in the integral (59) and kills it so that, for ε small, no new reflected signal is observed at time $t_0+\Delta t_0$. This is observed experimentally (see [9]).

If $\hat{f}(0) \neq 0$, which means that the pulse contains low frequencies, it is possible to use the low frequency limit

$$\lim_{\omega\downarrow 0} \frac{\Lambda(\omega,t)}{\omega^2}$$

in the inverse problem for $C_0(x)$ in (54).

5.3. BACK AND FORTH

Following an idea used in the context of target selection [10], it is possible to repeat the same procedure at another time t_1. The new pulse will be $S^\varepsilon_{t_0}$ whose Fourier transform $\widehat{S^\varepsilon_{t_0}}(\omega)$ is given by (61).

We now perform the following:

1. Send $S^\varepsilon_{t_0}$ back in the medium without time-reversal.
2. Get the reflected signal.
3. Cut it around time t_1 with the new window $H_\varepsilon(t) = H(t/\varepsilon^\delta)$.
4. Time-reverse it and send it back in the medium.
5. Get the new reflected signal denoted by $S^\varepsilon_{t_0,t_1}(\sigma)$.

Applying (61) with $\varepsilon^{-\gamma}\hat{f}$, t_0 and G_ε replaced respectively by $\widehat{S^\varepsilon_{t_0}}$, t_1 and H_ε we get

$$\widehat{S^\varepsilon_{t_0,t_1}}(\omega) = \varepsilon^\delta \int e^{iht_1} \overline{\widehat{S^\varepsilon_{t_0}}(\omega+\varepsilon h)}\, \overline{R^\varepsilon(0,\omega+\varepsilon h)} R^\varepsilon(0,\omega) \hat{H}(\varepsilon^\delta h) dh. \tag{62}$$

Replacing $\widehat{S^\varepsilon_{t_0}}(\omega+\varepsilon h)$ given in (61) we obtain

$$\begin{aligned} \widehat{S^\varepsilon_{t_0,t_1}}(\omega) = \quad & \varepsilon^{\delta+\rho-\gamma} \int\int e^{iht_1} e^{-ih't_0} \hat{f}(\omega+\varepsilon(h+h')) \\ & \times \left\{ R^\varepsilon(0,\omega) \overline{R^\varepsilon(0,\omega+\varepsilon h)}^2 R^\varepsilon(0,\omega+\varepsilon(h+h')) \right\} \\ & \times \hat{G}(\varepsilon^\rho h') \hat{H}(\varepsilon^\delta h) dh dh'. \end{aligned} \tag{63}$$

In this integral three frequencies are involved which is not good for an asymptotics. Rewriting

$$e^{iht_1} e^{-ih't_0} = e^{-i(h+h')t_0} e^{it(t_0+t_1)},$$

fixing $t = t_0 + t_1$ and integrating with respect to t_0 between 0 and t (instead of $-\infty$ and $+\infty$ by causality), we use

$$\int_0^t dt_0 e^{-i(h+h')t_0} \ldots = \int \delta_0(h+h') \ldots$$

to obtain

$$\begin{aligned} \int_0^t S^\varepsilon_{t_0,t-t_0}(\sigma) dt_0 = \quad & \varepsilon^{\delta+\rho-\gamma} \frac{1}{2\pi} \int e^{-i\omega\sigma} \hat{f}(\omega) \\ & \times \int e^{iht} \left\{ R^\varepsilon(0,\omega)^2 \overline{R^\varepsilon(0,\omega+\varepsilon h)}^2 \right\} \hat{G}(\varepsilon^\rho h) \hat{H}(\varepsilon^\delta h) dh d\omega. \end{aligned} \tag{64}$$

Using $\gamma = \delta + \rho$ (with $\rho = 0$ and $\delta = \gamma$ for instance), the Fourier transform of $\int_0^t S^\varepsilon_{t_0,t-t_0}(\sigma) dt_0$ gives a consistent estimator, as $\varepsilon \downarrow 0$, for $W^2(0,t)$ which can be used in the transport equations (54) to improve the estimation of the coefficients in the inverse problem. This procedure can be obviously carried out to higher order $W^N(0,t)$.

5.4. THE 3-D STRATIFIED CASE

With the notations introduced in Section 3.5, the scaled reflected velocity can be written

$$u^\varepsilon(t_0 + \varepsilon\sigma, x_0 + \varepsilon\Delta x, y_0 + \varepsilon\Delta y, 0)$$

where (x_0, y_0) is the offset center of observation, t_0 is the time window center and the last component is $z = 0$. A time-window of this reflected velocity will be used as a new pulse sent into the medium after time-reversal. The new pulse in the ε scale is

$$g(\sigma, \Delta x, \Delta y) = u^{\varepsilon}(t_0 - \varepsilon\sigma, x_0 + \varepsilon\Delta x, y_0 + \varepsilon\Delta y, 0)G_{\varepsilon}(-\sigma, \Delta x, \Delta y).$$

This new source will generates a new reflected velocity at the surface $u(t + \varepsilon\sigma, x + \varepsilon\Delta x, y + \varepsilon\Delta y, 0)$ for which an integral representation can be written. This representation involves the slowness (Section 3.5) and we refer to [11] for more details on this generalization of Section 2 to the three-dimensional case, as well as the asymptotic analysis as $\varepsilon \downarrow 0$.

The main result obtained in [11] says that this new reflected velocity is asymptotically 0 (for ε small) , unless $(x, y) = (-x_0, -y_0)$ and $t = t_0$. In other words we have refocusing at the right time and the right location.

6. Conclusion

We have presented in the simplest way, we believe, the technique of diffusion-approximation applied to waves propagating in randomly layered media. This has been done in the regime of separation of scales explained in Section 2.4.2. We have shown how to use this technique to study the shape of the transmitted front (Section 3) and the reflected signal (Section 4). In the last Section 5 we have presented recent results using time reversal mirror techniques.

References

1. Asch, M., Kohler, W., Papanicolaou, G., Postel, M., and White, B. (1991) Frequency content of randomly scattered signals, *SIAM Review* **33**, 519-626.
2. Asch, M., Papanicolaou, G., Postel, M., Sheng, P., and White, B. (1990) Frequency content of randomly scattered signals, Part I, *Wave Motion* **12**, 429-450.
3. Burridge, B., Papanicolaou, G., Sheng, P., and White, B. (1989) Probing a random medium with a pulse, *SIAM J. Appl. Math.* **49**, 582-607.
4. Burridge, B., Papanicolaou, G., and White, B. (1987) Statistics for pulse reflexion from a randomly layered medium, *SIAM J. Appl. Math.* **47** (1987, 146-168.
5. Chillan, J. and Fouque, J.P. (1998) Pressure fields generated by acoustical pulses propagating in randomly layered media, *SIAM J. Appl. Math.* **58**, 1532-1546.
6. Clouet, J.F. and Fouque, J.P. (1994) Spreading of a pulse travelling in random media, *Annals of Applied Probability* **4**, 1083-1097.
7. Clouet, J.F., Fouque, J.P., and Postel, M. (1995) Spectral analysis of ramdomly scattered signals using the wavelet transform, *Wave Motion* **22**, 145-170.
8. Clouet, J.F. and Fouque, J.P. (1997) A time-reversal method for an acoustical pulse propagating in randomly layered media, *Wave Motion* **25**, 361-368.
9. Fink, M. (1993) Time reversal mirrors, *J. Phys. D: Appl. Phys.* **26**, 1333-1350.
10. Fink, M. and Prada, (1994) Eigenmodes of the time reversal operator: a solution to selective focusing in multiple-target media, *Wave Motion* **20**, 151-163.
11. Fouque, J.P. and Ndzie, J. (in preparation) Refocusing of a time-reversed acoustical pulse propagating in randomly layered media.

12. Kohler, W., Papanicolaou, G., and White, B. (1996) Localization and Mode Conversion for Elastic Waves in Randomly Layered Media, *Wave Motion* **23**, 1-22 and 181-201.
13. Lewicki, P., Burridge, R., and De Hoop, M. (1996) Beyond effective medium theory: pulse stabilization for multimode wave propagation in high-contrast layered media, *SIAM J. Appl. Math.* **56**, 256-276.
14. Lewicki, P., Burridge, R., and Papanicolaou, G. (1994) Pulse stabilisation in a strongly heterogeneous layered medium, *Wave Motion* **20**, 177-195.
15. O'Doherty, R.F. and Anstey, N.A. (1971) Refections on amplitudes, *Geophysical Prospecting* **19**, 430-458.
16. Papanicolaou, G., Postel, M., Sheng, P., and White, B. (1990) Frequency content of randomly scattered signals, Part II, *Wave Motion* **12**, 527-549.
17. Solna, K. (1997) Stable spreading of acoustic pulses due to laminated microstructure, Ph.D. dissertation, Stanford University.

REFLECTION AND TRANSMISSION OF ACOUSTIC WAVES BY A LOCALLY-LAYERED RANDOM SLAB

W. KHOLER
Department of Mathematics
Virginia Polytechnic and State University
Blacksburg, VA 24061, USA
kohler@math.vt.edu

G. PAPANICOLAOU
Department of Mathematics
Stanford University
Stanford, CA 94305-2125, USA
papanicolaou@stanford.edu

AND

B. WHITE
Exxon Research and Engineering Company
Route 22 East
Annandale, NJ 08801, USA
bswhite@erenj.com

Abstract. We consider acoustic propagation through a slab in which rapidly-varying random plane layering has been subjected to smooth small amplitude undulations. The reflection and transmission properties of such a locally-layered slab are studied. Of principal interest is the question of the robustness of the plane layered theory. Do phenomena such as localization and the results of O'Doherty-Anstey theory persist under perturbations of plane layering? We establish some results affirming the robustness of these phenomena.

1. Introduction

The study of wave propagation in disordered or random media has been extensively pursued, being motivated by applications in many of the phys-

J.-P. Fouque (ed.), Diffuse Waves in Complex Media, 347–381.

ical sciences. References [1], [2] and [3] provide an overview of some recent activity. A sizable portion of this work has been focussed upon the study of one-dimensional models. Such models have relevance to geophysical applications; they are also the most tractable mathematically.

One-dimensional random wave propagation exhibits some striking features. Coherent multiple scattering, induced by the fluctuations in the underlying random medium, suppresses transmission and effectively localizes the wave (*c.f.* [4], [5]). If the random medium fluctuations are weak, the corresponding transmission length scale, or localization length, is large. O'Doherty and Anstey [6] discovered that in this regime, a pulse, observed in a Lagrangian frame moving with the appropriate random velocity, will appear to retain its shape up to a slow spreading; references [7, 8, 9, 14] provide further insight into this O'Doherty-Anstey phenomenon.

The question of robustness naturally arises. Do these phenomena persist under modest geometric perturbations of the underlying random layering? The difficulty inherent in providing an answer to this question lies in the fact that such geometric perturbations, no matter how modest, destroys transverse homogeneity and forces one to deal with stochastic partial, rather than ordinary, differential equations.

In [15] we studied extensively the problem of acoustic wave propagation in a rapidly-varying plane layered slab. A small parameter ε was used to delineate three relevant length scales. The largest scale or macroscale (*e.g.* the slab thickness L) was assumed to be $O(1)$. The acoustic wavelength defined an intermediate $O(\varepsilon)$ length scale while the correlation length of the random layering was assumed to vary on the smallest $O(\varepsilon^2)$ scale. On the one hand, therefore, the wavelength of the acoustic radiation was assumed to be small relative to the scale on which deterministic variations in the slab constitutive parameters occurred, so geometric acoustics might reasonably be expected to play an important role. On the other hand, the acoustic wavelength was assumed to span many correlation lengths and one might further expect this scaling to produce a meaningful probabilistic limit asymptotically as $\varepsilon \to 0$. These expectations were indeed realized in [15]. The multiple scattering induced by the random layering created localization phenomena which profoundly affected the reflection and transmission properties of the slab. The O'Doherty-Anstey phenomenon was also shown to follow from a weak-fluctuation variant of this model. In [17] we studied the reflection and transmission of elastic, time harmonic plane waves by randomly layered media, especially mode conversion by the inhomogeneities. In [18] we extended the statistical inverses theory for randomly layered media of [15] to the case of acoustic wave pulses generated by a point source.

In this paper our goal is to test the robustness of this theory with respect to perturbations in the layering geometry. In other words, will the essential features of the theory be preserved if we introduce gentle deformations (undulations) or other reasonably small perturbations into the plane layering structure? We shall refer to this deformed geometry as a locally layered medium. The O'Doherty-Anstey phenomenon is analyzed for a different class of locally layered random media by [14]. The deterministic background fluctuations are not small in amplitude but they are slowly varying.

In what follows, we extend much of the development of [15] to the more complicated locally layered setting, adhering closely to the setup and notation of that paper. In Section 2 the acoustic problem in a locally layered geometry is formulated, and scales are delineated in terms of the small parameter ε as described above. We assume a generalization of the plane-layered model that allows for small spatially-varying deviations in the direction of the random layering and also allows for some small, but fully three-dimensional nonrandom variations in the background. As will be shown, this generalization of the model does produce interesting leading order, three-dimensional spatial dependences in the final results.

In Sections 3 and 4 the quantities of interest, *i.e.* the reflected and transmitted pressure waves, are expressed in terms of reflection and transmission operators. In Section 5 the basic conservation relations for these operators are derived, and in Section 6, a limit for the operators is obtained, valid as $\varepsilon \downarrow 0$. This stochastic limit is conveniently expressed in terms of Ito "white noise" equations, and it is ¿from these governing Ito equations that all the relevant statistics are derived in the remainder of the paper. We check, in Section 7, that this stochastic limit for the operators does indeed reduce to the strictly plane layered theory of [15] when the appropriate expressions are set equal to zero.

In Sections 8, 9 and 10 we investigate, both for reflected and transmitted waves, the coherent field, the space and time correlation functions, and the localization length. A generalized O'Doherty-Anstey theory is presented in Section 11, also for both transmitted and reflected waves. For all these cases, the main qualitative features of the strictly plane-layered theory survive the three-dimensional perturbations, although some $O(1)$ three-dimensional effects can be observed.

2. Problem Formulation and Scaling

As in [15], let (x, y) be coordinates transverse to an initially unperturbed plane layering with the scattering region in the interval $-L \leq z \leq 0$. To define the locally layered geometry, we introduce the coordinate transformation

$$x' = x, \quad y' = y, \quad z' = z + \varepsilon\phi(x, y, z), \tag{1}$$

where ϕ is a smooth and bounded function. The undulations are defined by the surfaces z' = constant and exhibit an $O(\varepsilon)$ (wavelength scale) amplitude variation over $O(1)$ (macroscale) transverse distances.

We are primarily interested in the multiple scattering properties of the bulk locally layered material. Therefore, we shall model the problem so as to eliminate interface complications. We shall flatten the undulations at both ends of the slab by assuming

$$\phi(x, y, -L) = \phi(x, y, 0) = 0 \tag{2}$$

so that $z' = z$ at $-L$ and 0. The half-spaces $z < -L$ and $z > 0$ will be assumed to be constant and homogeneous acoustic media. In $z > 0$, the density, bulk modulus and sound speed will be denoted by ρ_0, K_0 and c_0, respectively (with $K_0 = \rho_0 c_0^2$). The subscript 2 will be used to index their constant counterparts in $z < -L$. In the randomly undulating slab region $-L \leq z \leq 0$, the subscript 1 will be used. As in [15], the density will be assumed to be constant, *i.e.* $\rho = \rho_1$. The bulk modulus will be modeled as

$$K^{-1} = K_1^{-1}(1 + \nu(z'/\varepsilon^2)) + \varepsilon K_{11}^{-1}(x', y', z') \, , \quad -L \leq z' \leq 0, \tag{3}$$

where K_1 is a constant, K_{11} is a deterministic spatially-varying perturbation and ν is a zero mean, stationary stochastic process bounded in modulus by a constant less than one. Since it is a function of z'/ε^2, the process ν decorrelates on the smallest $O(\varepsilon^2)$ spatial scale in a direction perpendicular to the undulations.

We shall assume that acoustic energy is incident upon the random slab from above. For the present discussion, we shall assume that this radiation is emitted by a point source located at $(0, 0, z_s)$ and having orientation defined by the unit vector $\mathbf{e}$. Then the governing acoustic equations for pressure p and particle velocity $\mathbf{u}$ are (*c.f.* equations (2.17)–(2.19) in [15])

$$\rho\partial_t \mathbf{u} + \nabla p = \varepsilon^{1/2} f(t/\varepsilon)\delta(x)\delta(y)\delta(z - z_s)\mathbf{e} \tag{4a}$$

$$K^{-1}\partial_t p + \nabla \cdot \mathbf{u} = 0 \tag{4b}$$

where

$$\rho = \begin{cases} \rho_0, & z > 0 \\ \rho_1, & -L < z < 0 \\ \rho_2, & z < -L \end{cases} \tag{5}$$

and

$$K^{-1} = \begin{cases} K_0^{-1}, & z > 0 \\ K_1^{-1}(1+\nu(z'/\varepsilon^2)) + \varepsilon K_{11}^{-1}(x',y',z'), & -L < z < 0 \\ K_2^{-1}, & z < -L \end{cases} \tag{6}$$

with x', y' and z' defined by (1). The function f in (4a) defines the pulsed temporal signal emitted by the point source; the $\varepsilon^{1/2}$ factor makes the total energy released by the pulsed point source independent of ε.

We begin by rewriting equations (4), within the (source-free) slab region, in terms of the primed variables. Let $u^{(j)}, j = 1, 2, 3$, denote the three components of particle velocity in the fixed, unprimed Cartesian system. Let a function ψ be defined in terms of the inverse map as follows

$$x = x' \;,\; y = y' \;,\; z = z' - \varepsilon\psi(x', y', z', \varepsilon) \;. \tag{7}$$

Note from (1) that ψ satisfies the functional equation:

$$\psi(x', y', z', \varepsilon) = \phi(x', y', z' - \varepsilon\psi(x', y', z', \varepsilon)) \;. \tag{8}$$

In the slab region $-L \leq z' \leq 0$, the component acoustic equations become

$$\begin{aligned} \rho_1 u_t^{(1)} + p_{x'} + \varepsilon\psi_{x'}(1-\varepsilon\psi_{z'})^{-1}p_{z'} &= 0 \\ \rho_1 u_t^{(2)} + p_{y'} + \varepsilon\psi_{y'}(1-\varepsilon\psi_{z'})^{-1}p_{z'} &= 0 \\ \rho_1 u_t^{(3)} + p_{z'} + \varepsilon\psi_{z'}(1-\varepsilon\psi_{z'})^{-1}p_{z'} &= 0 \\ \left[K_1^{-1}(1+\nu) + \varepsilon K_{11}^{-1}\right] p_t + u_{x'}^{(1)} + u_{y'}^{(2)} + u_{z'}^{(3)} & \\ + \varepsilon(1-\varepsilon\psi_{z'})^{-1}\left[\psi_{x'}u_{z'}^{(1)} + \psi_{y'}u_{z'}^{(2)} + \psi_{z'}u_{z'}^{(3)}\right] &= 0. \end{aligned} \tag{9}$$

Equations (9), in turn, can be used to obtain the following system of equations for p and $u^{(3)}$

$$\begin{aligned} p_{z'} &= -\rho_1(1-\varepsilon\psi_{z'})u_t^{(3)} \\ u_{z't}^{(3)} &= \frac{1-\varepsilon\psi_{z'}}{1+\varepsilon^2(\psi_{x'}^2+\psi_{y'}^2)}\Big[\rho_1^{-1}(p_{x'x'}+p_{y'y'}) - \left[K_1^{-1}(1+\nu)+\varepsilon K_{11}^{-1}\right]p_{tt} \\ &\quad -\varepsilon\left[2\partial_{x'}(\psi_{x'}u_t^{(3)}) + 2\partial_{y'}(\psi_{y'}u_t^{(3)}) - (\psi_{x'x'}+\psi_{y'y'})u_t^{(3)}\right]\Big], \\ &\quad -L \leq z' \leq 0 \,. \end{aligned} \tag{10}$$

For simplicity of notation, we shall now drop the primes and the (3) - superscript. As in equation (2.21) of [15], we introduce the following Fourier transforms

$$\hat{p}(\boldsymbol{\kappa},\omega,z) = \iiint p e^{i\omega(t-\boldsymbol{\kappa}\cdot\mathbf{x})/\varepsilon} dt d\mathbf{x}$$
$$\hat{u}(\boldsymbol{\kappa},\omega,z) = \iiint u e^{i\omega(t-\boldsymbol{\kappa}\cdot\mathbf{x})/\varepsilon} dt d\mathbf{x}\ , \ \boldsymbol{\kappa} = (\kappa_1,\kappa_2),\ \mathbf{x} = (x,y)\ . \quad (11)$$

¿From (8) we infer that

$$\psi = \phi - \varepsilon\phi_z\phi + O(\varepsilon^2)\ . \quad (12)$$

It then follows that the transformed pressure and particle velocity satisfy the following system of integro-differential equations

$$\begin{aligned}
\partial_z\hat{p} &= i\frac{\omega}{\varepsilon}\rho_1\hat{u} - i\omega\rho_1\hat{\phi}_z * \hat{u} + \dots \\
\partial_z\hat{u} &= i\frac{\omega}{\varepsilon}\left[K_1^{-1} - \rho_1^{-1}\kappa^2\right]\hat{p} + i\frac{\omega}{\varepsilon}K_1^{-1}\nu\hat{p} - i\omega\left[[K_1^{-1}\hat{\phi}_z - (\widehat{K_{11}^{-1}})] * \hat{p}\right] + \\
&\quad + i\omega\rho_1^{-1}\hat{\phi}_z * [\kappa^2\hat{p}] + 2\omega^2\left[\kappa_1[(\kappa_1\hat{\phi}) * \hat{u}] + \kappa_2[(\kappa_2\hat{\phi}) * \hat{u}]\right] + \dots \\
&\quad -L \le z \le 0,
\end{aligned} \quad (13)$$

where $\kappa^2 = \boldsymbol{\kappa}\cdot\boldsymbol{\kappa}$ and

$$\begin{aligned}
\hat{\phi}(\omega\boldsymbol{\kappa},z) &= \iint \phi e^{-i\omega\boldsymbol{\kappa}\cdot\mathbf{x}} d\mathbf{x} \\
(\hat{\phi} * \hat{u})(\boldsymbol{\kappa},\omega,z) &= \left(\frac{\omega}{2\pi}\right)^2 \iint \hat{\phi}(\omega\boldsymbol{\lambda},z)\hat{u}(\boldsymbol{\kappa}-\varepsilon\boldsymbol{\lambda},\omega,z)d\boldsymbol{\lambda},
\end{aligned} \quad (14)$$

and the ellipses in equations (13) indicate negligible terms which shall, in subsequent equations, simply be dropped. An examination of equations (13) reveals that the deformation function ϕ and the deterministic bulk modulus perturbation K_{11}^{-1} introduce basically the same level of complexity, *i.e.* $O(1)$ convolution terms.

We are ultimately interested in recasting equations (13) into equations for the amplitudes of upgoing and downgoing waves. Sound speeds c_j and acoustic impedances ζ_j for each of the three regions are defined as

$$c_j = (K_j/\rho_j)^{1/2}, \quad \zeta_j = \rho_j c_j/(1-\kappa^2 c_j^2)^{1/2}, \quad j = 0,1,2. \quad (15)$$

A travel time τ relative to $z = 0$ is then defined as

$$\tau(z,\kappa) = \begin{cases} (1-\kappa^2c_0^2)^{1/2}c_0^{-1}z, & z > 0 \\ (1-\kappa^2c_1^2)^{1/2}c_1^{-1}z, & -L \le z \le 0 \\ -(1-\kappa^2c_1^2)^{1/2}c_1^{-1}L + (1-\kappa^2c_2^2)^{1/2}c_2^{-1}(z+L), & z < -L\ . \end{cases} \quad (16)$$

(Recall that the depth variable in (16) is actually the primed variable. At points outside the slab, we simply equate primed and unprimed depths.) As in equation (2.26) of [15] , we introduce amplitudes A and B by the relations

$$\begin{aligned} \hat{p} &= \zeta^{1/2}[Ae^{i\frac{\omega}{\varepsilon}\tau} - Be^{-i\frac{\omega}{\varepsilon}\tau}] \\ \hat{u} &= \zeta^{-1/2}[Ae^{i\frac{\omega}{\varepsilon}\tau} + Be^{-i\frac{\omega}{\varepsilon}\tau}], \end{aligned} \tag{17}$$

where ζ is the piecewise-constant acoustic impedance taking on the values defined by (15). When equations (17) are substituted into (13), we obtain the following equations for A and B:

$$\begin{aligned} \frac{d}{dz}A &= i\frac{\omega}{\varepsilon}n\left[A - Be^{-i2\frac{\omega\tau}{\varepsilon}}\right] \\ &\quad -i\omega\frac{\rho_1}{\zeta_1}\left(\frac{\omega}{2\pi}\right)^2 \iint \partial_z\left[\hat{\phi}e^{-i\omega\tau_{\boldsymbol{\kappa}}\cdot\boldsymbol{\lambda}}\right] A(\boldsymbol{\kappa} - \varepsilon\boldsymbol{\lambda}, \omega, z)d\boldsymbol{\lambda} \\ &\quad +i\frac{\omega}{2}\zeta_1\left(\frac{\omega}{2\pi}\right)^2 \iint (\widehat{K_{11}^{-1}})e^{-i\omega\tau_{\boldsymbol{\kappa}}\cdot\boldsymbol{\lambda}} A(\boldsymbol{\kappa} - \varepsilon\boldsymbol{\lambda}, \omega, z)d\boldsymbol{\lambda} \\ \frac{d}{dz}B &= i\frac{\omega}{\varepsilon}n\left[Ae^{i2\frac{\omega}{\varepsilon}\tau} - B\right] \\ &\quad +i\omega\frac{\rho_1}{\zeta_1}\left(\frac{\omega}{2\pi}\right)^2 \iint \partial_z\left[\hat{\phi}e^{i\omega\tau_{\boldsymbol{\kappa}}\cdot\boldsymbol{\lambda}}\right] B(\boldsymbol{\kappa} - \varepsilon\boldsymbol{\lambda}, \omega, z)d\boldsymbol{\lambda} \\ &\quad -i\frac{\omega}{2}\zeta_1\left(\frac{\omega}{2\pi}\right)^2 \iint (\widehat{K_{11}^{-1}})\, e^{i\omega\tau_{\boldsymbol{\kappa}}\cdot\boldsymbol{\lambda}} B(\boldsymbol{\kappa} - \varepsilon\boldsymbol{\lambda}, \omega, z)d\boldsymbol{\lambda}, \qquad -L < z < 0, \end{aligned} \tag{18}$$

where (*c.f.* equation (2.28) of [15]):

$$\begin{aligned} n &= \frac{1}{2}K_1^{-1}\zeta_1\nu = \frac{\nu}{2c_1(1-\kappa^2c_1^2)^{1/2}} \\ \tau_{\boldsymbol{\kappa}} &= \nabla_{\boldsymbol{\kappa}}\tau = \frac{-c_1\boldsymbol{\kappa}z}{(1-\kappa^2c_1^2)^{1/2}}. \end{aligned} \tag{19}$$

The boundary conditions accompanying equations (18) follow from the requirement that pressure and the normal component of particle velocity be continuous across the slab interfaces. Because we have assumed that the undulations flatten out at these extremities (*c.f.* (1), (2)), it follows that $\hat{p}$ and $\hat{u}_3 = \hat{u}$ must be continuous at $z = -L$ and 0. Equations (17), in turn, transform these relations into corresponding interface relations for A and B.

In the homogeneous half-space below the slab, *i.e.* $z < -L$, the radiation condition requiring the waves to be downward-propagating leads to $A = 0$.

We shall assume that the slab effective medium is matched to the upper homogeneous half-space, *i.e.* that $\zeta_1 = \zeta_0$. This condition suppresses deterministic coherent reflections from the upper interface, $z = 0$. All energy returning to the upper half-space from the slab will do so because of multiple scattering by the undulating layers and/or reflection by the impedance mismatch at $z = -L$. For the point source excitation of (4a) with $\mathbf{e} = \mathbf{z}_0$, we obtain the boundary conditions (equations (2.30) - (2.32) and (A.1) of [15])

$$\begin{aligned} A(\boldsymbol{\kappa},\omega,-L^+) &= \Gamma_I(-L)e^{-i2\frac{\omega}{\varepsilon}\tau(-L)}B(\boldsymbol{\kappa},\omega,-L^+) \\ B(\boldsymbol{\kappa},\omega,0^-) &= B(\boldsymbol{\kappa},\omega,0^+) = \varepsilon^{3/2}\frac{\hat{f}(\omega)}{2}\zeta_0^{-1/2}e^{i\frac{\omega}{\varepsilon}\tau(z_s)}, \end{aligned} \tag{20}$$

where

$$\begin{aligned} \Gamma_I(-L) &= \frac{\zeta_1 - \zeta_2}{\zeta_1 + \zeta_2} \\ \hat{f}(\omega) &= \int f(t)e^{i\omega t}dt\,. \end{aligned} \tag{21}$$

In some of the subsequent discussions, we shall consider plane wave excitation. For those cases, the second of boundary conditions (20) will be appropriately changed.

The integral terms in (18) couple the amplitudes A and B for all values of slowness $\boldsymbol{\lambda}$. To emphasize this point, as well as to simultaneously recenter the equations, we make the change of dependent variables

$$\begin{aligned} \widetilde{A}(\boldsymbol{\kappa},\boldsymbol{\lambda},\omega,z) &= A(\boldsymbol{\kappa}+\varepsilon\boldsymbol{\lambda},\omega,z)e^{i\omega\tau_{\boldsymbol{\kappa}}\cdot\boldsymbol{\lambda}} \\ \widetilde{B}(\boldsymbol{\kappa},\boldsymbol{\lambda},\omega,z) &= B(\boldsymbol{\kappa}+\varepsilon\boldsymbol{\lambda},\omega,z)e^{-i\omega\tau_{\boldsymbol{\kappa}}\cdot\boldsymbol{\lambda}}\,. \end{aligned} \tag{22}$$

Equations (18) transform into

$$\begin{aligned} \frac{d}{dz}\widetilde{A} + i\omega\frac{\zeta_1}{\rho_1}\boldsymbol{\kappa}\cdot\boldsymbol{\lambda}\widetilde{A} &= i\frac{\omega}{\varepsilon}n\left[\widetilde{A} - \widetilde{B}e^{-i2\frac{\omega\tau}{\varepsilon}}\right] - i\omega\frac{\rho_1}{\zeta_1}\left(\frac{\omega}{2\pi}\right)^2\iint\Big[\hat{\phi}_z(\omega\boldsymbol{\lambda}',z) \\ &\quad + i\omega\frac{\zeta_1}{\rho_1}\boldsymbol{\kappa}\cdot\boldsymbol{\lambda}'\hat{\phi}(w\boldsymbol{\lambda}',z)\,\Big]\widetilde{A}(\boldsymbol{\kappa},\boldsymbol{\lambda}-\boldsymbol{\lambda}',\omega,z)d\boldsymbol{\lambda}' \\ &\quad + i\frac{\omega}{2}\zeta_1\left(\frac{\omega}{2\pi}\right)^2\iint(\widehat{K_{11}^{-1}})(\omega\boldsymbol{\lambda}',z)\widetilde{A}(\boldsymbol{\kappa},\boldsymbol{\lambda}-\boldsymbol{\lambda}',\omega,z)d\boldsymbol{\lambda}' \\ \frac{d}{dz}\widetilde{B} - i\omega\frac{\zeta_1}{\rho_1}\boldsymbol{\kappa}\cdot\boldsymbol{\lambda}\widetilde{B} &= i\frac{\omega}{\varepsilon}n\left[\widetilde{A}e^{i2\frac{\omega\tau}{\varepsilon}} - \widetilde{B}\right] + i\omega\frac{\rho_1}{\zeta_1}\left(\frac{\omega}{2\pi}\right)^2\iint\Big[\hat{\phi}_z(\omega\boldsymbol{\lambda}',z) \\ &\quad - i\omega\frac{\zeta_1}{\rho_1}\boldsymbol{\kappa}\cdot\boldsymbol{\lambda}'\hat{\phi}(\omega\boldsymbol{\lambda}',z)\,\Big]\widetilde{B}(\boldsymbol{\kappa},\boldsymbol{\lambda}-\boldsymbol{\lambda}',\omega,z)d\boldsymbol{\lambda}' \end{aligned}$$

$$-i\frac{\omega}{2}\zeta_1\left(\frac{\omega}{2\pi}\right)^2\iint(\widehat{K_{11}^{-1}})(\omega\boldsymbol{\lambda}',z)\widetilde{B}(\boldsymbol{\kappa},\boldsymbol{\lambda}-\boldsymbol{\lambda}',\omega,z)d\boldsymbol{\lambda}', \tag{23}$$

where all $\boldsymbol{\lambda}$ dependence in the coefficients, unless explicitly noted, has been collapsed.

The structure of equations (23) suggests that we view $\boldsymbol{\kappa}$ and $\boldsymbol{\lambda}$ as independent variables and introduce Fourier transforms with respect to slowness $\boldsymbol{\lambda}$. The correct conjugate spatial variable will, in fact, turn out to be the macroscopic transverse variable $\mathbf{x}$. Anticipating this fact, we define

$$\begin{aligned}\overline{A}(\boldsymbol{\kappa},\mathbf{x},\omega,z) &= \left(\frac{\omega}{2\pi}\right)^2\iint\widetilde{A}(\boldsymbol{\kappa},\boldsymbol{\lambda},\omega,z)e^{i\omega\boldsymbol{\lambda}\cdot\mathbf{x}}d\boldsymbol{\lambda}\\ \overline{B}(\boldsymbol{\kappa},\mathbf{x},\omega,z) &= \left(\frac{\omega}{2\pi}\right)^2\iint\widetilde{B}(\boldsymbol{\kappa},\boldsymbol{\lambda},\omega,z)e^{i\omega\boldsymbol{\lambda}\cdot\mathbf{x}}d\boldsymbol{\lambda}\,.\end{aligned} \tag{24}$$

Equations (23) transform into the following transport equations for $\overline{A}$ and $\overline{B}$

$$\begin{aligned}\frac{d}{dz}\overline{A}+\frac{\zeta_1}{\rho_1}\boldsymbol{\kappa}\cdot\nabla_{\mathbf{x}}\overline{A} &= i\frac{\omega}{\varepsilon}n\left[\overline{A}-\overline{B}e^{-i2\frac{\omega\tau}{\varepsilon}}\right]+i\frac{\omega}{2}\zeta_1K_{11}^{-1}(\mathbf{x},z)\overline{A}\\ &\quad -i\omega\frac{\rho_1}{\zeta_1}\left[\phi_z(\mathbf{x},z)+\frac{\zeta_1}{\rho_1}\boldsymbol{\kappa}\cdot\nabla_{\mathbf{x}}\phi(\mathbf{x},z)\right]\overline{A}\\ \frac{d}{dz}\overline{B}-\frac{\zeta_1}{\rho_1}\boldsymbol{\kappa}\cdot\nabla_{\mathbf{x}}\overline{B} &= i\frac{\omega}{\varepsilon}n\left[\overline{A}e^{i2\frac{\omega\tau}{\varepsilon}}-\overline{B}\right]-i\frac{\omega}{2}\zeta_1K_{11}^{-1}(\mathbf{x},z)\overline{B}\\ &\quad +i\omega\frac{\rho_1}{\zeta_1}\left[\phi_z(\mathbf{x},z)-\frac{\zeta_1}{\rho_1}\boldsymbol{\kappa}\cdot\nabla_{\mathbf{x}}\phi(\mathbf{x},z)\right]\overline{B}\,.\end{aligned} \tag{25}$$

Equations (23) and (25), along with corresponding boundary conditions, will form the basis for our discussion. The boundary conditions accompanying (25) are obvious adaptations of (20). The variables $\tilde{A},\tilde{B},\overline{A}$ and $\overline{B}$ have natural physical interpretations. Recall that the primes were dropped and that z actually corresponds to z'. Thus, $\overline{A}(\boldsymbol{\kappa},\mathbf{x},\omega,z')$ represents an upgoing wave amplitude corresponding to slowness $\boldsymbol{\kappa}$ and frequency ω at spatial position (x',y',z'); upgoing is interpreted locally in the sense of increasing z'.

3. Reflection and Transmission Operators

In the case of a plane layered slab, the scattering process is local in the slowness variable $\boldsymbol{\kappa}$. The upgoing amplitude can be equated to the product of the downgoing amplitude and a reflection coefficient – all evaluated at the same value of $\boldsymbol{\kappa}$ (equation (2.31) of [15]). The presence of undulations or spatially varying perturbations delocalizes the scattering process. Energy incident at one value of $\boldsymbol{\lambda}$ will be scattered into other values of $\boldsymbol{\lambda}$. Therefore, we formulate the following reflection relation

$$\widetilde{A}(\boldsymbol{\kappa},\boldsymbol{\lambda},\omega,z) = \iint \widetilde{\Gamma}(\boldsymbol{\kappa},\boldsymbol{\lambda},\boldsymbol{\lambda}',\omega,z)\widetilde{B}(\boldsymbol{\kappa},\boldsymbol{\lambda}',\omega,z)d\boldsymbol{\lambda}' . \tag{26}$$

The upgoing wave at slowness $\boldsymbol{\lambda}$ is thus the superposition of contributions ¿from downgoing waves incident at all values of slowness $\boldsymbol{\lambda}'$. The kernel of the integral operator, $\widetilde{\Gamma}$, determines how this slowness conversion occurs through scattering.

We can also define an analogous scattering relation in the spatial domain

$$\overline{A}(\boldsymbol{\kappa},\mathbf{x},\omega,z) = \iint \overline{\Gamma}(\boldsymbol{\kappa},\mathbf{x},\mathbf{x}',\omega,z)\overline{B}(\boldsymbol{\kappa},\mathbf{x}',\omega,z)d\mathbf{x}' . \tag{27}$$

The corresponding physical interpretation of (27) is that energy scattered upward at depth z and transverse location $\mathbf{x}$ is a linear superposition of multiple scattering contributions from energy incident at depth z and all transverse locations $\mathbf{x}'$.

¿From (24), it follows that the reflection kernels $\widetilde{\Gamma}$ and $\overline{\Gamma}$ are related as Fourier transforms.

$$\overline{\Gamma}(\boldsymbol{\kappa},\mathbf{x},\mathbf{x}',\omega,z) = \left(\frac{\omega}{2\pi}\right)^2 \int\cdot\cdot\int \widetilde{\Gamma}(\boldsymbol{\kappa},\boldsymbol{\lambda},\boldsymbol{\lambda}',\omega,z)e^{i\omega(\boldsymbol{\lambda}\cdot\mathbf{x}-\boldsymbol{\lambda}'\cdot\mathbf{x}')}d\boldsymbol{\lambda}d\boldsymbol{\lambda}' . \tag{28}$$

The defining relations (26), (27), when substituted into (23) and (25), respectively, lead to Riccati equations for the scattering kernels $\widetilde{\Gamma}$ and $\overline{\Gamma}$. The initial conditions for these equations follow from the first of boundary conditions (20). We obtain two related initial value problems.

$$\begin{aligned}
&\partial_z\widetilde{\Gamma} = \\
&i\frac{\omega}{\varepsilon}n\left[2\widetilde{\Gamma} - e^{-i2\frac{\omega}{\varepsilon}\tau}\delta(\boldsymbol{\lambda}-\boldsymbol{\lambda}') - e^{i2\frac{\omega}{\varepsilon}\tau}\iint \widetilde{\Gamma}(\boldsymbol{\kappa},\boldsymbol{\lambda},\boldsymbol{\lambda}'',\omega,z)\widetilde{\Gamma}(\boldsymbol{\kappa},\boldsymbol{\lambda}'',\boldsymbol{\lambda}',w,z)d\boldsymbol{\lambda}''\right] \\
&-i\omega\frac{\zeta_1}{\rho_1}\boldsymbol{\kappa}\cdot(\boldsymbol{\lambda}+\boldsymbol{\lambda}')\widetilde{\Gamma} - i\omega\frac{\rho_1}{\zeta_1}\left(\frac{\omega}{2\pi}\right)^2\iint \left[\hat{\phi}_z(\omega(\boldsymbol{\lambda}-\boldsymbol{\lambda}''),z)\right. \\
&\left.+i\omega\frac{\zeta_1}{\rho_1}\boldsymbol{\kappa}\cdot(\boldsymbol{\lambda}-\boldsymbol{\lambda}'')\hat{\phi}(\omega(\boldsymbol{\lambda}-\boldsymbol{\lambda}''),z)\,\right]\widetilde{\Gamma}(\boldsymbol{\kappa},\boldsymbol{\lambda}'',\boldsymbol{\lambda}',\omega,z)d\boldsymbol{\lambda}'' \\
&+i\frac{\omega}{2}\zeta_1\left(\frac{\omega}{2\pi}\right)^2\iint (\widehat{K_{11}^{-1}})(\omega(\boldsymbol{\lambda}-\boldsymbol{\lambda}''),z)\widetilde{\Gamma}(\boldsymbol{\kappa},\boldsymbol{\lambda}'',\boldsymbol{\lambda}',\omega,z)d\boldsymbol{\lambda}'' \\
&-i\omega\frac{\rho_1}{\zeta_1}\left(\frac{\omega}{2\pi}\right)^2\iint \widetilde{\Gamma}(\boldsymbol{\kappa},\boldsymbol{\lambda},\boldsymbol{\lambda}'',\omega,z)\left[\hat{\phi}_z(\omega(\boldsymbol{\lambda}''-\boldsymbol{\lambda}'),z)\right. \\
&\left.-i\omega\frac{\zeta_1}{\rho_1}\boldsymbol{\kappa}\cdot(\boldsymbol{\lambda}''-\boldsymbol{\lambda}')\hat{\phi}(\omega(\boldsymbol{\lambda}''-\boldsymbol{\lambda}'),z)\,\right]d\boldsymbol{\lambda}''
\end{aligned}$$

$$+i\frac{\omega}{2}\zeta_1\left(\frac{\omega}{2\pi}\right)^2\iint\widetilde{\Gamma}(\boldsymbol{\kappa},\boldsymbol{\lambda},\boldsymbol{\lambda}'',\omega,z)(\widehat{K_{11}^{-1}})(\omega(\boldsymbol{\lambda}''-\boldsymbol{\lambda}'),z)d\boldsymbol{\lambda}'', \tag{29}$$

with initial condition

$$\widetilde{\Gamma}(\boldsymbol{\kappa},\boldsymbol{\lambda},\boldsymbol{\lambda}',\omega,-L^+)=\Gamma_I(-L)e^{-i2\frac{\omega}{\varepsilon}\tau(-L)}\delta(\boldsymbol{\lambda}-\boldsymbol{\lambda}')$$

, and where (equation (2.30) of [15])

$$\Gamma_I(-L)=\frac{\zeta_1-\zeta_2}{\zeta_1+\zeta_2} \tag{30}$$

is an interface reflection coefficient. We also have

$$\begin{aligned}\partial_z\overline{\Gamma}+\frac{\zeta_1}{\rho_1}\boldsymbol{\kappa}\cdot(\nabla_{\mathbf{x}}-\nabla_{\mathbf{x}'})\overline{\Gamma} &= i\frac{\omega}{\varepsilon}n\Big[2\overline{\Gamma}-e^{-i2\frac{\omega}{\varepsilon}\tau}\delta(\mathbf{x}-\mathbf{x}')-e^{i2\frac{\omega}{\varepsilon}\tau}\cdot\\ &\quad\cdot\iint\overline{\Gamma}(\boldsymbol{\kappa},\mathbf{x},\mathbf{x}'',\omega,z)\overline{\Gamma}(\boldsymbol{\kappa},\mathbf{x}'',\mathbf{x}',\omega,z)d\mathbf{x}''\Big]\\ &\quad -i\omega\frac{\rho_1}{\zeta_1}\Big[\phi_z(\mathbf{x},z)+\frac{\zeta_1}{\rho_1}\boldsymbol{\kappa}\cdot\nabla_{\mathbf{x}}\phi(\mathbf{x},z)+\phi_z(\mathbf{x}',z)\\ &\quad -\frac{\zeta_1}{\rho_1}\boldsymbol{\kappa}\cdot\nabla_{\mathbf{x}'}\phi(\mathbf{x}',z)\Big]\overline{\Gamma}\\ &\quad +i\frac{\omega}{2}\zeta_1\left(K_{11}^{-1}(\mathbf{x},z)+K_{11}^{-1}(\mathbf{x}',z)\right)\overline{\Gamma},\end{aligned} \tag{31}$$

with initial condition

$$\overline{\Gamma}(\boldsymbol{\kappa},\mathbf{x},\mathbf{x}',\omega,-L^+)=\Gamma_I(-L)e^{-i2\frac{\omega}{\varepsilon}\tau(-L)}\delta(\mathbf{x}-\mathbf{x}')\ .$$

Transmission operators are introduced as linear integral operators relating the downgoing wave at the slab bottom to the downgoing wave at the variable depth z, *i.e.*

$$\widetilde{B}(\boldsymbol{\kappa},\boldsymbol{\lambda},\omega,-L^-)=\iint\widetilde{T}(\boldsymbol{\kappa},\boldsymbol{\lambda},\boldsymbol{\lambda}',\omega,z)\widetilde{B}(\boldsymbol{\kappa},\boldsymbol{\lambda}',\omega,z)d\boldsymbol{\lambda}', \tag{32}$$

and

$$\overline{B}(\boldsymbol{\kappa},\mathbf{x},\omega,-L^-)=\iint\overline{T}(\boldsymbol{\kappa},\mathbf{x},\mathbf{x}',\omega,z)\overline{B}(\boldsymbol{\kappa},\mathbf{x}',\omega,z)d\mathbf{x}'\ . \tag{33}$$

Equations for the transmission kernels are obtained by differentiating (32) and (33) with respect to z and using (23) – (27). We obtain the following initial value problems

$$\partial_z\widetilde{T}(\boldsymbol{\kappa},\boldsymbol{\lambda},\boldsymbol{\lambda}',\omega,z)\ +i\omega\frac{\zeta_1}{\rho_1}\boldsymbol{\kappa}\cdot\boldsymbol{\lambda}'\widetilde{T}(\boldsymbol{\kappa},\boldsymbol{\lambda},\boldsymbol{\lambda}',\omega,z)=-i\frac{\omega}{\varepsilon}n\left[e^{i2\frac{\omega}{\varepsilon}\tau}\iint\widetilde{T}(\boldsymbol{\kappa},\boldsymbol{\lambda},\boldsymbol{\lambda}'',\omega,z)\cdot\right.$$

$$\cdot \widetilde{\Gamma}(\boldsymbol{\kappa}, \boldsymbol{\lambda}'', \boldsymbol{\lambda}', \omega, z) d\boldsymbol{\lambda}'' - \widetilde{T}(\boldsymbol{\kappa}, \boldsymbol{\lambda}, \boldsymbol{\lambda}', \omega, z)\Big]$$

$$-i\omega \frac{\rho_1}{\zeta_1} \left(\frac{\omega}{2\pi}\right)^2 \iint \widetilde{T}(\boldsymbol{\kappa}, \boldsymbol{\lambda}, \boldsymbol{\lambda}'', \omega, z) \Big[\hat{\phi}_z(\omega(\boldsymbol{\lambda}'' - \boldsymbol{\lambda}'), z)$$

$$-i\omega \frac{\zeta_1}{\rho_1} \boldsymbol{\kappa} \cdot (\boldsymbol{\lambda}'' - \boldsymbol{\lambda}') \hat{\phi}(\omega(\boldsymbol{\lambda}'' - \boldsymbol{\lambda}'), z)\Big] d\boldsymbol{\lambda}''$$

$$+i\frac{\omega}{2}\zeta_1 \left(\frac{\omega}{2\pi}\right)^2 \iint \widetilde{T}(\boldsymbol{\kappa}, \boldsymbol{\lambda}, \boldsymbol{\lambda}'', \omega, z)(\widehat{K_{11}^{-1}})(\omega(\boldsymbol{\lambda}'' - \boldsymbol{\lambda}'), z) d\boldsymbol{\lambda}''$$

$$\widetilde{T}(\boldsymbol{\kappa}, \boldsymbol{\lambda}, \boldsymbol{\lambda}', \omega, -L^+) = \frac{2(\zeta_1\zeta_2)^{1/2}}{\zeta_1 + \zeta_2} \delta(\boldsymbol{\lambda} - \boldsymbol{\lambda}'), \tag{34}$$

and

$$\partial_z \overline{T}(\boldsymbol{\kappa}, \mathbf{x}, \mathbf{x}', \omega, z) - \frac{\zeta_1}{\rho_1} \boldsymbol{\kappa} \cdot \nabla_{\mathbf{x}'} \overline{T}(\boldsymbol{\kappa}, \mathbf{x}, \mathbf{x}', \omega, z) =$$

$$-i\frac{\omega}{\varepsilon} n \left[e^{i2\frac{\omega}{\varepsilon}\tau} \iint \overline{T}(\boldsymbol{\kappa}, \mathbf{x}, \mathbf{x}'', \omega, z) \overline{\Gamma}(\boldsymbol{\kappa}, \mathbf{x}'', \mathbf{x}', \omega, z) d\mathbf{x}'' - \overline{T}(\boldsymbol{\kappa}, \mathbf{x}, \mathbf{x}', \omega, x)\right]$$

$$-\overline{T}(\boldsymbol{\kappa}, \mathbf{x}, \mathbf{x}', \omega, z) \left[i\omega \frac{\rho_1}{\zeta_1} \left[\phi_z(\mathbf{x}', z) - \frac{\zeta_1}{\rho_1} \boldsymbol{\kappa} \cdot \nabla_{\mathbf{x}'} \phi(\mathbf{x}', z)\right] - i\frac{\omega}{2} \zeta_1 K_{11}^{-1}(\mathbf{x}', z)\right],$$

$$\overline{T}(\boldsymbol{\kappa}, \mathbf{x}, \mathbf{x}', \omega, -L^+) = \frac{2(\zeta_1\zeta_2)^{1/2}}{\zeta_1 + \zeta_2} \delta(\mathbf{x} - \mathbf{x}') . \tag{35}$$

Note that the two transmission kernels $\widetilde{T}$ and $\overline{T}$ satisfy transform relation (28).

4. Quantities of Interest

The reflected pressure at $z = 0$ and the transmitted pressure, *i.e.* the total pressure at $z = -L$, are the quantities of interest. Consider first the reflected pressure. Recall that we have assumed that the slab effective medium is matched to the upper half-space ($\zeta_1 = \zeta_0$). From (11), (17) and (22), noting that $\tau_{\boldsymbol{\kappa}}(0) = \mathbf{0}$, we obtain

$$p_{\text{refl}}(t, \mathbf{x}, 0) = (2\pi\varepsilon)^{-3} \iiint e^{i\omega(\boldsymbol{\kappa}\cdot\mathbf{x} - t)/\varepsilon} \zeta_0^{1/2} \omega^2 \Big[\iint \widetilde{\Gamma}(\boldsymbol{\kappa}, \mathbf{0}, \boldsymbol{\lambda}', \omega, 0) \cdot$$

$$\cdot B(\boldsymbol{\kappa} + \varepsilon\boldsymbol{\lambda}', \omega, 0) d\boldsymbol{\lambda}'\Big] d\omega d\boldsymbol{\kappa} . \tag{36}$$

Equation (36) can be recast into a form that makes its physical content more transparent. We make the change of variables $\overline{\boldsymbol{\kappa}} = \boldsymbol{\kappa} + \varepsilon\boldsymbol{\lambda}', \overline{\boldsymbol{\lambda}} = \boldsymbol{\lambda}'$ (and subsequently drop the overbar); we use the fact that $\widetilde{\Gamma}(\boldsymbol{\kappa} - \varepsilon\boldsymbol{\lambda}, \mathbf{0}, \boldsymbol{\lambda}, \omega, 0) =$

$\widetilde{\Gamma}(\boldsymbol{\kappa}, -\boldsymbol{\lambda}, \mathbf{0}, \omega, 0)$. With these steps and the use of transform relation (28), (36) can be rewritten as

$$p_{\text{refl}}(t, \mathbf{x}, 0) = (2\pi\varepsilon)^{-3} \iiint e^{i\omega(\boldsymbol{\kappa}\cdot\mathbf{x}-t)/\varepsilon} \left[\iint \overline{\Gamma}(\boldsymbol{\kappa}, \mathbf{x}, \mathbf{x}', \omega, 0) d\mathbf{x}'\right] \zeta_0^{1/2} \cdot \\ \cdot B(\boldsymbol{\kappa}, \omega, 0)\omega^2 d\omega d\boldsymbol{\kappa} . \tag{37}$$

The physical content of (37) is clear. The spectral synthesis involves a product of two terms. The first term, $\overline{\overline{\Gamma}}(\boldsymbol{\kappa}, \mathbf{x}, \omega) \equiv \iint \overline{\Gamma}(\boldsymbol{\kappa}, \mathbf{x}, \mathbf{x}', \omega, 0) d\mathbf{x}'$, is an input reflection coefficient whose value at transverse position $\mathbf{x}$ involves a superposition of scatter contributions from all other transverse points. The second term, $\zeta_0^{1/2} B(\boldsymbol{\kappa}, \omega, 0)$, is the spectral representation of the incident pressure; for the case of a point source, it is given by (20).

Now consider the transmitted pressure. From (17), (32) and the radiation condition in $z < -L$, we obtain the representation

$$\begin{aligned} p_{\text{trans}}(t, \mathbf{x}, -L) &= -(2\pi\varepsilon)^{-3} \iiint e^{i\frac{\omega}{\varepsilon}(\boldsymbol{\kappa}\cdot\mathbf{x}-t)} \zeta_2^{1/2} B(\boldsymbol{\kappa}, \omega, -L^-) e^{-i\frac{\omega}{\varepsilon}\tau(-L)} \omega^2 d\omega d\boldsymbol{\kappa} \\ &= -(2\pi\varepsilon)^{-3} \iiint e^{i\frac{\omega}{\varepsilon}(\boldsymbol{\kappa}\cdot\mathbf{x}-t)} \zeta_2^{1/2} \cdot \\ &\quad \cdot \left[\iint \widetilde{T}(\boldsymbol{\kappa}, \mathbf{0}, \boldsymbol{\lambda}', \omega, 0) B(\boldsymbol{\kappa} + \varepsilon\boldsymbol{\lambda}', \omega, 0) d\boldsymbol{\lambda}'\right] \omega^2 d\omega d\boldsymbol{\kappa} . \end{aligned} \tag{38}$$

To recast (38) into a more physically meaningful form, we again make the change of variables $\overline{\boldsymbol{\kappa}} = \boldsymbol{\kappa} + \varepsilon\boldsymbol{\lambda}', \overline{\boldsymbol{\lambda}} = \boldsymbol{\lambda}'$ and use the fact that $\widetilde{T}(\boldsymbol{\kappa} - \varepsilon\boldsymbol{\lambda}, \mathbf{0}, \boldsymbol{\lambda}, \omega, 0) = e^{-i\omega\tau_{\boldsymbol{\kappa}}(-L)\cdot\boldsymbol{\lambda}}\, \widetilde{T}(\boldsymbol{\kappa}, -\boldsymbol{\lambda}, \mathbf{0}, \omega, 0)$. Since $\overline{T}$ and $\widetilde{T}$ also satisfy transform relation (28), we obtain

$$p_{\text{trans}}(t, \mathbf{x}, -L) = -(2\pi\varepsilon)^{-3} \iiint e^{i\frac{\omega}{\varepsilon}(\boldsymbol{\kappa}\cdot\mathbf{x}-t-\tau(-L))} \left[\iint \overline{T}(\boldsymbol{\kappa}, \mathbf{x}, \mathbf{x}', \omega, 0) d\mathbf{x}'\right] \zeta_2^{1/2} \\ \cdot B(\boldsymbol{\kappa}, \omega, 0)\omega^2 d\omega d\boldsymbol{\kappa} . \tag{39}$$

Thus, the transmitted pressure is likewise a spectral synthesis involving the product of two terms. $\overline{\overline{T}}(\boldsymbol{\kappa}, \mathbf{x}, \omega) \equiv \iint \overline{T}(\boldsymbol{\kappa}, \mathbf{x}, \mathbf{x}', \omega, 0) d\mathbf{x}'$ is a slab transmission coefficient whose value at $\mathbf{x}$ is obtained by superposing contributions from all transverse starting points $\mathbf{x}'$. The term $-\zeta_2^{1/2} B(\boldsymbol{\kappa}, \omega, 0)$ equals $\sqrt{\frac{\zeta_2}{\zeta_0}}$ times the spectrally-resolved incident pressure (20). Note also that $-\tau(-L) = (1 - \kappa^2 c_1^2)^{1/2} c_1^{-1} L$ is the transit time required for a ray to travel a linear path, characterized by slowness κ, through the slab (*c.f.* 16).

5. Conservation Relations

We examine the conservation relations present in the model that we have developed. By direct computation, using equations (23), one can show that

$$\frac{d}{dz}\iint \left[|\widetilde{B}(\boldsymbol{\kappa},\boldsymbol{\lambda},\omega,z)|^2 - |\widetilde{A}(\boldsymbol{\kappa},\boldsymbol{\lambda},\omega,z)|^2\right] d\boldsymbol{\lambda} = 0 \,. \tag{40}$$

Use is made of the fact that ϕ and K_{11} are real-valued; thus, for example, $\hat{\phi}^*(\omega\boldsymbol{\lambda},z) = \hat{\phi}(-\omega\boldsymbol{\lambda},z)$. Noting (22), it immediately follows that

$$\frac{d}{dz}\iint \left[|B(\boldsymbol{\kappa}+\varepsilon\boldsymbol{\lambda},\omega,z)|^2 - |A(\boldsymbol{\kappa}+\varepsilon\boldsymbol{\lambda},\omega,z)|^2\right] d\boldsymbol{\lambda} = 0 \,. \tag{41}$$

Similarly, using Parseval's identity (or direct computation and equations (25)), we obtain

$$\frac{d}{dz}\iint \left[|\overline{B}(\boldsymbol{\kappa},\mathbf{x},\omega,z)|^2 - |\overline{A}(\boldsymbol{\kappa},\mathbf{x},\omega,z)|^2\right] d\mathbf{x} = 0 \,. \tag{42}$$

Thus, the resulting conservation law exists in a transversely integrated sense. The undulations and mean bulk modulus perturbations couple different slowness values; only the total integrated acoustic flux is conserved. From (17) and (41) it follows that

$$\begin{aligned}\iint Re\{\hat{p}(\boldsymbol{\kappa}+\varepsilon\boldsymbol{\lambda},\omega,0)\hat{u}^*(\boldsymbol{\kappa}+\varepsilon\boldsymbol{\lambda},\omega,0)\}d\boldsymbol{\lambda} \;=\; & \iint Re\{\hat{p}(\boldsymbol{\kappa}+\varepsilon\boldsymbol{\lambda},\omega,-L)\cdot \\ & \cdot\hat{u}^*(\boldsymbol{\kappa}+\varepsilon\boldsymbol{\lambda},\omega,-L)\}d\boldsymbol{\lambda}\end{aligned} \tag{43}$$

Conservation relations (40) and (42) lead to analogous results for the reflection and transmission operators introduced in Section 3. Let $\widetilde{\mathbf{\Gamma}},\overline{\mathbf{\Gamma}},\widetilde{\mathbf{T}}$ and $\overline{\mathbf{T}}$ denote the integral operators defined by (26), (27), (32) and (33), respectively. Thus, for example (26) now becomes $\widetilde{A} = \widetilde{\mathbf{\Gamma}}\widetilde{B}$. Let $\langle\cdot,\cdot\rangle$ and $||\cdot||$ denote the (complex) inner product and norm on $L_2(d\boldsymbol{\lambda})$. Then, it follows from (26), (32) and (40) that

$$\begin{aligned}\frac{d}{dz}\langle\widetilde{B}(z),(\mathbf{I}-\widetilde{\mathbf{\Gamma}}^{\dagger}(z)\widetilde{\mathbf{\Gamma}}(z))\widetilde{B}(z)\rangle \;=\; & \frac{d}{dz}\langle\widetilde{B}(-L^-),(\widetilde{\mathbf{T}}^{-1}(z))^{\dagger}(\mathbf{I}-\widetilde{\mathbf{\Gamma}}^{\dagger}(z)\widetilde{\mathbf{\Gamma}}(z))\cdot \\ & \cdot\widetilde{\mathbf{T}}^{-1}(z)\widetilde{B}(-L^-)\rangle = 0 \,.\end{aligned} \tag{44}$$

However, from (29), (30) and (34)

$$(\widetilde{\mathbf{T}}^{-1}(-L^+))^{\dagger}(\mathbf{I}-\widetilde{\mathbf{\Gamma}}^{\dagger}(-L^+)\widetilde{\mathbf{\Gamma}}(-L^+))\widetilde{\mathbf{T}}^{-1}(-L^+) = \mathbf{I} \,. \tag{45}$$

Therefore, we obtain

$$\langle\widetilde{B}(-L^-),[(\widetilde{\mathbf{T}}^{-1}(z)^{\dagger}(\mathbf{I}-\widetilde{\mathbf{\Gamma}}^{\dagger}(z)\widetilde{\mathbf{\Gamma}}(z))\widetilde{\mathbf{T}}^{-1}(z)-\mathbf{I}]\widetilde{B}(-L^-)\rangle = 0 \tag{46}$$

for all possible transmitted fields $\widetilde{B}(-L^-)$. We thus obtain the conservation law

$$\widetilde{\mathbf{\Gamma}}^\dagger(z)\widetilde{\mathbf{\Gamma}}(z) + \widetilde{\mathbf{T}}^\dagger(z)\widetilde{\mathbf{T}}(z) = \mathbf{I}, \tag{47}$$

and an analogous argument leads to the following spatial domain counterpart of (47)

$$\overline{\mathbf{\Gamma}}^\dagger(z)\overline{\mathbf{\Gamma}}(z) + \overline{\mathbf{T}}^\dagger(z)\overline{\mathbf{T}}(z) = \mathbf{I}. \tag{48}$$

6. The Stochastic Limit; Ito Equations

Our goal is to describe the asymptotic behavior of quantities of interest, such as the reflected and transmitted pressure, in the limit as $\varepsilon \to 0$. Therefore, we need to characterize the limit reflection and transmission processes. A convenient way to describe these limit processes is by the use of Ito equations.

We shall first formulate operator versions of equations (23) and (25). We then pass to the limit for these linear equations, obtaining linear Ito equations for the operators which propagate the fields $\widetilde{A}, \widetilde{B}, \overline{A}$ and $\overline{B}$ upward from initial values at the slab bottom. Once these Ito equations have been obtained, the defining relations and Ito calculus enable us to readily derive corresponding Ito equations for the reflection and transmission operators $\widetilde{\mathbf{\Gamma}}, \overline{\mathbf{\Gamma}}, \widetilde{\mathbf{T}}$ and $\overline{\mathbf{T}}$.

We define the operators $\widetilde{\mathbf{A}}, \widetilde{\mathbf{B}}$ by the relations

$$\widetilde{\mathbf{A}}(z)\,(\widetilde{B}(-L^-)) = \widetilde{A}(z) \quad , \quad \widetilde{\mathbf{B}}(z)\,(\widetilde{B}(-L^-)) = \widetilde{B}(z) \quad , \quad -L^- \leq z \leq 0\,. \tag{49}$$

These operators propagate the transmitted field at $-L^-$ upward through the slab. If $\zeta_2 \neq \zeta_1$ (*i.e.* $\Gamma_I(-L) \neq 0$), the operators are discontinuous at $-L$. However, on the interval $-L^+ < z \leq 0$, equations (23) imply that (prior to taking the ε-limit), the operators satisfy

$$\frac{d}{dz}\widetilde{\mathbf{A}} = i\frac{\omega}{\varepsilon}n\left[\widetilde{\mathbf{A}} - e^{-i2\frac{\omega}{\varepsilon}\tau}\widetilde{\mathbf{B}}\right] + i\omega\widetilde{\mathbf{Q}}_A\widetilde{\mathbf{A}}$$

$$\frac{d}{dz}\widetilde{\mathbf{B}} = i\frac{\omega}{\varepsilon}n\left[e^{i2\frac{\omega}{\varepsilon}\tau}\widetilde{\mathbf{A}} - \widetilde{\mathbf{B}}\right] + i\omega\widetilde{\mathbf{Q}}_B\widetilde{B} \quad , \quad -L < z < 0 \tag{50}$$

where, for brevity, we define

$$\widetilde{\mathbf{Q}}_A(\widetilde{A})(z) \equiv -\frac{\zeta_1}{\rho_1}\boldsymbol{\kappa}\cdot\boldsymbol{\lambda}\widetilde{A}(\boldsymbol{\kappa},\boldsymbol{\lambda},\omega,z)$$

$$-\frac{\rho_1}{\zeta_1}\left(\frac{\omega}{2\pi}\right)^2\iint\left[\hat{\phi}_z(\omega\boldsymbol{\lambda}',z) + i\omega\frac{\zeta_1}{\rho_1}\boldsymbol{\kappa}\cdot\boldsymbol{\lambda}'\cdot\hat{\phi}(\omega\boldsymbol{\lambda}',z)\right]\widetilde{A}(\boldsymbol{\kappa},\boldsymbol{\lambda}-\boldsymbol{\lambda}',\omega,z)d\boldsymbol{\lambda}'$$

$$+\frac{\zeta_1}{2}\left(\frac{\omega}{2\pi}\right)^2 \iint (\widehat{K_{11}^{-1}})(\omega\boldsymbol{\lambda}',z)\widetilde{A}(\boldsymbol{\kappa},\boldsymbol{\lambda}-\boldsymbol{\lambda}',\omega,z)d\boldsymbol{\lambda}' ,$$

$$\widetilde{\mathbf{Q}}_B(\widetilde{B})(z) \equiv \frac{\zeta_1}{\rho_1}\boldsymbol{\kappa}\cdot\boldsymbol{\lambda}\widetilde{B}(\boldsymbol{\kappa},\boldsymbol{\lambda},w,z)$$

$$+\frac{\rho_1}{\zeta_1}\left(\frac{\omega}{2\pi}\right)^2 \iint \left[\hat{\phi}_z(\omega\boldsymbol{\lambda}',z) - i\omega\frac{\zeta_1}{\rho_1}\boldsymbol{\kappa}\cdot\boldsymbol{\lambda}'\cdot\hat{\phi}(\omega\boldsymbol{\lambda}',z)\right]\widetilde{B}(\boldsymbol{\kappa},\boldsymbol{\lambda}-\boldsymbol{\lambda}',\omega,z)d\boldsymbol{\lambda}'$$

$$-\frac{\zeta_1}{2}\left(\frac{\omega}{2\pi}\right)^2 \iint \left(\widehat{K_{11}^{-1}}\right)(\omega\boldsymbol{\lambda}',z)\widetilde{B}(\boldsymbol{\kappa},\boldsymbol{\lambda}-\boldsymbol{\lambda}',\omega,z)d\boldsymbol{\lambda}' . \tag{51}$$

If we apply the $\varepsilon \to 0$ limit to equations (50), we obtain operator-valued Markov processes which we can characterize in terms of linear Ito equations. Applying this limit is tantamount to making the following substitutions

$$\varepsilon^{-1}n\left(z/\varepsilon^2\right)dz \to \sqrt{\alpha_{nn}}\,d\beta_1$$
$$\varepsilon^{-1}n\left(z/\varepsilon^2\right)e^{\pm i2\frac{\omega}{\varepsilon}\tau(z)}dz \to \sqrt{\alpha_{nn}}\left(\frac{d\beta_2 \pm id\beta_3}{\sqrt{2}}\right), \tag{52}$$

where the β_j, $j = 1,2,3$, are real-valued, independent Brownian motions ($E\{d\beta_i d\beta_j\} = dz\delta_{ij}$) and (*cf.* (2.19), (2.28) and (3.13) in [15])

$$\alpha_{nn} \equiv \frac{\int_0^\infty E\{\nu(s)\nu(0)\}ds}{4c_1^2(1-c_1^2\kappa^2)} \equiv \frac{\alpha}{c_1^2(1-c_1^2\kappa^2)} . \tag{53}$$

The limit, roughly speaking, combines the dynamics of the Central Limit Theorem and the Method of Averaging. Therefore, the noise terms in (52) that are multiplied by the rapidly-varying phase factor become asymptotically independent of the noise term that has no such factor. Taking the $\varepsilon \to 0$ limit in (50) leads to the following operator-valued linear Ito equations

$$d\widetilde{\mathbf{A}} = i\omega\sqrt{2\alpha_{nn}}\left[\widetilde{\mathbf{A}}d\beta_1 - \widetilde{\mathbf{B}}\left(\frac{d\beta_2 - id\beta_3}{\sqrt{2}}\right)\right] + i\omega\widetilde{\mathbf{Q}}_A\widetilde{\mathbf{A}}dz$$
$$d\widetilde{\mathbf{B}} = i\omega\sqrt{2\alpha_{nn}}\left[\widetilde{\mathbf{A}}\left(\frac{d\beta_2 + id\beta_3}{\sqrt{2}}\right) - \widetilde{\mathbf{B}}d\beta_1\right] + i\omega\widetilde{\mathbf{Q}}_B\widetilde{\mathbf{B}}dz . \tag{54}$$

Having representations (54), we can use the operator relations

$$\widetilde{\mathbf{\Gamma}} = \widetilde{\mathbf{A}}\widetilde{\mathbf{B}}^{-1} \quad , \quad \widetilde{\mathbf{T}} = \widetilde{\mathbf{B}}^{-1} \tag{55}$$

and Ito calculus to derive the following initial value problems for the reflection and transmission operators

$$d\widetilde{\mathbf{\Gamma}} = i\omega\sqrt{2\alpha_{nn}}\left[2\widetilde{\mathbf{\Gamma}}d\beta_1 - \mathbf{I}\left(\frac{d\beta_2 - id\beta_3}{\sqrt{2}}\right) - \widetilde{\mathbf{\Gamma}}^2\left(\frac{d\beta_2 + id\beta_3}{\sqrt{2}}\right)\right]$$

$$+\left[-6\omega^2\alpha_{nn}\widetilde{\mathbf{\Gamma}}+i\omega\left(\widetilde{\mathbf{Q}}_A\widetilde{\mathbf{\Gamma}}-\widetilde{\mathbf{\Gamma}}\widetilde{\mathbf{Q}}_B\right)\right]dz,$$

$$\widetilde{\mathbf{\Gamma}}|_{z=-L^+}=\Gamma_I(-L)e^{-i2\frac{\omega}{\varepsilon}\tau(-L)}\mathbf{I} \tag{56}$$

$$d\widetilde{\mathbf{T}}=i\omega\sqrt{2\alpha_{nn}}\left[\widetilde{\mathbf{T}}d\beta_1-\widetilde{\mathbf{T}}\widetilde{\mathbf{\Gamma}}\left(\frac{d\beta_2+id\beta_3}{\sqrt{2}}\right)\right]+\left[-2\omega^2\alpha_{nn}\widetilde{\mathbf{T}}-i\omega\widetilde{\mathbf{T}}\widetilde{\mathbf{Q}}_B\right]dz,$$

$$\widetilde{\mathbf{T}}|_{z=-L^+}=\frac{2\sqrt{\zeta_1\zeta_2}}{\zeta_1+\zeta_2}\mathbf{I}\,. \tag{57}$$

Note that the two Ito equations contain the drift terms $-6\omega^2\alpha_{nn}\widetilde{\mathbf{\Gamma}}dz$ and $-2\omega^2\alpha_{nn}\widetilde{\mathbf{T}}dz$, respectively. Since α_{nn}, defined by (53), is positive, these terms, arising ¿from the random layering, introduce exponential decay into the evolution of the reflection and transmission operators. From a physical point of view, such behavior is to be expected; it arises from the loss of coherence and localization induced by the random multiple scattering. We shall develop these ideas further in Sections 8 and 10.

Counterpart Ito equations for corresponding spatial domain operators $\overline{\mathbf{A}},\overline{\mathbf{B}},\overline{\mathbf{\Gamma}}$ and $\overline{\mathbf{T}}$ can be obtained by applying Fourier transforms (24) and (28) to (54), (56) and (57), or by repeating the development of this section, starting with equations (25). The defining relations are

$$\overline{\mathbf{A}}(z)(\overline{B}(-L^-))=\overline{A}(z)\,,\ \overline{\mathbf{B}}(z)(\overline{B}(-L^-))=\overline{B}(z),\ \overline{\mathbf{\Gamma}}=\overline{\mathbf{A}}\,\overline{\mathbf{B}}^{-1}\,,\ \overline{\mathbf{T}}=\overline{\mathbf{B}}^{-1}, \tag{58}$$

and the corresponding Ito equations become

$$d\overline{\mathbf{A}}=i\omega\sqrt{2\alpha_{nn}}\left[\overline{\mathbf{A}}d\beta_1-\overline{\mathbf{B}}\left(\frac{d\beta_2-id\beta_3}{\sqrt{2}}\right)\right]+\left[-\frac{\zeta_1}{\rho_1}\boldsymbol{\kappa}\cdot\nabla_{\mathbf{x}}+i\omega\overline{q}_A(\mathbf{x},z)\mathbf{I}\right]\overline{\mathbf{A}}dz$$

$$d\overline{\mathbf{B}}=i\omega\sqrt{2\alpha_{nn}}\left[\overline{\mathbf{A}}\left(\frac{d\beta_2+id\beta_3}{\sqrt{2}}\right)-\overline{\mathbf{B}}d\beta_1\right]+\left[\frac{\zeta_1}{\rho_1}\boldsymbol{\kappa}\cdot\nabla_{\mathbf{x}}+i\omega\overline{q}_B(\mathbf{x},z)\mathbf{I}\right]\overline{\mathbf{B}}dz, \tag{59}$$

where

$$\overline{q}_A(\mathbf{x},z)\equiv-\frac{\rho_1}{\zeta_1}\left[\phi_z(\mathbf{x},z)+\frac{\zeta_1}{\rho_1}\boldsymbol{\kappa}\cdot\nabla_{\mathbf{x}}\phi(\mathbf{x},z)\right]+\frac{\zeta_1}{2}K_{11}^{-1}(\mathbf{x},z)$$

$$\overline{q}_B(\mathbf{x},z)\equiv\frac{\rho_1}{\zeta_1}\left[\phi_z(\mathbf{x},z)-\frac{\zeta_1}{\rho_1}\boldsymbol{\kappa}\cdot\nabla_{\mathbf{x}}\phi(\mathbf{x},z)\right]-\frac{\zeta_1}{2}K_{11}^{-1}(\mathbf{x},z), \tag{60}$$

and

$$d\overline{\mathbf{\Gamma}}=i\omega\sqrt{2\alpha_{nn}}\left[2\overline{\mathbf{\Gamma}}d\beta_1-\mathbf{I}\left(\frac{d\beta_2-id\beta_3}{\sqrt{2}}\right)-\overline{\mathbf{\Gamma}}^2\left(\frac{d\beta_2+id\beta_3}{\sqrt{2}}\right)\right]+\left[-6\omega^2\alpha_{nn}\overline{\mathbf{\Gamma}}\right.$$

$$\left.-\frac{\zeta_1}{\rho_1}[(\boldsymbol{\kappa}\cdot\nabla)\,\overline{\mathbf{\Gamma}}+\overline{\mathbf{\Gamma}}(\boldsymbol{\kappa}\cdot\nabla)]+i\omega[\overline{q}_A\mathbf{I}\overline{\mathbf{\Gamma}}-\overline{\mathbf{\Gamma}}\overline{q}_B\mathbf{I}]\right]dz,$$

$$\overline{\mathbf{\Gamma}}|_{z=-L^+}=\Gamma_I(-L)e^{-i2\frac{\omega}{\varepsilon}\tau(-L)}\mathbf{I}, \tag{61}$$

$$d\overline{\mathbf{T}} = i\omega\sqrt{2\alpha_{nn}}\left[\overline{\mathbf{T}}d\beta_1 - \overline{\mathbf{T}}\overline{\mathbf{\Gamma}}\left(\frac{d\beta_2 + id\beta_3}{\sqrt{2}}\right)\right] + \left[-2\omega^2\alpha_{nn}\overline{\mathbf{T}} - \frac{\zeta_1}{\rho_1}\overline{\mathbf{T}}(\boldsymbol{\kappa}\cdot\nabla)\right.$$
$$\left. -i\omega\overline{\mathbf{T}}\overline{q}_B\mathbf{I}\right]dz,$$
$$\overline{\mathbf{T}}|_{z=-L^+} = \frac{2\sqrt{\zeta_1\zeta_2}}{\zeta_1+\zeta_2}\mathbf{I}\ . \tag{62}$$

From these operator equations, one can readily obtain Ito equations for the fields themselves (*i.e.* $\widetilde{A}(z), ..., \overline{B}(z)$) and the reflection and transmission integral kernels. One can also verify that the conservation relations determined in Section 5 remain valid in the limit Ito setting.

7. Reduction to the Plane-Layered Case

Now that the locally layered model for the slab has been developed, we shall examine how this model reduces to the plane-layered description when the undulations and perturbations are removed.

If we set $\phi = 0$ and $K_{11}^{-1} = 0$, then the operators $\widetilde{\mathbf{Q}}_A$ and $\widetilde{\mathbf{Q}}_B$, defined by (51), reduce to $-\frac{\zeta_1}{\rho_1}(\boldsymbol{\kappa}\cdot\boldsymbol{\lambda})\mathbf{I}$ and $\frac{\zeta_1}{\rho_1}(\boldsymbol{\kappa}\cdot\boldsymbol{\lambda})\mathbf{I}$, respectively. In this case, the Ito equation and initial condition for $\widetilde{\Gamma}$, the kernel of the reflection operator $\widetilde{\mathbf{\Gamma}}$, reduce to (*c.f.* (56))

$$d\widetilde{\Gamma}(\boldsymbol{\kappa},\boldsymbol{\lambda},\boldsymbol{\lambda}',\omega,z) = i\omega\sqrt{2\alpha_{nn}}\left[2\widetilde{\Gamma}(\boldsymbol{\kappa},\boldsymbol{\lambda},\boldsymbol{\lambda}',\omega,z)d\beta_1 - \delta(\boldsymbol{\lambda}-\boldsymbol{\lambda}')\left(\frac{d\beta_2 - id\beta_3}{\sqrt{2}}\right)\right.$$
$$\left. - \iint \widetilde{\Gamma}(\boldsymbol{\kappa},\boldsymbol{\lambda},\boldsymbol{\lambda}'',\omega,z)\widetilde{\Gamma}(\boldsymbol{\kappa},\boldsymbol{\lambda}'',\boldsymbol{\lambda}',\omega,z)d\boldsymbol{\lambda}''\left(\frac{d\beta_2 + id\beta_3}{\sqrt{2}}\right)\right]$$
$$+\left[-6\omega^2\alpha_{nn} - i\omega\frac{\zeta_1}{\rho_1}\boldsymbol{\kappa}\cdot(\boldsymbol{\lambda}+\boldsymbol{\lambda}')\right]\widetilde{\Gamma}(\boldsymbol{\kappa},\boldsymbol{\lambda},\boldsymbol{\lambda}',\omega,z)dz,$$
$$\widetilde{\Gamma}(\boldsymbol{\kappa},\boldsymbol{\lambda},\boldsymbol{\lambda}',\omega,-L^+) = \Gamma_I(-L)e^{-i2\frac{\omega}{\varepsilon}\tau(-L)}\delta(\boldsymbol{\lambda}-\boldsymbol{\lambda}')\ . \tag{63}$$

The solution of (63) has the form

$$\widetilde{\Gamma}(\boldsymbol{\kappa},\boldsymbol{\lambda},\boldsymbol{\lambda}',\omega,z) = \widetilde{\Gamma}(\boldsymbol{\kappa},\boldsymbol{\lambda},\omega,z)\delta(\boldsymbol{\lambda}-\boldsymbol{\lambda}'), \tag{64}$$

and, noting (28), the kernel of the corresponding spatial domain operator assumes the form

$$\overline{\Gamma}(\boldsymbol{\kappa},\mathbf{x},\mathbf{x}',\omega,z) = \overline{\Gamma}(\boldsymbol{\kappa},\mathbf{x}-\mathbf{x}',\omega,z)\ . \tag{65}$$

The reflection coefficient appearing in the expression for the reflected pressure (37), *i.e.* $\overline{\overline{\Gamma}} = \iint \overline{\Gamma}(\boldsymbol{\kappa},\mathbf{x}-\mathbf{x}',\omega,0)d\mathbf{x}'$, thus becomes independent of $\mathbf{x}$. If we define

$$\overline{\overline{\Gamma}}(\boldsymbol{\kappa},\omega,z) \equiv \iint \overline{\Gamma}(\boldsymbol{\kappa},\mathbf{x},\omega,z)d\mathbf{x}, \tag{66}$$

one can show that $\overline{\overline{\Gamma}}$ satisfies the initial value problem

$$d\overline{\overline{\Gamma}} = i\omega\sqrt{2\alpha_{nn}}\left[2\overline{\overline{\Gamma}}d\beta_1 - \left(\frac{d\beta_2 - id\beta_3}{\sqrt{2}}\right) - \overline{\overline{\Gamma}}^2\left(\frac{d\beta_2 + id\beta_3}{\sqrt{2}}\right)\right] - 6\omega^2\alpha_{nn}\overline{\overline{\Gamma}}dz,$$

$$\overline{\overline{\Gamma}}|_{z=-L^+} = \Gamma_I(-L)e^{-i2\frac{\omega}{\varepsilon}\tau(-L)} . \tag{67}$$

This problem is, in fact, the limit problem for the plane-layered random slab ((2.32) and (3.1) of [15]).

The same argument can be used in the case of transmission. With $\phi = 0$ and $K_{11}^{-1} = 0$, we note from (34), (64) that the transmission operator kernel has the form

$$\widetilde{T}(\boldsymbol{\kappa}, \boldsymbol{\lambda}, \boldsymbol{\lambda}', \omega, z) = \widetilde{T}(\boldsymbol{\kappa}, \boldsymbol{\lambda}, \omega, z)\delta(\boldsymbol{\lambda} - \boldsymbol{\lambda}') . \tag{68}$$

The kernel of the spatial domain transmission kernel thus assumes the form

$$\overline{T}(\boldsymbol{\kappa}, \mathbf{x}, \mathbf{x}', \omega, z) = \overline{T}(\boldsymbol{\kappa}, \mathbf{x} - \mathbf{x}', \omega, z) \tag{69}$$

and the transmission coefficient $\overline{\overline{T}} = \iint T(\boldsymbol{\kappa}, \mathbf{x}, \mathbf{x}', \omega, 0)d\mathbf{x}'$ appearing in the expression for the transmitted pressure (39) is no longer a function of $\mathbf{x}$. As in (66), let

$$\overline{\overline{T}}(\boldsymbol{\kappa}, \omega, z) \equiv \iint \overline{T}(\boldsymbol{\kappa}, \mathbf{x}, \omega, z)d\mathbf{x} . \tag{70}$$

Then, it can be shown that $\overline{\overline{T}}$ satisfies the following limit problem for the plane-layered slab

$$d\overline{\overline{T}} = i\omega\sqrt{2\alpha_{nn}}\left[\overline{\overline{T}}d\beta_1 - \overline{\overline{T}}\,\overline{\overline{\Gamma}}\left(\frac{d\beta_2 + id\beta_3}{\sqrt{2}}\right)\right] - 2\omega^2\alpha_{nn}\overline{\overline{T}}dz$$

$$\overline{\overline{T}}|_{z=-L^+} = \frac{2\sqrt{\zeta_1\zeta_2}}{\zeta_1 + \zeta_2}I . \tag{71}$$

8. Coherent Fields

We begin our evaluation of the locally layered model by determining the mean or coherent reflected and transmitted pressure. From (37) and (39), it is clear that we must compute $E\{\overline{\Gamma}(\boldsymbol{\kappa}, \mathbf{x}, \mathbf{x}', \omega, 0)\}$ and $E\{\overline{T}(\boldsymbol{\kappa}, \mathbf{x}, \mathbf{x}', \omega, 0)\}$.

¿From (61), we note that the mean reflection operator satisfies the initial value problem

$$\begin{aligned} d\langle\overline{\mathbf{\Gamma}}\rangle &= -\left[6\omega^2\alpha_{nn}\langle\overline{\mathbf{\Gamma}}\rangle + \frac{\zeta_1}{\rho_1}[(\boldsymbol{\kappa}\cdot\nabla)\langle\overline{\mathbf{\Gamma}}\rangle + \langle\overline{\mathbf{\Gamma}}\rangle(\boldsymbol{\kappa}\cdot\nabla)] - i\omega\left[\overline{q}_A\mathbf{I}\langle\overline{\mathbf{\Gamma}}\rangle\right.\right. \\ &\quad \left.\left. -\langle\overline{\mathbf{\Gamma}}\rangle\overline{q}_B\mathbf{I}\ \right]\right] dz \\ \langle\overline{\mathbf{\Gamma}}\rangle|_{z=-L^+} &= \Gamma_I(-L)e^{-i2\frac{\omega}{\varepsilon}\tau(-L)}\mathbf{I} . \end{aligned} \tag{72}$$

The corresponding kernel of this mean reflection operator, i.e. $\langle\overline{\Gamma}(\boldsymbol{\kappa}, \mathbf{x}, \mathbf{x}', \omega, z)\rangle$, is thus a solution of

$$\partial_z\langle\overline{\Gamma}\rangle + \frac{\zeta_1}{\rho_1}\left[\boldsymbol{\kappa}\cdot\nabla_{\mathbf{x}}\langle\overline{\Gamma}\rangle - \boldsymbol{\kappa}\cdot\nabla_{\mathbf{x}'}\langle\overline{\Gamma}\rangle\right] = -\left[6\omega\alpha_{nn} - i\omega\left(\overline{q}_A(\mathbf{x}, z) - \overline{q}_B(\mathbf{x}', z)\right)\right]\langle\overline{\Gamma}\rangle$$
$$\langle\overline{\Gamma}\rangle|_{z=-L^+} = \Gamma_I(-L)e^{-i2\frac{\omega}{\varepsilon}\tau(-L)}\delta(\mathbf{x}-\mathbf{x}') \ . \tag{73}$$

The solution of this initial value problem at $z = 0$ is

$$\langle\overline{\Gamma}(\boldsymbol{\kappa}, \mathbf{x}, \mathbf{x}', \omega, 0)\rangle = \Gamma_I(-L)e^{-i2\frac{\omega}{\varepsilon}\tau(-L)}\exp\Big\{-6\omega^2\alpha_{nn}L + i\omega\frac{\zeta_1}{2}\int_{-L}^0\Big[K_{11}^{-1}(\mathbf{x}+\frac{\zeta_1}{\rho_1}\boldsymbol{\kappa}\sigma, \sigma) + K_{11}^{-1}\left(\mathbf{x}' - \frac{\zeta_1}{\rho_1}\boldsymbol{\kappa}\sigma, \sigma\right)\Big]d\sigma\Big\}\delta\left(\mathbf{x}-\mathbf{x}'-2\frac{\zeta_1}{\rho_1}\boldsymbol{\kappa}L\right) \tag{74}$$

and the coherent reflected pressure, assuming point source excitation, becomes (*c.f.* (20), (37))

$$\langle p_{\text{refl}}(t, \mathbf{x}, 0)\rangle = \frac{\varepsilon^{-3/2}}{2(2\pi)^3}\iiint e^{i\frac{\omega}{\varepsilon}[\boldsymbol{\kappa}\cdot\mathbf{x} - t + \tau(z_s) - 2\tau(-L)]}\exp\Big\{-6\omega^2\alpha_{nn}L + i\omega\frac{\zeta_1}{2}\int_{-L}^0\Big[K_{11}^{-1}(\mathbf{x}+\frac{\zeta_1}{\rho_1}\boldsymbol{\kappa}\sigma, \sigma) + K_{11}^{-1}(\mathbf{x} - \frac{\zeta_1}{\rho_1}\boldsymbol{\kappa}(2L+\sigma), \sigma)\Big]d\sigma\Big\}\cdot \Gamma_I(-L)\hat{f}(\omega)\omega^2 d\omega d\boldsymbol{\kappa} \ . \tag{75}$$

Because we have assumed $\zeta_1 = \zeta_0$, the slab effective medium is matched to the upper half-space. Coherent energy emitted by the source experiences no interface reflection at $z = 0$. It enters the slab, is partially reflected by the mismatch at $z = -L$ (if $0 < \Gamma_I(-L) < 1$), and the reflected portion travels upward to $z = 0$. The factor $\tau(z_s) - 2\tau(-L) = (1-\kappa^2c_0^2)^{1/2}\ c_0^{-1}z_s + (1-\kappa^2c_1^2)^{1/2}c_1^{-1}(2L)$ accounts for the time delay associated with this two-way transit. The factor $e^{-6\omega^2\alpha_{nn}L}$ represents attenuation arising from loss of coherence. Multiple scattering by the random microstructure strips energy from the coherent field, converting it into incoherent energy. Because of our assumption that $\phi(x, y, 0) = \phi(x, y, -L) = 0$, the function ϕ defining the undulations does not contribute to the phase shift in (75). This phase shift amounts to an integration of the reciprocal bulk modulus perturbation along the piecewise-linear ray path associated with the reflection ¿from the interface at $z = -L$.

We can make these observations more explicit by using a Stationary Phase approximation to evaluate (75). For simplicity, assume that $z_s = 0$, i.e. that the point source lies on the interface at the origin. In that case we obtain

$$\langle p_{\text{refl}}(t,\mathbf{x},0)\rangle \approx \frac{\sqrt{\varepsilon}\Gamma_I(-L)\sin^2\theta}{8\pi L c_1}\frac{\partial}{\partial t}\left[\frac{1}{2\pi}\int d\omega \hat{f}(\omega) e^{-i\frac{\omega}{\varepsilon}[t-2L(c_1\sin\theta)^{-1}]}e^{-\omega^2\beta_1(L)}\cdot\right.$$
$$\cdot\exp\left\{i\omega\rho_1 c_1(2\sin\theta)^{-1}\int_{-L}^{0}\left[K_{11}^{-1}(\mathbf{x}[1+\sigma/(2L)],\sigma)\right.\right.$$
$$\left.\left.\left.+K_{11}^{-1}(\mathbf{x}[-\sigma/(2L)],\sigma)\right]d\sigma\right\}\right], \tag{76}$$

where

$$\kappa c_1 \equiv \cos\theta \equiv |\mathbf{x}|/(|\mathbf{x}|^2+4L^2)^{1/2} \quad , \quad \beta_1(L) \equiv 6\alpha L/(c_1\sin\theta)^2 \ . \tag{77}$$

The interface reflection coefficient $\Gamma_I(-L)$ in (76) likewise depends upon θ. Let angles $\theta=\theta_1$ and θ_2 be related by Snell's Law, *i.e.*

$$\frac{c_1}{c_2} = \frac{\cos\theta_1}{\cos\theta_2} \ . \tag{78}$$

Then $\Gamma_I(-L)$ in (76) is given by

$$\Gamma_I(-L) = \frac{\rho_1 c_1\sin\theta_2 - \rho_2 c_2\sin\theta_1}{\rho_1 c_1\sin\theta_2 + \rho_2 c_2\sin\theta_1} \tag{79}$$

(*c.f.* (3.22)–(3.26) and Fig. 3.1 in [15]). The only difference between (76) and the comparable expression for the plane-layered coherent return is the phase shift caused by the bulk modulus perturbation K_{11}. In particular, the Gaussian spreading factor $\gamma(L)$ is not affected by the three dimensional deterministic perturbations.

¿From (62), we infer that the mean transmission operator satisfies the initial value problem

$$\begin{aligned} d\langle\overline{\mathbf{T}}\rangle &= \left[-2\omega^2\alpha_{nn}\langle\overline{\mathbf{T}}\rangle - \frac{\zeta_1}{\rho_1}\langle\overline{\mathbf{T}}\rangle(\boldsymbol{\kappa}\cdot\nabla) - i\omega\langle\overline{\mathbf{T}}\rangle\overline{q}_B\mathbf{I}\right]dz \\ \langle\overline{\mathbf{T}}\rangle|_{z=-L^+} &= \frac{2\sqrt{\zeta_1\zeta_2}}{\zeta_1+\zeta_2}\mathbf{I} \ . \end{aligned} \tag{80}$$

The corresponding kernel of this mean transmission operator, i.e. $\langle\overline{T}(\boldsymbol{\kappa},\mathbf{x},\mathbf{x}',\omega,z)\rangle$, therefore is a solution of

$$\begin{aligned} \frac{\partial}{\partial z}\langle\overline{T}\rangle - \frac{\zeta_1}{\rho_1}\boldsymbol{\kappa}\cdot\nabla_{\mathbf{x}'}\langle\overline{T}\rangle &= -\left[2\omega^2\alpha_{nn} + i\omega\overline{q}_B(\mathbf{x}',z)\right]\langle\overline{T}\rangle \\ \langle\overline{T}\rangle|_{z=-L^+} &= \frac{2\sqrt{\zeta_1\zeta_2}}{\zeta_1+\zeta_2}\delta(\mathbf{x}-\mathbf{x}') \ . \end{aligned} \tag{81}$$

The solution of this initial value problem at $z = 0$ is

$$\langle \overline{T}(\boldsymbol{\kappa}, \mathbf{x}, \mathbf{x}', \omega, 0)\rangle = \frac{2\sqrt{\zeta_1\zeta_2}}{\zeta_1+\zeta_2} \exp\left\{-2\omega^2\alpha_{nn}L + i\omega\frac{\zeta_1}{2}\cdot \right.$$
$$\left. \cdot \int_{-L}^{0} K_{11}^{-1}(\mathbf{x} - \frac{\zeta_1}{\rho_1}\boldsymbol{\kappa}(L+\sigma), \sigma)d\sigma\right\}\delta(\mathbf{x}-\mathbf{x}'-\frac{\zeta_1}{\rho_1}\boldsymbol{\kappa}L)\,. \quad (82)$$

The factor $e^{-2\omega^2\alpha_{nn}L}$ accounts for the attenuation of the coherent transmitted wave due to scattering by the random microstructure. The function ϕ, defining the undulations, again does not contribute to the phase integral, due to the fact that ϕ vanishes at both interfaces. The phase integral itself amounts to an integration of perturbation K_{11}^{-1} along a straight line path beginning at $\mathbf{x}' = \mathbf{x} - \boldsymbol{\kappa}\frac{\zeta_1}{\rho_1}L$ on the upper face ($z = 0$) and ending at $\mathbf{x}$ when $z = -L$.

For point source excitation (20) (with $z_s = 0$), we obtain from (39) that

$$\langle p_{\mathrm{trans}}(t, \mathbf{x}, -L)\rangle = \frac{-\varepsilon^{-3/2}}{2(2\pi)^3}\iiint e^{i\frac{\omega}{\varepsilon}(\boldsymbol{\kappa}\cdot\mathbf{x}-t-\tau(-L))}\frac{2\zeta_2}{\zeta_1+\zeta_2}e^{-2\omega^2\alpha_{nn}L}\exp\left\{i\omega\frac{\zeta_1}{2}\cdot\right.$$
$$\left.\cdot\int_{-L}^{0} K_{11}^{-1}(\mathbf{x} - \frac{\zeta_1}{\rho_1}\boldsymbol{\kappa}(L+\sigma), \sigma)d\sigma\right\}\hat{f}(\omega)\omega^2 d\omega d\boldsymbol{\kappa}, \quad (83)$$

and a Stationary Phase approximate evaluation of (82) leads to

$$\langle p_{\mathrm{trans}}(t, \mathbf{x}, -L)\rangle \approx \frac{-\varepsilon^{-1/2}(sin\theta_1)^2}{4\pi L c_1}\left(\frac{2\rho_2 c_2\sin\theta_1}{\rho_1 c_1 \sin\theta_2 + \rho_2 c_2 \sin\theta_1}\right)$$
$$\cdot\frac{\partial}{\partial t}\left[\frac{1}{2\pi}\int d\omega \hat{f}(\omega)\exp\left\{-i\frac{\omega}{\varepsilon}[t - L(c_1\sin\theta_1)^{-1}]\right\}\exp\left\{-\frac{1}{3}\omega^2\beta_1(L)\right\}\right.$$
$$\left.\cdot\exp\left\{i\omega\rho_1 c_1(2\sin\theta_1)^{-1}\int_{-L}^{0}K_{11}^{-1}(\mathbf{x}[-\sigma/L], \sigma)d\sigma\right\}\right], \quad (84)$$

where

$$\kappa c_1 \equiv \cos\theta_1 \equiv |\mathbf{x}|^2/(|\mathbf{x}|^2 + L^2)^{1/2}\,, \quad (85)$$

$\beta_1(L)$ is defined by (77), and θ_2 is defined by (78). Note that in (83), the phase integral is taken along the linear ray path connecting the source at the origin and the exiting point at $(\mathbf{x}, -L)$.

9. Pressure Correlations

We now consider two-point, two-time pressure correlations. To simplify matters, we shall, as in [16], assume plane wave excitation. In this case, the incident pressure at $z = 0$ has the form

$$p_{\mathrm{inc}}(t, \mathbf{x}, 0) = \frac{-1}{\sqrt{\varepsilon}} f\left(\frac{t - \boldsymbol{\kappa}_0\cdot\mathbf{x}}{\varepsilon}\right), \quad (86)$$

where $\boldsymbol{\kappa}_0$ is the transverse slowness vector for the incident acoustic wave. The counterpart to (20) becomes

$$B(\boldsymbol{\kappa},\omega,0)=\left(\frac{2\pi}{\omega}\right)^2\varepsilon^{5/2}\,\zeta_0^{-1/2}\hat{f}(\omega)\delta(\boldsymbol{\kappa}-\boldsymbol{\kappa}_0), \tag{87}$$

and, when substituted into (37), we obtain the following expression for the reflected pressure

$$p_{\text{refl}}(t,\mathbf{x},0)=\frac{1}{2\pi\sqrt{\varepsilon}}\int e^{i\omega(\boldsymbol{\kappa}_0\cdot\mathbf{x}-t)/\varepsilon}\left[\iint\overline{\Gamma}(\boldsymbol{\kappa}_0,\mathbf{x},\mathbf{x}',\omega,0)d\mathbf{x}'\right]\hat{f}(\omega)d\omega\,. \tag{88}$$

In the random plane-layered problem, the reflected pressure would also be a plane wave. In our case, the undulations and bulk modulus perturbations impart a complex transverse spatial structure to the reflected pressure through the presence of $\overline{\Gamma}$ in (88). A similar statement can be made about the transmitted pressure. Substitution of (87) into (39) leads to

$$\begin{aligned}p_{\text{trans}}(t,\mathbf{x},-L)\;&=\;\frac{-1}{2\pi\sqrt{\varepsilon}}\int e^{i\omega[\boldsymbol{\kappa}_0\cdot\mathbf{x}-t+(1-\kappa_0^2c_1^2)^{1/2}c_1^{-1}L]/\varepsilon}\,\cdot\\&\cdot\left[\iint\overline{T}(\boldsymbol{\kappa}_0,\mathbf{x},\mathbf{x}',\omega,0)d\mathbf{x}'\right]\sqrt{\frac{\zeta_2}{\zeta_0}}\hat{f}(\omega)d\omega\,.\end{aligned} \tag{89}$$

The two-point, two-time correlation function for the reflected pressure assumes the form

$$\begin{aligned}&\langle p_{\text{refl}}(t+\tfrac{\varepsilon}{2}\overline{t},\,\mathbf{x}+\tfrac{\varepsilon}{2}\overline{\mathbf{x}},0)p_{\text{refl}}(t-\tfrac{\varepsilon}{2}\overline{t},\mathbf{x}-\tfrac{\varepsilon}{2}\overline{\mathbf{x}},0)\rangle=\\&\quad\frac{1}{(2\pi)^2}\iint e^{i\omega(\boldsymbol{\kappa}_0\cdot\overline{\mathbf{x}}-\overline{t})}e^{-ih(\boldsymbol{\kappa}_0\cdot\mathbf{x}-t)}\left[\int\cdots\int\langle\overline{\Gamma}(\boldsymbol{\kappa}_0,\mathbf{x}+\tfrac{\varepsilon}{2}\overline{\mathbf{x}},\mathbf{x}',\omega-\tfrac{\varepsilon}{2}h,0)\right.\\&\quad\left.\cdot\overline{\Gamma}^*(\boldsymbol{\kappa}_0,\mathbf{x}-\tfrac{\varepsilon}{2}\overline{\mathbf{x}},\mathbf{x}'',\omega+\tfrac{\varepsilon}{2}h,0)\rangle d\mathbf{x}'d\mathbf{x}''\right]|\hat{f}(\omega)|^2d\omega dh\,.\end{aligned} \tag{90}$$

From (90) it's clear that we must determine the asymptotic (*i.e.* ε-limit) behavior of $\langle\overline{\Gamma}_0(\boldsymbol{\kappa},\mathbf{x},\mathbf{x}',\omega-\frac{\varepsilon}{2}h,0)\overline{\Gamma}^*(\boldsymbol{\kappa}_0,\overline{\mathbf{x}},\overline{\mathbf{x}}',\omega+\frac{\varepsilon}{2}h,0)\rangle$. In [16] and [15], this was accomplished by solving an infinite system of transport equations (W-equations). In the present problem, for general oblique plane wave incidence, this approach is complicated by the presence of the $\boldsymbol{\kappa}_0\cdot\nabla$ drift operators present in the $\overline{\Gamma}$ Riccati equation. (*c.f.* (31), (61)). To obtain a closed system of moment equations, we must consider the asymptotic behavior of $\{\langle\overline{\Gamma}^{NM}\rangle\}_{N,M=0}^{\infty}$, where $\overline{\Gamma}^{NM}\equiv\prod_{n=0}^{N}\prod_{m=0}^{M}\overline{\Gamma}(\boldsymbol{\kappa}_0,\mathbf{x}_n,\mathbf{x}_n',\omega-$

$\frac{\varepsilon}{2}h, 0)\overline{\Gamma}^*(\boldsymbol{\kappa}_0, \overline{\mathbf{x}}_m, \overline{\mathbf{x}}'_m, \omega + \frac{\varepsilon}{2}h, 0)$. Thus, as N and M increase, the transverse spatial domain of $\overline{\Gamma}^{NM}$, *i.e.* $R^{4(N+M)}$, likewise increases.

Considerable simplification can be achieved, however, in the important special case of normal incidence, when $\boldsymbol{\kappa}_0 = \mathbf{0}$. In this case, we infer ¿from (61) that the limiting Ito equation for the integral operator kernel $\overline{\Gamma}(\mathbf{0}, \mathbf{x}, \mathbf{x}', \omega \pm \frac{\varepsilon}{2}h, z) \equiv \overline{\Gamma}^{(\pm)}$ is

$$d\overline{\Gamma}^{(\pm)} = i\omega\sqrt{2\alpha_{nn}}\Big[2\overline{\Gamma}^{(\pm)}d\beta_1 - e^{\mp ih\tau}\delta(\mathbf{x}-\mathbf{x}')\left(\frac{d\beta_2 - id\beta_3}{\sqrt{2}}\right)$$
$$-e^{\pm ih\tau}\iint \overline{\Gamma}^{(\pm)}(\mathbf{0}, \mathbf{x}, \mathbf{x}'', \omega \pm \frac{\varepsilon}{2}h, z)\overline{\Gamma}^{(\pm)}(\mathbf{0}, \mathbf{x}'', \mathbf{x}', \omega \pm \frac{\varepsilon}{2}h, z)d\mathbf{x}''\left(\frac{d\beta_2 + id\beta_3}{\sqrt{2}}\right)\Big]$$
$$+\left[-6\omega^2\alpha_{nn}\overline{\Gamma}^{(\pm)} + i\omega(\overline{q}_A(\mathbf{x}, z) - \overline{q}_B(\mathbf{x}', z))\overline{\Gamma}^{(\pm)}\right]dz, \tag{91}$$

with

$$\tau = c_1^{-1}z \quad , \quad \alpha_{nn} = \alpha c_1^{-2}$$
$$\overline{q}_A(\mathbf{x}, z) = -c_1^{-1}\phi_z(\mathbf{x}, z) + \frac{\rho_1 c_1}{2}K_{11}^{-1}(\mathbf{x}, z) = -\overline{q}_B(\mathbf{x}, z)\ . \tag{92}$$

Also

$$\overline{\Gamma}^{(\pm)}|_{z=-L^+} = \left[\frac{\rho_1 c_1 - \rho_2 c_2}{\rho_1 c_1 + \rho_2 c_2}\right] e^{-i(2\frac{\omega}{\varepsilon} \pm h)c_1^{-1}L}\delta(\mathbf{x} - \mathbf{x}')\ . \tag{93}$$

The solution of (91)–(93) has the form

$$\overline{\Gamma}(\mathbf{0}, \mathbf{x}, \mathbf{x}', \omega \pm \frac{\varepsilon}{2}h, z) \equiv \overline{\Gamma}(\mathbf{x}, \omega, \pm h, z)\delta(\mathbf{x} - \mathbf{x}') \equiv \overline{\Gamma}^{(\pm)}\delta(\mathbf{x} - \mathbf{x}')\ . \tag{94}$$

Substitution of (94) leads to

$$d\overline{\Gamma}^{(\pm)} = i\omega\sqrt{2\alpha_{nn}}\left[2\overline{\Gamma}^{(\pm)}d\beta_1 - e^{\mp ih\tau}\left(\frac{d\beta_2 - id\beta_3}{\sqrt{2}}\right) - e^{\pm ih\tau}\overline{\Gamma}^{(\pm)2}\left(\frac{d\beta_2 + id\beta_3}{\sqrt{2}}\right)\right]$$
$$+\left[-6\omega^2\alpha_{nn} + i2\omega\overline{q}_A(\mathbf{x}, z)\right]\overline{\Gamma}^{(\pm)}dz,$$
$$\overline{\Gamma}^{(\pm)}|_{z=-L^+} = \left[\frac{\rho_1 c_1 - \rho_2 c_2}{\rho_1 c_1 + \rho_2 c_2}\right] e^{-i(2\frac{\omega}{\varepsilon} \pm h)c_1^{-1}L} \tag{95}$$

and correlation function (90) reduces to

$$\langle p_{\text{refl}}(t + \frac{\varepsilon}{2}\overline{t}, \mathbf{x} + \frac{\varepsilon}{2}\overline{\mathbf{x}}, 0)p_{\text{refl}}(t - \frac{\varepsilon}{2}\overline{t}, \mathbf{x} - \frac{\varepsilon}{2}\overline{\mathbf{x}}, 0)\rangle =$$
$$\frac{1}{(2\pi)^2}\iint e^{-i\omega(\omega\overline{t} - ht)}\langle\overline{\Gamma}(\mathbf{x} + \frac{\varepsilon}{2}\overline{\mathbf{x}}, \omega, -h, 0)\overline{\Gamma}^*(\mathbf{x} - \frac{\varepsilon}{2}\overline{\mathbf{x}}, \omega, h, 0)\rangle|\hat{f}(\omega)|^2 d\omega dh\ . \tag{96}$$

In problem (95), transverse spatial dependence enters parametrically. We can obtain a system of moment equations very similar to those in [16] by defining

$$W^N(\mathbf{x},\mathbf{x}',\omega,h,z) \equiv \langle\left(\overline{\Gamma}(\mathbf{x},\omega,-h,z)\overline{\Gamma}^*(\mathbf{x}',\omega,h,z)\right)^N\rangle \quad N=0,1,2,\ldots\,. \tag{97}$$

Using (95) and the Ito calculus, we obtain

$$\begin{aligned}\frac{d}{dz}W^N &= 2\omega^2\alpha_{nn}N^2\left[e^{i2h\tau}W^{N-1}-2W^N+e^{-i2h\tau}W^{N+1}\right]\\ &\quad +i2\omega N\left[\overline{q}_A(\mathbf{x},z)-\overline{q}_A(\mathbf{x}',z)\right]W^N, \quad -L<z\leq 0\\ W^N|_{z=-L^+} &= \left[\frac{\rho_1c_1-\rho_2c_2}{\rho_1c_1+\rho_2c_2}\right]^{2N}e^{i2Nhc_1^{-1}L}\,. \end{aligned}\tag{98}$$

The phase term involving $\overline{q}_A$ in (98) plays a role only when $\mathbf{x}\neq\mathbf{x}'$. If we collapse the spatial offset in (96), *i.e.* set $\overline{\mathbf{x}}=\mathbf{0}$ and consider $\langle p_{\text{refl}}(t+\frac{\varepsilon}{2}\overline{t},\mathbf{x},0)p_{\text{refl}}(t-\frac{\varepsilon}{2}\overline{t},\mathbf{x},0)\rangle$, the phase term vanishes and we obtain a system of W-equations identical to that arising in the plane-layered case (*c.f.* [16]). This observation is important since it identifies a robustness present in the plane-layered theory.

Similar observations can be made in the case of transmitted normal plane wave pressure correlations. From (89), with $\boldsymbol{\kappa}_0=\mathbf{0}$

$$\begin{aligned}&\langle p_{\text{trans}}(t+\frac{\varepsilon}{2}\overline{t},\mathbf{x}+\frac{\varepsilon}{2}\overline{\mathbf{x}},-L)p_{\text{trans}}(t-\frac{\varepsilon}{2}\overline{t},\mathbf{x}-\frac{\varepsilon}{2}\overline{\mathbf{x}},-L)\rangle=\\ &\quad\frac{1}{(2\pi)^2}\iint e^{-i(\omega\overline{t}-h[t-c^{-1}L])}\\ &\quad\cdot\left[\int\cdots\int\langle\overline{T}(\mathbf{0},\mathbf{x}+\frac{\varepsilon}{2}\overline{\mathbf{x}},\mathbf{x}',\omega-\frac{\varepsilon}{2}h,0)\overline{T}^*(\mathbf{0},\mathbf{x}-\frac{\varepsilon}{2}\overline{\mathbf{x}},\mathbf{x}'',\omega+\frac{\varepsilon}{2}h,0)\rangle d\mathbf{x}'d\mathbf{x}''\right]\\ &\quad\cdot\frac{\zeta_2}{\zeta_0}\left|\hat{f}(\omega)\right|^2 d\omega dh\,.\end{aligned}\tag{99}$$

As with the reflection coefficient, in the case of normal plane wave incidence, we infer from (62) and (94) that the kernel of the transmission integral operator has the form

$$\overline{T}(\mathbf{0},\mathbf{x},\mathbf{x}',\omega\pm\frac{\varepsilon}{2}h,z)\equiv\overline{T}(\mathbf{x},\omega,\pm h,z)\delta(\mathbf{x}-\mathbf{x}')\equiv\overline{T}^{(\pm)}\delta(\mathbf{x}-\mathbf{x}') \tag{100}$$

and $\overline{T}^{(\pm)}$ satisfies

$$d\overline{T}^{(\pm)}=i\omega\sqrt{2\alpha_{nn}}\left[\overline{T}^{(\pm)}d\beta_1-e^{\pm ih\tau}\overline{T}^{(\pm)}\overline{\Gamma}^{(\pm)}\left(\frac{d\beta_2+id\beta_3}{\sqrt{2}}\right)\right]-\left[2\omega^2\alpha_{nn}+\right.$$

$$i\omega\overline{q}_B\Big]\overline{T}^{(\pm)}dz$$

$$\overline{T}^{(\pm)}|_{z=-L^+} = \frac{2\sqrt{\rho_1 c_1 \rho_2 c_2}}{\rho_1 c_1 + \rho_2 c_2} \ . \tag{101}$$

where τ, α_{nn} and $\overline{q}_B$ are given by (92). The transmitted pressure correlation becomes

$$\langle p_{\text{trans}}(t + \frac{\varepsilon}{2}\overline{t}, \mathbf{x} + \frac{\varepsilon}{2}\overline{\mathbf{x}}, -L) p_{\text{trans}}(t - \frac{\varepsilon}{2}\overline{t}, \mathbf{x} - \frac{\varepsilon}{2}\overline{\mathbf{x}}, -L)\rangle =$$
$$\frac{1}{(2\pi)^2}\iint e^{-i(\omega\overline{t} - h[t - c^{-1}L])}\langle\overline{T}(\mathbf{x} + \frac{\varepsilon}{2}\overline{\mathbf{x}}, \omega, -h, 0)\overline{T}^*(\mathbf{x} - \frac{\varepsilon}{2}\overline{\mathbf{x}}, \omega, h, 0)\rangle \frac{\zeta_2}{\zeta_0}\Big|\hat{f}(\omega)\Big|^2 d\omega dh \ . \tag{102}$$

The counterpart of (97) becomes the system of moments:

$$\begin{aligned} Z^N(\mathbf{x}, \mathbf{x}', \omega, h, z) &\equiv \langle\Big(\overline{\Gamma}(\mathbf{x}, \omega, -h, z)\overline{\Gamma}^*(\mathbf{x}', \omega, h, z)\Big)^N \overline{T}(\mathbf{x}, \omega, -h, z)\cdot \\ &\quad \cdot\overline{T}^*(\mathbf{x}', \omega, h, z)\rangle \quad , \quad N = 0, 1, 2, \ldots \end{aligned}$$

which are solutions of the following initial value problem

$$\begin{aligned} \frac{d}{dz}Z^N &= 2\omega^2\alpha_{nn}\Big[N^2 e^{i2h\tau}Z^{N-1} - (2N^2 + 2N + 1)Z^N \\ &\quad + e^{-i2h\tau}(N^2 + 2N + 1)Z^{N+1}\Big] \\ &\quad + i\omega(2N + 1)\left[\overline{q}_A(\mathbf{x}) - \overline{q}_A(\mathbf{x}')\right] Z^N \\ Z^N|_{z=-L^+} &= \frac{4\rho_1 c_1 \rho_2 c_2 (\rho_1 c_1 - \rho_2 c_2)^{2N}}{(\rho_1 c_1 + \rho_2 c_2)^{2N+2}} e^{i2Nhc_1^{-1}L} \ . \end{aligned} \tag{103}$$

As with the case of the reflected pressure correlations, problem (103) reduces to the plane-layered problem if we collapse the spatial offset.

10. Localization

We now show that for time harmonic, normally incident plane wave excitation, the limiting behavior of the locally-layered perturbed random slab localizes the energy in the same manner as that of the plane-layered slab. Let $\boldsymbol{\kappa}_0 \to \mathbf{0}, (2\pi\sqrt{\varepsilon})^{-1}\hat{f}(\omega) \to \delta(\omega - \omega_0)$ and use (100) in (89). Then

$$p_{\text{trans}}(t, \mathbf{x}, -L) = -\sqrt{\frac{\rho_2 c_2}{\rho_1 c_1}}\, e^{-i\omega_0(t - c_1^{-1}L)/\varepsilon}\overline{T}(\mathbf{x}, \omega_0, 0). \tag{104}$$

In the discussion of localization, it is convenient to relocate the origin at $z = -L$. Therefore, let $\zeta \equiv z + L$ and let $\overline{T}(\mathbf{x}, \omega_0, z) \to \overline{T}(\mathbf{x}, \omega_0, \zeta)$. It follows from (101) and (92) that

$$d\left[|\ln \overline{T}(\mathbf{x},\omega_0,\zeta)|\right] = \omega\sqrt{\alpha_{nn}}\left[Im\{\overline{\Gamma}\}d\beta_2 + Re\{\overline{\Gamma}\}d\beta_3\right] - \omega^2\alpha_{nn}d\zeta. \quad (105)$$

Using the boundedness of $\overline{\Gamma}$, we conclude that

$$\lim_{\zeta\to\infty} \zeta^{-1}|\ln \overline{T}(\mathbf{x},\omega_0,\zeta)| = -\omega^2\alpha_{nn} \quad (106)$$

with probability one, which is identical to the result obtained in the plane-layered case.

11. O'Doherty-Anstey Theory

In this section, we shall establish the robustness of O'Doherty-Anstey theory for both transmission and reflection. We restrict attention to plane wave pulses. The point source case is analyzed by [13] for plane layered random media and in [14] for a different class of locally layered random media. Related work on the O'Doherty-Anstey theory can be found in [10, 11, 12].

Recall that (*c.f.* (62)

$$\begin{aligned} d\overline{\mathbf{T}} &= i\omega\sqrt{2\alpha_{nn}}\left[\overline{\mathbf{T}}d\beta_1 - \overline{\mathbf{T}}\overline{\mathbf{\Gamma}}\left(\frac{d\beta_2 + id\beta_3}{\sqrt{2}}\right)\right] + \left[-2\omega^2\alpha_{nn}\overline{\mathbf{T}} - \frac{\zeta_1}{\rho_1}\overline{\mathbf{T}}(\boldsymbol{\kappa}\cdot\nabla)\right. \\ &\quad \left. -i\omega\overline{\mathbf{T}}\overline{q}_B\mathbf{I}\right]dz, \\ \overline{\mathbf{T}}|_{z=-L^+} &= \frac{2\sqrt{\zeta_1\zeta_2}}{\zeta_1+\zeta_2}\mathbf{I}\,. \end{aligned} \quad (107)$$

Define

$$\widehat{\mathbf{T}} = e^{-i\omega\sqrt{2\alpha_{nn}}\beta_1(z)}\overline{\mathbf{T}}, \quad (108)$$

where β_1 is normalized so that $\beta_1(-L) = 0$. Then

$$\begin{aligned} d\widehat{\mathbf{T}} &= -i\omega\sqrt{2\alpha_{nn}}\widehat{\mathbf{T}}\overline{\mathbf{\Gamma}}\left(\frac{d\beta_2 + id\beta_3}{\sqrt{2}}\right) + \left[-\omega^2\alpha_{nn}\widehat{\mathbf{T}} - \frac{\zeta_1}{\rho_1}\widehat{\mathbf{T}}(\boldsymbol{\kappa}\cdot\nabla) - i\omega\widehat{\mathbf{T}}\overline{q}_B\mathbf{I}\right]dz, \\ \widehat{\mathbf{T}}|_{z=-L^+} &= \frac{2\sqrt{\zeta_1\zeta_2}}{\zeta_1+\zeta_2}\mathbf{I}\,. \end{aligned} \quad (109)$$

Equation (89) is an expression for the transmitted pressure. We shall now consider $\sqrt{\varepsilon}p_{\mathrm{trans}}$ and show that this quantity has an asymptotic limit. From (108), we write a corresponding relation between kernels as

$$\overline{T} = e^{-i\omega\sqrt{2\alpha_{nn}}\beta_1(z)}\widehat{T}. \quad (110)$$

Inserting (110) into (89) yields

$$\sqrt{\varepsilon}p_{\text{trans}}(t,\mathbf{x},-L) = \frac{-1}{2\pi}\int e^{i\omega[\boldsymbol{\kappa}_0\cdot\mathbf{x}-t+(1-\kappa_0^2c_1^2)^{1/2}c_1^{-1}L+\varepsilon\sqrt{2\alpha_{nn}}\beta_1(0)]/\varepsilon}\cdot$$

$$\cdot\left[\iint \widehat{T}(\boldsymbol{\kappa}_0,\mathbf{x},\mathbf{x}',\omega,0)d\mathbf{x}'\right]\sqrt{\frac{\zeta_2}{\zeta_0}}\hat{f}(\omega)d\omega\,. \tag{111}$$

Define

$$\widehat{T}_0 \quad = \quad E\{\widehat{T}(\boldsymbol{\kappa}_0,\mathbf{x},\mathbf{x}',\omega,z)|\beta_1(z)\}. \tag{112}$$

Then

$$E\{\sqrt{\varepsilon}p_{\text{trans}}(t,\mathbf{x},-L)|\beta_1(0)\} = \frac{-1}{2\pi}\int e^{i\omega[\boldsymbol{\kappa}_0\cdot\mathbf{x}-t+(1-\kappa_0^2c_1^2)^{1/2}c_1^{-1}L+\varepsilon\sqrt{2\alpha_{nn}}\beta_1(0)]/\varepsilon}\cdot$$

$$\cdot\left[\iint \widehat{T}_0(\boldsymbol{\kappa}_0,\mathbf{x},\mathbf{x}',\omega,0)d\mathbf{x}'\right]\sqrt{\frac{\zeta_2}{\zeta_0}}\hat{f}(\omega)d\omega\,. \tag{113}$$

Note in particular that $E\{\left(\frac{d\beta_2(z)+id\beta_3(z)}{\sqrt{2}}\right)|\beta_1(z)\} = 0$, so that

$$\begin{aligned} d\widehat{\mathbf{T}}_0 &= \left[-\omega^2\alpha_{nn}\widehat{\mathbf{T}}_0 - \frac{\zeta_1}{\rho_1}\widehat{\mathbf{T}}_0(\boldsymbol{\kappa}\cdot\nabla) - i\omega\widehat{\mathbf{T}}_0\overline{q}_B\mathbf{I}\right]dz \\ \widehat{\mathbf{T}}_0|_{z=-L^+} &= \frac{2\sqrt{\zeta_1\zeta_2}}{\zeta_1+\zeta_2}\mathbf{I}\,. \end{aligned} \tag{114}$$

We can solve (114) by simply noting that replacing α_{nn} by $2\alpha_{nn}$ in this equation gives equation (80) for the mean transmission operator. Therefore, the kernel $\widehat{T}_0$ can be obtained by replacing α_{nn} by $\alpha_{nn}/2$ in (82) to obtain

$$\begin{aligned} \widehat{T}_0 &= \frac{2\sqrt{\zeta_1\zeta_2}}{\zeta_1+\zeta_2}\exp\left\{-\omega^2\alpha_{nn}L + i\omega\frac{\zeta_1}{2}\int_{-L}^{0} K_{11}^{-1}(\mathbf{x}-\frac{\zeta_1}{\rho_1}\boldsymbol{\kappa}(L+\sigma),\sigma)d\sigma\right\} \\ &\cdot\delta(\mathbf{x}-\mathbf{x}'-\frac{\zeta_1}{\rho_1}\boldsymbol{\kappa}L)\,. \end{aligned} \tag{115}$$

The δ-function reduces integration over $\mathbf{x}'$ to an evaluation. Therefore, substituting (115) into (113) yields

$$\begin{aligned} &E\{\sqrt{\varepsilon}p_{\text{trans}}(t,\mathbf{x},-L)|\beta_1(0)\} = \\ &\frac{-1}{2\pi}\int \exp\{i\omega[\boldsymbol{\kappa}_0\cdot\mathbf{x}-t+(1-\kappa_0^2c_1^2)^{1/2}c_1^{-1}L+\varepsilon\sqrt{2\alpha_{nn}}\beta_1(0) \\ &+\varepsilon\frac{\zeta_1}{2}\int_{-L}^{0} K_{11}^{-1}(\mathbf{x}-\frac{\zeta_1}{\rho_1}\boldsymbol{\kappa}(L+\sigma),\sigma)d\sigma]/\varepsilon - \omega^2\alpha_{nn}L\}\cdot \\ &\cdot\frac{2\zeta_2}{\zeta_1+\zeta_2}\hat{f}(\omega)d\omega \end{aligned} \tag{116}$$

It is instructive to compare the mean transmitted pressure field (116) with the random time shift $\varepsilon\sqrt{2\alpha_{nn}}\beta_1(0)$ and the mean transmitted pressure field without the shift (83). For the comparison we have to convert (83) to the plane wave case by ignoring the $\boldsymbol{\kappa}$ integration, setting $\boldsymbol{\kappa} = \boldsymbol{\kappa}_0$ and adjusting the constants in front of the ω integral appropriately (compare (20) and (87)). We hind that when we average (116) with respect to the shift we recover the coherent pressure field (83). Thus, when the pulse is observed in a random time frame that is defined by $t - \varepsilon\sqrt{2\alpha_{nn}}\beta_1(0)$, then the Gaussian spreading factor is one half that of the coherent field. Moreover, in this time frame the pulse stabilizes, that is its variance tends to zero.

We now show that the conditional variance of $\sqrt{\varepsilon}p_{\text{trans}}$, given $\beta_1(0)$, is zero. From (111) we have that

$$
\begin{aligned}
&\left(\sqrt{\varepsilon}p_{\text{trans}}(t,\mathbf{x},-L)\right)^2 = \\
&\quad \frac{1}{4\pi^2}\int\int \exp\{i(\omega_1+\omega_2)[\boldsymbol{\kappa}_0\cdot\mathbf{x} - t + (1-\kappa_0^2 c_1^2)^{1/2}c_1^{-1}L \\
&\quad +\varepsilon\sqrt{2\alpha_{nn}}\beta_1(0)]/\varepsilon\}\left[\int\int\int\int \widehat{T}(\boldsymbol{\kappa}_0,\mathbf{x},\mathbf{x}_1',\omega_1,0)\widehat{T}(\boldsymbol{\kappa}_0,\mathbf{x},\mathbf{x}_2',\omega_2,0)d\mathbf{x}_1'd\mathbf{x}_2'\right]\cdot \\
&\quad \cdot\frac{\zeta_2}{\zeta_0}\hat{f}(\omega_1)d\omega_1\hat{f}(\omega_2)d\omega_2.
\end{aligned}
\tag{117}
$$

We will show that, for $\omega_1 \neq \omega_2$,

$$
\begin{aligned}
&E\{\widehat{T}(\boldsymbol{\kappa}_0,\mathbf{x}_1,\mathbf{x}_1',\omega_1,z)\widehat{T}(\boldsymbol{\kappa}_0,\mathbf{x}_2,\mathbf{x}_2',\omega_2,z)|\beta_1(z)\} = \\
&\quad E\{\widehat{T}(\boldsymbol{\kappa}_0,\mathbf{x}_1,\mathbf{x}_1',\omega_1,z)|\beta_1(z)\}E\{\widehat{T}(\boldsymbol{\kappa}_0,\mathbf{x}_2,\mathbf{x}_2',\omega_2,z)|\beta_1(z)\}.
\end{aligned}
\tag{118}
$$

From (117) and (118) it will follow that

$$
E\{(\sqrt{\varepsilon}p_{\text{trans}}(t,\mathbf{x},-L))^2|\beta_1(0)\} = (E\{\sqrt{\varepsilon}p_{\text{trans}}(t,\mathbf{x},-L)|\beta_1(0)\})^2, \tag{119}
$$

and the vanishing of the conditional variance of $\sqrt{\varepsilon}p_{\text{trans}}$ given $\beta_1(0)$ will have been shown.

It remains to show (118). From (109), we can write equations for the kernels as

$$
\begin{aligned}
d\widehat{T}(\boldsymbol{\kappa}_0,\mathbf{x},\mathbf{x}',\omega,z) &= -i\omega\sqrt{2\alpha_{nn}}\int \widehat{T}(\boldsymbol{\kappa}_0,\mathbf{x},\mathbf{x}'',\omega,z)\overline{\Gamma}(\boldsymbol{\kappa}_0,\mathbf{x}'',\mathbf{x}',\omega,z)\left(\frac{d\beta_2+id\beta_3}{\sqrt{2}}\right) \\
&\quad -\omega^2\alpha_{nn}\widehat{T}dz + \frac{\zeta_1}{\rho_1}(\boldsymbol{\kappa}\cdot\nabla_{x'})\widehat{T}dz - i\omega\overline{q}_B(\mathbf{x}',z)\widehat{T}dz.
\end{aligned}
\tag{120}
$$

Let

$$
\widehat{T}^{(j)} = \widehat{T}((\boldsymbol{\kappa}_0,\mathbf{x}_j,\mathbf{x}_j',\omega_j,z), \quad j = 1,2. \tag{121}
$$

We shall use the fact that $\frac{\beta_2(\omega_1,z)+i\beta_3(\omega_1,z)}{\sqrt{2}}$ is independent of $\frac{\beta_2(\omega_2,z)+i\beta_3(\omega_2,z)}{\sqrt{2}}$ for $\omega_1 \neq \omega_2$ to obtain that

$$d(\widehat{T}^{(1)}\widehat{T}^{(2)}) = \widehat{T}^{(1)}(d\widehat{T}^{(2)}) + (d\widehat{T}^{(1)})\widehat{T}^{(2)}. \tag{122}$$

There is no Ito term since $d\left[\frac{\beta_2(\omega_1,z)+i\beta_3(\omega_1,z)}{\sqrt{2}}\right] d\left[\frac{\beta_2(\omega_2,z)+i\beta_3(\omega_2,z)}{\sqrt{2}}\right] = 0$. It follows that

$$\begin{aligned} dE\{\widehat{T}^{(1)}\widehat{T}^{(2)}|\beta_1(z)\} = \{&-(\omega_1^2+\omega_2^2)\alpha_{nn} + \frac{\zeta_1}{\rho_1}(\boldsymbol{\kappa}\cdot\nabla_{\mathbf{x}_1'} + \boldsymbol{\kappa}\cdot\nabla_{\mathbf{x}_2'}) \\ &-i[\omega_1\overline{q}_B(\mathbf{x}_1',z) + \omega_1\overline{q}_B(\mathbf{x}_2',z))]\}E\{\widehat{T}^{(1)}\widehat{T}^{(2)}|\beta_1(z)\}dz. \end{aligned} \tag{123}$$

However, from (120)

$$dE\{\widehat{T}^{(j)}|\beta_1(z)\} = \left[-\omega_j^2\alpha_{nn} + \frac{\zeta_1}{\rho_1}(\boldsymbol{\kappa}\cdot\nabla_{\mathbf{x}_j'}) - i\omega_j\overline{q}_B(\mathbf{x}_j',z)\right]E\{\widehat{T}^{(j)}|\beta_1(z)\}dz \tag{124}$$

so that

$$\begin{aligned} d[E\{\widehat{T}^{(1)}|\beta_1(z)\}E\{\widehat{T}^{(2)}|\beta_1(z)\}] = \Big[&-(\omega_1^2+\omega_2^2)\alpha_{nn} + \frac{\zeta_1}{\rho_1}(\boldsymbol{\kappa}\cdot\nabla_{\mathbf{x}_1'} + \boldsymbol{\kappa}\cdot\nabla_{\mathbf{x}_2'}) \\ &-i[\omega_1\overline{q}_B(\mathbf{x}_1',z) + \omega_1\overline{q}_B(\mathbf{x}_2',z))]\Big]E\{\widehat{T}^{(1)}|\beta_1(z)\}E\{\widehat{T}^{(2)}|\beta_1(z)\}dz. \end{aligned} \tag{125}$$

A comparison of (123) and (125), together with the fact that both equations have the same initial conditions, verifies (118) and establishes the O'Doherty-Anstey result for the case of transmission.

The argument establishing the O'Doherty-Anstey result for reflection proceeds in much the same way (see [11, 10] for the plane layered case). We show that the conditional variance of $\sqrt{\varepsilon}p_{\text{refl}}(t,\mathbf{x},0)$, given $\beta_1(0)$, is zero. Noting (61), let

$$\widehat{\mathbf{\Gamma}} = e^{-i2\omega\sqrt{2\alpha_{nn}}\beta_1(z)}\overline{\mathbf{\Gamma}}. \tag{126}$$

Then, the initial value problem for $\widehat{\mathbf{\Gamma}}$ becomes

$$\begin{aligned} d\widehat{\mathbf{\Gamma}} = &-i\omega\sqrt{2\alpha_{nn}}\left[e^{-i2\omega\sqrt{2\alpha_{nn}}\beta_1}\mathbf{I}\left(\frac{d\beta_2 - id\beta_3}{\sqrt{2}}\right) + e^{i2\omega\sqrt{2\alpha_{nn}}\beta_1}\widehat{\mathbf{\Gamma}}^2\left(\frac{d\beta_2 + id\beta_3}{\sqrt{2}}\right)\right] \\ &+\left[-2\omega^2\alpha_{nn}\widehat{\mathbf{\Gamma}} - \frac{\zeta_1}{\rho_1}[(\boldsymbol{\kappa}\cdot\nabla)\widehat{\mathbf{\Gamma}} + \widehat{\mathbf{\Gamma}}(\boldsymbol{\kappa}\cdot\nabla)] + i\omega[\overline{q}_A\mathbf{I}\widehat{\mathbf{\Gamma}} - \widehat{\mathbf{\Gamma}}\overline{q}_B\mathbf{I}\,]\,dz \right. \\ &\widehat{\mathbf{\Gamma}}|_{z=-L^+} = \Gamma_I(-L)e^{-i2\frac{\omega}{\varepsilon}\tau(-L)}\mathbf{I}. \end{aligned} \tag{127}$$

Noting (88), the equation for $\sqrt{\varepsilon}p_{\text{refl}}$ can be written as

$$\sqrt{\varepsilon}p_{\text{refl}}(t,\mathbf{x},0) = \frac{1}{2\pi}\int e^{i\omega(\boldsymbol{\kappa}_0\cdot\mathbf{x}-t)/\varepsilon}\left[\iint\overline{\Gamma}(\boldsymbol{\kappa}_0,\mathbf{x},\mathbf{x}',\omega,0)d\mathbf{x}'\right]\hat{f}(\omega)d\omega \tag{128}$$

From (126), it follows that the relationship between kernels is

$$\overline{\Gamma} \;=\; e^{i2\omega\sqrt{2\alpha_{nn}}\beta_1}\widehat{\Gamma}. \tag{129}$$

If we define

$$\widehat{\Gamma}_0 = E\{\widehat{\Gamma}(\boldsymbol{\kappa}_0, \mathbf{x}, \mathbf{x}', \omega, z)|\beta_1(z)\}, \tag{130}$$

then it follows from (128) - (130) that

$$\begin{aligned} E\{\sqrt{\varepsilon}p_{\text{refl}}(t,\mathbf{x},0)|\beta_1(0)\} \;&=\; \frac{1}{2\pi}\int e^{i\omega(\boldsymbol{\kappa}_0\cdot\mathbf{x}-t+2\varepsilon\sqrt{2\alpha_{nn}}\beta_1(0))/\varepsilon}\;\cdot \\ &\cdot\left[\iint \widehat{\Gamma}_0(\boldsymbol{\kappa}_0,\mathbf{x},\mathbf{x}',\omega,0)d\mathbf{x}'\right]\hat{f}(\omega)d\omega \end{aligned} \tag{131}$$

Using (130), (127) and the fact that $(\beta_2 + i\beta_3)/\sqrt{2}$ is independent of β_1, we obtain the following initial value problem for the operator $\widehat{\mathbf{\Gamma}}_0$ (having kernel $\widehat{\Gamma}_0$).

$$\begin{aligned} d\widehat{\mathbf{\Gamma}}_0 \;&=\; \Bigg[-2\omega^2\alpha_{nn}\widehat{\mathbf{\Gamma}}_0 - \frac{\zeta_1}{\rho_1}[(\boldsymbol{\kappa}\cdot\nabla)\,\widehat{\mathbf{\Gamma}}_0 + \widehat{\mathbf{\Gamma}}_0(\boldsymbol{\kappa}\cdot\nabla)] \\ &\quad + i\omega[\overline{q}_A\mathbf{I}\widehat{\mathbf{\Gamma}}_0 - \widehat{\mathbf{\Gamma}}_0\overline{q}_B\mathbf{I}]\Bigg]dz \\ \widehat{\mathbf{\Gamma}}_0|_{z=-L^+} \;&=\; \Gamma_I(-L)e^{-i2\frac{\omega}{\varepsilon}\tau(-L)}\mathbf{I}. \end{aligned} \tag{132}$$

A solution for the kernel $\widehat{\Gamma}_0$ can be obtained by observing that (132) becomes identical to problem (72) for $\langle\mathbf{\Gamma}\rangle$ if α_{nn} is replaced by $3\alpha_{nn}$. Therefore, replacing α_{nn} by $\alpha_{nn}/3$ in (74) yields the following expression for $\widehat{\Gamma}_0$

$$\begin{aligned} \langle\widehat{\Gamma}_0(\boldsymbol{\kappa},\mathbf{x},\mathbf{x}',\omega,0)\rangle = \Gamma_I(-L)e^{-i2\frac{\omega}{\varepsilon}\tau(-L)}\exp\Bigg\{-2\omega^2\alpha_{nn}L + i\omega\frac{\zeta_1}{2}\int_{-L}^{0}\Bigg[K_{11}^{-1}\Big(\mathbf{x}+ \\ \frac{\zeta_1}{\rho_1}\boldsymbol{\kappa}_0\sigma,\sigma\Big) + K_{11}^{-1}\left(\mathbf{x}'-\frac{\zeta_1}{\rho_1}\boldsymbol{\kappa}_0\sigma,\sigma\right)\Bigg]d\sigma\Bigg\}\delta\left(\mathbf{x}-\mathbf{x}'-2\frac{\zeta_1}{\rho_1}\boldsymbol{\kappa}L\right) \end{aligned} \tag{133}$$

When (133) is substituted into (131), the δ-function reduces the $\mathbf{x}'$ integration to an evaluation and we obtain

$$\begin{aligned} E\{\sqrt{\varepsilon}p_{\text{refl}}(t,\mathbf{x},0)|\beta_1(0)\} = \frac{1}{2\pi}\int\exp\Bigg\{i\frac{\omega}{\varepsilon}\Bigg[\boldsymbol{\kappa}_0\cdot\mathbf{x}-t-2\tau(-L)+2\varepsilon\sqrt{2\alpha_{nn}}\beta_1(0) \\ +\frac{\zeta_1}{2}\int_{-L}^{0}\left[K_{11}^{-1}\left(\mathbf{x}+\frac{\zeta_1}{\rho_1}\boldsymbol{\kappa}_0\sigma,\sigma\right)+K_{11}^{-1}\left(\mathbf{x}'-\frac{\zeta_1}{\rho_1}\boldsymbol{\kappa}_0\sigma,\sigma\right)\right]d\sigma\Bigg] \\ -2\omega^2\alpha_{nn}L\Bigg\}\Gamma_I(-L)\hat{f}(\omega)d\omega. \end{aligned} \tag{134}$$

It is instructive here again to compare the mean reflected field (134) with the coherent reflected field (75), specialized to plane wave pulses as explained above for the transmitted pressure. When (134) is averaged over the random time shift $2\varepsilon\sqrt{2\alpha_{nn}}\beta_1(0)$ we obtain exactly the coherent reflected field (75). The Gaussian spreading factor is now three times bigger. The reflected pulse stabilizes as well.

We next show that the conditional variance of $\sqrt{\varepsilon}p_{\text{refl}}$, given $\beta_1(0)$, is zero. From (128)

$$(\sqrt{\varepsilon}p_{\text{refl}}(t,\mathbf{x},0))^2 = \frac{1}{4\pi^2}\int\int \exp\left\{i(\omega_1+\omega_2)(\boldsymbol{\kappa}_0\cdot\mathbf{x}-t+2\varepsilon\sqrt{2\alpha_{nn}}\beta_1(0))/\varepsilon\right\}\cdot \left[\int\int\int\int \widehat{\Gamma}(\boldsymbol{\kappa}_0,\mathbf{x},\mathbf{x}_1',\omega_1,0)\widehat{\Gamma}(\boldsymbol{\kappa}_0,\mathbf{x},\mathbf{x}_2',\omega_2,0)d\mathbf{x}_1'd\mathbf{x}_2'\right]\hat{f}(\omega_1)\hat{f}(\omega_2)d\omega_1 d\omega_2 \tag{135}$$

We will show that, for $\omega_1 \neq \omega_2$,

$$E\left\{\widehat{\Gamma}(\boldsymbol{\kappa}_0,\mathbf{x},\mathbf{x}_1',\omega_1,z)\widehat{\Gamma}(\boldsymbol{\kappa}_0,\mathbf{x},\mathbf{x}_2',\omega_2,z)|\beta_1(z)\right\} = E\left\{\widehat{\Gamma}(\boldsymbol{\kappa}_0,\mathbf{x},\mathbf{x}_1',\omega_1,z)|\beta_1(z)\right\}E\left\{\widehat{\Gamma}(\boldsymbol{\kappa}_0,\mathbf{x},\mathbf{x}_2',\omega_2,z)|\beta_1(z)\right\} \tag{136}$$

It will then follow from (136) and (135) that

$$E\{(\sqrt{\varepsilon}p_{\text{refl}}(t,\mathbf{x},0))^2|\beta_1(0)\} = \left(E\{\sqrt{\varepsilon}p_{\text{refl}}(t,\mathbf{x},0)|\beta_1(0)\}\right)^2, \tag{137}$$

and the vanishing of the conditional variance of $\sqrt{\varepsilon}p_{\text{refl}}$, given $\beta_1(0)$, will thus have been shown.

It remains to show (136). From (127), we can write the equation of the kernel $\widehat{\Gamma}$ as

$$\begin{aligned} d\widehat{\Gamma}(\boldsymbol{\kappa},\mathbf{x},\mathbf{x}',\omega,z) &= -i\omega\sqrt{2\alpha_{nn}}\left(e^{-i2\omega\sqrt{2\alpha_{nn}}\beta_1}\delta(\mathbf{x}-\mathbf{x}')\left(\frac{d\beta_2(z)-d\beta_3(z)}{\sqrt{2}}\right)\right. \\ &\quad\left.+\,e^{i2\omega\sqrt{2\alpha_{nn}}\beta_1}\left[\int\int\widehat{\Gamma}(\mathbf{x},\mathbf{x}'',\omega,z)\widehat{\Gamma}(\mathbf{x}'',\mathbf{x}',\omega,z)d\mathbf{x}''\right]\delta(\mathbf{x}-\mathbf{x}')\left(\frac{d\beta_2(z)+d\beta_3(z)}{\sqrt{2}}\right)\right) \\ &\quad+\left[-2\alpha_{nn}\omega^2\widehat{\Gamma}-\frac{\zeta_1}{\rho_1}[\boldsymbol{\kappa}\cdot\nabla_{\mathbf{x}}\widehat{\Gamma}-\boldsymbol{\kappa}\cdot\nabla_{\mathbf{x}'}\widehat{\Gamma}]+i\omega[\overline{q}_A(\mathbf{x},z)-\overline{q}_B(\mathbf{x}',z)]\widehat{\Gamma}\right]dz. \end{aligned} \tag{138}$$

Let

$$\widehat{\Gamma}^{(j)} = \widehat{\Gamma}(\boldsymbol{\kappa}_0,\mathbf{x}_j,\mathbf{x}_j',\omega_j,z), \quad j=1,2. \tag{139}$$

We use the fact that $\frac{\beta_2(\omega_1,z)+i\beta_3(\omega_1,z)}{\sqrt{2}}$ is independent of $\frac{\beta_2(\omega_2,z)+i\beta_3(\omega_2,z)}{\sqrt{2}}$, for $\omega_1\neq\omega_2$, so that $d\frac{\beta_2(\omega_1,z)+i\beta_3(\omega_1,z)}{\sqrt{2}}d\frac{\beta_2(\omega_2,z)+i\beta_3(\omega_2,z)}{\sqrt{2}}=0$ to obtain

$$d\left(\widehat{\Gamma}^{(1)}\widehat{\Gamma}^{(2)}\right) = \widehat{\Gamma}^{(1)}d\widehat{\Gamma}^{(2)}+\left(d\widehat{\Gamma}^{(1)}\right)\widehat{\Gamma}^{(2)}. \tag{140}$$

It follows that

$$dE\left\{\widehat{\Gamma}^1\widehat{\Gamma}^2|\beta_1(z)\right\} =$$
$$\Big[-2\alpha_{nn}(\omega_1^2+\omega_2^2) - \frac{\zeta_1}{\rho_1}[\boldsymbol{\kappa}\cdot\nabla_{\mathbf{x}_1} + \boldsymbol{\kappa}\cdot\nabla_{\mathbf{x}_2} - \boldsymbol{\kappa}\cdot\nabla_{\mathbf{x}_1'} + \boldsymbol{\kappa}\cdot\nabla_{\mathbf{x}_2'}]$$
$$+[\omega_1\overline{q}_A(\mathbf{x}_1,z) + \omega_2\overline{q}_A(\mathbf{x}_2,z) - \omega_1\overline{q}_B(\mathbf{x}_1',z) - \omega_2\overline{q}_B(\mathbf{x}_2',z)\Big] E\left\{\widehat{\Gamma}^1\widehat{\Gamma}^2|\beta_1(z)\right\} dz. \tag{141}$$

However, from (138)

$$dE\left\{\widehat{\Gamma}^{(j)}|\beta_1(z)\right\} =$$
$$\Big[-2\alpha_{nn}\omega_j^2 - \frac{\zeta_1}{\rho_1}[\boldsymbol{\kappa}\cdot\nabla_{\mathbf{x}_j} - \boldsymbol{\kappa}\cdot\nabla_{\mathbf{x}_j'}]$$
$$+i\omega_j\left(\overline{q}_A(\mathbf{x}_j,z) - \overline{q}_B(\mathbf{x}_j',z)\right)\Big] E\left\{\widehat{\Gamma}^j|\beta_1(z)\right\} dz, \tag{142}$$

so that

$$d\Big[E\left\{\widehat{\Gamma}^1|\beta_1(z)\right\} E\left\{\widehat{\Gamma}^2|\beta_1(z)\right\}\Big] =$$
$$\Big[-2\alpha_{nn}(\omega_1^2+\omega_2^2) - \frac{\zeta_1}{\rho_1}[\boldsymbol{\kappa}\cdot\nabla_{\mathbf{x}_1} + \boldsymbol{\kappa}\cdot\nabla_{\mathbf{x}_2} - \boldsymbol{\kappa}\cdot\nabla_{\mathbf{x}_1'} + \boldsymbol{\kappa}\cdot\nabla_{\mathbf{x}_2'}]$$
$$+i[\omega_1\overline{q}_A(\mathbf{x}_1,z) + \omega_2\overline{q}_A(\mathbf{x}_2,z) - \omega_1\overline{q}_B(\mathbf{x}_1',z) - \omega_2\overline{q}_B(\mathbf{x}_2',z)]\Big]\cdot$$
$$\cdot E\left\{\widehat{\Gamma}^1|\beta_1(z)\right\} E\left\{\widehat{\Gamma}^2|\beta_1(z)\right\} dz. \tag{143}$$

Comparing (141) and (143), and noting that both equations have the same initial conditions, establishes (136).

If proper allowance is made for the differences between point source and plane wave excitation (*c.f.* (20) and (87)), the conditional means considered in this section can be seen to agree with their counterparts in Section 8. Using the fact that $E\{e^{i\omega\sqrt{2\alpha_{nn}}\beta_1(0)}\} = e^{-\omega^2\alpha_{nn}L}$, expressions (116) and (134) can be put into agreement with (83) and (75), respectively.

12. Conclusions

We have confirmed the robustness of the random plane layered theory of acoustics by obtaining qualitatively similar results from a model that has three dimensional perturbations. This perturbed model allows for small spatial dependence of the background bulk modulus, and for small spatial deviations in the normal to the random layers. These deviations from plane layers produce $O(1)$ effects on many of the statistics of interest.

For the generalized O'Doherty-Anstey theory, and for the reflected and transmitted coherent fields, only the variations of the background have an effect, to leading order and not on the Gaussian spreading rate. However, all the spatial variations have a leading order effect in computing the space and time correlation functions of the reflected and transmitted pressure waves at non-normal incidence. For these correlations, a system of moment equations was derived. These moment equations are qualitatively similar to what is obtained in the plane layered case, but with new terms that are dependent on the three-dimensional effects.

However, when the incident wave is normal to the layering, a remarkable robustness is obtained for the coherent field and for the time autocorrelation function of reflected or transmitted pressure at a single point in space. In these cases all three-dimensional effects vanish, and leading order results identical to the plane layered case are reproduced exactly. A related robustness result is obtained for the localization length, which, at normal incidence, is identical to that in the strictly plane layered theory. Thus despite the nontrivial deviation from a one-dimensional model, the localization phenomenon, at normal incidence, is preserved quantitatively as well as qualitatively.

Acknowledgement

The work of G. Papanicolaou was suppoted by NSF grant DMS-9622854 and AFOSR grant F49620-98-1-0211.

References

1. P. Sheng, Editor, *Scattering and Localization of Classical Waves in Random Media*, World Scientific, Singapore (1990).
2. I. M. Lifshits, S. A. Gredeskul and L. A. Pastur, *Introduction to the Theory of Disordered Systems*, J. Wiley, New York (1988).
3. A. Crisanti, G. Paladin and A. Vulpiani, *Products of Random Matrices in Statistical Physics*, Springer, New York (1993).
4. I. J. Goldsheid, S. A. Molchanov and L. A. Pastur, *A Random One-Dimensional Schroedinger Operator Has Pure Point Spectrum*, Functional Anal. Appl. 11, 1-10 (1977).
5. V. I. Klyatskin, *Stochastic Equations and Waves in Randomly Inhomogeneous Media*, Nauka, Moscow (1980) (in Russian).
6. R. F. O'Doherty and N. A. Anstey, *Reflections on Amplitudes*, Geophysical Prospecting 19, 430-458 (1971).
7. R. Burridge, B. White and G. Papanicolaou, *One-Dimensional Wave Propagation in a Highly Discontinuous Medium*, Wave Motion 10, 19 - 44 (1988).
8. J.F. Clouet and J.P. Fouque, *Spreading of a pulse travelling in a random medium*, Annals of Applied Probability, 4, 1083-1097, 1994.

9. P. Lewicki, R. Burridge and G. Papanicolaou, *Pulse Stabilization in a Strongly Heterogeneous Layered Medium*, Wave Motion 20, 177-196 (1994).
10. P. Lewicki, *Long time evolution of wavefronts in random media*, SIAM J. Appl. Math. 54, 907-934, 1994.
11. P. Lewicki and G. Papanicolaou, *Reflection of wavefronts by randomly layered media*, Wave Motion 20, 245-266, 1994.
12. P. Lewicki, R. Burridge and M.V. de Hoop, *Beyond effective medium theory: Pulse stabilization for multimode wave propagation in high contrast media*, SIAM J. Appl. Math. 56, 256-276, 1996.
13. J. Chillan and J.P. Fouque, *Pressure fields generated by acoustical pulses propagating in randomly layered media*, SIAM J. Appl. Math. 58, 1532-1546, 1998.
14. K. Solna, *Stable spreading of acoustic pulses due to laminated microstructure*, Ph.D. thesis, Stanford University, 1997.
15. M. Asch, W. Kohler, G. Papanicolaou, M. Postel, and B. White, *Frequency Content of Randomly Scattered Signals*, SIAM Review 33, 519-625 (1991).
16. R. Burridge, G. Papanicolaou, P. Sheng and B. White, *Probing a Random Medium with a Pulse*, SIAM J. Appl. Math. 49, 582-607 (1989).
17. W. Kohler, G. Papanicolaou and B. White, *Localization and Mode Conversion for Elastic Waves in Randomly Layered Media*, Wave Motion 23, 1-22 and 181-201, 1996.
18. M. Asch, W. Kohler, G. Papanicolaou M. Postel and B. White, *Statistical Inversion from Reflections of Spherical Waves By A Randomly Layered Medium*, Waves in Random Media 6, 293-334, 1996.

SEISMIC WAVE DIFFUSION IN THE EARTH LITHOSPHERE

M. CAMPILLO AND L. MARGERIN
Laboratoire de Géophysique Interne et Tectonophysique,
Observatoire de Grenoble,
Université Joseph Fourier, BP 53,
38041 Grenoble Cedex, France
E. mail: Michel.Campillo@ujf-grenoble.fr,
Ludovic.Margerin@ujf-grenoble.fr

AND

N. M. SHAPIRO
Instituto de Geofisica,
Universidad National Autonoma de Mexico,
Mexico
E. mail: shapiro@ollin.igeofcu.unam.mx

Abstract. This paper is devoted to the study of the time decay of the coda of seismograms. We consider a conceptual model of the Earth upper layers: a diffractive crust overlying an almost homogeneous mantle. We simulate the multiple scattering of the seismic waves using the classical radiative transfer equation in a scalar approximation. We solve the equation using the Monte Carlo method and give a particular attention to the asymptotics of the solution. Under the condition that the ratio mean free path/layer thickness is less than one , we can give an analytical form of the time decay of the coda energy as the solution of a diffusive equation. Otherwise, our transcription of the boundary condition is not valid. The asymptotic form is similar to the one that was proposed by Aki and Chouet (1975) to fit their observations. We checked numerically that, even when the mean free path is larger than the layer thickness, the asymptote of the radiative transfer solution has the same functional form than the one obtained when we are able to compute an analytical solution of the diffusion equation. We show a direct comparison of observations made in Mexico with the results of our model in which we included the effect of a weak absorption. The measurements are well predicted by the model, both for the absolute level of apparent attenuation and for its frequency dependence. We stress the

J.-P. Fouque (ed.), Diffuse Waves in Complex Media, 383–404.

importance of the leakage of diffuse energy in the mantle and propose to define a time of residence of diffuse waves in the crust to characterize the temporal decay of the coda energy. An important aspect of our interpretation is the fact that the coda decay corresponds to the decay of scattered waves in the diffusive regime. To demonstrate the diffuse character of the coda, we look for the energy partitionning that is expected for elastic waves in the diffusive regime. Considering a series of records from earthquakes at different epicentral distances, we show that the equilibration between the two modes (that is the energy partitioning) appears very early in the coda. The ratio between compressional and shear energies is very stable while the energy level changes by several orders of magnitude. The energy ratio is independent of the earthquake considered. This experiment tends to confirm that the seismic coda corresponds to waves in the diffusive regime and therefore supports our interpretation that the decay observed in Mexico at low frequency is strongly governed by the rate of leakage of diffuse energy in the mantle.

1. Introduction

The seismograms consist of records of the 3 component motion of the free surface of the Earth during earthquakes. The seismograms begin with a series of arrivals of compressional, shear and surface waves that can be interpreted in terms of the ray theory. The very deep structure of our planet has been explored by using this type of waves. On short period seismograms (in the range 1-10 Hz), after these arrivals the envelope decays slowly to reach the level of seismic noise after a delay that can be currently several tens of times the travel time of direct waves. This long tail is called the seismic coda. The coda consists of waves that are arriving to the receiver from almost all directions (see for example Campillo at al, 1998 for recent references). Applying a frequency-wave number analysis to the data of the NORSAR array, Dainty and Toksoz (1990) have shown that the coda is dominated by waves with apparent velocities less then 4 km/s, that correspond in the Earth crust to the shear elastic waves called S waves in seismology.

In their pioneering work, Aki and Chouet (1975) showed that the time decay of energy in the coda is a regional characteristic, independent of the source depth or magnitude. They measured this constant decay rate after a lapse time that is larger than two times the travel time of direct shear waves. They found that the coda envelope can be fitted by the formula:

$$\rho(f,t) = \frac{S_0(f)g_\pi(f)}{2\pi\beta^2 t^{2\gamma}} \exp\left(-\frac{2\pi f t}{Q_c}\right), \tag{1}$$

where $S_0(f)$ is the shear energy emitted by the source, $g_\pi(f)$ is the backscattering coefficient and β the shear wave velocity. γ is a factor of spreading chosen to be 1 since it is assumed that the coda is made up of body waves. For large lapse time the decay is governed essentially by the exponential term. Q_c, the "coda quality factor" characterizes the decay rate (through the exponential term) and was introduced by analogy with the classical quality factor associated with absorption.

The parameter Qc (coda Q) has been measured in a large number of regions and the different authors reported essentially the Aki and Chouet conclusions about the stability of the decay rate. Qc appeared to be the easiest amplitude parameter to measure while the peak amplitude or attenuation of direct waves are in most cases very difficult to evaluate in the Earth due to focusing and defocusing effects.

It was found that Qc is increasing with frequency with a rate that is larger when its value is small at low frequency. At high frequency, the values observed are very high almost everywhere while around 1 Hz Q_c exhibits regional variations correlated with the tectonic style.

Aki and Chouet interpreted their result using either the single scattering approximation or the diffusion approximation for scalar waves in a full space. Later, Wu (1985) introduced the stationary radiative transfer equation for scalar waves. Abubikarov and Gusev (1987) and Hoshiba (1991) used the Monte Carlo method to solve the radiative transfer equation for scalar waves in the time domain. In all these studies, a constant background velocity and a homogeneous distribution of scatterers were assumed. The use of this type of model in the interpretation of the observed coda decay rate suggests a strong absorption that would be the dominant process in the apparent attenuation of seismic waves in the crust. On the other hand, the study of the decay of direct waves with distance led to the inverse conclusion (see Campillo and Plantet, 1991): when Qs the quality factor of S waves is small (significantly less than 1000), it has a strong frequency dependence characteristic of the scattering effect on primary waves. The measurements of Qs in tectonically active regions suggest that the scattering effect is dominant over the absorption. This apparent contradiction between these two interpretations can be solved by considering a more realistic model of the Earth structure to interpret the coda decay.

2. Radiative transfer in the crust

The major characteristic of the structure of the first tens of kilometers of the solid Earth is the existence of the crust overlying the mantle. The crust and the mantle have very different physical properties and chemical composition. The mantle has a very homogeneous composition with mostly olivine (Mg,Fe)SiO4 while the crust is the result of the differentiation of light elements during the Earth existence. In geologically active provinces, the compositionnal heterogeneity of the crust makes it possible the progressive build up of mechanical heterogeneity during the long history of the deformations associated with the tectonic processes. Besides the mantle has a ductile mechanical behavior while the upper crust is brittle. This results in the development of crack networks in the crust. Experiments of deep reflection seismic soundings in the continental domain provided images of a reflective crust above a transparent mantle (see for example Allmendinger et al., 1987 or Meissner, 1989).

It is therefore reasonable to set up a model in which an upper layer containing numerous scatterers (impedance fluctuations, cracks) overlies a half space that is weakly diffusive. There is another special feature of this structure that cannot be neglected. The wave velocities are significantly higher in the mantle than in the crust (typically $4700m/s$ and $3500m/s$ for S waves). As a consequence, the wavefield produced by a source located in the crust comprises strong guided waves due to post critical reflections on the boundary. This particularity of the Green function has to be incorporated in our treatment of the multiple diffraction. A simple conceptual model that accounts for the specificities of the Earth was studied by Margerin et al. (1998a). The configuration is shown in Fig 1. Let us begin with a purely elastic model. Considering the observation that coda waves are mostly S waves, we neglect the compressional P waves and assume that, after a few diffractions the S waves are depolarized. We therefore limit our analysis to a scalar problem. This model may appear too simple but our goal at this stage is to investigate the consequences of the existence of the layering. After a series of numerical studies using Monte Carlo simulation, it was shown that the energy decay in such a model varies with time according to the regime of diffraction.

We present in Fig 2 an example of the results for a configuration in which the crustal thickness is assumed to be 30 km while the mean free path in the crust is 50 km. For very short lapse time, the solution is asymptotic to the single scattering solution. Nevertheless, the divergence of the two curves is almost instantaneous, demonstrating the importance of multiple scattering and the extreme limitation of the single scattering approximation in this problem. Rapidly, the solution of the radiative transfer equation

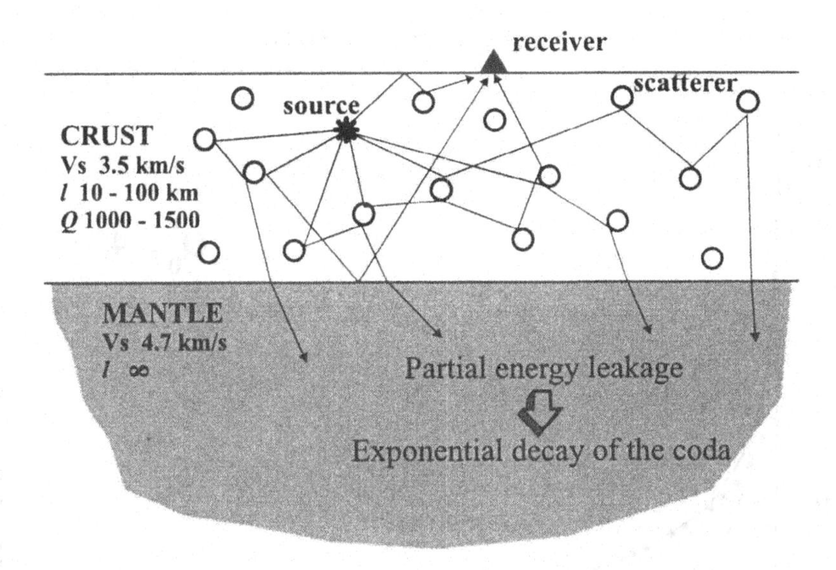

Figure 1. Sketch of our simplified model with the physical parameters used in the computations.

becomes asymptotic to a curve which has the expression proposed by Aki and Chouet from their data analysis. We fit the curve using equation (1) with Qc being the free parameter. This result indicates that the simple purely elastic model we propose is able to reproduce this very strong property of the observation, that is the existence of an exponential decay term. Margerin et al. (1998a,b) showed that the fonctional form of the decay is identical for a broad range of mean free path and crustal thickness.

We interpret this exponential behavior as the consequence of the leakage of the diffuse energy from the crust to the mantle. This leakage is nevertheless governed by the reflection coefficient at the base of the crust. In the following we shall use the diffusion approximation to understand this behavior and to give a formal interpretation of Q_c.

3. Diffusion equation and asymptotics

The asymptote of the energy envelope of the multiply scattered waves is given, for large lapse times, by the solution of the diffusion equation. Although this is a general statement, the manner of writing the diffusion equation with its initial conditions is far to be obvious. Margerin et al. (1998a) studied explicitely a particular case where the mean free path is smaller than the crustal thickness ($l < H$). The source is assumed to be at the surface. In this case, one can assume that the wavefield is almost diffusive when it reaches the base of the crust. It is therefore possible to write

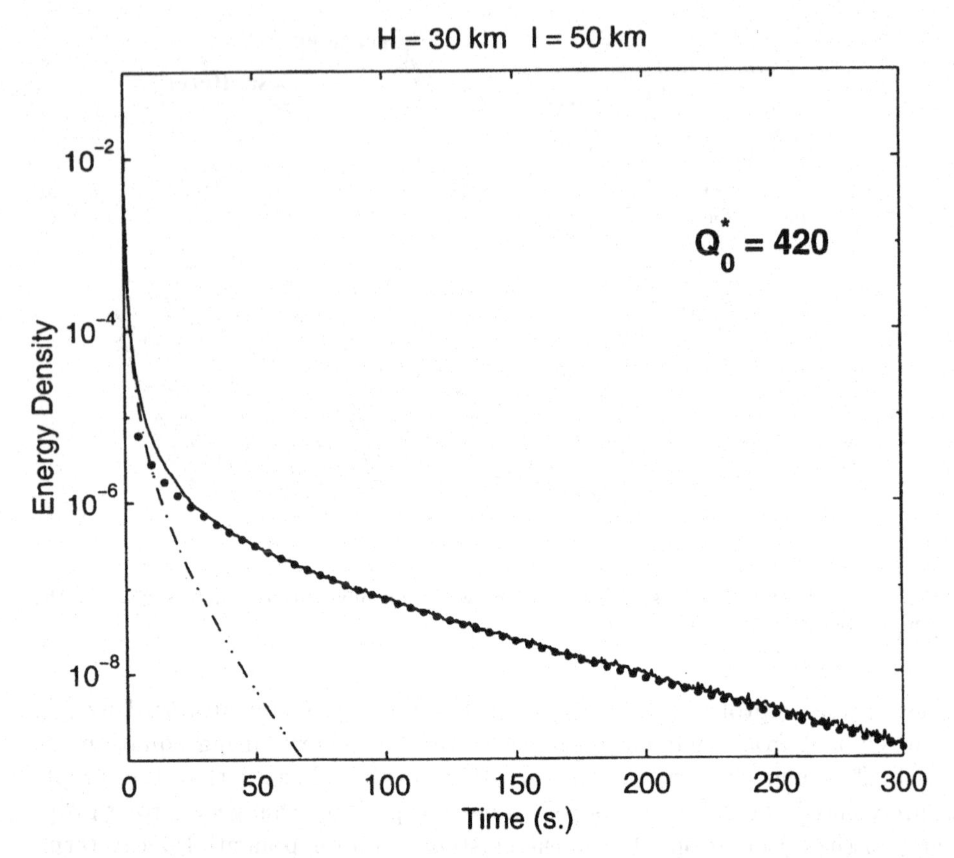

Figure 2. Coda decay obtained in our model with $H = 30$ km and the mean free path of the crust l equal to 50 km. Solid lines show the numerical solution of the Radiative Transfer Equation obtained by Monte Carlo modeling. The thin dotted lines shows the result of the single scattering approximation. The black circles indicate an approximation of the radiative transfer solution obtained with the formula $\frac{1}{t}\exp(-\frac{2\pi t}{Q_0^*})$. The Q_0^* value corresponding to the best approximation is 420. The maximum standard deviation of the Q_0^* is $\Delta Q_0^* \approx 30$.

directly a boundary condition for diffuse waves and to solve the diffusion equation analytically. It was checked that, under the condition $l < H$, the agreement between the solution of the diffusion and radiative transfer equations is perfect for large lapse times. The complete solution corresponding to the conceptual model presented above is:

$$\rho(r,t) = \frac{\exp(-\frac{r^2}{4D_1 t})}{2\pi H D_1 t} \sum_n \frac{\sin\xi_n + \frac{t_n\gamma}{H}\cos\xi_n}{(1+\frac{\gamma}{H})\sin\xi_n + \frac{t_n\gamma}{H}\cos\xi_n} \exp(-D_1 \frac{\xi_n^2}{H^2} t) \quad (2)$$

where ρ is the energy density in the coda, t, the time elapsed since the energy release at the source, f, the frequency of waves. The ξ_n are the roots of the equation :

$$\xi_n \tan\xi_n = \frac{H}{\gamma}, \qquad \xi_n \in]n\pi, n\pi + \frac{\pi}{2}[\qquad n \in N. \quad (3)$$

In this expression γ is a variable that depends on the differential cross section of the scatterer in the crust, their density and the reflection coefficient at the boundary (Margerin et al, 1998a).

For a configuration where the source and the receiver are close, the leading term of the solution has the form:

$$\rho(t) = \frac{1}{t}\exp(-\frac{t}{\tau_d}) \quad (4)$$

and

$$\tau_d = \frac{Q_c^\star}{2\pi f} \simeq \frac{3H^2}{vl\xi^2}, \quad (5)$$

where ξ is the root of the equation :

$$\xi \tan\xi = \frac{H}{\gamma}, \qquad \xi \in \left]0, \frac{\pi}{2}\right[. \quad (6)$$

$Q_c^\star$ is a parameter that describes the decay rate of the coda in our model. We denote by $Q_0^\star$ the value of $Q_c^\star$ at the frequency $f = 1$ Hz. We maintain in the notation the analogy with the classical quality factor because of the wide practice of this parametrization in experimental seismology. It allows to compare directly the output of our model with the measurements made on actual records. However, the use of τ_d, the time of residence of diffuse energy in the crust would be more appropriate to characterize the leakage effect. Whatever the parametrization chosen, the important result is the fact that the solution of the diffusion equation in a model without absorption gives a functional decay identical to the one that has been observed by Aki and Chouet and widely confirmed later on. The solution of the diffusion equation makes it possible to give an explicit expression of $Q_c^\star$ or τ_d in terms of the physical parameters of the model, again under the condition $l < H$. We are not able to write down a similar problem of diffusion when $l > H$. In this case the field that reaches initially the boundary is not diffusive and therefore we cannot incorporate simply the boundary conditions in our model. Nevertheless, the numerical simulation of the radiative transfer equation for different ratios l/H indicates that the same functional form of the asymptotic decay is expected when $l < H$ or $l > H$ (Margerin et al., 1998b).

4. Comparison with observations

In the previous sections we studied a simple conceptual model in which we assume no absorption. Indeed two effects can govern the actual decay of the coda of the seismograms. One is the leakage that we just discussed and another is the anelastic absorption which can be represented by the quality factor Q_i. In order to test the relative importance of these effects, we present a comparison between the decays observed at a series of seismic stations and the results of our simulations using realistic values for the parameters of the model. The stations are located along the Pacific Coast of Mexico, a region which is tectonically active.

We use the records at the seismological stations PNIG, HUIG, CAIG and ZIIG of the Mexican Seismological National Network (see Singh et al., 1997). We consider the records of N earthquakes which occured at epicentral distances between 30 and 150 km. At each station, the decay at a given frequency f is parametrized through the measure of $Q_c(f)$ in the way proposed by Aki and Chouet (1975) (equation 1). The results are shown in Fig 3.

Between 1 and 5 Hz, they exhibit a strong frequency dependence as expected from our model. In the region where these data are collected, the crustal thickness is between 20 and 30 km (Kostoglodov et al., 1996). The mean free path is not objectively determined. Therefore, we consider values of l between 20 and 70 km, a range that covers the values proposed for the Earth crust in the frequency range considered here.

The results of the numerical simulations in absence of absorption are plotted in continuous lines in Fig 3 for the extreme values of l 20 and 70 km. The results obtained at low frequency (1Hz) are close to the observed values, while at high frequency Q_c is widely overestimated by our purely elastic model. We interpret this discrepancy by the effect of anelasticity. According to laboratory studies, the quality factor inherent to dry rocks is independent of frequency (Johnston, 1981). For crustal rocks at depth and the frequency range considered here, it is difficult to evaluate directly a possible frequency dependence of Q_i. In absence of further evidence, we consider the model with the smallest number of free parameters: a constant Q_i. We add to our model a small absorption term described by $Q_i = 1200$. This value corresponds to the high frequency limit of the measured Q_c. This

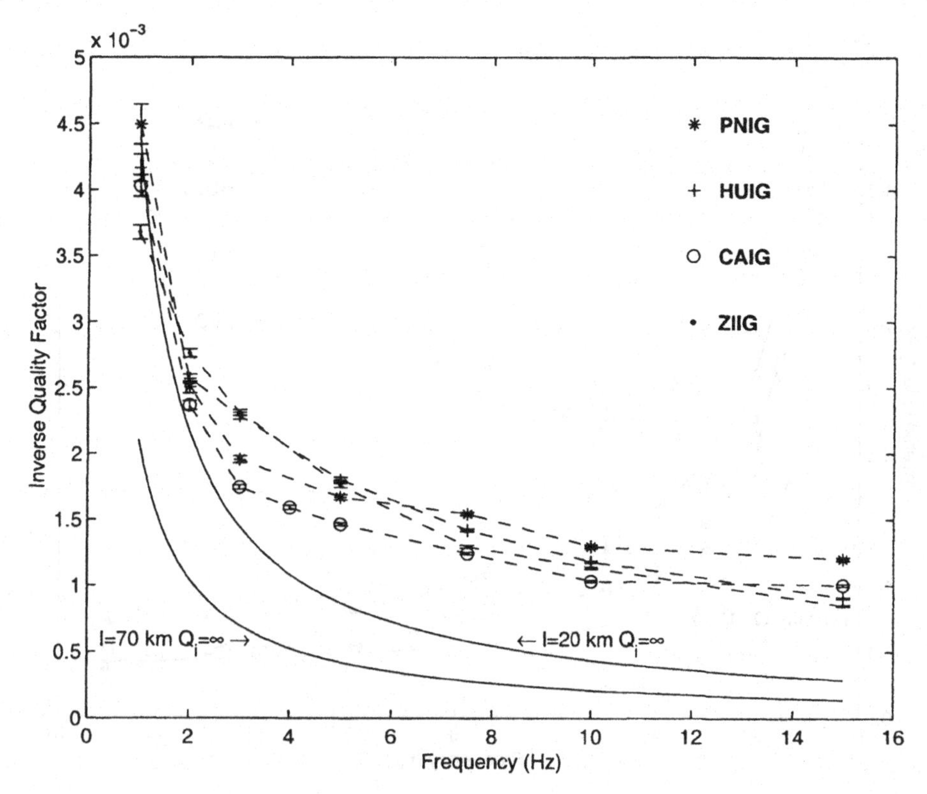

Figure 3. The symbols correspond to Q_c as a function of frequency measured at the different stations. The continuous lines correspon d to the solution of the radiative transfer equation in a purely elastic medium for the extreme values of mean free path 20 km and 70 km.

is a reasonable guess since this value is similar to the one measured in geologically stable regions (Singh and Herrmann, 1983, Hasegawa, 1985) where the crust was not affected by tectonic motions since the early stages of the Earth history and is therefore probably very homogeneous. The apparent attenuation of direct waves measured in these regions is a good evaluation of the intrinsic attenuation of rocks that has to be taken into account in the modelling. After adding to our model a weak attenuation term described by $Q_i = 1200$, we compute again the coda decay parameter Q_c^* for different frequencies and for the extreme values of l of 20 km and 70 km. The results are compared with the measurements in Fig 4.

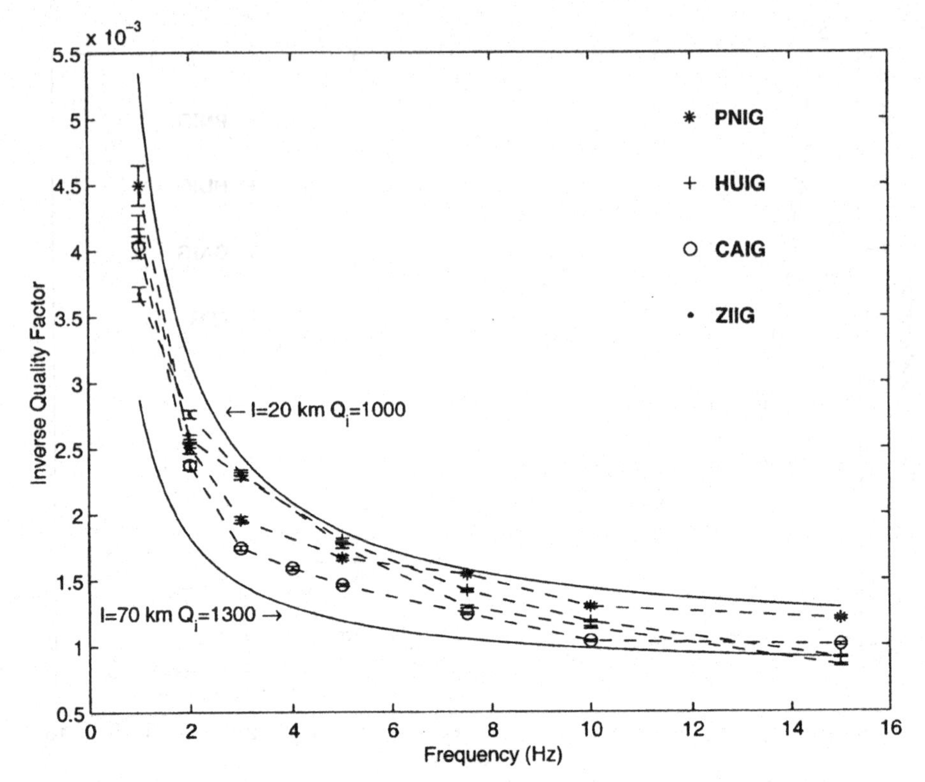

Figure 4. A comparison of the measured values of Q_c with the ones deduced from our model including a weak intrinsic absorption.

The numerical results are now in agreement with the observations in the complete range of frequency.

This agreement obtained with a very simple model shows that we catch the essential of the physical processes that govern the coda decay in this example: the leakage of diffuse energy in the mantle, that dominates at low frequency and leads to a linear frequency dependence of $Q_c^{\star}$ and the intrinsic absorption of rocks that dominates the attenuation in the high frequency limit. Indeed such a simple conceptual model is not expected to represent the various aspects of the Earth structure but it illustrates the importance of the leakage, a process that was neglected in previous interpretations of Q_c. This effect has to be considered in tectonically active regions where the scattering in the crust is strong (l of the order of H) while in stable regions, we expect a weak scattering in the crust (l large) and therefore a predominance of the intrinsic attenuation over the leakage effect in the exponential decay of the coda energy.

In the example of the records in Mexico, the conclusion of our analysis is that the decay rate we measure at low frequency is governed by the leakage into the mantle of diffuse waves of the crust. A strong support to this interpretation should be to demonstrate the diffusive character of coda waves. It is difficult to assess explicitely in what regime is the wavefield from conventional measurements. In the case of seismic waves, we can use the properties of the elastic waves. In the following we use an energy balance property of the elastic waves as a marker of the onset of the diffusive regime.

5. Partition of P and S energy

The understanding of the regime of the scattering of seismic waves in the Earth is fundamental for their interpretation in terms of the properties of the Earth materials. Most of previous theoretical studies of the seismic coda (including our study of the effect of the layering) used the acoustic approximation and did not consider the different polarizations of the elastic waves and the mode conversions that occur at each scattering event. As a consequence, the experimental studies were essentially concerned with the measurement of the decay rate of the coda envelopes. However, such kind of measurements is insufficient to distinguish unambiguously between the different scattering regimes.

Recently, the elastic radiative transfer equation has been derived (Weaver, 1982, 1990, Ryzhik et al., 1996; Papanicolaou et al., 1996; Turner, 1998). This equation takes into account the wave polarization and the mode conversion between P and S waves in an infinite elastic body. A fundamental property expected from these studies is that, in the diffusion regime, R, the ratio of energy densities of S and P waves, becomes constant:

$$R = \frac{W_s}{W_p} = \frac{2\alpha^3}{\beta^3} \tag{7}$$

where α is the P-wave velocity and β is the S-wave velocity. This property can be regarded simply as a complete randomization of the field in the phase space. For a medium with roughly $\alpha/\beta = \sqrt{3}$, which is the case in the lithosphere, this ratio is 10.39. Therefore, the elastic diffusion approximation predicts that the seismic coda is dominated by S waves as it was observed with actual records. A constant ratio between P and S energy densities is not expected for deterministic arrivals for which this ratio depends on the nature of the source and of the path. The measurement of the repartition of the S and P wave energies and its possible equilibration can give a very strong indication on the regime of seismic wave scattering. The problem is that S and P waves cannot be separated using a record of ground displacement at one point. Particularly in the coda, the wavefield consists of numerous simultaneous arrivals from unknown directions, making the separation between P and S waves impossible using standard signal processing techniques.

Shapiro et al. (1998) present a new approach to separate P and S wave energies. This technique is based on the processing of the data from a small-aperture array. It implies the measurements of the spatial derivatives and the calculation of the curl and divergence of the displacement $\vec{u}$. We have:

$$\vec{curl}\vec{u} = 0 \tag{8}$$

for P waves and

$$div\vec{u} = 0 \tag{9}$$

for S waves. We use these fundamental properties to estimate P and S waves energies separately.

Following Aki and Richards (1980) we can write the density of energy associated with the deformation in elastic medium:

$$W = \frac{1}{2}\sigma_{ij}\chi_{ij} \tag{10}$$

where χ is the deformation tensor and σ is the stress tensor. In an isotropic medium, the equation (10) can be rewritten as:

$$W = (\frac{\lambda}{2} + \mu)(div\vec{u})^2 + \frac{\mu}{2}(\vec{curl}\vec{u})^2 + \mu I \tag{11}$$

where

$$I = 2(\frac{\partial u_x}{\partial y}\frac{\partial u_y}{\partial x} + \frac{\partial u_x}{\partial z}\frac{\partial u_z}{\partial x} + \frac{\partial u_y}{\partial z}\frac{\partial u_z}{\partial y}) - 2(\frac{\partial u_x}{\partial x}\frac{\partial u_y}{\partial y} + \frac{\partial u_x}{\partial x}\frac{\partial u_z}{\partial z} + \frac{\partial u_y}{\partial y}\frac{\partial u_z}{\partial z}) \tag{12}$$

and λ and μ are Lamé constants. In this study we are looking for a property of a diffuse wavefield, i.e. a random and almost isotropic field. We calculate the average energy $\overline{W}$ in a sufficiently large time window. In this case, average of cross-products of a-priory non correlated functions become zero and we obtain:

$$\overline{I} = 0 \tag{13}$$

and

$$\overline{W} = \overline{W_P} + \overline{W_S} \tag{14}$$

where:

$$\overline{W_P} = \overline{(\frac{\lambda}{2} + \mu)(div\vec{u})^2} \tag{15}$$

$$\overline{W_S} = \overline{\frac{\mu}{2}(curl\vec{u})^2} \tag{16}$$

According to the theory (7) of the infinite body, if the wavefield is diffusive, the ratio between these two energies is a constant given by the Lamé parameters, whatever the total energy of the field is and for any type of source. For non-random field, this ratio evolves rapidly according to the arrivals of deterministic waves. For deterministic arrivals, the ratio indeed looses its signification since we neglect the interference term I in the evaluation of energy. We will evaluate this ratio with actual data using the equations (15) and (16). The direct way to do it would be to install seismic receivers in closely located boreholes at different depths relatively far from the earth surface. This would allow to calculate the partial derivatives with respect to the three spatial coordinates, to estimate the curl and the divergence and hence the S and P wave energies. However, installing numerous seismic stations in bore-holes is extremely expensive and the actual boreholes are not deep enough to allow to neglect the effect of the free surface. So far, there is no theoretical result to describe the equilibration at the surface of a half-space. The problem is indeed more difficult than for the infinite body because the free surface implies deterministic reflection and conversion.

At a depth below the surface larger than the transport mean free path, and because of the scatterings, the reflected wavefield is diffuse and therefore presents the same equilibration between the modes as for the infinite medium. The effect of the surface will be confined to a limit layer. At the surface, the effect of reflection and conversion affects the ratio of energies. This effect is independent of the amplitude of the incident diffuse field. Therefore, as long as an equilibration at a constant ratio occurs in the bulk, a form of equilibration is expected at the free surface. The value of the ratio of the energies at the surface can be different from the one in the bulk. Our goal here is to investigate the existence of the equilibration rather than to discuss the value of the ratio.

We set up an experiment to estimate the S-to-P energy ratio and its temporal evolution using only receivers located at the surface.

We installed a temporary small aperture array close to the city of Chilpancingo (the capital of the Guerrero state, Mexico). The high seismicity rate in Mexico made it possible to record rapidly a series of local earthquakes with magnitudes larger than 4. The location of the array is shown in Fig 5.

The geometry of the array is a 50 m side square (Fig 6). The sensors are CMG-40T seismometers connected to Reftek digitizers. The absolute time was provided by the radio signals of the GPS satellites.

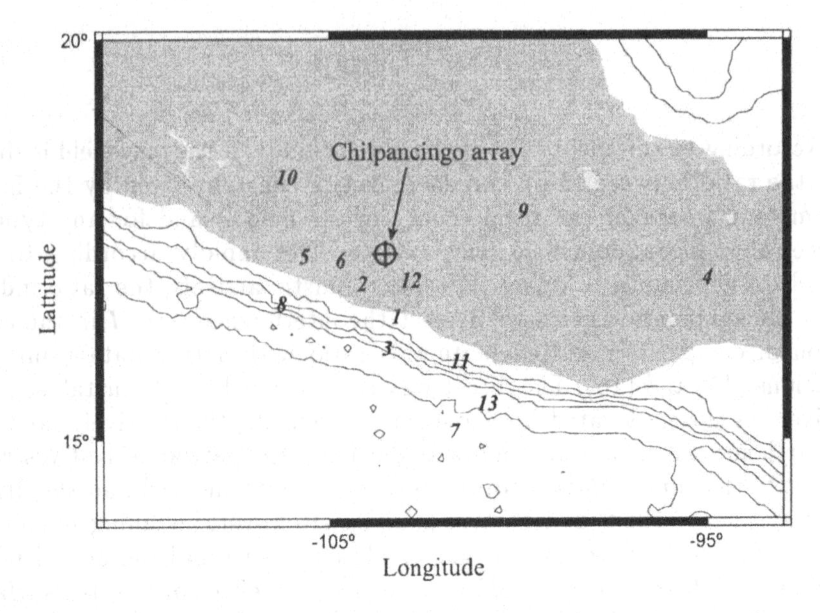

Figure 5. Map of Southern Mexico. The crossed circle shows the location of the Chilpancingo array. The italic numbers indicate the epicenters of the events used in Shapiro et al., 1998.

The spatial derivatives of the displacement with respect to horizontal coordinates were estimated by the relations:

$$\frac{\partial u_i}{\partial x} = \frac{u_i^3 - u_i^1}{d} \qquad \frac{\partial u_i}{\partial y} = \frac{u_i^2 - u_i^4}{d} \qquad i = x, y, z \tag{17}$$

where u_i^n is the displacement recorded at station n of the array (see Fig 4) and d is the distance between the two stations.

We estimated spatial derivatives of the displacement with respect to the vertical coordinate using the boundary condition at the surface. The free stress condition is:

$$\begin{pmatrix} \sigma_{xx} & \sigma_{xy} & \sigma_{xz} \\ \sigma_{yx} & \sigma_{yy} & \sigma_{yz} \\ \sigma_{zx} & \sigma_{zy} & \sigma_{zz} \end{pmatrix} \begin{pmatrix} 0 \\ 0 \\ 1 \end{pmatrix} = 0 \tag{18}$$

In an isotropic medium, it leads to the relations between the derivatives:

$$\left\{ \begin{array}{lcl} \frac{\partial u_y}{\partial z} & = & -\frac{\partial u_z}{\partial y} \\ \frac{\partial u_x}{\partial z} & = & -\frac{\partial u_z}{\partial x} \\ \frac{\partial u_z}{\partial z} & = & -\frac{\lambda}{\lambda + 2\mu}(\frac{\partial u_x}{\partial x} + \frac{\partial u_y}{\partial y}) \end{array} \right. \tag{19}$$

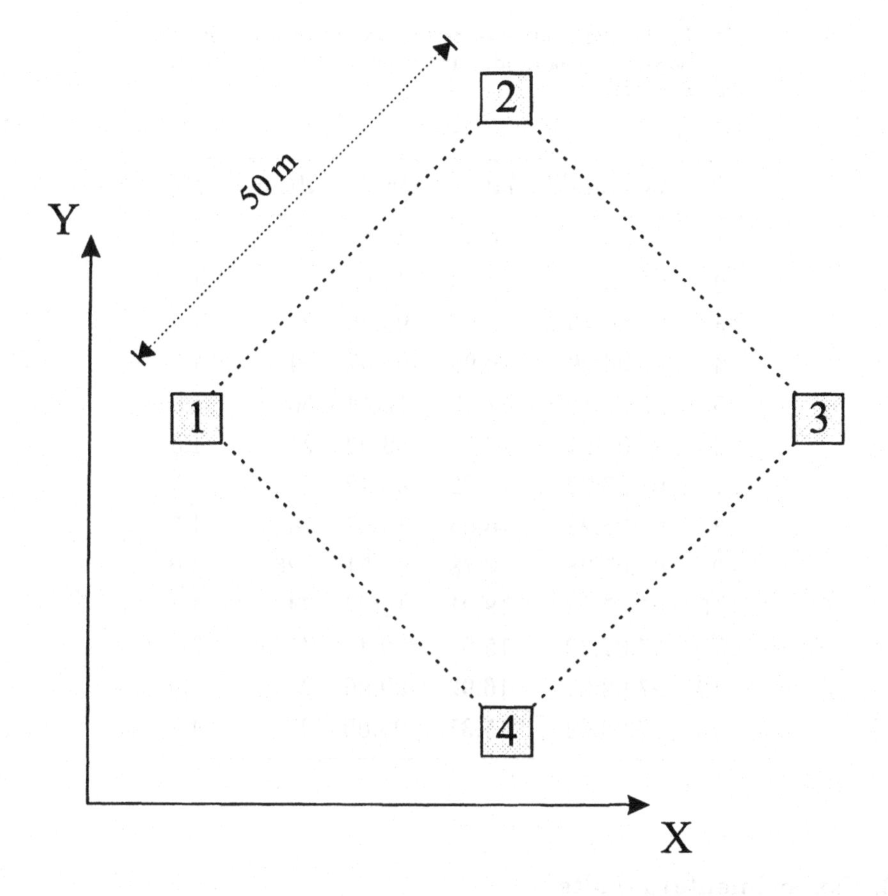

Figure 6. Configuration of the Chilpancingo array.

This last equation can be written:

$$\frac{\partial u_z}{\partial z} = (2(\beta/\alpha)^2 - 1)(\frac{\partial u_x}{\partial x} + \frac{\partial u_y}{\partial y}) \tag{20}$$

Using equations (17), (19), and (20) we can now estimate the whole set of spatial derivatives of the displacement. We can therefore rewrite equations (15) and (16) and estimate the ratio of S and P wave energy densities:

$$R = \frac{1}{4}(\frac{\alpha}{\beta})^2 \frac{4\frac{\partial u_z}{\partial x}^2 + 4\frac{\partial u_z}{\partial y}^2 + (\frac{\partial u_x}{\partial y} - \frac{\partial u_y}{\partial x})^2}{(\frac{\partial u_x}{\partial x} + \frac{\partial u_y}{\partial y})^2} \tag{21}$$

TABLE 1. Parameters of earthquake used in the study. Locations and magnitudes are given by Mexican Seismological Survey.

N	yy:mm:dd	lat	lon	H(km)	M
1	97:06:21	16.47	-99.18	5	4.5
2	97:06:28	16.88	-99.63	7	4.1
3	97:06:29	16.07	-99.30	23	4.4
4	97:06:29	16.96	-95.07	64	4.5
5	97:07:19	17.22	-100.4	56	4.9
6	97:07:21	17.17	-99.92	24	4.5
7	97:07:22	15.02	-98.42	5	5.1
8	97:07:24	16.63	-100.7	16	4.5
9	97:07:28	17.78	-97.51	126	4.0
10	97:07:29	18.21	-100.7	74	4.4
11	97:07:30	15.9	-98.4	?	?
12	97:08:01	16.92	-99.06	35	4.3
13	97:08:03	15.37	-98.05	27	4.7

6. Experimental results

During the three months of the experiment we recorded 13 earthquakes located in southern Mexico and with magnitudes large enough to produce a coda with a good signal-to-noise ratio. Event locations are listed in Table 1 and are shown in Fig 5.

We selected events not too far away in order to have late arrivals associated with backscattered waves rather than deterministic low velocity guided modes or critical deep reflections. The energy ratio was computed for each record according to equation (21). We calculated the average S-to-P energy ratio (R) in a moving window of width 16 s. The results of the processing for two events are shown in Fig 7. The two events have different locations. The time window corresponding to the coda with a signal to noise ratio greater than 4 is indicated by grey shading.

The time evolution of the energy ratio shows that the ratio tends to stabilize in the coda. The ratio is widely variable in the noise and at the onset of direct waves. In the coda, in spite of some fluctuations, it is impor-

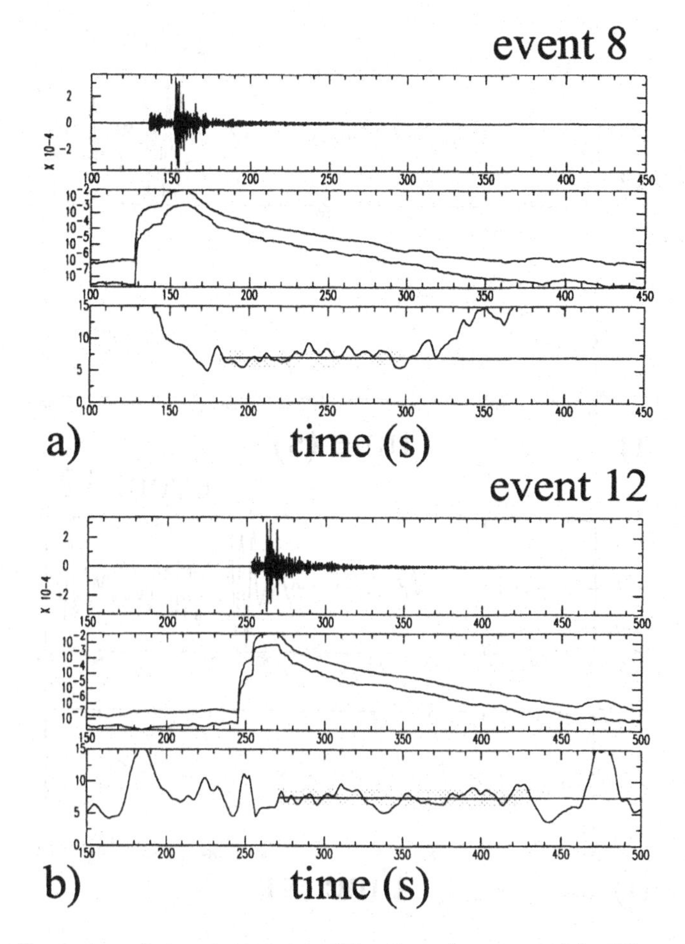

Figure 7. Examples of measurements . We selected events with different epicentral distances and backazimuths: a) event 8; b) event 12. For each event, we present the vertical component seismogram band-passed between 1 and 3 Hz (upper part), the P and S energies smoothed in a 16 s moving window (central part) and the S-to-P energy ratio (lower part).

tant to note that the ratio varies of about 25% of the mean value for the largest fluctuations while, in the same time, the energy varies by a factor of 10,000. An average value of R for the whole coda window is about 7.5 with a standard deviation of 1. We present in Fig 8 the results of essentially the same analysis but limited to the time window corresponding to the arrivals of the body waves.

The grey shading indicates the arrival times of the direct P and S waves. The mean value of the energy ratio R is calculated on 1s intervals. Its evolution is now plotted on a logarithmic scale. The energy ratio shows very

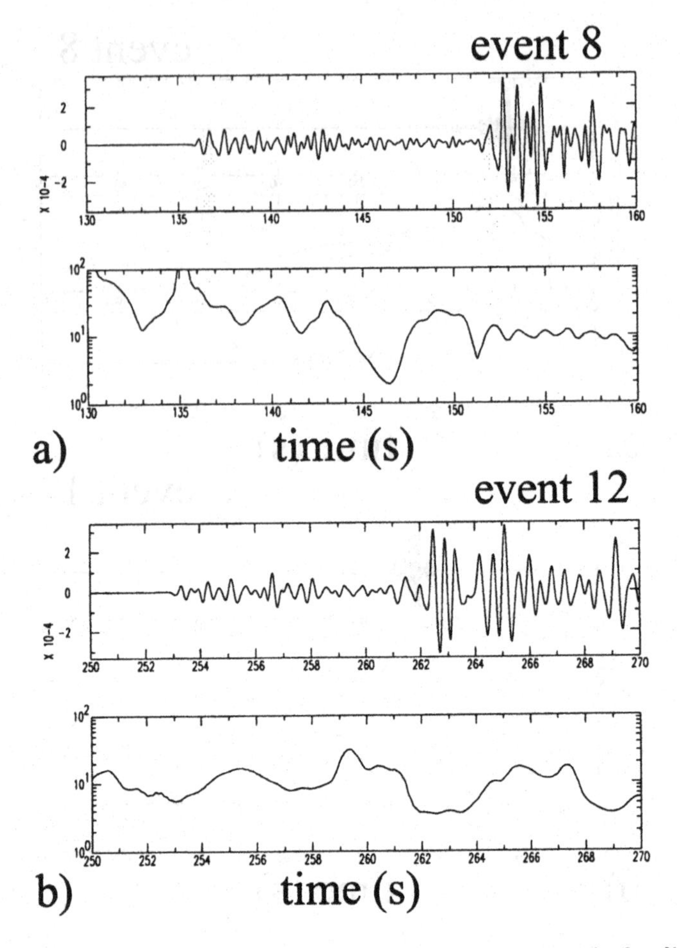

Figure 8. Examples of the measurement of the S-to-P energy ratio for direct waves: a) event 8; b) event 12. For each event, we present the vertical component seismogram band-passed between 1 and 3 Hz (upper part), the P and S energies smoothed in a 1 s moving window (central part) and the S-to-P energy ratio (lower part). Note the difference of scale in the plot of the energy ratio with respect to Fig 7.

rapid and huge variations during the deterministic arrivals. It is expected from the properties of reflection and conversion of body waves at a free surface that the energy ratio has great variations depending upon the angle of incidence and the polarization of the incident wave. The presence of a low velocity layer at the surface provokes strong P to S conversions that can explain the surprisingly high values of R observed at the onset of P waves. We performed the same measurements for the 13 earthquakes. Concerning the coda, we found in every case that the two modes equilibrate very rapidly after the arrival of the direct waves. We extract the values of R for the coda and for the direct P and S waves in order to compare how these measures

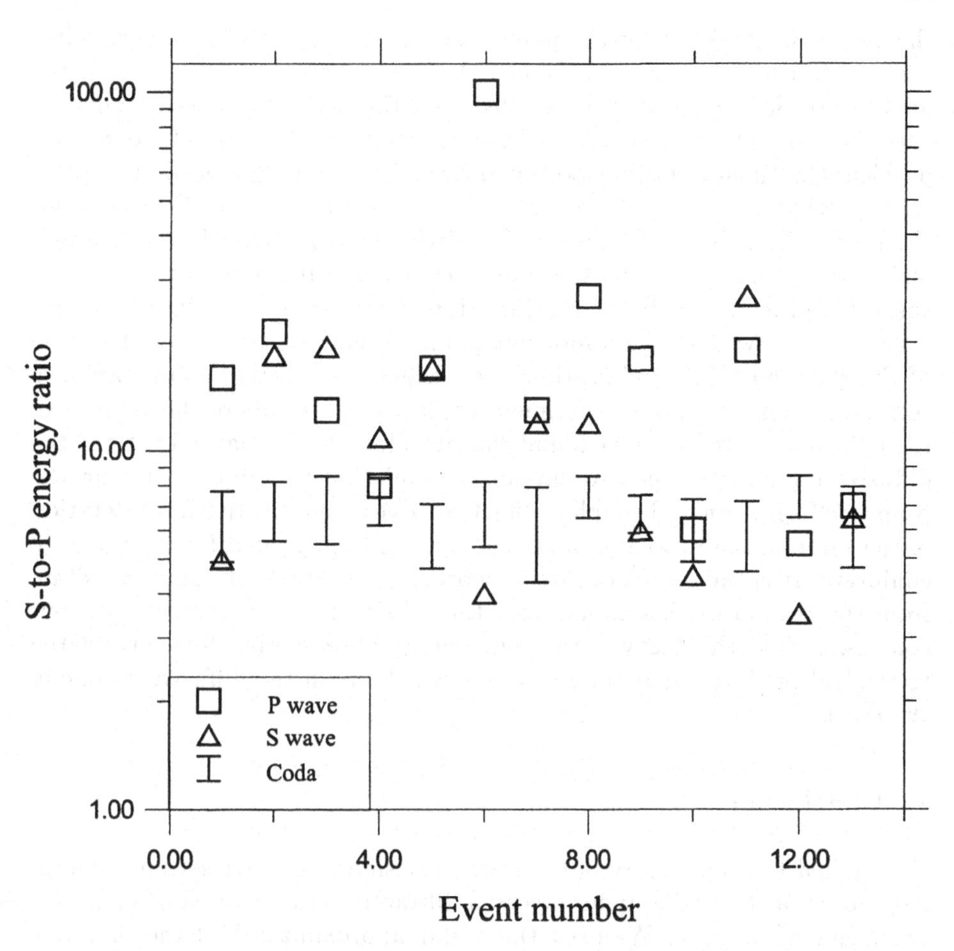

Figure 9. S-to-P energy ratios for the 13 events. The squares and the triangles indicate the ratios measured for P and S waves, respectively. Vertical bars show average values plus minus standard deviation of the S-to-P energy ratio measured in the coda window.

vary from one event to another. The results of the individual measurements are shown in Fig 9. For all events we obtain almost the same R value in the coda: about 7 in average. It means that the S-to-P energy ratio in the coda stabilizes at a values that does not depend on the seismic source nor on the earthquake location. The picture is completely different for direct waves. The ratio is much more variable from an event to the other with mean values which are significantly higher than for the coda.

The stabilization of R in the coda at the same level for all earthquakes is a good indication that the diffusion regime is reached. The theoretical value expected for diffuse waves in an infinite body is given by the equation (7). The value which we have observed at the free surface is lower than the one theoretically predicted for the poissonian solid (7 and 10.39, respectively). This difference can be due to the fact that we did not take into account the presence of the reflective surface. Obviously the reflection is a deterministic effect that breaks the isotropy of the diffuse field. This is a fundamental problem that is not yet fully solved as already noticed and we do not know exactly what is the equilibration ratio at the free surface. One can also claim that in a real situation a part of the energy of the P and S waves which are diffuse in the bulk are diffracted into surface waves at the free surface, a process that is not considered in the theory so far. These different arguments show that the important point in our analysis is not the value of the ratio but the equilibration itself which is a marker of the diffusive regime. In this sense, our experiment confirms the results of the analysis of the coda decay for which we found that it follows a law characteristic of the diffusive regime. We need to develop a simulation technique including the proper polarization and coupling effects to give a quantitative interpretation of the value of the ratio and of the short lapse time needed for the waves to equibrate. It is the object of further works. At this present stage, we retain from the results of this experiment that their are in agreement with our conclusion that the decay of the coda energy observed in Mexico is related to the leakage into the mantle of waves which are in the diffusive regime in the crust.

7. Conclusion

We studied the time decay of the coda of seismograms. We set up a simplified model of the Earth upper layers: a diffractive crust overlying an almost homogeneous mantle. We used the scalar approximation of the radiative transfer equation to model the multiple scattering of the seismic waves. We solved the equation using the Monte Carlo method. When the ratio mean free path/layer thickness is less than one, we can give an analytical form of the time decay of the coda energy as the solution of a diffusive equation. Otherwise, our transcription of the boundary condition is not valid. The asymptotic form is similar to the one that was proposed by Aki and Chouet (1975) to fit the observations. When the mean free path is larger than the layer thickness, we found numerically that the asymptote of the radiative transfer solution has the same functional form as the one obtained when our diffusion approximation is valid. The decays computed in our model in which we included the effect of a weak absorption are in a very good agreement with the observations. The leakage of diffuse energy into the mantle plays a prominent part at low frequency and controls the frequency dependence of Q_c when, as in our case, the mean free path is of the order of the crustal thickness. We propose to characterize the temporal decay of the coda energy by the time of residence of diffuse waves in the crust. We showed the diffuse character of the coda by looking for the energy partitionning that is expected for elastic waves in the diffusive regime. Considering a series of records from earthquakes at different epicentral distances, we show that the equilibration between the two modes (that is the energy partitioning) appears very early in the coda. The ratio between compressional and shear energies is very stable while the energy level changes by several orders of magnitude. The energy ratio is independent of the earthquake considered. This experiment tends to confirm that the seismic coda corresponds to waves in the diffusive regime and therefore that the decay is strongly governed by the rate of leakage of diffuse energy in the mantle.

Acknowledgments

We acknowledge financial support from Program "Interieur de la Terre" of INSU/CNRS (France), from the CONACYT project 0974-PT (Mexico) and from the European Union Contract CH*-CT92-0025. We thank B. van Tiggelen, R. Maynard and S.K. Singh for their help and suggestions.

References

1. Abubakirov, I.R. and A. A. Gusev, 1990. Estimation of scattering properties of li thosphere of Kamchatka based on Monte-Carlo simulation of record envelope of a n ear earthquake, *Physics of the Earth and Planetary Interior*, **64**, 52 -67.
2. Aki, K. and B. Chouet, 1975. Origin of coda waves: Source, attenuation, and scattering effects, *J. Geophys. Res.*, **80**, 3322-3342.
3. Allmendinger, R.W., K.D. Nelson, C.J. Otter, M. Barazangi, L.D. Brown and J.E. Oliver, 1987. Deep seismic reflection characteristics of the continental crust, *Geology*, **15**, 304-310.
4. Campillo, M. and J.L. Plantet, 1991. Frequency dependence and spatial distribut ion of seismic attenuation in France: experimental results and possible interpretati ons, *Phys. Earth Planet. Interiors*, **67**, 48-64.
5. Campillo, M., L. Margerin and K. Aki 1998 Seismology, in New Aspects of Electromagnetic and Acoustic Waves Diffusion, POAN Research Group (ed), Springer tracts in modern physics; vol 144, 87-99.
6. Dainty, A.M. and M.N. Toksoz 1990 Array analysis of seismic scattering, *Bull. Seism. Soc. Am.*, **80**, 2248-2260
7. Hasegawa, H. S., 1985. Attenuation of *Lg* waves in the canadian shield, *Bull. Seismol. Soc. Am.*, **75**, 1569-1582.
8. Hoshiba, M., 1991. Simulation of multiple scattered coda wave excitation based on the energy conservation law, *Phys. Earth and Planet. Interiors*, **67**, 123-136.
9. Johnston, D.H. Attenuation: A state of the art summary. in: Seismic wave attenuation, M.N. Toksoz and D.H. Johnston editors, Geophysics Reprint Series No. 2, Society Of Exploration Geophysicists, Tulsa, 1981.

10. Kostoglodov, V., Bandy, W., Domingez, J., and Mena, M., 1996. Gravity and seismicity over the Guerrero seismic gap, Mexico. *Geophys. Res. Lett.*, **23**, 3385-3388.
11. Margerin, L., M. Campillo, B.A. Van Tiggelen, 1998a Radiative transfer and diff usion of waves in a layered medium: a new insight into coda Q, *Geophys. J. Int.*, **134**, 596-612.
12. Margerin, L., M. Campillo,, N.M. Shapiro and B. van Tiggelen 1998b The time of residence of diffuse waves in the crust and the physical interpretation of coda Q. Application to seismograms recorded in Mexico, submitted to Geophys. J. Int.
13. Meissner, R., 1989. Rupture, creep, lamellae and crocodiles: happenings in the continental crust, *Terra Nova*, **1**, 17-28.
14. Papanicolaou, G.C., L.V. Ryzhik and J.B. Keller 1996 Stability of the P to S energy ratio in the diffusive regime, *Bull. Seism. Soc. Am.*, **86**, 1107-1115.
15. Ryzhik, L.V., G.C. Papanicolaou and J.B. Keller 1996 Tansport equations for elastic and other waves in random media, *Wave Motion*, **24**, 327-370.
16. Shapiro, N.M., M. Campillo, L. Margerin, S.K. Singh, V. Kostoglodov and J. Pacheco 1998 The energy partitioning between P and S waves and the diffusive character of the seismic coda. submitted to Bull. Seism. Soc. Am.
17. Singh, S., & Herrmann, R.B., 1983. Regionalization of crustal coda Q in the c ontinental United States, *Journal of Geophysical Research*, **88**, 527-5 38.
18. Singh, S. K., Pacheco, J., Courboulex, F., and Novelo, D. A. 1997 Source parameters of the Pinotepa Nacional, Mexico, earthquake of 27 March, 1996 ($Mw = 5.4$) estimated from near-field recording of a single station., *J. Seismology*, **1**, 39-45.
19. Turner, J.A. 1998 Scattering and Diffusion of Seismic Waves. *Bull. Seism. Soc. Am.*, **88**, 276-283.
20. Weaver, R.L. 1982 On diffuse waves in solid media. *J. acoust. soc. Am.*, **71**, 1608-1609.
21. Weaver, R.L. 1990 Diffusivity of ultrasound in polycrystals, *J. Mech. Phys. Solids*, **38**, 55-86.
22. Wu R.S., 1985 Multiple scattering and energy transfer of seismic wa ves-separation of scattering effect from intrinsic attenuation-I Theoretical modeling. *Geophys. J. R. Astr. Soc*, **82**, 57-80.

IMAGING AND AVERAGING IN COMPLEX MEDIA

ROEL SNIEDER
Dept. of Geophysics
Utrecht University
P.O. Box 80.021
3508 TA Utrecht
The Netherlands
email snieder@geo.uu.nl

1. Introduction, complex wave propagation in the earth

Media can be complex in at least two different ways. The first kind of complexity is that a medium has small-scale inhomogeneities that allow waves to travel along a multitude of paths. For such a medium geometrical optics cannot be used to account for the wave propagation. The second kind of complexity is that the inhomogeneities are so strong that multiple scattering effects are operative. In this work the effects of both small-scale perturbations of the medium are treated (section 3), as well as the effects of multiple scattering (section 4). Imaging of wave-fields is very different in complex media than in simple media. The effects of complex wave propagation on imaging techniques is discussed in the sections 2 and 5.

One does not have to look far for a complex medium, because the propagation of elastic waves through the earth can be extremely complex. As an example, consider the top panel of figure 1 in which the ground motion recorded by a seismic station in the Netherlands after an earthquake in Greece is shown. The signal has a very complex appearance. The frequency spectrum of the recorded signal is shown in figure 2. At a frequency of around $0.2Hz$ (corresponding to a period of $5s$) the spectrum shows a pronounced peak. This peak is also present in the absence of waves generated by earthquakes. Instead of the earth response to the earthquake, this peak is due to seismic noise that is generated at the oceans when water waves with near-opposite wavenumber interact [*Longuet-Higgins*, 1950; *Webb*, 1998]. It turns out that the seismic signal at frequencies greater than the frequency

J.-P. Fouque (ed.), Diffuse Waves in Complex Media, 405–454.

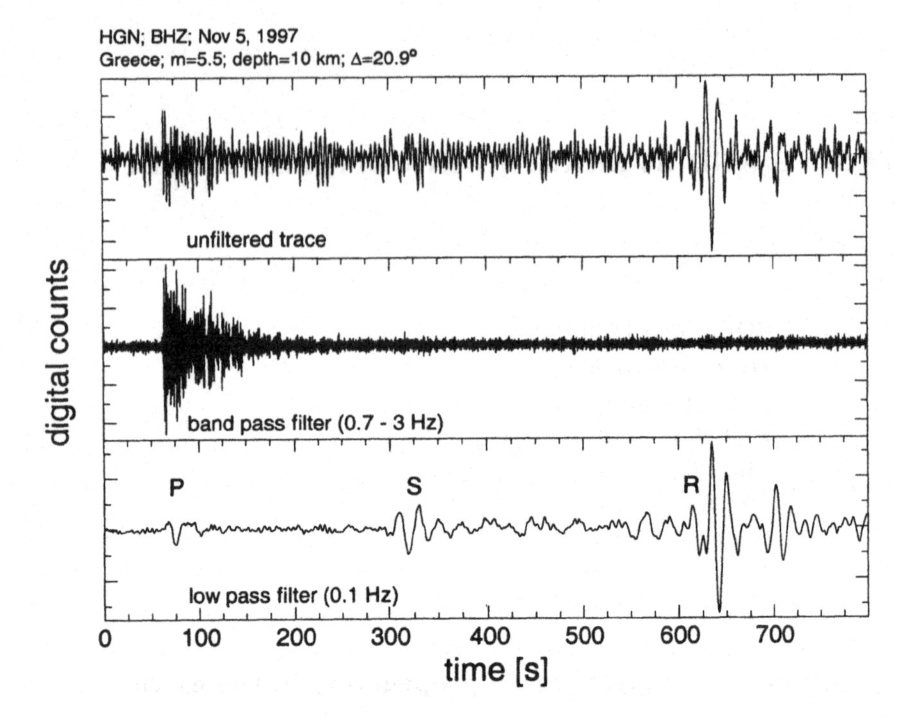

Figure 1. Vertical ground motion recorded in the Netherlands after an earthquake in Greece (top panel). Also shown are the band-passed filtered seismogram containing frequencies from 0.7 Hz to 3 Hz (middle panel) and the low-passed filtered seismogram with frequencies less than 0.1 Hz (bottom panel).

of the microseismic noise is of a completely different character than the signal at lower frequencies.

This can be seen by comparing the seismic signal that contains only frequency components between 0.7 and 3.0 Hz in the middle panel of figure 1 with the low-pass filtered seismogram in the bottom panel of figure 1 that contains only frequencies lower than 0.1 Hz. The high-frequency signal shown in the middle panel starts with an impulsive arrival that directly merges is a long wave-train that consists of multiple scattered waves, this wave-train is called the *coda.*[1] The high-frequency signal shown in the middle panel in figure 1 is of a completely different nature from the low-frequency components shown in the bottom panel. The low-frequency signal does not show a strong coda after the first arriving wave around $90s$. The main features in this signal are distinct arrivals that correspond to the longitudinal and transverse waves that propagate through an elastic

[1]The term coda comes from music, where it refers to the closing part of a piece of music.

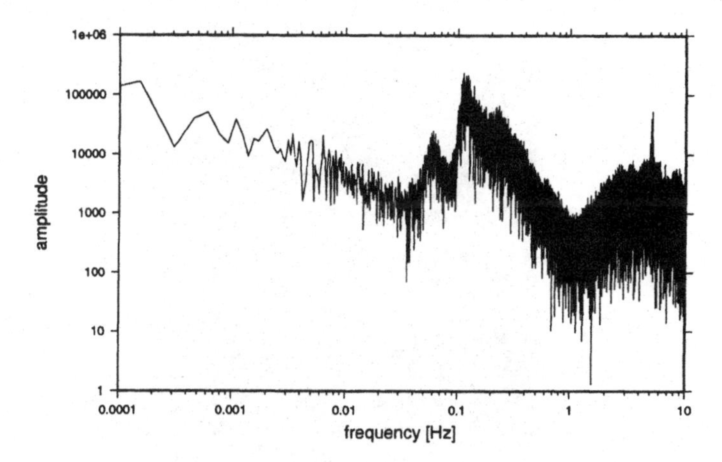

Figure 2. The amplitude spectrum of the seismogram in the top panel of the previous figure.

medium. These arrivals are labelled with "P" and "S" according to the common seismological nomenclature of P(rimary)- and S(econdary)-waves. The other main arrival is the surface wave that arrives around 640s after the earthquake, this wave is denoted by the label "R."

It is striking that the frequency-band of the microseismic noise separates for this epicentral distance a wave regime that is dominated by strong scattering (the middle panel) from a regime (in the lower panel) where there is little scattering and where one essentially only observes the direct waves that are explained well by geometrical optics. The waves in both frequency bands travelled through the same medium, the earth. The lesson to be learned from this is that one cannot simply speak of complex media or simple media, because depending on the frequency, the waves may interact in very different ways with the medium. This is due to the frequency-dependence of the relevant scattering and mode-conversion coefficients for body waves [*Aki and Richards*, 1980], surface waves [*Snieder*, 1986; *Snieder and Nolet*, 1987] and for normal modes [*Woodhouse and Dahlen*, 1978; *Snieder and Romanowicz*, 1988]. The name *complex media* is therefore is misnomer, because depending on the frequency the wave propagation in a medium can either be very simple or extremely complex.

On a smaller scale, the material comprising the earth's crust can lead to very complex wave propagation. This is beautifully illustrated by the laboratory measurements made by *Nishizawa et al.* [1997]. Using a laser-doppler interferometer they measured elastic waves that propagate through samples made of granite and of steel. Cross-sections of the three granites that were used are shown in figure 3. The waves that have propagated through these granite samples are shown in the figures 4 and 5. As a reference, the waves

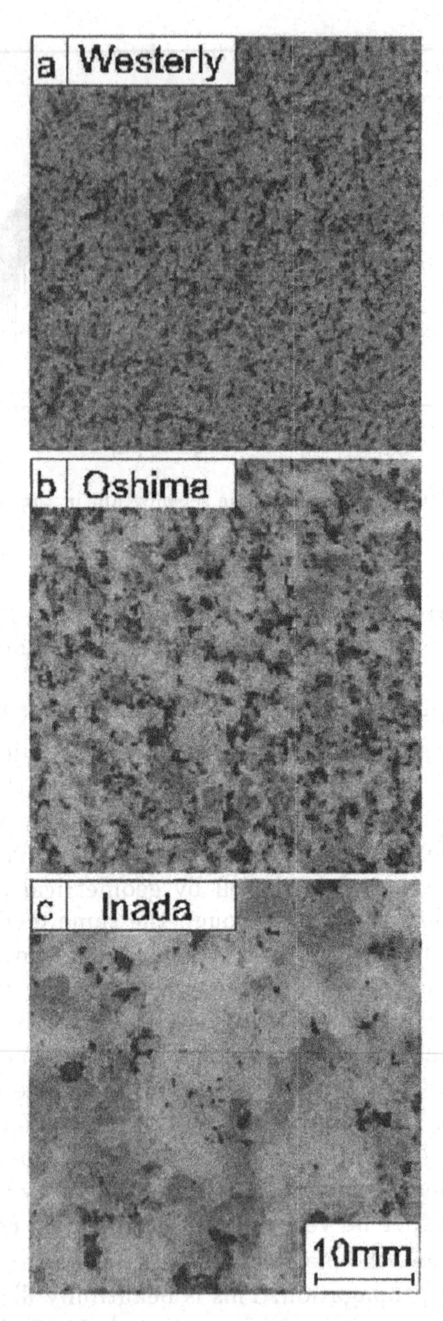

Figure 3. Cross section through the granite samples used for the wavefield measurements shown in the next figures. (Courtesy of Osamu Nishizaza)

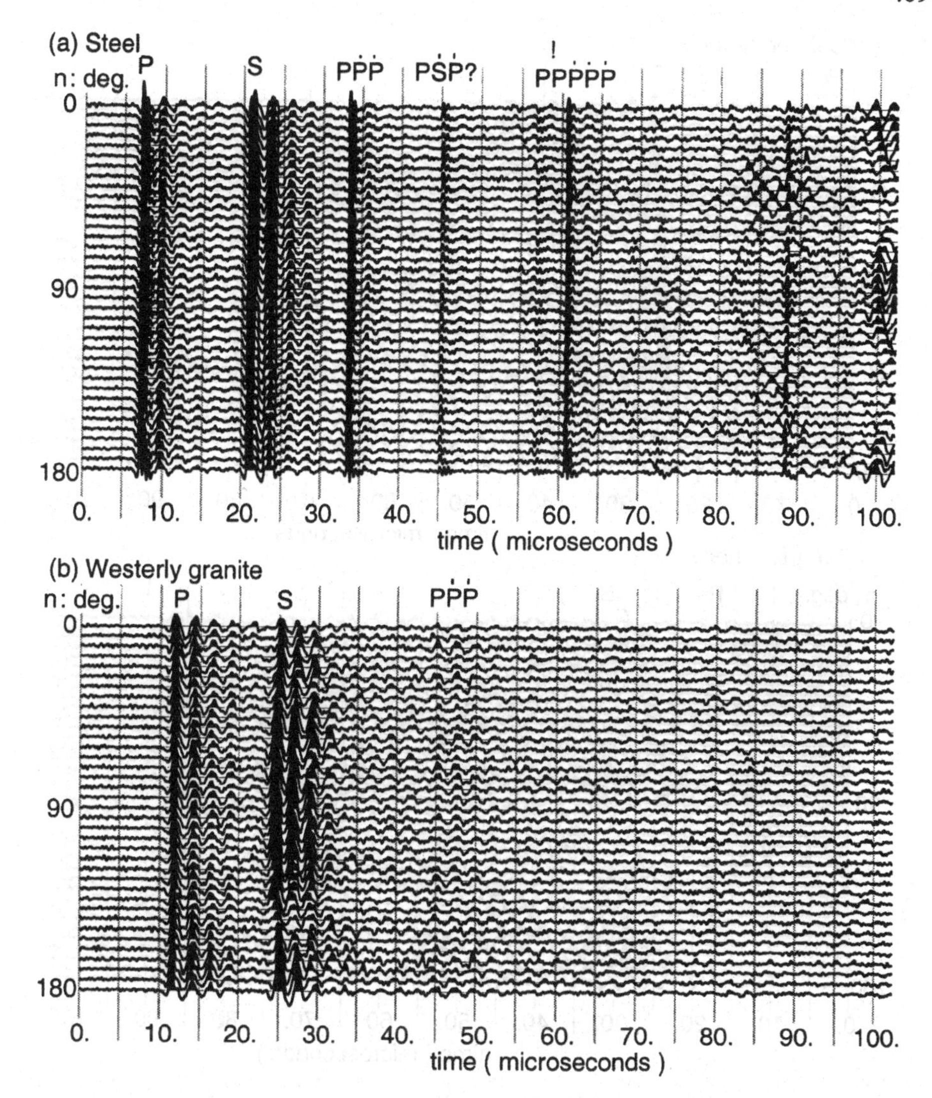

Figure 4. Wavefield that have propagated through a steel block and the Westerley granite sample of the previous figure. (Courtesy of Osamu Nishizawa.)

that propagated through a steel block is shown in figure 4; the steel block can be considered to be a nearly homogeneous medium. Note that in the steel sample (figure 4) the wave-field is dominated by coherent arrivals that are the P- and S-waves that propagate through the sample and are reflected and converted at the boundaries of the sample. In the most homogeneous granite sample, the Westerey granite, one can still observe the direct P- and S-waves. However, in the more inhomogeneous granite, notably the Inada

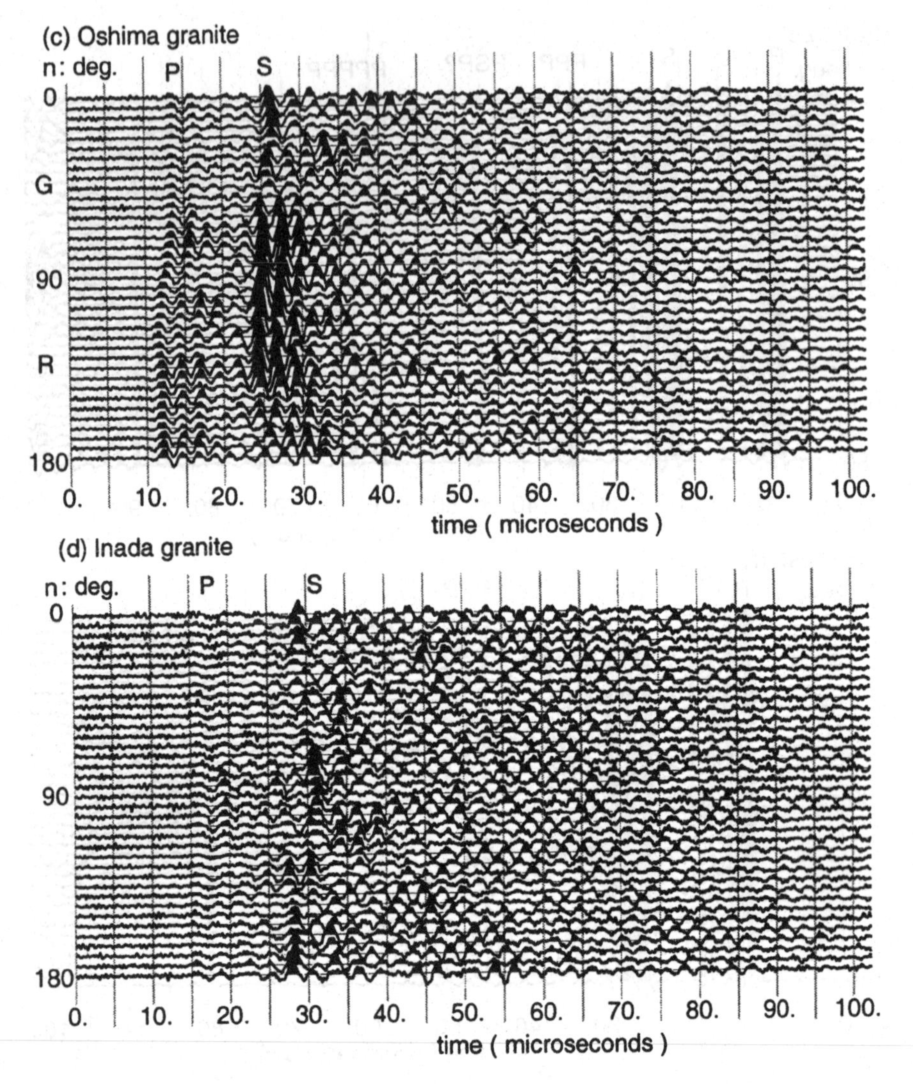

Figure 5. Wavefield that have propagated through the Oshima and Inada granite samples. (Courtesy of Osamu Nishizawa.)

granite in figure 5, one cannot identify these direct wave arrivals anymore, and the wave-field makes an incoherent and noisy impression. In this sample the wave scattering is so strong that the direct ballistic arrivals have disappeared.

It is tempting to speak of "noisy signals." However, one should be careful of using the phrase "noise" when referring to wave-fields that appear to show little organisation [*Scales and Snieder*, 1998]. This is illustrated

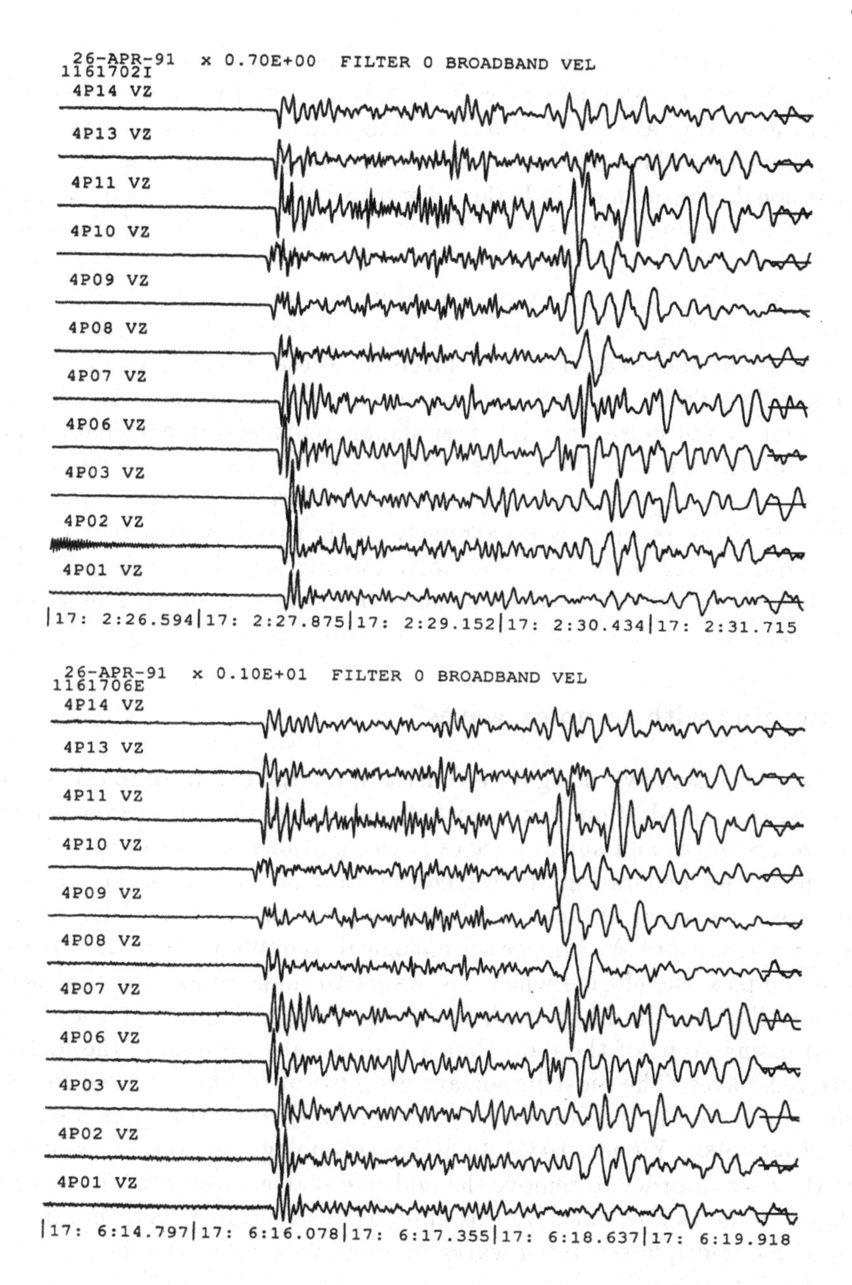

Figure 6. Wave-field recorded by an array in California after two nearby earthquakes (top panel and bottom panel respectively). Note the extreme resemblance of each trace in the top panel with the corresponding trace in the bottom panel. (Courtesy of Peggy Hellweg.)

by the recordings of the ground motion of an array in California that is shown in the top panel of figure 6. This ground motion was excited by a weak local earthquake, and the wave propagated mainly through the earth's crust. Note that the wave-field is extremely complex and makes a "noisy" impression because there is little coherence between the signals recorded at different stations. By a stroke of luck, a second very similar earthquake occurred at almost the same location of the first earthquake, the ground motion for this second event recorded by the same array is shown in the bottom panel of figure 6. Note that the signals shown in the panels in figure 6 are virtually identical. The waveforms for these two earthquakes can be matched "wiggle by wiggle." This means that the signals shown in figure 6 are definitely not noise. Instead these signals represent an extremely reproducible earth-response. That these signals are not due to ambient random noise can be seen by considering the signal before the first arriving wave. This part of the recordings is extremely weak; this indicates that the true noise level is extremely low. The noisy looking signals shown in figure 6 therefore should not be labelled as noise but rather as deterministic earth response.

2. Imaging with complex waves?

The wave-fields shown in figure 6 consist to a large extent of multiple scattered waves. In addition to pure wave scattering, the conversion between P-waves, S-waves and surface waves is an additional physical effect that is responsible for generating the waveforms that are recorded. The question then arises: to what extent can these complex waves be used for imaging the earth's interior? Within the seismological community one tends to avoid these complex waveforms when one wants to make images of the earth's interior. The standard approach to seismic prospecting includes a number of processing steps of the data that are aimed at suppressing the multiple scattered waves. The most important example of techniques used for *multiple suppression* is the normal move out (NMO) correction plus stack [e.g. *Claerbout*, 1985; *Yilmaz*, 1987]. In global seismology one often low-pass filters the data in order to remove the multiple scattered waves. The example of figure 1 shows how effective this approach can be. However, the elimination of the multiple scattered waves by preprocessing of the data entails a loss of information.

The main motivation for accepting this loss of information is that the retrieval of earth structure from multiple scattered waves is assumed to be an inherently unstable process so that random noise or unwarranted assumptions used in the imaging process magnify under the nonlinear inversion steps needed to map the multiple reflected waves onto a model for the earth's interior. It is for these reasons that *Claerbout* [1985] notes on p. 363 that:

> *"Multiple reflections is a good subject for nuclear physicists, astrophysicists and mathematicians who enter our field. Those who are willing to take up the challenge of trying to carry theory through to industrial practice are rewarded by some humility."*

At this point it is not known whether this pessimistic conclusion about the possibility to use strongly scattered waves is justified.

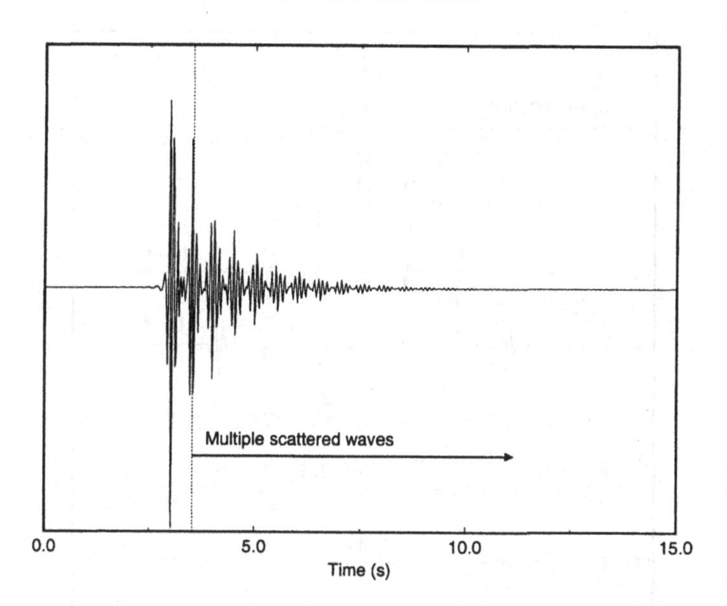

Figure 7. Wave-field at one of the two receivers in the 1D numerical experiment.

In any case, this conclusion appears to be add odds with the experiments of *Derode et al.* [1995] who used a Time Reversal Mirror (TRM) to image acoustic waves that have been strongly scattered back onto the source position. In their experiment, waves emitted by a source propagate through a dense assemblage of 2000 metal rods that strongly scatter acoustic waves. The waves are digitally recorded at an array of 128 receivers. The recorded waves are time-reversed (i.e. the time t is replaced by $-t$) and these time-reversed waves are send back from the receivers into the medium. Since the equation for the propagation of acoustic waves is invariant for time-reversal, the waves should revert along their original path and at time $t = 0$ focus on the source that originally excited the waves at $t = 0$. The surprise of the experiment of *Derode et al.* [1995] is not that the system exhibits invariance for time reversal, but the surprise is that this time-reversed imaging works even when the employed waves are strongly scattered and when the waves that are send back into the medium are contaminated with errors due to discretization and instrumental effects. The stability of the time-reversed imaging of *Derode et al.* [1995] seems to be at odds with the pessimistic conclusions of *Claerbout* [1985]. So who is right, *Claerbout* [1985] or *Derode et al.* [1995]? Or in other words, can one use multiple scattered waves for imaging purposes or is it impossible to do this in a robust fashion?

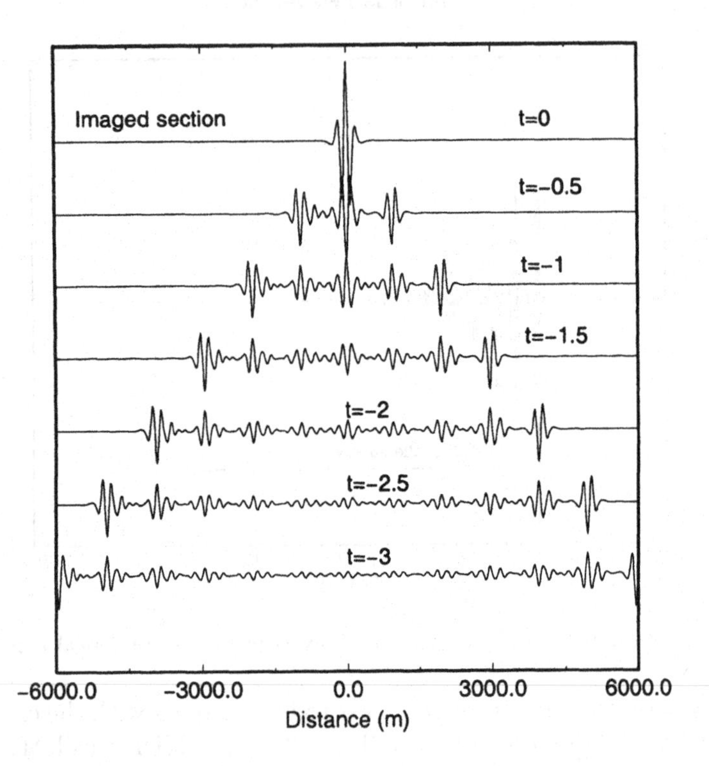

Figure 8. Wave-field after time reversal at different times.

In order to resolve this paradox an example from *Scales and Snieder* [1997] is shown that is based on the multiple scattering theory for isotropic point scatterers that is described in section 4.1. At this point the details of this scattering theory are not yet important, it suffices to note that the full multiple scattering response is used in the numerical simulation. The examples shown are for wave propagation in one dimension, i.e. the waves propagate along a line. The source is located at $x = 0$, on each side of the source two strong scatterers are positioned within a distance of $500m$.

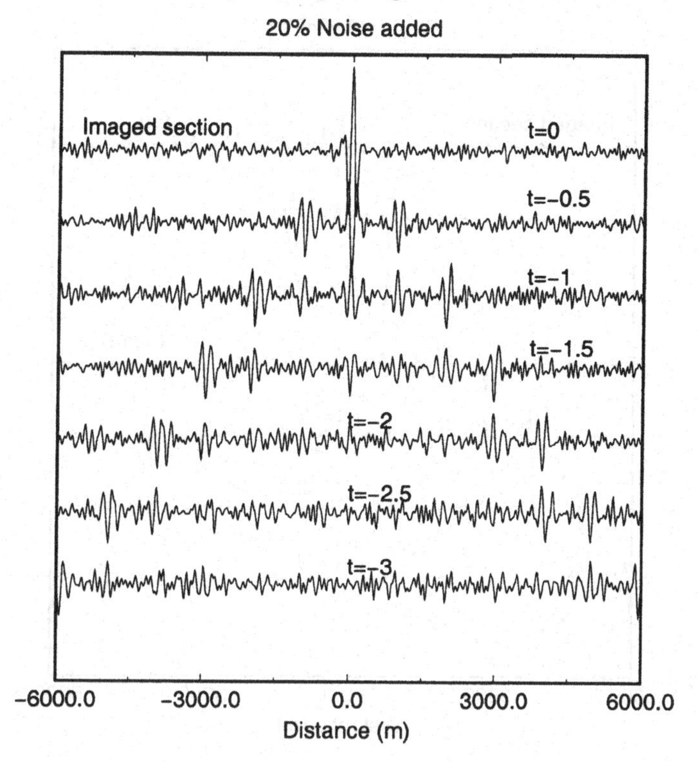

Figure 9. Time-reversed wave-field when 20% noise has been added to the signals that are emitted from the receivers.

Two receivers located at $\pm 6000m$ record the wave-field. The wave-field recorded by one of the receivers is shown in figure 7. After time-reversal the waves recorded at the two receivers are sent back from the receivers into the medium (which includes the scatterers). In figure 8 one can see how the waves propagate through the medium. Note that the wave-field is confined to a "light-cone" with the source at $x = 0$ and $t = 0$ as apex. As a consequence of the time-reversal invariance of the underlying wave equation, the wave field collapses at $t = 0$ onto the source at $x = 0$.

The time-reversed wave-field for data that are contaminated with 20% noise is shown in figure 9. Despite the background noise in the back-propagating waves, the wave-field still collapses well at $t = 0$ onto the source at $x = 0$. This implies that the back-propagation is fairly stable for the addition of random noise, despite the fact that the waves propagate through a strongly scattering medium.

Figure 10 shows the time-reversed wave-field propagating through the

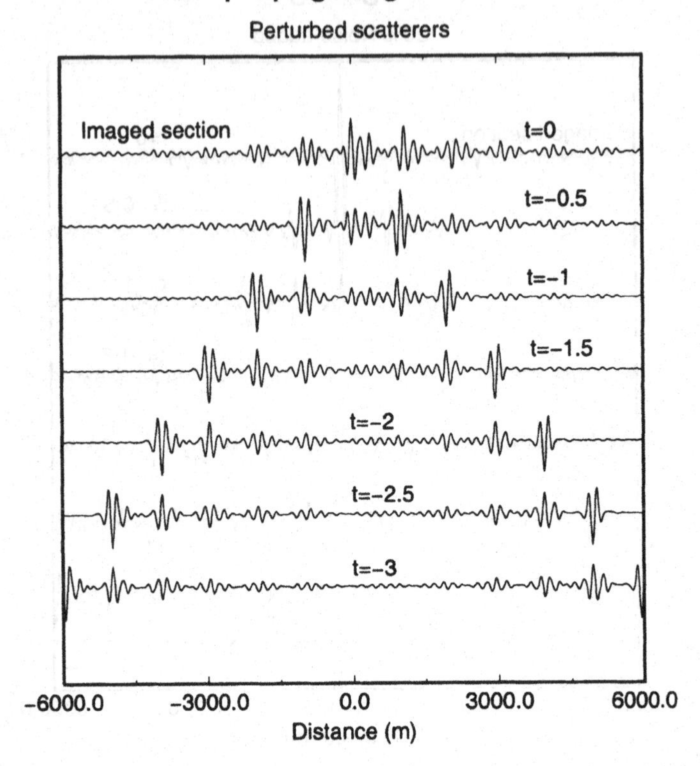

Figure 10. Time-reversed wavefield when the scatterers are perturbed.

medium, but now the position of the scatterers have been randomly perturbed with an rms perturbation of about $25m$. (This is $1/4$ of the dominant wavelength.) Note that now the waves poorly focus at $t = 0$ onto the source at $x = 0$ and that energy leaks out off the "light-cone", notably near $t = 0$. The small perturbation in the location of the scatterers has destroyed the delicate interference process that produces the focusing of the wave-field onto the source, which suggests that focusing through multiple scattering media is an inherently unstable process.

Let us now return to the issue: is *Claerbout* [1985] right in his statement about the futility of using multiple scattering data for imaging the earth's interior, or are *Derode et al.* [1995] right with their apparently stable time-reversed experiments? The answer is: both are right! We are considering two different kind of instabilities. In the physical experiment of *Derode et al.* [1995], the time-reversed wave-field is slightly in error because of the 8-bit digitization that they used. However, the time-reversed wave-field acts as a linear boundary condition on the complete wave-field. This implies that since the medium used by *Derode et al.*[1995] for the forward and reverse propagation is identical, their experiments are not prone to any nonlinear error magnification process. In contrast, *Claerbout* [1985] refers to the fact that imaging is extremely difficult in a strongly scattering medium which is known with only a limited accuracy. His pessimistic conclusion is supported by the example shown in figure 10.

The upshot of this issue is: when one discusses instability of multiple scattering processes one should carefully state whether one refers to (i) instability of the wave-field to perturbations of the boundary conditions, (ii) instability of the wave-field to perturbations of the medium, or (iii) instability of the inverse problem to perturbations of the data. These are different issues that should not be confused.

Note that we have not fully addressed the problem whether imaging of structures is possible on the basis of strongly scattered waves. In fact, this is still an open question. However, strong scattering is not the only complexity seismologists have to deal with. The presence of small-scale structure in the earth imposes an important restriction on the way wave propagation in the earth should be described. The imprint on small-scale variations of the medium on wave propagation is the subject of the next chapter.

3. Averaging in complex media

3.1. RAY THEORY AS WORKHORSE

As shown in the lower panel of figure 1, the elastic response of the earth is for low frequencies characterized by a limited number of wave arrivals. This does not imply that the interior of the earth is a smooth quasi-homogeneous medium. In fact, the coda waves for the high-frequency waves shown in the middle panel of figure 1 show that the interior of the earth is not simple and smooth at all depths. In general, seismologists have followed the strategy to pre-process the data such that the complex wave propagation phenomena are suppressed. The seismic industry relies almost exclusively on single scattered waves for imaging oil reservoirs. Within global seismology, virtually all tomographic models are based on transmission data. Both in seismic imaging and in global seismology one usually employs geometric ray theory to account for wave propagation in the earth. In this sense ray theory can be called the workhorse of seismology.

Ray theory is an approximate theory that is valid when the length-scale a of the inhomogeneity is much larger than:

[1] The wavelength λ of the waves: $a \gg \lambda$

[2] The width of the Fresnel zone: $a \gg L_F$

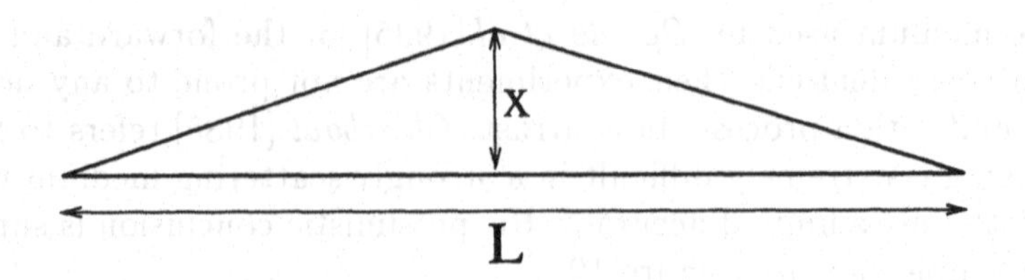

Figure 11. Geometry for the calculation of the Fresnel zone in a homogeneous medium.

A detailed discussion of these requirements is given by *Kravtsov* [1988].

The definition of the Fresnel zone warrants some further discussion. The (first) Fresnel zone is defined as the region of space that when a wave is scattered once at a point within the Fresnel zone the scattered wave arrives almost in phase with the direct wave. This idea can be based in a more rigorous analysis on Kirchhoff integrals [*Kravtsov,* 1988]. As a first example, consider a homogeneous medium. In such a medium the ray is a straight line. Consider a ray of length L as shown in figure 11, and consider the wave that is scattered once at a point at a distance x from the ray. Let us for simplicity consider a scattering point at the middle of the ray. For small values of x the detour of the scattered wave compared to the direct wave is given by $2\sqrt{(L/2)^2 + x^2} - L \approx 2x^2/L$. When this detour is equal to a quarter of a wavelength one is at the edge of the Fresnel zone. The width of the Fresnel zone is for a homogeneous medium thus given by:

$$L_F = \sqrt{\frac{\lambda L}{8}} . \tag{1}$$

One should note that for a fixed wavelength the width of the Fresnel zone grows without limit with the path length L. This means that for a fixed wavelength λ and scale-length a of the heterogeneity the second criterion will always be violated when the path length is of the order of $8a^2/\lambda$.

Let us now consider whether the requirements for the validity of ray theory are satisfied for wave propagation in the earth. As a first example let us consider a P-wave of a period of $1s$ that travels over a teleseismic distance, say $10^4 km$. With a P-velocity of $8km/s$ this leads to a Fresnel zone with a width of $L_F = 100km$. Note that this is really the half-width if the Fresnel zone so that the total width of the Fresnel zone is $200km$. This is about one third of the thickness of the upper mantle, and one may question whether the earth has little variations on that scale.

As a next example consider the Fresnel zone for a PP-wave at the earth's surface that is shown in figure 12. In this example the period of the wave is $20s$. A PP-wave travels downward from a source, is bend towards the earth's surface, reflects at the earth's surface and then travels to the receiving point.

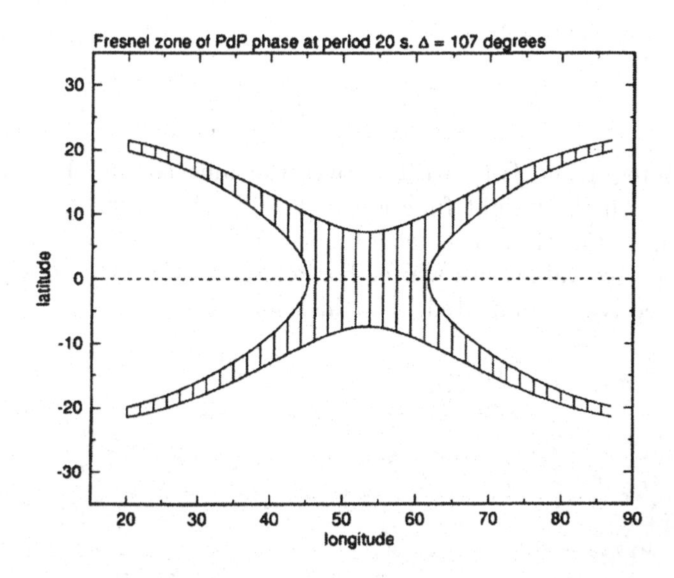

Figure 12. Horizontal cross-section through the Fresnel zone of a PP wave at the bounce point at the Earth's surface for a epicentral distance of 107 degrees and a period of 20s.

Note that the Fresnel zone for this wave does not have an ellipsoidal cross-section. This is due to the fact that this wave is a minimax phase [*Choy and Richards*, 1975]; this is a different way of saying that the travel time is not a minimum along the ray but that the travel-time surface is a saddle-point surface. However, the important point is not the shape of the Fresnel zone but the size; the total width of the Fresnel zone is about 20 degrees, or about $2000km$. It is not likely that the earth is smooth on this length scale. Other examples of exotic Fresnel zone are shown by *Marquering et al.* [1998].

Fresnel zones can not only be defined for travelling body waves, but also for the normal modes of the earth. Just as an atom with spherical symmetry, a spherically symmetric earth model has normal modes whose horizontal dependence are described by spherical harmonics $Y_{lm}(\theta, \varphi)$. The normal modes with the same order l but different degree m are degenerate and have the same eigen-frequency [*Aki and Richards*, 1980]. One thus has multiplets of normal modes that have a $(2l+1)$-fold degeneracy. Horizontal variations in the earth structure lead to line splitting, and one can show that the centroid-frequency shift $\delta\omega_l$ of a split multiplet observed at a given station for a given earthquake can be written as an integral of the local perturbation δm of the earth's structure [*Tanimoto*, 1984; *Snieder*, 1993;

Dahlen and Tromp, 1998]:

$$\delta\omega_l = \iint K(\theta,\varphi)\delta m(\theta,\varphi)d\Omega\,, \tag{2}$$

where the integration $\iint(\cdots)d\Omega$ is over the unit sphere. For the mode ${}_0S_{20}$ with $l = 20$ the weight function $K(\theta,\varphi)$ is shown in figure 13 for an earthquake in Indonesia that is recorded in Western Europe. Positive values are grey, negative values are white. For reference, the great circle joining source and receiver is indicated by a thick line.

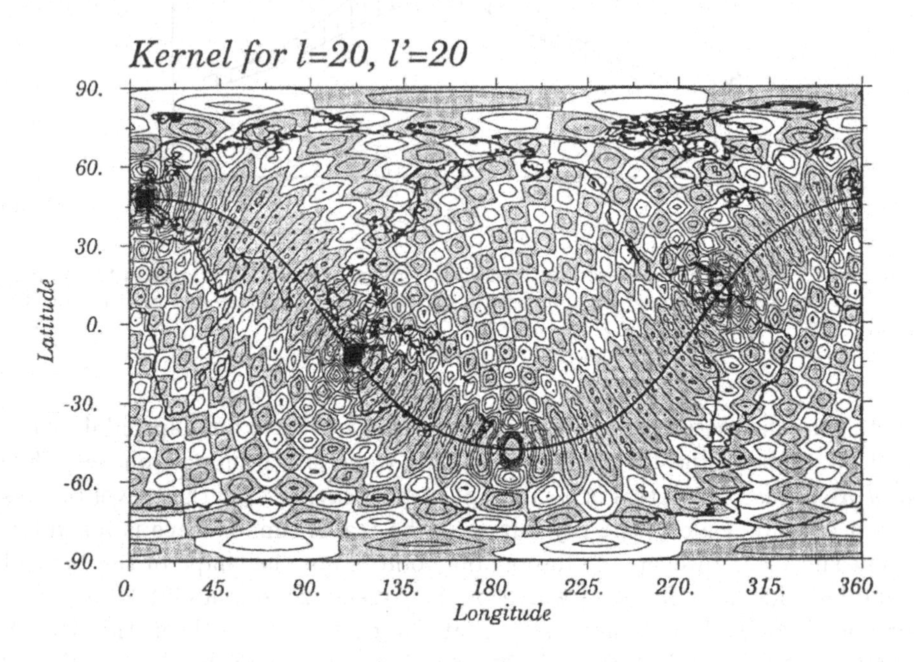

Figure 13. Contour diagram of the sensitivity function $K(\theta,\varphi)$ for mode ${}_0S_{20}$ for an earthquake in Indonesia that is recorded in Europe for mode . Positive values are shown in grey, negative values in white.

Away from the great circle the weight function is oscillatory and changes sign. When the earth structure is smooth on the scale of these oscillations the contribution of this region integrates to zero. The only net contribution then comes from the grey region that straddles the great circle. This region can be seen as the Fresnel zone for this normal mode frequency shift. Note that the width of the Fresnel zone is about $6000km$; the Fresnel zone extends from the north of Tibet to the spreading ridge in the Indian Ocean. It is hard to sustain that the earth structure is smooth on that length scale.

The discussion up to this point assumed that there was a characteristic length-scale of the velocity perturbations in the earth. It is debatable whether such a length scale indeed exists. In-situ measurements taken in boreholes provide evidence that the structure of the earth's crust is self-similar [*Dolan et al.*, 1998; *Herrmann*, 1997]. Such a medium has perturbations on every length scale, and the criteria for the validity of ray theory are thus surely violated.

The same argument can be used for the propagation of waves through a turbulent medium. Turbulence is characterized by a coupling of the different length-scales of the heterogeneity within the medium by the nonlinear terms in the equation of motion. A turbulent gas or fluid thus displays perturbations on every length scale. This means that ray theory should strictly speaking not be used to describe wave propagation in turbulent media such as the ocean (as used in ocean tomography) or the atmosphere (as used in ionospheric tomography).

It is interesting to consider how we deal with the propagation of light in the atmosphere. We explain the twinkling of the stars using ray theory by stating that thermal fluctuations lead to focussing and defocussing of light that is perceived as twinkling. The argument is ray-geometric. In fact, geometrical optics was originally developed to describe the propagation of light. However, we explain the fact that the sky is blue using scattering theory. In both problems we are considering light of the same wavelength propagating through the same medium but apparently:

At night we believe in ray theory while during the day we believe in scattering theory.

The examples of this section indicate that one often uses ray theory rather carelessly. It is of interest to study the propagation of waves when the requirements for the validity of ray theory are violated. For the moment we will consider media that are complex because they have perturbations on small-scales that violate the requirements for the validity of theory. However, it is assumed in this chapter that the perturbations are weak. For many practical applications such as mantle tomography or light propagation in the atmosphere, this condition is justified. We will return to the issue of wave complexity by multiple scattering by strong perturbations of the medium in section 4.

3.2. PERTURBATIONS OF A HOMOGENEOUS MEDIUM

For simplicity we consider the Helmholtz equation

$$\nabla^2 u + \frac{\omega^2}{v^2(\mathbf{r})} \left(1 + n(\mathbf{r})\right) u = 0 \,. \tag{3}$$

The velocity $v(\mathbf{r})$ is a reference velocity that is assumed to be so smooth that ray theory can be used to account for wave propagation through this medium. The unperturbed wave-field that propagates through this medium is denoted by $u_0(\mathbf{r})$. The velocity is perturbed by the quantity $n(\mathbf{r})$, this quantity is not necessarily smooth. The first order perturbation of the phase of the wave-field follows from the Rytov approximation [*Rytov et al.*, 1989] and is given by

$$\delta\varphi = \Im m\left(\frac{u_B}{u_0}\right) , \tag{4}$$

with u_B given by

$$u_B(\mathbf{r}) = -\int \frac{\omega^2}{v^2(\mathbf{r}')} G(\mathbf{r}, \mathbf{r}') n(\mathbf{r}') u_0(\mathbf{r}') dV' , \tag{5}$$

details can be found in *Snieder and Lomax* [1996]. In this expression $G(\mathbf{r}, \mathbf{r}')$ is the Green's fucntion for the reference medium $v(\mathbf{r})$.

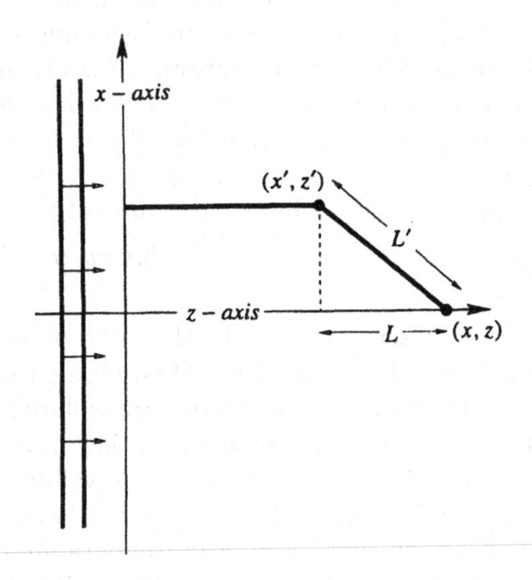

Figure 14. Definition of the geometric variables for a plane incident wave in a homogeneous reference medium.

Let us specialize to the special case of a homogeneous medium in two dimensions with an incoming plane wave that propagates in the z-direction, see figure 14. The wavenumber is given by $k = \omega/v$. The Green's function $G(\mathbf{r}, \mathbf{r}')$ in (5) is equal to $(-i/4)\, H_0^{(1)}(k\,|\mathbf{r} - \mathbf{r}'|)$ [*Morse and Ingard,* 1968]. Using the far-field approximation for the Hankel function one can rewrite the phase shift from expressions (4) and (5) as:

$$\delta\varphi(\mathbf{r}) = \Im m\left(\frac{k^2 e^{i\pi/4}}{\sqrt{8\pi k}} \int \left\{ \int \frac{n(\mathbf{r}') e^{ikD(\mathbf{r}')}}{\sqrt{|\mathbf{r} - \mathbf{r}'|}} dx' \right\} dz' \right) . \tag{6}$$

In this expression D is the detour of the scattered wave compared to the direct wave, in the notation of figure 14:

$$D(\mathbf{r}') = L - L' \ . \tag{7}$$

If the perturbation $n(\mathbf{r}')$ would be sufficiently smooth, one could use a stationary phase evaluation of the x'-integral in (6) [*Bleistein*, 1984]. This would lead to the ray-geometric solution [*Snieder*, 1987]. However, the perturbation is not assumed to be smooth, hence the x'-integral cannot be solved with a stationary phase evaluation. Instead, consider the stationary phase integral without the perturbation $n(\mathbf{r}')$ and without the geometrical spreading $\sqrt{|\mathbf{r} - \mathbf{r}'|}$:

$$\int e^{ikD(\mathbf{r}')} dx' = \sqrt{\frac{2\pi\,|z - z'|}{k}}\; e^{i\pi/4} \ . \tag{8}$$

Next, divide the x-integral in (6) by the left hand side of this expression and multiply with the right hand side of (6), and ignoring the variation of the geometrical spreading of $|\mathbf{r} - \mathbf{r}'|$ with x', the result can be written as:

$$\delta\varphi(\mathbf{r}) = \Re e \left(\frac{k}{2} \int \frac{\int n(\mathbf{r}') e^{ikD(\mathbf{r}')} dx'}{\int e^{ikD(\mathbf{r}')} dx'} dz' \right) \ . \tag{9}$$

This is an interesting expression because the x'-integration can be seen as a weigthed average of the perturbation $n(\mathbf{r}')$ with a weight function $\exp(ikD)$. This weight function allows for the relative timing of the scattered waves compared to the direct wave.

When one changes the geometry of the problem the unperturbed wave u_0 changes, and when one considers the problem in 3 dimensions the Green's function changes. This means that for a different geometry the integral (6) has a different form. However, as shown by *Snieder and Lomax* [1996] the averaging integral (9) holds in both 2 and 3 dimensions as well as for a point source instead of a plane incoming wave. The only difference is that in 3 dimensions one needs to integrate over two transverse coordinates x and y. For this reason expression (9) is more general than the original integral (6).

3.3. PERTURBATIONS OF AN INHOMOGENEOUS MEDIUM

The derivation for a homogeneous medium can be generalized for a background medium that is inhomogeneous with velocity variations that are so smooth that it is justified to use ray theory. Perturbations of this medium are denoted by $n(\mathbf{r})$, these perturbations can have any scale. The corresponding derivation is given by *Snieder and Lomax* [1996] with some corrections in *Snieder and Chapman* [1998]. For such an inhomogeneous reference medium the detour D needs to be replaced by the delay time T of a scattered wave compared to the unperturbed wave. Using ray-centered coordinates [*Červený and Hron,* 1980] the phase shift can for such a medium be written as

$$\delta\varphi = \frac{\omega}{2}\Re e\left(\int ds \frac{1}{v(s)} \frac{\int n(\mathbf{r})e^{i\omega T} h(s, q_1, q_2) dq_1 dq_2}{\int e^{i\omega T} dq_1 dq_2}\right) . \tag{10}$$

In this expression s is the integration variable along the reference ray while $q_1$and q_2 denote the coordinates perpendicular to the reference ray. The Jacobian of the transformation to ray-centered coordinates is given by $h(s, q_1, q_2) = 1 + (1/v)(\mathbf{q}\cdot\nabla\mathbf{v})$. The derivation of (10) is much more complex than the equivalent expression (9) for a homogeneous reference medium because in an inhomogeneous medium the relation between geometrical spreading and wave-front curvature is nontrivial, and one needs to invoke the equations of dynamic ray tracing to derive (10).

However, this expression has a similar interpretation as an averaging integral as (9). The quantity that is averaged in (10) is effectively given by $\omega n(\mathbf{r})/2v(\mathbf{r})$, and the weight function for the function is given by $\exp(i\omega T)$. It follows from the Helmholtz equation (3) that the local wavenumber is given by $k^2(\mathbf{r}) = (\omega^2/v^2(\mathbf{r}))\ \ (1 + n(\mathbf{r}))$. To first order in $n(\mathbf{r})$ the local perturbation of the wavenumber is thus given by

$$\delta k(\mathbf{r}) = \frac{\omega}{v(\mathbf{r})}\left\{\sqrt{1 + n(\mathbf{r})} - 1\right\} \approx \frac{\omega n(\mathbf{r})}{2v(\mathbf{r})} \tag{11}$$

This means that according to (10) for a general variable reference medium the phase shift is given by a weighted average of the local wavenumber as defined in (11) with a weight function $\exp(i\omega T)$.

3.4. THE SIGNIFICANCE OF WAVE-FIELD AVERAGING

The equations (9) and (10) state that the perturbation of the phase shift is a weighted average of the slowness perturbation over the Fresnel zone. This implies that the continuity of the wave-field smooths out the perturbations of the medium in the sense that small-scale perturbations of the medium only affect in an average sense the properties of the transmitted wave. As shown in *Snieder and Lomax* [1996] a similar result can be derived for the amplitude of the transmitted wave.

This averaging property plays a crucial role in seismic tomography. As argued in section 3.1 the requirements for the validity of ray theory are not very justifiable for the earth. Nevertheless, the majority of techniques for seismic tomography are based on ray theory [e.g. *Iyer and Hirahara*, 1993]. Let us now divide the perturbations of the medium in small-scale perturbations that violate the requirements for the validity of ray theory, and large-scale perturbations for which ray theory is valid. The averaging integrals (9) or (10) imply that the small-scale perturbations are averaged out. In contrast, the large-scale perturbations lead to a change in the phase that is identical to the phase shift predicted by ray theory. This means that a tomographic inversion for the large-scale perturbations is not affected by the presence of small-scale perturbations. Only when one aims at resolving features that are smaller than the width of the Fresnel zone, one needs to replace ray theory by the averaging integrals (9) or (10).

Of course, the theory of this chapter is only applicable for perturbations that are so weak that the relation between the perturbation of the phase and the perturbation of the medium can be linearized. However, virtually all schemes for seismic tomography are based on such a linearization, hence this is no additional complexity as far as the application in seismic tomography is concerned.

3.5. FOR THOSE WHO DON'T LIKE THE RYTOV APPROXIMATION

Readers may not feel comfortable with the use of the Rytov approximation. The validity of the Rytov approximation is a topic of heated debate for at least 30 years. An important restriction of the Rytov approximation is that there is no known generalization to vector waves. The results obtained in section 3.2 can, however, also be obtained in a completely different way by making a coordinate transformation as shown in figure 15 from the Cartesian coordinates (x, y, z) to a new coordinate system (x, y, Z) where the lines of constant values of the new Z-coordinate coincide with the perturbed wavefronts [*Snieder*, 1998]. One can view this technique as a higher-dimensional version of the method of strained coordinates [*Nayfeh*, 1973] Solving the wave equation in this new coordinate system using a parabolic approximation leads to the same averaging integral (9) as was obtained from the Rytov approximation.

4. Multiple scattering by isotropic point scatterers

As argued in section 1, complex variations in the structure of the earth can give rise to very complex waveforms through the process of multiple scattering, see the examples in the middle panel of figure 1 and in the figures 6. Formulating a multiple scattering theory for the general elastic problem is a task of formidable proportions. However, for the special case of

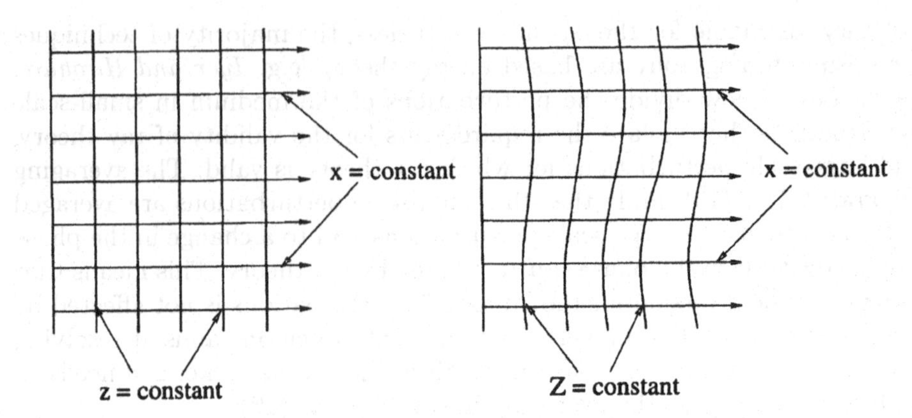

Figure 15. Definition of the coordinate systems (x, y, z) and (x, y, Z) where the planes $z = const$, coincide with the unperturbed wavefront (left panel) and where the planes $Z = const$, coincide with the perturbed wavefront (right panel).

isotropic point scatterers the multiple scattering problem has a very simple and elegant solution.

4.1. DERIVATION OF THE THEORY

Let us consider a system with point scatterers at locations $\mathbf{r}_j$. The complex scattering coefficient of scatterer j is denoted by A_j. This coefficient contains the full nonlinear interaction of the wave that is incident on the scatterer and the scattered wave. It is shown in appendix A that this scattering coefficient contains the superposition of all consecutive multiple scattering interactions with the same scatterer. Let the total wave-field that is incident on scatterer j be denoted by u_j. The wave that is scattered by this scatterer is then given by $G(\mathbf{r}, \mathbf{r}_j)A_j u_j$, where $G(\mathbf{r}, \mathbf{r}')$ is the Green's function for the medium in which the scatterers are embedded. Since the scattering is assumed to be isotropic, there is no dependence on the scattering angle. The total wave-field can be written as the superposition of the unperturbed wave $u^{(0)}(\mathbf{r})$ and the waves emanating from all the scatterers:

$$u(\mathbf{r}) = u^{(0)}(\mathbf{r}) + \sum_j G(\mathbf{r}, \mathbf{r}_j) A_j u_j \ . \tag{12}$$

The wave-field that is incident on scatterer i follows from this expression by setting $\mathbf{r} = \mathbf{r}_i$, and by omitting the term $j = i$ from the sum in (12) because the wave incident wave on scatterer i only has contributions from the unperturbed wave $u^{(0)}(\mathbf{r})$ and from the waves coming from the *other* scatterers:

$$u_i = \mathbf{u}^{(0)}(\mathbf{r}_i) + \sum_{j \neq i} G(\mathbf{r}_i, \mathbf{r}_j) A_j u_j \ . \tag{13}$$

This is all that is needed to solve the scattering problem. Equation (13) constitutes a linear system of equations for the complex coefficients u_j. This system can be solved numerically. Once the u_j are determined one can compute the wave-field at any location $\mathbf{r}$ by inserting the u_j in expression (12).

For convenience the system of equations (13) can also be written in vector form:

$$(\mathbf{I} - \mathbf{S})\,\mathbf{u} = \mathbf{u}^{(0)} \ , \tag{14}$$

where $\mathbf{u}$ is the vector with u_j as the j-th component and the vector $\mathbf{u}^{(0)}$ has $u^{(0)}(\mathbf{r}_j)$ as the j-th component. The identity matrix is denoted by $\mathbf{I}$ and the components of the matrix $\mathbf{S}$ are given by:

$$S_{ij} = \begin{cases} 0 & for\ i = j \\ G(\mathbf{r}_i, \mathbf{r}_j) A_j & for\ i \neq j \end{cases} \tag{15}$$

When energy is conserved, the optical theorem imposes a constraint on the imaginary component of the forward scattering amplitude and the scattered power averaged over all directions [*van der Hulst*, 1949; *Ishimaru*, 1997]. The generalization of the optical theorem for elastic surface wave modes is given by *Snieder* [1988]. For isotropic scatterers the optical theorem imposes the following constraint on the scattering coefficient in different dimensions:

$$\Im m(A) = \begin{cases} -\dfrac{1}{2k}|A|^2 & in\ 1D \\[2ex] -\dfrac{1}{4}|A|^2 & in\ 2D \\[2ex] -\dfrac{k}{4\pi}|A|^2 & in\ 3D \end{cases} \tag{16}$$

Note that the scattering formalism can be applied to any number of dimensions and that the numerical implementation is very similar in different number of dimensions.

4.2. THE FEYNMAN PATH SUMMATION

The scattering equations of the previous section can be rewritten in a different form that is useful for a number of applications. The linear system of equations (14) can be solved by matrix inversion: $\mathbf{u} = (\mathbf{I} - \mathbf{S})^{-1}\,\mathbf{u}^{(0)}$. Using an expansion of the inverse $(\mathbf{I}-\mathbf{S})^{-1}$ this can also be written as:

$$\mathbf{u} = \mathbf{u}^{(0)} + \mathbf{S}\mathbf{u}^{(0)} + \mathbf{S}^2\mathbf{u}^{(0)} + \mathbf{S}^3\mathbf{u}^{(0)} + \cdots \tag{17}$$

Inserting this expression in (12) and using the definition (15) for $\mathbf{S}$, the total wave-field is given by

$$u(\mathbf{r}) = u^{(0)}(\mathbf{r}) + \sum_i G(\mathbf{r},\mathbf{r}_i)A_i u^{(0)}(\mathbf{r}_i) + \sum_{i\neq j}\sum_j G(\mathbf{r},\mathbf{r}_i)A_i G(\mathbf{r}_i,\mathbf{r}_j)A_j u^{(0)}(\mathbf{r}_j) + \ldots \tag{18}$$

This result can be seen as the Neumann series solution of the scattering problem. The series does have a clear physical meaning. The first term denotes the arrival of the unperturbed wave. The second term consists of all the single scattered waves. The second term contains all the waves that are scattered twice by different scatterers. The series (18) can thus be seen as a summation of waves that have travelled along all possible paths between scatterers, *but that never visit the same scatterers on two consecutive scattering events.* One can therefore view expression (18) as a discrete version of the Feynman path integral [*Feynman and Hibbs*, 1965] where the total wave-field is expressed as a path-integral over all possible trajectories. Since (18) is the discrete equivalent of the Feynman path integral it will be called the "Feynman path summation." Note that no approximations have been made in the deriving the Feynman path summation.

With a slight change of notation one can write the Feynman path summation also as

$$u(\mathbf{r}) = \sum_P e^{ikL_P}\left(\Pi C\right) u^{(0)} . \tag{19}$$

In this expression the sum $\sum_P$ denotes a sum over all paths starting from the source and travelling along all possible combinations of scatterers with the same scatterer not encountered on consecutive scattering events. The path length for path P is denoted by L_P. The term (ΠC) gives the product of scattering coefficients for that path and the geometrical spreading and other constants that appear in the Green's function. Note that for 2 dimensions, one can only split-off the term $\exp(ikL_P)$ when the distance between the scatterers is much larger than a wavelength so that the far-field approximation for the Hankel function can be used.

4.3. THE EFFECTIVE WAVENUMBER FOR AN ENSEMBLE OF SCATTERERS

Let us consider a medium of isotropic point scatterers as described in section 4.1. The scatterers have three effects: (1) the ballistic wave that travels through the medium will experience a phase shift, (2) the ballistic wave will be reduced in amplitude due to scattering losses and (3) a coda of scattered waves is generated. In this section we will not account for the last effect. The first two effects can be described by introducing an effective local wavenumber k_{eff} for the medium with the scatterers. This effective wavenumber was derived by *Groenenboom and Snieder* [1995] using an approximate invariant embedding technique. In this section the effective wavenumber is derived using the averaging integral that is derived in section 3.2.

As a starting point consider the Helmholtz equation (3) for the special case that the reference medium is homogeneous:

$$\nabla^2 u(\mathbf{r}) + k^2 (1 + n(\mathbf{r}))\, u(\mathbf{r}) = 0 \,. \tag{20}$$

In order to relate this equation to the scattering theory of section 4.1 we need to establish the connection between the perturbation $n(\mathbf{r})$ and the scattering coefficients A_j in (12). This can be achieved by letting the operator $(\nabla^2 + k^2)$ act on (12). Using the fact that the unperturbed wave satisfies $(\nabla^2 + k^2)\, u^{(0)}(\mathbf{r}) = 0$ and that the Green's function satisfies $(\nabla^2 + k^2)\, G(\mathbf{r}, \mathbf{r}') = \delta(\mathbf{r} - \mathbf{r}')$ one finds that

$$\left(\nabla^2 + k^2\right) u(\mathbf{r}) = \sum_j \delta(\mathbf{r} - \mathbf{r}_j) A_j u_j \,. \tag{21}$$

Comparing this with (20) one readily finds that

$$n(\mathbf{r}) = -\frac{1}{k^2} \sum_j \delta(\mathbf{r} - \mathbf{r}_j) A_j \,. \tag{22}$$

In order to obtain an effective wavenumber, let us write the phase change due to the perturbation in equation (9) as $\delta\varphi = \int \delta k \; dz$, where δk is the perturbation in the wavenumber. Inserting (22) in the averaging integral (9) and using the above expression for δk one obtains the following expression for the wavenumber perturbation:

$$\delta k = \frac{1}{2} k \langle n \rangle_F = -\frac{1}{2k} \left\langle \sum_j \delta(\mathbf{r} - \mathbf{r}_j) A_j \right\rangle_F \,. \tag{23}$$

The brackets $\langle \cdots \rangle_F$ do *not* denote an ensemble average, but a weighted average over the Fresnel zone:

$$\langle \cdots \rangle_F \equiv \frac{\int (\cdots)\, e^{ikD(\mathbf{r}')} dx'}{\int e^{ikD(\mathbf{r}')} dx'} \tag{24}$$

Note that we have tacitly removed the operation of taking the real part of (23). Since the scattering coefficients A_j are complex, the wavenumber perturbation is also complex. This describes the amplitude decay of the ballistic wave due to scattering losses. One can derive this rigorously by using that the logarithmic perturbation of the amplitude is the imaginary part of the averaging integral in (9), see *Snieder and Lomax* [1996] for details. Adding the wavenumber perturbation δk to the unperturbed wavenumber k one obtains the complex effective wavenumber:

$$k_{eff} = k - \frac{1}{2k} \left\langle \sum_j \delta(\mathbf{r} - \mathbf{r}_j) A_j \right\rangle_F \tag{25}$$

Let us now consider the special case that the scatterers have an identical scattering coefficient A and that the weight function in the averaging integral is replaced by its value on the reference ray: $\exp(ikD) \rightarrow 1$. Under these assumptions the wavenumber perturbation integrated in the z-direction can be rewritten in the following way:

$$\begin{aligned} \int \left\langle \textstyle\sum_j \delta(\mathbf{r} - \mathbf{r}_j) A_j \right\rangle_F dz &= A \int \left\langle \textstyle\sum_j \delta(\mathbf{r} - \mathbf{r}_j) \right\rangle_F dz \\ &= A \, \frac{\text{Number of scatterers in Fresnel zone}}{\text{Volume of Fresnel zone}} \\ &= A \langle N \rangle \end{aligned} \tag{26}$$

where $\langle N \rangle$ is the number of scatterers per unit volume area over the Fresnel zone. The corresponding effective wavenumber is then given by

$$k_{eff} = k - \frac{A \langle N \rangle}{2k} \tag{27}$$

This expression agrees with the effective wavenumber that is derived using other techniques [*Lax*, 1951; *Waterman and Truell*, 1961].

4.4. A NUMERICAL EXAMPLE

As an illustration of the theory of this chapter a numerical example of wave propagation in two dimensions is presented. The geometry of the problem is shown in figure 16. An isotropic source (shown by a square) is placed in the middle of 12 receivers (shown by triangles) that are located on a circle. Between the source and the receivers point scatterers are present, these are indicated by dots. Details on the parameters used in the simulation can be found in *Groenenboom and Snieder* [1995].

The wave-field in the absence of scatterers is indicated by the dashed line in figure 17. Since the source is isotropic and the reference medium is

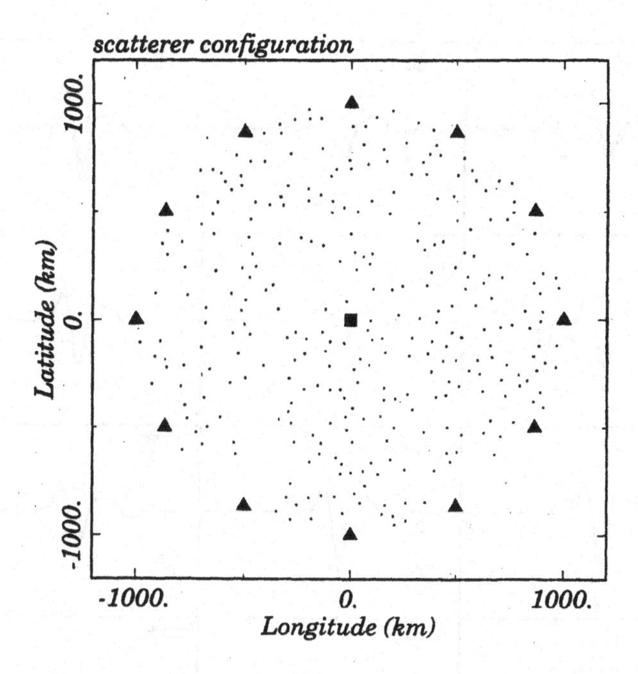

Figure 16. Geometry of the scattering experiment with a source (square) surrounded by 12 receivers (triangles) and scatterers (dots).

homogeneous the unperturbed wave is identical for all the receivers. The exact scattering response is indicated by the thin solid line in figure 17. This response was computed using the expressions (12) and (13) in the frequency domain and by carrying out a Fourier transform. Note that the amplitude of the ballistic wave has been reduced severely (about a factor 3) due to scattering losses. This strong scattering loss is consistent with the presence of strong scattered waves after the arrival of the ballistic wave. This means that in this problem strong multiple scattering is operative.

The thick solid line in figure 17 gives the wave-field computed with the effective wavenumber given in expression (27). For every source-receiver pair this quantity was computed by counting the number of scatterers within the Fresnel zone for that source-receiver pair. Note the good agreement between the ballistic wave in the exact response and the wave-field computed with the effective wavenumber (27). The scattered waves arriving after the ballistic wave are not reproduced in the wave-forms computed with the effective wavenumber technique. This is due to the fact that in this

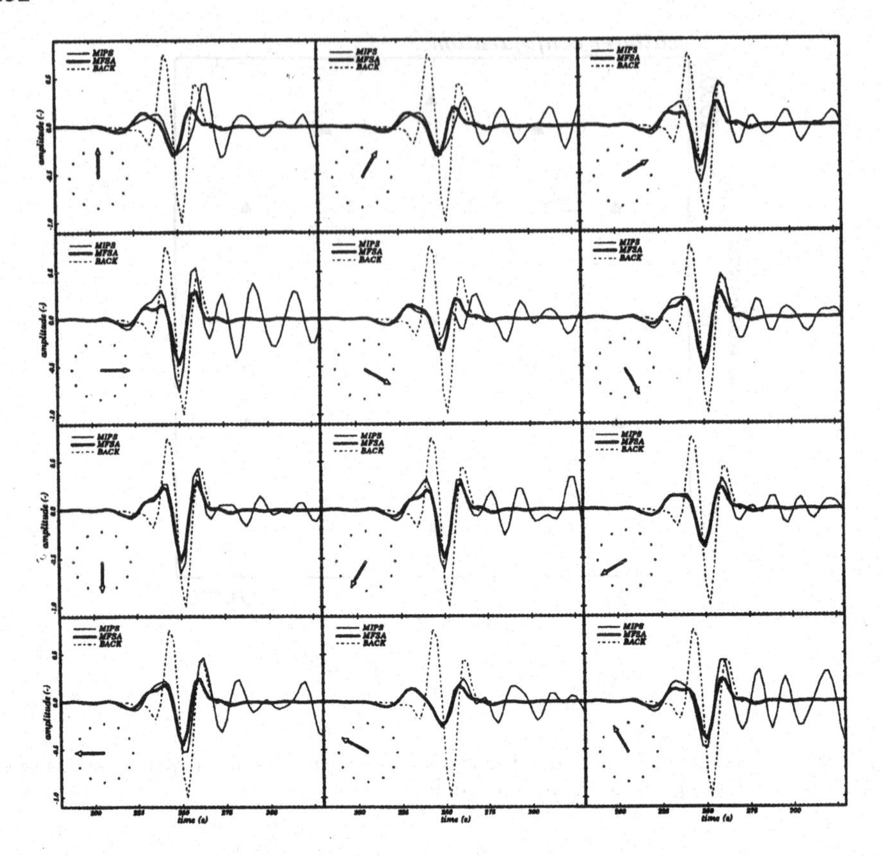

Figure 17. Waves at the 12 receivers of the previous figure. The clock indicates the receiver position. Shown is the wave-field in the absence of scatterers(dashed line), the complete wavefield in the presence of scatterers (thin solid line) and the wavefield computed with the effective wavenumber technique (thick solid line).

approach only the scatterers within the Fresnel are accounted for, whereas the coda is generated by the scatterers outside the Fresnel zone as well.

The ballistic wave varies considerably for the different receivers due the variations in the scatterer density. Notice that the waveforms computed with the effective wavenumber technique follow this variability quite well. It can be seen that the waveforms in the presence of scatterers exhibit a pronounced dispersion that is reproduced well with the effective wavenumber approach. This example shows that the effective wavenumber can be a powerful tool for describing the effects of scattering on the ballistic wave, even in the case when this wave is strongly perturbed.

5. Imaging waves or particles in strongly scattering media

In this section we return to the issue to what extent imaging in strongly scattering media is possible. It was shown in section 2 that the random addition of noise enters only the linear boundary conditions of the problem, hence it does not need to be considered as a source of instability due to nonlinear effects. In classical mechanics, the instability of trajectories of particles to perturbations of the initial conditions is the signature of chaotic behavior [e.g. *Tabor*, 1989]. In order to use waves for imaging purposes, the temporal evolution of the wave-field may or may not be unstable for perturbations of the location of the source of the wave-field or for perturbations of the medium. In this sense one may wonder how the *wave-field* in a strongly scattering system behaves when it is known that *particles* propagating through the same system exhibit chaotic behavior. The imprint of chaos on wave systems is usually studied in the context of quantum chaos [e.g. *Tabor*, 1989; *Gutzwiller*, 1990]. However, in that field of research the attention usually is focused on the spectral properties of a closed wave-system rather than the stability of the temporal evolution of the wave-field.

The issue of the stability of the temporal evolution of wave or particle propagation can be studied using a Time-Reversal-Mirror. This device has been used in laboratory experiments [*Derode et al.*, 1995; *Fink*, 1997]. In this technique wave propagate through a medium and are recorded at receivers. The recorded wave-field is reversed in time, which is equivalent by replacing t by $-t$, and this time-reversed wave-field is send back into the medium. Since the system is invariant for time reversal, the wave propagate back along their original path and focus back on the original source position at $t = 0$. Since the waves are imaged in this method on the original source position this technique will be referred to as Time-Reversed-Imaging (TRI). This principle can obviously also be applied to particles. Whether the waves (or particles) really return to the original source position when the system is perturbed depends on the stability properties of the wave or particle propagation.

The system used here is similar to the one used by *Derode et al.* [1995] and is shown in figure 18. The numerical experiments shown in this section are from *Snieder and Scales* [1998]. Particles or waves are emitted from a source and propagate through a system of 200 strong isotropic point scatterers. For the waves, 96 receivers are located on the line indicated in figure 18. The recorded wave-forms are time-reversed and then send back into the medium from the receivers that now act as sources. A particle is recorded for the time-reversed imaging when it traverses the receiver line in figure 18 at a certain time t. TRI of the particle is realized by reversing

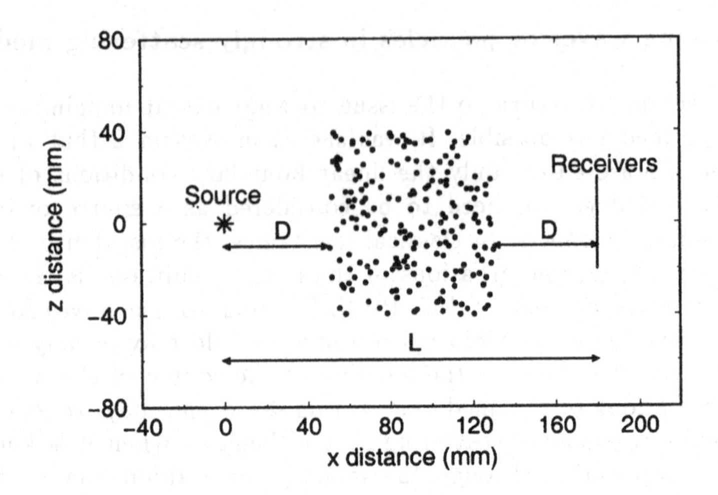

Figure 18. Geometry of the numerical experiment with TRI. The scatterers are shown by thick dots.

the velocity of the particle $\mathbf{v} \rightarrow -\mathbf{v}$ and by sending the particle back into the medium at time $-t$.

When comparing the stability of wave or particle propagation, one can either specify the medium (e.g. a quantum mechanical potential) or the scattering properties of the waves and particles. Here the latter approach is taken by using isotropic point scatterers for both waves and particles. This choice ensures that the only difference between the waves and particles lies in the dynamics of propagation, rather than in a different interaction with the scatterers.

5.1. SCATTERING OF PARTICLES AND WAVES

Isotropic scattering of particles that is invariant under time reversal is ensured by requiring that both the velocity v (=*1500 m/s*) and the impact parameter b of the particles are conserved during scattering and that (in two dimensions) the scattering angle Θ is linear in the impact parameter:

$$\Theta = \pi \left(\frac{\sigma - 2b}{\sigma} \right) \qquad for \quad |b| \leq \sigma/2 \,, \tag{28}$$

where σ is the scattering cross-section. See figure 19 for the definition of variables. For larger values of the impact parameter the particle is not

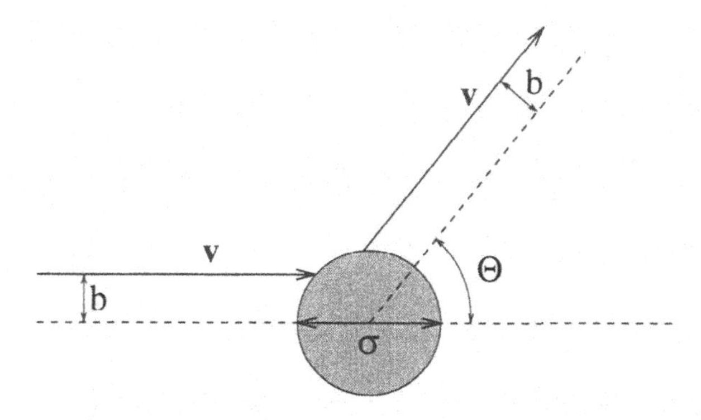

Figure 19. Definition of the impact parameter b and the scattering angle Θ.

scattered (i.e. $\Theta = 0$). Table 1 shows the values of the relevant length

TABLE 1. Numerical values of parameters in the numerical experiment.

Symbol	Property	Value
σ	Scattering cross section	1.592 mm
l	Mean free path	15.56 mm
λ	Dominant wavelength	2.5 mm

scales in the numerical experiment. Figure 20 shows the mean number of encounters n with scatterers for the particles that cross the receiver line as a function of time t. A least-squares fit of the line in figure 20 gives the mean time between consecutive encounters with scatterers, this can be used to infer that the mean free path l is given by $l = 15.56mm$. This quantity is much less than the size of the scattering region ($80mm$), which implies that the particles are strongly scattered.

For the scattering of waves, the scattering theory for isotropic point scatterers of section 4.1 is used. The wave-field for both the forward and reverse propagation is computed with equations (12) and (13). The wave propagation can only be invariant for time-reversal when energy is conserved. It is thus crucial that the scattering coefficients satisfy the optical theorem given in equation (16). The wavefield recorded by a receiver in the middle of the receiver array is shown in figure 21. The wave-field has the same character as the seismogram shown in the middle panel of figure 1; the direct wave merges with a coda of strongly scattered waves. This reflects

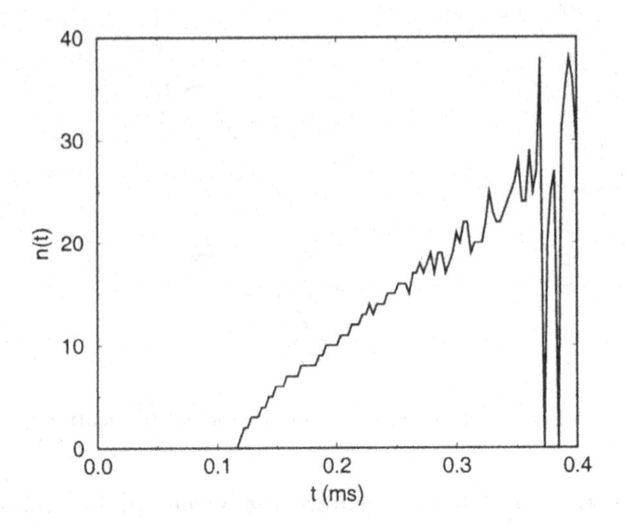

Figure 20. **Mean number of scatterers encountered by particles as a function of the arrival time at the receiver line.**

the fact that for the employed parameter setting strong multiple scattering is operative.

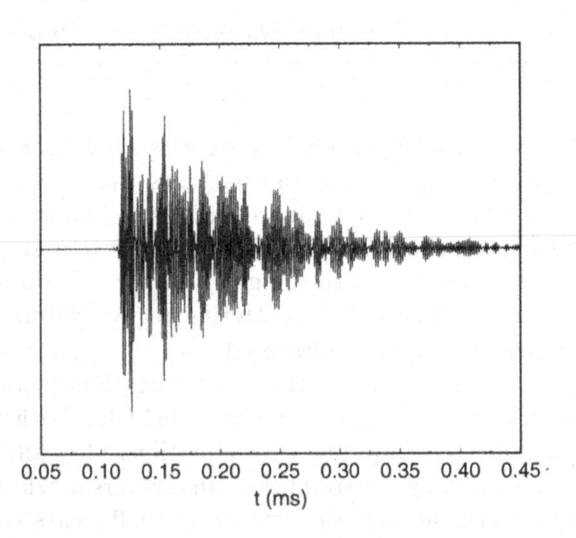

Figure 21. **Wave-field at a receiver located in the middle of the receiver array.**

TRI of the waves is carried out by recording the wave-field at 96 equidistant receivers on the receiver line, and by using the complex conjugate of the wave-field in the frequency domain as source signals that are emitted from the receivers that now act as sources. An example of the wave-field after TRI is shown by the thick solid line in figure 22 where the time-reversed wave-field at time $t = 0$ is shown along the line $x = 0$. In the ideal case the wave-field focusses on the source at location $z = 0$. Due to the limited aperture of the receiver array and the finite bandwidth of the waves the focussing peak has a finite width and has sidelobes. For the TRI only the waves arriving between $0.25s$ and $0.30s$ have been used. It can be seen in figure 21 that this time window only contains multiple scattered waves in the coda.

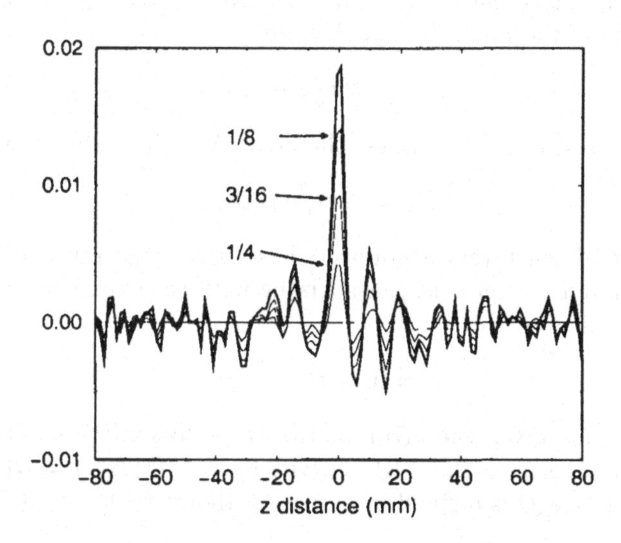

Figure 22. Time reversed wavefield at time $t = 0$ along the line $x = 0$. The source location is at $z = 0$. The thick solid lines is for TRI with the unperturbed receiver positions. The thin solid lines are for the wavefield after TRI with perturbed receiver positions. The numbers denote the rms value of the perturbation as a fraction of the dominant wavelength.

5.2. STABILITY ANALYSIS FOR SCATTERED PARTICLES

Consider the trajectories of a particle that is scattered with impact parameter b and with a perturbed impact parameter $b + \Delta$. The divergence of the trajectories is shown in figure 23. It follows from this figure and expression (28) that the divergence of the trajectories is given by $|\mathbf{r}_{b+\Delta}(t) - \mathbf{r}_b(t)| \approx vt\left(\Theta(b+\Delta) - \Theta(b)\right) = 2\pi vt\Delta/\sigma$. This implies that the error Δ_{out} at time t

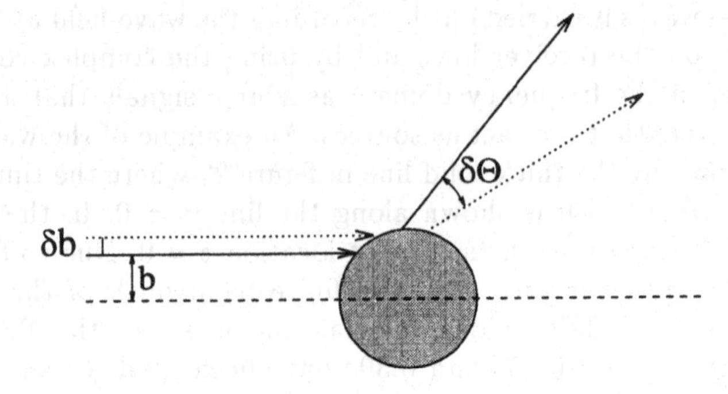

Figure 23. Divergence of the trajectories of the particles with impact parameter b and $b + \delta b$.

since the scattering is related to the initial error Δ_{in} by $\Delta_{out} = 2\pi\,(vt/\sigma)\,\Delta_{in}$. On average, vt is the mean-free path l, hence

$$\Delta_{out} = 2\pi\,(l/\sigma)\,\Delta_{in}\ . \tag{29}$$

When a wave is scattered n times, the error Δ_n follows by recursion:

$$\Delta_n = (2\pi l/\sigma)^n\,\Delta_0\ . \tag{30}$$

The number of scatterers encountered is on average given by $n = vt/l$, hence the Lyapunov exponent μ associated with the exponential divergence of trajectories is given by

$$\mu = \ln\left(2\pi l/\sigma\right)\ v/l\ . \tag{31}$$

Equation (30) gives the error in the trajectory after n scattering encounters. The error δ in the TRI is given by $\delta = D\,(\Delta\Theta) = D(d\Theta/db)\Delta_n$, where the distance D is defined in figure 18. Hence with (30) it follows that

$$\delta = \frac{2\pi D}{\sigma}\left(\frac{2\pi l}{\sigma}\right)^n \Delta_0\ . \tag{32}$$

When the error in the trajectory is of the order of $\sigma/2$ the trajectory will be completely different because the particle then encounters different scatterers. The associated critical perturbation δ_c follows from (30):

$$\delta_c = \left(\frac{\sigma}{2\pi l}\right)^n \frac{\sigma}{2}\ . \tag{33}$$

Using expression (31) for the Lyapunov exponent and using the relation $n = vt/l$ to eliminate n one finds that this critical length scale decreases exponentially with time:

$$\delta_c = \frac{\sigma}{2}\,e^{-\mu t}\ . \tag{34}$$

For the numerical experiments the critical length scale is shown in table 2. Also indicated is the precision with which the numerical simulations have been carried out. (All calculations were done in 64 bit arithmetic on an SGI Power Challenge.) Since the mean free path l is much larger than the scattering cross-section σ (table 1) the critical length scale decreases dramatically with the number of scattering encounters.

TABLE 2. Critical error δ_c for different numbers of scattering encounters. Also indicated is the employed machine precision.

n	$\delta_c(mm)$
1	0.0129
2	2.11×10^{-4}
3	3.43×10^{-6}
4	5.60×10^{-8}
5	9.11×10^{-10}
6	1.48×10^{-11}
7	2.41×10^{-13}
8	3.93×10^{-15}
Machine precision	0.22×10^{-15}
9	6.41×10^{-17}

The previous analysis applies for a perturbation of the starting point of a particle. When the scatterer locations are perturbed over a distance δ, a term δ should be added to the right-hand side of (29). The error after n scattering encounters is then given by

$$\Delta_n = \left(\frac{\left(\frac{2\pi l}{\sigma}\right)^{n+1} - 1}{\left(\frac{2\pi l}{\sigma}\right) - 1} \right) \Delta_0 \,. \tag{35}$$

However, given the high numerical value of $2\pi l/\sigma$ (≈ 61) in the numerical experiments this result is similar to (30) for the perturbation of initial conditions. The associated critical length scale is shown in table 3.

It follows that for the particles the critical length scale depends on the scattering cross section and the mean free path, and that this quantity depends exponentially on the number of scattering encounters (and thus

TABLE 3. Critical length scales δ_c for different perturbations.

	Scatterer location	Source location
Particles	$\frac{1}{2}\left(\frac{\left(\frac{2\pi l}{\sigma}\right)-1}{\left(\frac{2\pi l}{\sigma}\right)^{n+1}-1}\right)\sigma$	$\frac{1}{2}\left(\frac{\sigma}{2\pi l}\right)^{n}\sigma$
Ballistic wave	$\frac{\sqrt{\lambda L}}{\sqrt{12(n+1)}}$	$\frac{\lambda}{4}$
Coda wave	$\frac{1}{\sqrt{2n}}\frac{\lambda}{4}$	$\frac{\lambda}{4}$

on time). Due to this dependence the critical length scale δ_c is dramatically smaller than the scattering cross section σ.

5.3. STABILITY ANALYSIS FOR SCATTERED WAVES

A fundamental difference between time reversed imaging of waves and particles is that TRI of particles occurs because a trajectory returns to the source at $t = 0$ whereas for waves TRI is achieved because the waves interfere at $t = 0$ constructively only near the source. The Feynman path summation given in expression (19) provides the basis for quantifying when the interference process that leads to TRI of waves is destroyed by the perturbations. When either the sources or the scatterers are perturbed for the waves, the dominant effect on the wave-field is the perturbation of the path length L_P in (19). When the variance σ_L of the path length is of the order of a quarter wavelength the resulting interference pattern is destroyed. Hence TRI of waves will break down when $\sigma_L \approx \lambda/4$.

The effect of the perturbation in the i-th component of the position vector of scatterer j on the path length L_P follows from the derivative

$$\frac{\partial L_P}{\partial x_i^{(j)}} = \frac{x_i^{(j)} - x_i^{(j-1)}}{\left|\mathbf{r}^{(j)} - \mathbf{r}^{(j-1)}\right|} - \frac{x_i^{(j+1)} - x_i^{(j)}}{\left|\mathbf{r}^{(j+1)} - \mathbf{r}^{(j)}\right|}, \tag{36}$$

which implies that

$$\sum_i \left(\frac{\partial L_P}{\partial x_i^{(j)}} \right)^2 = 2 - 2 \frac{\left(\mathbf{r}^{(j)} - \mathbf{r}^{(j-1)}\right) \cdot \left(\mathbf{r}^{(j+1)} - \mathbf{r}^{(j)}\right)}{\left|\mathbf{r}^{(j)} - \mathbf{r}^{(j-1)}\right| \left|\mathbf{r}^{(j+1)} - \mathbf{r}^{(j)}\right|} = 2\left(1 - \cos\Theta_j\right) , \tag{37}$$

where Θ_j is the scattering angle at scatterer j. This angle is related to the angle φ_j in figure 30 by the relation $\Theta_j = \varphi_j - \varphi_{j-1}$. When the perturbations of the locations of different scatterers are independent, the total variance in the path length is thus given by

$$\sigma_L^2 = \sum_j 2\left(1 - \cos\Theta_j\right) \delta^2 , \tag{38}$$

where δ is the variance in the x and z coordinates of the scatterers.

For the coda the cosine of the scattering angle has zero mean because all scattering angles are equally likely: $\langle \cos\Theta_j \rangle = 0$. Using this, it follows from (38) that the variance in the path length joining n scatterers is given by

$$\sigma_L^{coda} = \sqrt{2n}\, \delta . \tag{39}$$

TRI of the coda breaks down when this quantity equals $\lambda/4$. The critical length scale for perturbations of the scatterer locations is thus given by

$$\delta_c^{coda} = \lambda / \left(4\sqrt{2n}\right) , \tag{40}$$

see Table 3. Note that in contrast to the situation for particles this critical length scale does not depend exponentially on n. Using the fact that the number of encountered scatterers increases linearly with time ($n = vt/l$) one finds that the critical length scale for the coda waves varies with time as $\delta_c^{coda} \sim 1/\sqrt{t}$. This time-dependence of the critical length scale for the coda waves is in stark contrast with the exponential decrease of the critical length scale for the particles with time given in (34). The $1/\sqrt{t}$ time-dependence of the critical length scale was also obtained by *Ballentine and Zibin* [1996] who show that for a periodically kicked rotator with a kick-strength that leads to classical solutions that are chaotic, the critical rotation angle $\delta\theta_{1/2}$ varies with time as $1/\sqrt{t}$ for large time.

For the ballistic wave (the wave that propagates along the line of sight from source to receiver) only forward scattering is of relevance. For forward scattering the term $(1 - \cos\Theta_j) \approx \Theta_j^2/2$ is small because the scattering angle is small. This leads to a reduction of the variance with a factor λ/L. The detailed analysis in appendix B shows that

$$\sigma_L^{ball} \approx \sqrt{\frac{3(n+1)}{4}} \sqrt{\frac{\lambda}{L}}\, \delta . \tag{41}$$

When σ_L^{ball} is about a quarter wavelength the interference is destroyed, hence the critical length scale is given by

$$\delta_c^{ball} \approx \sqrt{\lambda L}/\sqrt{12(n+1)}\,, \tag{42}$$

(see Table 3). Note that this length scale is proportional to the width $L_F = \sqrt{\lambda L/8}$ of the first Fresnel zone given in equation (1); when a scatterer is moved over the width of the Fresnel zone it contributes in a fundamentally different way to the ballistic wave.

This is consistent with the theory of section 4.3 where it was argued that only the average scattering properties over the Fresnel zone affect the ballistic wave. The ballistic wave is only affected by the perturbation when a scatterer is moved out off the Fresnel zone or when a new scatterer is moved into the Fresnel zone. The relevant length scale for the ballistic wave thus is the width of the Fresnel zone rather than the wavelength, see expression (1).

When the source locations are perturbed over a distance δ but the scatterers remain fixed, only the length of the trajectory to the first scatterer is perturbed. This means that for this perturbation for both the coda and the ballistic wave $\sigma_L = \delta$. Thus, the critical length scale for perturbation of the source locations for both the coda and the ballistic wave is given by $\delta_c^{source} = \lambda/4$, see table 3.

5.4. NUMERICAL SIMULATIONS

In the TRI of particles 20,000 particles are propagated from the source to the receiver line and after time-reversal back-propagated to the source. For the case when the receivers and the scatterers are not perturbed, the only relevant error is the error in the numerical calculations. It follows from table 2 that with the employed machine precision particles with more than 8 scattering encounters will not be focused on the source during TRI. The numerical experiments confirm this conclusion. In the top panel of figure 24 the particles that have had 6 or fewer scatterer encounters are shown after TRI at time $t = 0$. In the ideal case all particles should be imaged at the source at $x = z = 0$. This is indeed the case, several thousand particles are imaged at the source, and only a few particles are far from the source position. The location of the particles that have had between 7 and 9 scatterer encounters is shown after TRI in the middle panel of figure 24. One can observe a clustering of particles near the source position, but the imaging is clearly degraded. The imaging further degrades for the particles that have encountered 10 or more scatterers, see the bottom panel of figure 24. In that case most particles are far from the source location. figure 24 confirms the analytical estimates of table 2 that with the employed numeri-

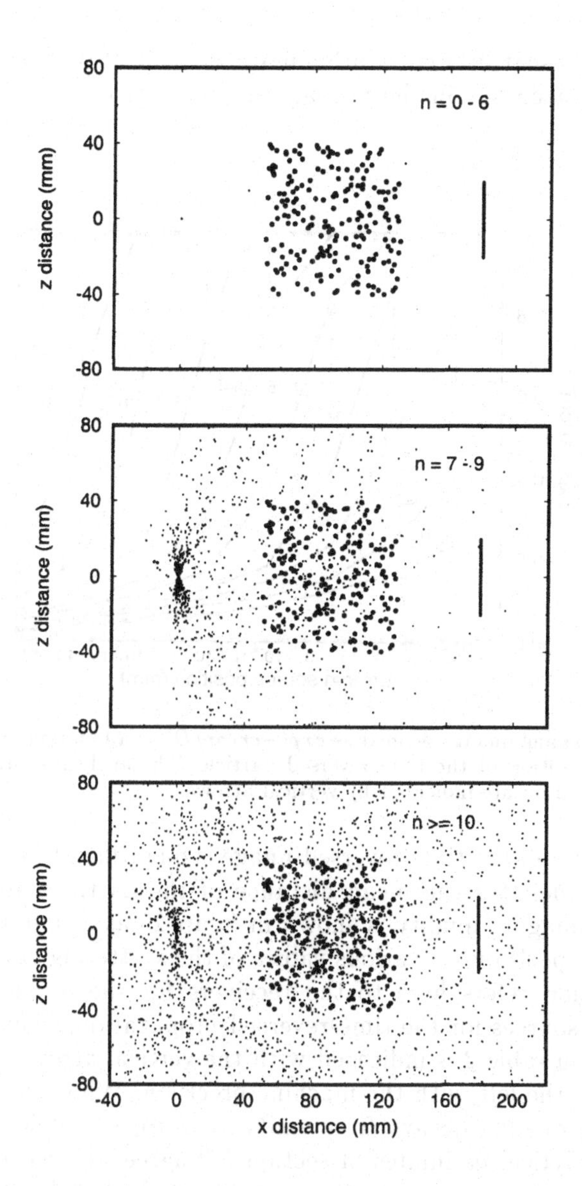

Figure 24. Locations of the particles (thin dots) at $t = 0$ after TRI for particles that have had 6 or less scattering encounters (top panel), between 7 and 9 scattering encounters (middle panel) and 10 or more scattering encountes (bottom panel). In the top panel several thousand particles are imaged on the source position at $x = z = 0$.

cal precision, 8 scatterer encounters mark the transition from accurate TRI to the loss of accuracy in the propagation of particles.

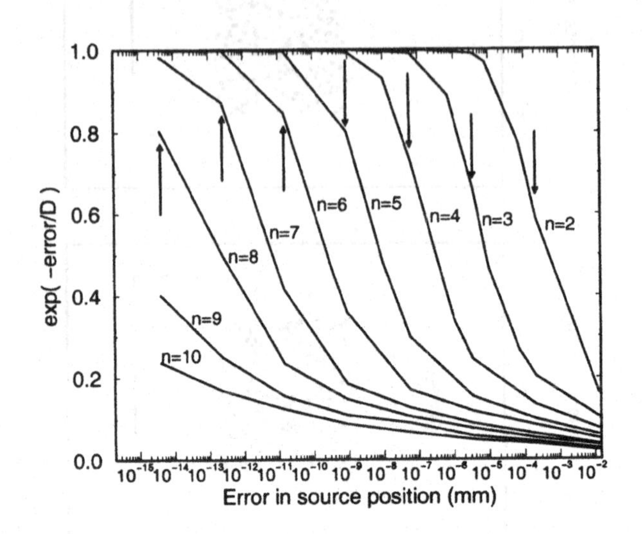

Figure 25. Imaging quality defined as $exp(-error/D)$ as a function of the perturbation of the initial position of the time reversed particle. The analytical stimates of the the critical perturbation are indicated by vertical arrows.

The quality of the time-reversed-image is quantified by $\exp(-error/D)$, where *error* denotes the mean distance of the particles to the source at $t = 0$. This imaging quality is shown in figure 25 as a function of the error in the source position for various values of the scattering encounter n. (The source position refers here to the receivers for the forward propagation that acts as sources for the time-reversed propagation.) The critical length scale shown in table 2 is indicated with the vertical arrows. The horizontal scale ends at the left with the machine precision. When the TRI degrades, the imaging quality decays from unity to zero, it follows from figure 25 that the analytical estimates of section 5.2 agree well with the numerical results. When the scatterer locations are perturbed rather than the source locations, the results are virtually the same. This is due to the fact that for large values of $2\pi l/\sigma$ the expressions (30) and (35) are almost identical.

For the waves, TRI has been carried out for several time windows, see table 4. The imaged section along the line $z = 0$ of figure 18 is shown in figure 22 by the thick solid line. In this example, a short time window of the coda from $0.25s$ to $0.30s$ has been used. This section compares favorably with the experimental results of *Derode et al.* [1995]. The thin lines in figure 22 give the imaged section for various values of the perturbation in

TABLE 4. Time windows used in the different numerical experiments with waves and number of scattering encounters.

Wave	Time window (s)	Number of encounters
Ballistic	0.11-0.13	2
Coda 1	0.20-0.25	13
Coda 2	0.30-0.35	22
Coda 3	0.40-0.45	32

the position of the receivers that act as sources in the TRI; the number indicates the variance in the perturbation of the source location measured in wavelengths. It can be seen that TRI indeed breaks down when the source locations are perturbed over about a quarter wavelength. The quality of the TRI can be quantified by computing the ratio of the amplitudes of the imaging peak of TRI with perturbation to the imaging peaks without perturbation.

The resulting relative peak heights are shown in figure 26 as a function of the error in source or scatterer locations for TRI experiments with the time windows shown in table 4. The critical length scales shown in table 3 are for each case indicated by vertical arrows. The curves for the perturbation of the source position are the four solid lines in the middle. These curves are identical for the four employed time windows and show a decay when the perturbation is of the order $\lambda/4$ (which has the numerical value $0.625mm$). For the perturbation of scatterers for TRI of the ballistic wave, the critical length scale is significantly larger, and agrees will with the critical length scale shown in table 3. For TRI of the waves in the three coda intervals the critical length scale is appreciably less than a wavelength. The reason is that the number of scattering encounters is large for these waves, see table 4. The agreement between numerical simulations and the estimates shown in table 3 is very good. This confirms the assumption that the dominant effect of the perturbation of the time reversed imaging of waves is the perturbation of the path length.

5.5. WHY DO WAVES AND PARTICLES BEHAVE DIFFERENT?

It follows from the TRI of particles and waves that the stability of particle and wave motion for the perturbation of initial conditions or scatterer locations is fundamentally different. In the numerical experiments particles that have encountered 8 or more scatterers do not return to the source after

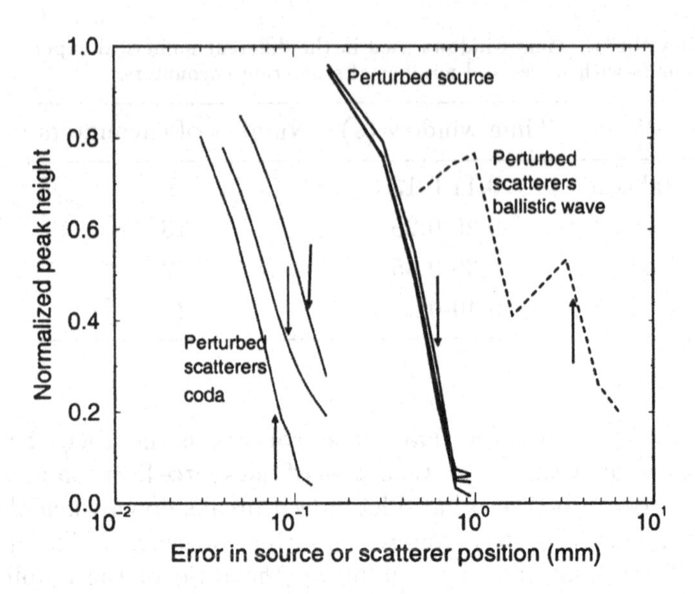

Figure 26. Quality of TRI of waves measured as the ratio of the peak height of the imaged section for the experiment with perturbed conditions compared to the peak height for the unperturbed imaged section. The dashed line represents the ballistic wave with perturbed scatterers. The dotted lines on the left are for the three coda intervals for peturbed scatterers with the latest coda interval on the left. The critical length scales from the theory are shown by vertical arrows.

TRI, whereas waves in the coda that have encountered up to at least 30 scatterers (table 4) focus well on the source after TRI.

The physical reason for this difference is that *particles* follow a single trajectory. When the initial condition of a particle or a scatterer along the trajectory are perturbed, the whole trajectory is perturbed, often in a dramatic fashion. Because of the chaotic nature of trajectories, the critical length scale is significantly less than the scattering cross-section by a factor that depends exponentially on the number of encountered scatterers and hence *exponentially* on time ($e^{-\mu t}$). This pertains both to the perturbation of the source position as well as to the perturbation of the scatterer position.

For the *waves* when the source or scatterer location is perturbed, the different wave-paths are not perturbed fundamentally; only the length of the wave-paths is changed. However, this perturbation only leads to appreciable effects when this perturbation is approximately a quarter-wavelength, because it is the interference of the waves along all possible wave-paths that determines the total wave-field. For both the coda and the ballistic wave the critical perturbation of the source location is a quarter-wavelength. For the perturbation of the scatterers the critical length-scale for the coda is proportional to the wavelength, but much smaller with a factor $1/\sqrt{2n}$. In contrast, the ballistic wave is only sensitive to perturbations of the scatterer position that are of the order of the width of the Fresnel zone. Since the number of scatterers encountered increases linearly with time, the critical length scale varies with time as $1/\sqrt{t}$ rather than the exponential time-dependence for the ciritical length scale for particles.

This implies that waves and particles react in fundamentally different ways to perturbations of the initial conditions or of the medium. The reason for this is that particles "select" a certain trajectory whereas waves travel along all possible trajectories visiting all the scatterers in all possible combinations. It is the "selection process" of a particle trajectory that creates the fundamentally larger instability of particle propagation than of wave propagation.

6. Discussion, imaging in complex media?

Imaging can only be carried out when the medium is known reasonably well because one needs to know the position of the waves in order to determine where the image is created. In general, imaging is achieved by an interference process. When a medium without scatterers is used as a reference model for an imaging experiment, then the medium must be known with such an accuracy that the error in the location of the wavefronts is less than $\lambda/4$. A larger error will destroy the interference process through which the imaging is achieved.

As shown in section 5.3, when multiple scattered waves are used for to carry out imaging the scatterer locations must be known with an error less than $(\lambda/4)/\sqrt{2n}$, where n is the number of scatterers encountered along the path. This suggest that multiple scattering does not fundamentally change the requirements that the medium needs to satisfy for successful imaging. The only change is the additional factor $1/\sqrt{2n}$ in the required precision, which suggest a quantitative difference in the required accuracy rather than a qualitative difference. Given the fact that the number of scatterer encounters n can be quite large this quantitative difference can be significant.

However, there is a more important qualitative issue at stake. When one carries out imaging in a medium without scatterers, one needs to know the location of the *wavefronts* within an accuracy $\lambda/4$. As shown in the sections 3.2 and 4.3, the location of the wavefronts depends predominantly on the average properties of the medium over the Fresnel zone. This means that a detailed knowledge of the medium is not needed, as long as the properties averaged over the Fresnel zone are known. In contrast to this, for imaging with multiple scattered waves the location of the *scatterers* must be known with an accuracy $(\lambda/4)/\sqrt{2n}$. It is for this reason that multiple scattered waves often are not used in imaging experiments where the medium is known with only limited precision.

However, the requirements for imaging with waves are favorable compared to the requirements for imaging with particles. There are two reasons for this:

[1] Waves travel along *all* possible trajectories, and remain doing this when the scatterers are perturbed. In contrast to this, particle travel along a *single* trajectory. As shown in the sections 5.4 and 5.5 this makes the propagation of particles fundamentally more sensitive to perturbations of the scatterers than the propagation of waves.

[2] For the ballistic wave the wave-field depends only on the average properties of the medium over the Fresnel zone, see the sections 3.2 and 4.3. The averaging carried out by the wave-field obviates the need for a detailed description of the structure within the Fresnel zone. There is no counterpart of this averaging property for particles.

Both factors are favorable for the stability of wave propagation compared to particle propagation in strongly scattering media. This is not only of importance for the possibility (or impossibility) of imaging experiments, but this difference should also be a crucial element in the comparison between classical chaos and quantum chaos.

Acknowledgments: I thank Peggy Hellweg, Axel Roehm and Osamu Nishizawa for providing figures. Numerous discussions with John Scales have been invaluable for the work on wave and particle chaos. Mathias Fink is thanked for making me aware of the robustness of time-reversed imaging in the presence of strong scattering.

A. The nonlinear scattering coefficient of a single scatterer

As a simple prototype of a system that exhibits multiple scattering by a single scatterer consider a delta function scatterer with strength g in one dimension:

$$L^{(0)}u(x) = g\delta(x)u(x)\,. \tag{43}$$

The operator $L^{(0)}$ characterizes the dynamics of the unperturbed system. For the Schrödinger equation this operator is given by $L^{(0)} = \nabla^2 + k^2$. Let the solution in the absence of the scatterer be denoted by $u^{(0)}$. The field then satisfies the Lippmann-Schwinger equation

$$u(x) = u^{(0)}(x) + \int G(x, x')g\delta(x')u(x')dx' , \tag{44}$$

which is equivalent to

$$u(x) = u^{(0)}(x) + G(x, 0)gu(0) . \tag{45}$$

This expression appears to be the same as equation (12) for the special case of a single scatterer. However, one cannot make this comparison because the wave-field u_j in (12) denotes the waves that are incident on the scatterer while $u(0)$ denotes the *total* wave-field at the scatterer.

These equations can be reconciled by setting $x = 0$ in (45), which gives: $u(0) = u^{(0)}(0) + G(x, 0)gu(0)$. The wave-field at the scatterer can be obtained from this result and is given by

$$u(0) = \frac{1}{1 - gG(0, 0)} u^{(0)}(0) \tag{46}$$

Inserting $u(0)$ in the right hand side of (45) then gives

$$u(x) = u^{(0)}(x) + G(x, 0)\frac{g}{1 - gG(0, 0)} u^{(0)}(0) . \tag{47}$$

This expression can be compared with equation (12) for the special case of one scatterer because the incident wave-field is given by $u^{(0)}$. Note that the scattering coefficient $g/(1 - gG(0, 0))$ depends nonlinearly on the scattering strength g.

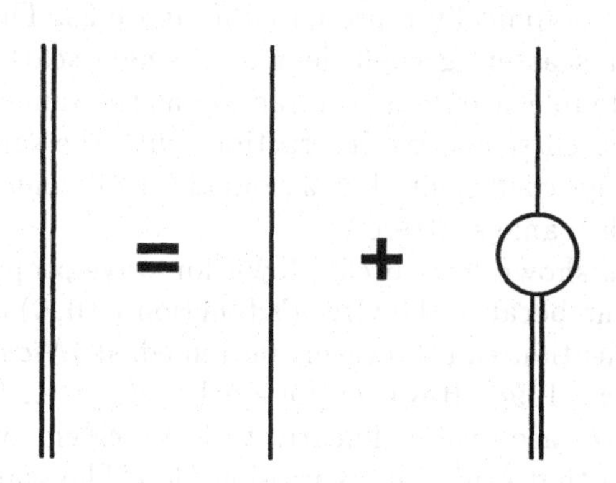

Figure 27. Diagrammatic representation of the wave-field $u = u^{(0)} + Ggu$. The single line denotes the unperturbed wave $u^{(0)}$ and G, the double line denotes the perturbed wave u and the open circle denotes the linear scattering coefficient g.

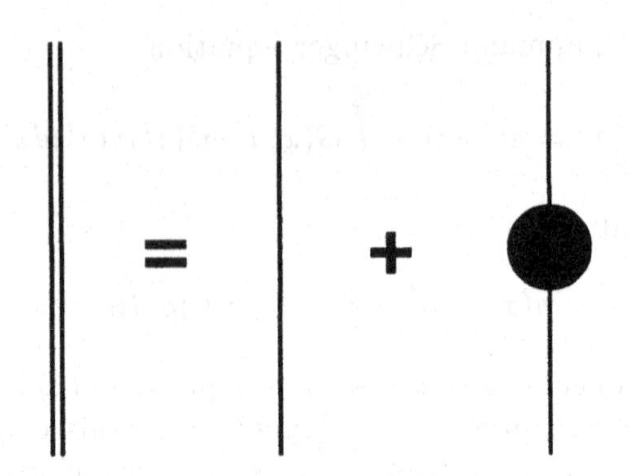

Figure 28. Alternative diagrammatic representation of the wave-field $u = u^{(0)} + G(1 - gG)^{-1} g u^{(0)}$. The single line denotes the unperturbed wave $u^{(0)}$ and G, the black circle denotes the nonlinear scattering coefficient $(1 - gG)^{-1} g$.

The scattering process can thus be described by the two alternative representations (45) and (47). These representations are shown graphically in the figures 27 and 28 respectively. In these figures a single line denotes the unperturbed wave-field and the double line the perturbed wave-field. The single scattering interaction g is denoted with an open circle while the nonlinear scattering coefficient $g/\left(1 - gG(0,0)\right)$ is shown with a black circle.

It is instructive to consider the following expansion of the nonlinear scattering coefficient:

$$\frac{g}{1 - gG} = g + gGg + gGgGg + \cdots \tag{48}$$

This expression is graphically represented in figure 29. The series expresses the full nonlinear scattering coefficient as a single scattering interaction, plus a repeated double scattering interaction at the same scatterer plus all higher order repeated scattering interactions with the same scatterer. The nonlinear scattering coefficient thus accounts for all consecutive scattering encounter with the same scatterer.

The equations shown here break down for wave-propagation problems in more dimensions because the Green's function $G(0,0)$ is singular in that case. This complication can, however, be remedied [*Nieuwenhuizen et al.*, 1992; *de Vries et al.*, 1998]. However, for any linear system in any dimension, the scattered waves are related linearly to the incident wave-field. This is the only condition that is used in expression (12). The scattering coefficient are in general related in a nonlinear way to the perturbation that causes the scattering.

As an example one can consider scattering of acoustic waves by a spherical scatterer with a perturbed bulk modulus with a radius much smaller

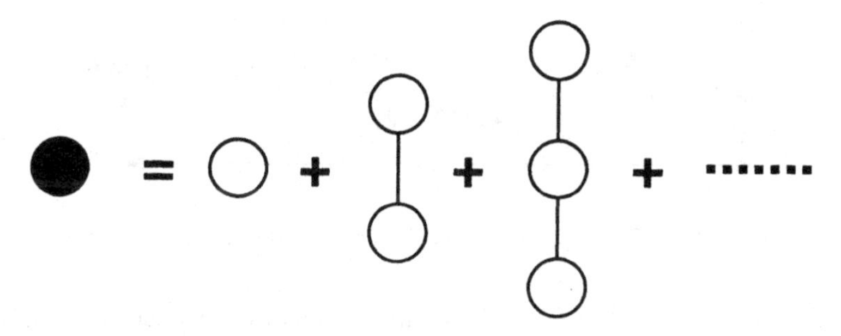

Figure 29. Diagrammatic representation of the nonlinear scattering coefficient. The single line denotes the unperturbed Green's function $G(0,0)$. The open circle denotes the scattering g and the black circle stands for the nonlinear scattering coefficient.

than a wavelength. As shown in *Morse and Ingard* [1968] the scattering is isotropic and the scattered waves depend linearly on the incident waves. Thus the conditions required for equation (12) are satisfied for this system.

B. The variance in the path length for the ballistic wave

For small scattering angles Θ_j the variance in the path length given in (38) reduces to

$$\sigma_L^2 = \sum_{j=1}^{n} \left\langle \Theta_j^2 \right\rangle \delta^2 . \tag{49}$$

As a simplified model for the ballistic wave we assume that n scatterers are separated with the same spacing Δ along the source-receiver path, this separation is given by

$$\Delta = \frac{L}{(n+1)} , \tag{50}$$

see figure 30. The fixed positions of the source and receiver are denoted by (x_0, z_0) and (x_{n+1}, z_{n+1}) respectively, hence $x_0 = x_{n+1} = 0$. The positions of the scatterers are prescribed by the distance x_j from the source-receiver line, the angle between the path from scatterer j to scatterer $j+1$ is denoted by φ_j. Since this angle is small $\varphi_j = (x_{j+1} - x_j)/\Delta$, so that the scattering angle is given by

$$\Theta_j = \varphi_j - \varphi_{j-1} = (x_{j+1} - 2x_j + x_{j-1})/\Delta . \tag{51}$$

The variance in the path length follows by inserting (51) in (49). The cross-terms that appear vanish on average because the positions of the scatterers are independent: $\langle x_{j+1} x_j \rangle = 0$. When all the scatterers have the same rms distance $\langle x^2 \rangle$ to the source-receiver line (with the exception of the terms $j = 0$ and $j = n+1$ for which $x_0 = x_{n+1} = 0$) it follows

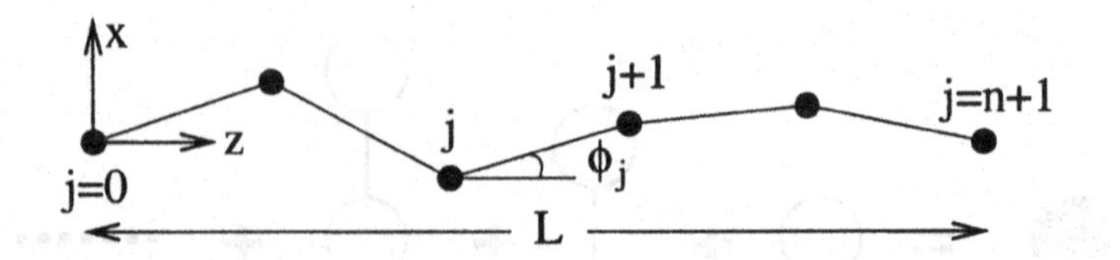

Figure 30. Geometry of the path of the scattered ballistic wave.

that $\sigma_L^2 = (6n-2)\langle x^2 \rangle \delta^2/\Delta$. Given the crudeness of the scattering model, the factor -2 in this expression is ignored so that the scattering angle is approximately equal to

$$\sigma_L^2 = \frac{6\delta^2}{\Delta^2} n \left\langle x^2 \right\rangle \tag{52}$$

The rms distance $\langle x^2 \rangle$ is at this point unknown, this quantity follows from the requirement that the scatterers are located within the first Fresnel zone. The length L_j of the path from scatterer j to scatterer $j+1$ is given by $L_j = \sqrt{\Delta^2 + (x_{j+1} - x_j)^2} \approx \Delta + (x_{j+1} - x_j)^2/2\Delta$. The detour d of the scattered wave compared to the direct wave is thus given by:

$$d = \sum_{j=0}^{n} L_j - L = \frac{1}{2\Delta} \sum_{j=0}^{n} (x_{j+1} - x_j)^2 \tag{53}$$

Using the fact that the scatterer positions are uncorrelated one finds with $x_0 = x_{n+1} = 0$ that on average

$$\langle d \rangle = \frac{n \langle x^2 \rangle}{\Delta} \,. \tag{54}$$

The scatterers contribute to the ballistic wave when the detour is less than a quarter wavelength. Assuming that this corresponds to a mean detour $\langle d \rangle$ of $\lambda/8$ one finds the corresponding variance $\langle x^2 \rangle$ from (54):

$$n \left\langle x^2 \right\rangle = \frac{\lambda \Delta}{8} \,. \tag{55}$$

Using this in (52) and using (50) to eliminate Δ then gives the variance of the path length of the ballistic wave:

$$\sigma_L^{ball} = \sqrt{\frac{3(n+1)}{4}} \sqrt{\frac{\lambda}{L}}\, \delta \,. \tag{56}$$

References

1. Aki, K., and P.G. Richards, *Quantitative seismology*, Freeman Co, San Fransisco, 1980.

2. Alsina, D., R.L. Woodward, and R.K. Snieder, Shear-Wave Velocity Structure in North America from Large-Scale Waveform Inversions of Surface Waves, *J. Geophys. Res., 101*, 15969-15986, 1996.
3. Ballentine, L.E. and J.P. Zibin, Classical state sensitivity from quantum mechanics, *Phys. Rev. A, 54*, 3813-3819, 1996.
4. Bleistein, N., *Mathematical methods for wave phenomena*, Academic Press, Orlando, 1984.
5. Červený, V., and F. Hron, The ray series method and dynamical ray tracing system for three-dimensional inhomogeneous media, *Bull. Seismol. Soc. Am., 70*, 47-77, 1980.
6. Choy, G.L., and P.G. Richards, Pulse distortion and Hilbert transformation in multiply relfected and refracted body waves, *Bull. Seismol. Soc. Am., 65*, 55-70, 1975.
7. Claerbout, J.F., *Imaging the Earth's interior*, Blackwell, Oxford, 1985.
8. Dahlen, F.A., and J. Tromp, *Theoretical global seismology*, Princeton University Press, Princeton, 1998.
9. Derode, A., P. Roux, and M. Fink, Robust acoustic time reversal with high-order multiple scattering, *Phys. Rev. Lett., 75*, 4206-4209, 1995.
10. de Vries, P., D.V. van Coevoorden, and A. Lagendijk, Point scatterers for clasical waves, *Rev. Mod. Phys., 70*, 447-466, 1998.
11. Dolan, S.S., C.J. Bean, and B. Riollet, The broad-band fractal nature of heterogeneity in the upper crust from petrophysical logs, *Geophys. J. Int., 132*, 489-507, 1998.
12. Draeger, C., and M. Fink, One-channel time reversal of elastic waves in a chaotic 2D-silicon cavity, *Phys. Rev. Lett., 79*, 407-410, 1997.
13. Feynman, R.P., and A.R. Hibbs, *Quantum mechanics and path integrals*, McGraw-Hill, New York, 1965.
14. Fink, M., Time Reversed Acoustics, *Physics Today, 50(3)*, 34-40, 1997.
15. Groenenboom, J., and R. Snieder, Attenuation, dispersion and anisotropy by multiple scattering of transmitted waves through distributions of scatterers, *J. Acoust. Soc. Am., 98*, 3482-3492, 1995.
16. Gutzwiller, M.C., *Chaos in classical and quantum mechanics*, Springer Verlag, New York, 1990.
17. Herrmann, F., A scaling medium representation; a discussion on well-logs, fractals and waves, *PhD thesis, Delft University of Technology*, pp. 298, 1997.
18. Ishimaru, A., *Wave propagation and scattering in random media*, Oxford University Press, Oxford, 1997.
19. Iyer, H.M., and K. (Eds.) Hirahara, *Seismic tomography; theory and practice*, Chapman and Hall, London, 1993.
20. Kravtsov, Ya.A., Rays and caustics as physical objects, in *Prog. in Optics, XXVI*, edited by E. Wolf, 227-348, Elsevier, Amsterdam, 1988.
21. Lax, M., Multiple scattering of waves, *Rev. Mod. Phys., 23*, 287-310, 1951.
22. Lomax, A., and R. Snieder, Estimation of finite-frequency waveforms through wavelength-dependent averaging of velocity, *Geophys. J. Int., 126*, 369-381, 1996.
23. Longuet-Higgins, M.S., A theory for for the generation of microseisms, *Phil. Trans. R. Soc. London Ser. A, 243*, 1-35, 1950.
24. Marquering, H., G Nolet, and F.A. Dahlen, Three-dimensional waveform sensitivity kernels, *Geophys. J. Int., 132*, 521-534, 1998.
25. Morse, P., and H. Feshbach, *Methods of theoretical physics, Part 1*, McGraw-Hill, New York, 1953.
26. Morse, P.M., and K.U. Ingard, *Theorerical acoustics*, McGraw-Hill, New York, 1968.
27. Nayfeh, A.H., *Perturbation methods*, John Wiley and Sons, New York, 1973.
28. Neele, F., and R. Snieder, Are long-period body wave coda caused by lateral heterogeneity?, *Geoph. J. Int., 107*, 131-153, 1991.
29. Niewenhuizen, T.M., A. Lagendijk, and B.A. van Tiggelen, Resonant point scatterers in multiple scattering of classical waves, *Phys. Lett. A., 169*, 191-194, 1992.

30. Nishizawa, O., T. Satoh, Xinglin Lei, and Y. Kuwahara, Laboratory studies of seismic wave propagation in inhomogeneous media using a laser doppler vibrometer, *Bull. Seismol. Soc. Am., 87*, 809-823, 1997.
31. Rytov, S.M., Yu. A. Kravtsov, and V.I. Tatarskii, *Principles of statistical radiophysics 4; Wave propagation through random media*, Springer-Verlag, Berlin, 1989.
32. Scales, J., and R. Snieder, Humility and nonlinearity, *Geophysics, 62*, 1355-1358, 1997.
33. Scales, J., and R. Snieder, What is noise?, *Geophysics, 63*, 1122-1124, 1998.
34. Snieder, R., 3D Linearized scattering of surface waves and a formalism for surface wave holography, *Geophys. J. R. astron. Soc., 84*, 581-605, 1986.
35. Snieder, R., On the connection between ray theory and scattering theory for surface waves, in *Mathematical Geophysics, a survey of recent developments in seismology and geodynamics*, edited by Vlaar, N.J., Nolet, G., Wortel, M.J.R. and Cloetingh, S.A.P.L., 77-83, Reidel, Dordrecht, 1987.
36. Snieder, R., The optical theorem for surface waves and the relation with surface wave attenuation, *Geophys. J., 95*, 293-302, 1988.
37. Snieder, R., Global inversions using normal modes and long-period surface waves, in *Seismic tomography*, edited by H.M. Iyer and K. Hirahara, 23-63, Prentice-Hall, London, 1993.
38. Snieder, R., The evolution of phase fronts and the method of strained coordinates, *J. Acoust. Soc. Am., 103*, 3180-3183, 1998.
39. Snieder, R., and C. Chapman, The reciprocity properties of geometrical spreading, *Geophys. J. Int., 132*, 89-95, 1998.
40. Snieder, R., and A. Lomax, Wavefield smoothing and the effect of rough velocity perturbations on arrival times and amplitudes, *Geophys. J. Int., 125*, 796-812, 1996.
41. Snieder, R. and G. Nolet, Linearized scattering of surface waves on a spherical Earth, *J. Geophys., 61*, 55-63, 1987.
42. Snieder, R. and B. Romanowicz, A new formalism for the effect of lateral heterogeneity on normal modes and surface waves -I: Isotropic perturbations, perturbations of interfaces and gravitational perturbations *Geophys. J. R. astron. Soc., 92*, 207-222, 1988.
43. Snieder, R., and J.A. Scales, Time reversed imaging as a diagnostic of wave and particle chaos, *Phys. Rev. E, 58*, 5668-5675, 1998.
44. Tabor, M., *Chaos and integrability in nonlinear dynamics*, Wiley-Interscience, New York, 1989.
45. Tanimoto, T., A simple derivation of the formula to calculate synthetic long-period seismograms in a heterogeneous earth by normal mode summation, *Geophys. J.R. Astron. Soc.,* 77, 275-278, 1984.
46. van der Hulst, H.C., On the attenuation of plane waves by obstacles of arbitrary size and form, *Physica, 15*, 740-746, 1949.
47. Waterman, P.C., and R. Truell, Multiple scattering of waves, *J. Math. Phys., 2*, 512-537, 1961.
48. Webb, S.C., Broadband seismology and noise under the ocean, *Rev. Geophys., 36*, 105-142, 1998.
49. Woodhouse J.H. and F.A. Dahlen, The effect of a general aspherical perturbation on the free oscillations of the Earth, *Geophys. J.R. astr. Soc., 53*, 335-354, 1978.
50. Yilmaz, O., Seismic data processing, *Investigations in geophysics, 2*, Society of Exploration Geophysicists, Tulsa, 1987.

Subject Index